Textile and Clothing Design Technology

Textile and Clothing Design Technology

Edited by

Tom Cassidy

Parikshit Goswami

CRC Press
Taylor & Francis Group
Boca Raton London New York

CRC Press is an imprint of the
Taylor & Francis Group, an **Informa** business

CRC Press
Taylor & Francis Group
6000 Broken Sound Parkway NW, Suite 300
Boca Raton, FL 33487-2742

First issued in paperback 2020

ISBN 13: 978-0-367-57258-7 (pbk)
ISBN 13: 978-1-4987-9639-2 (hbk)

Library of Congress Cataloging-in-Publication Data

Names: Cassidy, Tom, Professor of design, editor. | Goswami, Parikshit, editor.
Title: Textile and clothing design technology / [edited by] Tom Cassidy, Parikshit Goswami.
Description: Boca Raton : Taylor & Francis, a CRC title, part of the Taylor & Francis imprint, a member of the Taylor & Francis Group, the academic division of T&F Informa, plc, [2018] | Includes bibliographical references and index.
Identifiers: LCCN 2017026567 | ISBN 9781498796392 (hardback :acid-free paper) | ISBN 9781315156163 (ebook)
Subjects: LCSH: Textile design. | Clothing and dress.
Classification: LCC TS1475 .T45 2018 | DDC 667/.022--dc23
LC record available at https://lccn.loc.gov/2017026567

**Visit the Taylor & Francis Web site at
http://www.taylorandfrancis.com**

**and the CRC Press Web site at
http://www.crcpress.com**

Contents

Editors

Professor Tom Cassidy holds the chair of design in the University of Leeds, Leeds, United Kingdom. He holds four degrees (ATI, MSc, MBA, and PhD) and is a fellow of the Design Research Society (FRDS). He has been in education for 38 years following 9 years in the industry. He has carried out many consultancies, both nationally and internationally since the late 1970s, for agencies such as the UN, ODA, and Hong Kong Polytechnic University, King's Park, Hong Kong. Professor Cassidy is a regular reviewer for the *Textile Research Journal, Measurement Science and Technology*, the *Journal of the Textile Institute*, and many other academic journals. He is on the editorial board of the *Journal of Fiber Bioengineering and Informatics, Textile Progress*, and the *Journal of Global Fashion Marketing*. He has supervised 25 PhD students and 3 MPhil students to completion and is currently supervising 10 students. Professor Cassidy's research interests are wide and span most areas of design technology and management with particular strengths in textile and clothing design technology and education. He has held grants from the EPSRC and the AHRC for projects in design technology and design management.

Professor Parikshit Goswami (BSc (Tech), MSc, PhD, MRSC, ASDC, C Col, ATI, CText) obtained a bachelor's degree in textile technology and then worked within different industries, including a world leading third-party textile testing and quality certification company. Subsequently, he did his MSc in Advanced Textile and Performance Clothing and earned his PhD from the University of Leeds. Professor Goswami is presently the Head of Department of Fashion and Textiles at University of Huddersfield. Prior to joining University of Huddersfield, Professor Goswami was the Director of Research and Innovation at the School of Design, University of Leeds and he lead the Fibre and Fabric Functionalisation Research Group.

Professor Goswami's domains of research are product development using flexible materials and application of chemistry for functionalising textiles. Professor Goswami manages a large profile of research concerned with fibre/polymer science, nonwovens, medical textiles (implantable materials and non-implantable materials), sustainable materials (fundamentally understand the properties of new material), application and development of nano/submicron fibres for novel applications, and plasma treatment for functionalising textiles and textile chemistry. He is a member of Society of Dyers and Colourists (SDC), Education, Qualifications and Accreditation Board (EQAB), and was a member of EDANA's Board Working Group on Sustainability.

Contributors

Mirela Blaga
Gheorghe Asachi Technical University
    of Iaşi
Iaşi, Romania

Tom Cassidy
School of Design
University of Leeds
Leeds, United Kingdom

Tracy Cassidy
University of Huddersfield
Huddersfield, United Kingdom

Arobindo Chatterjee
Department of Textile Technology
Dr. B. R. Ambedkar National Institute of
    Technology Jalandhar
Jalandhar, India

Hugh Gong
School of Materials
University of Manchester
Manchester, United Kingdom

Parikshit Goswami
Head of Department
Department of Fashion and Textiles
University of Huddersfield
Huddersfield, United Kingdom

Tom O'Haire
Fibre and Fabric Functionalisation
    Research Group
School of Design
University of Leeds
Leeds, United Kingdom

Andrew J Hebden
Fibre and Fabric Functionalisation
    Group
School of Design
University of Leeds
Leeds, United Kingdom

Aileen Jefferson
Manchester Metropolitan University
Manchester, United Kingdom

Dian Li
School of Design
University of Leeds
Leeds, United Kingdom

John McLoughlin
Manchester Metropolitan University
Manchester, United Kingdom

Sandip Mukherjee
National Institute of Fashion Technology
Kolkata, India

Riikka Räisänen
Craft Studies
University of Helsinki
Helsinki, Finland

Sadhan Chandra Ray
University of Calcutta
Kolkata, India

Asim Kumar Roy Choudhury
KPS Institute of Polytechnic
Government College of Engineering &
    Textile Technology Serampore
Serampore, India

Muhammad Tausif
Fibre and Fabric Functionalisation
    Research Group
School of Design
University of Leeds
Leeds, United Kingdom

Lindsey Waterton Taylor
School of Design
University of Leeds
Leeds, United Kingdom

1

Introduction

Tom Cassidy and Parikshit Goswami

Technology, before the word was hijacked by the computer community in the 1970s, was defined as the science of the useful arts. This definition is particularly relevant to what this book is about. In the past, the education of textile technologists tended to be process oriented. The technologist today is better prepared for the textile and related industries by being product oriented and by being more able to work closely with and facilitate the ideas of designers. Designers use elements like color, shape, form, silhouette, fabric, and so on combined with principles such as harmony, rhythm, symmetry, balance, to create new and exciting products (garments, fabrics, yarns, fibers, etc.) for the market. In this book, we will learn about different product types as perceived and understood by the design technologist.

In the textile industry, there is a pressing need for people who can enable the creative solutions of designers to be interpreted into manufacturing language and data. The design technologist has to understand the elements and principles employed by designers and how these change for various textile media. He or she also has to have a good understanding of the processes, materials, and products for which the textile designer is required to produce creative solutions. The reader will find occasional overlaps between various chapters, this is beneficial because it reflects the nature of the textile and clothing industries and it will be advantageous for the reader to appreciate more than one approach to some areas. This book will suit designers wishing to improve their technological knowledge, technologists wishing to understand the design process, and to anyone who wants to work at this design/technology interface or to go on to R&D study in this field.

In Chapter 2, Goswami and O'Haire discuss the various natural and synthetic fibers and what are the implications of their shapes, surface characteristics, and mechanical properties on their use as design elements. These implications will include cost, handle, yarn spinnability, coloration, availability, and sustainability.

Chapter 3 is written by Cassidy, T and Li and describes the characteristics of the conventional/traditional staple yarns and how these are influenced by the technologies used to produce them. The chapter explains how the components of the yarns and their structural values (e.g., fiber type, fineness and length, twist content, and direction yarn count and structure) can be considered as design elements and thus manipulated to produce the design principles desired such as balance, symmetry/asymmetry, rhythm, and harmony. This chapter will also discuss new and alternative staple yarn spinning methods, some of which have become established, whereas others are striving to establish their place in the market.

Gong has contributed Chapter 5, which builds on the foundations laid by Chapters 3 and 4 and provides the reader with descriptions of various techniques and equipment used to produce fancy yarns, sometimes also called effect yarns. The market for these yarns tends to be subject to cyclic trends, but the handknitting yarn market provides a stable demand for fancy yarns. It is important that the textile and clothing design technologist is aware of the different types of yarn available and what possible innovations can be brought about by using or developing this technology.

The field of textile dyeing is a constantly evolving area and presents opportunities that can be addressed through innovative thinking. The primary objective of textile dyeing is to improve the aesthetic properties of textiles. In Chapter 6, Choudhury discusses different methods of dyeing fibers and filaments and different methods of quantifying the measured color.

In Chapter 7, Taylor describes how the interlacement patterns for warp and weft yarns can be controlled by manipulating the loom operations, which include shedding, picking, beating up, and taking up. The classification of weave designs and applications are discussed in detail. A particularly attractive feature of this chapter is the learning activities, which will be extremely valuable for a budding design technologist.

Chapter 8 on weft knitting technology by Mukherjee explains how the adjustable moving cams are used on weft knitting machines to control the movement of beard, latch, or compound needles to form normal stitches, tuck stitches, and floats. The various knit structures and their applications are described and circular knitting including sock making. The developing technology of whole garment knitting is also explained.

In Chapter 9, Ray discusses the development of warp knitting and the differences between raschel and tricot machines and their resultant fabric structures. He goes on to show how these fabrics tend to be much stiffer than weft knits and how they were originally intended for upholstery but are now used quite extensively in apparels, particularly for children's wear often in conjunction with fibers and yarns having elastic properties.

In Chapter 10, Tausif and Goswami discuss nonwoven fabric and how they are produced. They discuss key methods for web formation and web bonding to produce a myriad of nonwoven structures. The web formation methods include dry-, wet-, and polymer-laid methods, and web bonding methods cover chemical, thermal, and mechanical means to provide structural integrity to webs. They discuss the applications of nonwoven fabrics and potential for nonwovens to be used for apparel and fashion applications.

In Chapter 11, Choudhury discusses different methods for dyeing and printing fabrics.

Chapter 12 on color knowledge by Cassidy, T.D. discusses and develops the use of the 12-hue color wheel. It will facilitate a good understanding of color terminology, color composition, and color schemes essential for the compilation of color stories. An understanding of color mixing, both additive and subtractive, is also provided. Color is arguably the most important design element and, therefore, the knowledge provided by this chapter is essential for a textile and clothing design technologist.

Hebden and Goswami in Chapter 13 discuss textile finishing and how it can be used as a valuable tool to significantly enhance the value of finished textile products.

Jefferson, in her Chapter 14 on clothing technology, tackles the other important aspects of the clothing industry such as pattern construction, cutting and grading, and lay planning. She explains how modern technology has been applied to problem areas such as body sizing and the design process and then goes onto show examples of how new designers are making use of new technologies.

In Chapter 15, McLoughlin explains various types of seams and stitches used to join various sections of a garment and where and when they should be used. He also goes on to describe the sewing machines used to form the different stitch types. This is an area that has often been neglected in the textile and clothing technology, engineering, and design literature, but which is essential for a well-rounded design technologist.

In Chapter 16 on knitwear technology, Cassidy, T.D. discusses the differences between cut and sewn and fully fashioned garments and hybrids using techniques from both. There is a valuable section on knitted patterns and structures. In particular, attention is paid to linking, mocklinking, overlocking, and so on and how to recognize how a knitwear garment has been styled and formed. The importance of choosing the right style dependent on the value of the materials used (e.g., Cashmere versus acrylic) to provide the optimum quality and value for money is explained.

Raisenan, in her Chapter 17 on the measurement of textile material properties, concentrates on the practical reasons for testing rather than getting the reader lost in mathematical considerations. It is a holistic approach that is long overdue. The reader is invited to learn about the various standards that are applied to the many aspects of the textile and clothing supply chain by the international standards bodies. The importance of standard conditions and sampling methods is also discussed, but with reference given to pragmatic issues. The reader will still have to go to other publications to get details of the test methods outlined in this chapter, but the importance of this chapter is that at least the reader will now understand clearly what they are looking for and why.

Chapter 18 on how to become a textile consultant by Tom Cassidy describes some of the skills and experience required and discusses how the choice of appropriate technology is important. He discusses this with relation to case studies carried out, which involved remote, rural communities. It must, however, be emphasized that appropriate technology or perhaps the choice of appropriate levels of technology is important in much larger scale projects. For example, automatic doffing systems for spinning equipment is impressive and works very well but may not be appropriate if part of a company's and country's aims is to employ more people.

Before you set off to read, learn from, and enjoy this book, the editors would like you to consider the words of Aldred Barker, who was an early professor in the Department of Textile Industries at the University of Leeds: "If someone wishes to learn a new language they will only be successful when they learn to think in that language. If someone wants to learn about textiles they should think in fabric."

2

Fibers and Filaments

Tom O'Haire and Parikshit Goswami

CONTENTS

2.1 Introduction

Fibers are the fundamental building blocks of all products defined as a textile, regardless of application or construction. Fibers come in various forms and guises, which can have an impact on the performance, aesthetics, handle of a textile material. Natural fibers were used for millennia to create garments and tools that contributed to the expansion and development of humanity. More recently, science and industry has created a series of fibers and filaments derived from man-made polymeric material through chemical and extrusion processing, most notably nylon and polyester. This chapter will introduce the most common textile fibers from both natural and synthetic sources along with the properties of such fiber and filaments. Providing a comprehensive list of the subtle relationships between form and function for each fiber is beyond the scope of the text, but it should provide an overview of the fundamentals. This will enable the reader to assess critically how fiber type will impact on the properties of the final fabric. It is essential that the reader appreciates that appropriate selection of fiber is of fundamental importance when considering a desired design, application, or visual effect. To this end, the influence of fiber type, form, shape, diameter, and finish on properties and design is appraised by considering the major fiber types, the common production route, and how the manufacture can influence the types of yarns and fabrics available. The economic and societal implications are also of importance with the volumes, costs, and environmental impact of each class of fiber being considered.

2.2 Fundamental Fiber Properties

A fiber is simply considered to be a linear strand with flexibility and a length many times its width. This differentiates it from other assemblies such as tapes, films, and rods. For the designer, fibers and filaments could be considered the smallest element in a textile construction. The properties of a fiber will determine how it appears, how it drapes, how it conforms, and how it stretches. The designer should be aware that there is a full gamut of aesthetic finishes that can be generated by variation of this simple element.

Fibers are typically available in either staple or continuous filament. Staple fibers are elements with a length that is limited via a natural or man-made process. In contrast, continuous filaments are considered to be uncut and could be as long as many hundreds of kilometers. A continuous filament product can be easily converted in to a staple form with a length cutter. The natural materials (except silk) are available exclusively as spun staple products, whereas the synthetic filaments are often available in more formats, with monofilament yarns, multifilament yarns, and staple being the principal categories.

2.2.1 Fiber Structure

Whether staple or filament, natural or synthetic, the fundamental principles of physics apply to each fiber type. These properties are determined by the dimensions and molecular structure of the fiber. Fibers are constructed from long chains of atoms known as polymers. These structures can be formed naturally (cellulose, keratin, collagen, asbestos) or may be formed through synthetic chemical processing (polyethylene, nylon, polyester). Regardless of the method of synthesis, it is the length, molecular structure, and net orientation of these polymers that will govern not only the mechanical properties (strength, stiffness, elasticity) but also the processing in to fabrics (filament manufacture, fabric manufacture, dyeing) and the behavior of such materials (crease resistance, water adsorption, wicking). For example, cotton and flax are comprised largely of cellulose, a chain of carbon, oxygen, and hydrogen linked in the structure shown in Figure 2.1.

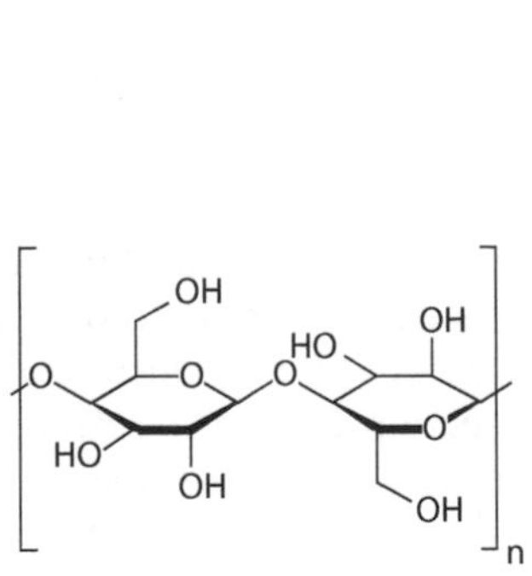

FIGURE 2.1
Chemical structure of cellulose (n is the degree of polymerization).

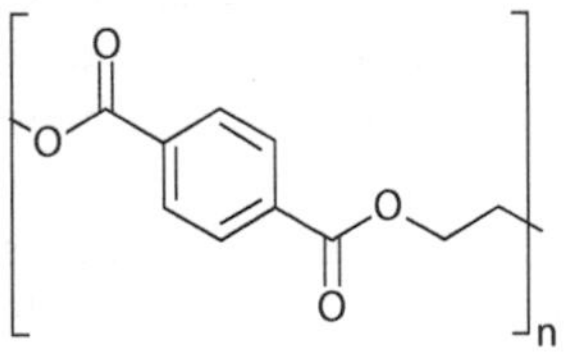

FIGURE 2.2
Chemical repeat of poly(ethylene terephthalate) (n is the degree of polymerization).

The length and arrangement of these chains determine the strength and stiffness of the fiber. Cellulose chains group together and arrange into linear fibrils, which, like a bunch of twigs, is strong and difficult to bend. It is the hydroxyl (–OH) chemical groups in the cellulose chain that allow for the adsorption and wetting of cellulose (hydrophilicity), which adds to the comfort and moisture response of cotton and linen products. The cellulose chains will also readily form hydrogen bonds between hydroxyl groups and with water. The hydroxyl–hydroxyl bond arranges the chains into a regular, patterned arrangement known as a crystal. There will also be regions of disorder where chains terminate or cannot form a regular structure. This combination of order and disorder (semicrystallinity) provides textile fibers with the strength and ductility necessary to form yarns and fabrics. The chemical groups in cellulose are also responsible for the phenomenon of creasing in cellulose where the hydroxyl groups form hydrogen bonds in new positions, breaking the initial conformation and creating a crease. The original structure can be recovered by resetting the hydrogen bonds with the addition of heat, water, and pressure (ironing). In contrast, poly(ethylene terephthalate) (PET) is a linear chain constructed of aromatic and ester linkages (Figure 2.2). The chemical groups in PET are less likely to form hydrogen bonds and thus polyester sportswear garments show excellent crease recovery and may not need ironing after washing. The incorporation of PET fibers into a cotton yarn increases its crease recovery, as the PET fibers are inclined to return to original state, thus creating easy-care fabrics.

2.2.2 Fiber Dimensions

The principle property that often determines the markets and application for a fiber is the diameter or fineness. The fineness of a fiber determines the flexibility of a fiber and the fiber assembly. Fiber fineness is typically expressed in microns when referring to diameter and in decitex or denier when referring to linear density. The finer the fiber, the more readily it will bend and is one of the key factors in determining softness and comfort levels. Textile fibers have to be sufficiently fine that they can be formed into yarns and fabrics and do not itch or demonstrate a *prickle effect* on the skin. Short segments of fiber do not bend upon skin contact until they reach a threshold value beyond which they buckle. For coarse fibers, this threshold force exceeds the force required to activate the nerve sensation (Naylor et al., 1997). Fiber fineness will determine the number of fibers within a yarn, with more fibers within a cross section generating more frictional contact and allowing for the creation of stronger yarns. There are established methods for quickly measuring the fineness of natural materials: the cotton industry uses the micronaire system and large segments of the woolen trade use the CSIRO laserscan (Gordon, 2007; Sommerville, 2009).

Length is another key metric in determining the processability of a material. This is often one of the pricing considerations with natural materials, with length being ranked the most important parameter for ring and air jet spun cotton yarns (Gordon, 2007). Longer fibers can make yarn processing easy and could be used in a finer and stronger yarn. In contrast, short fibers can not only increase yarn hairiness and bulk but also can significantly reduce the processing yield as fibers are lost during spinning (Ureyen and Kadoglu, 2006). For natural fibers, any supply is a distribution of lengths from long to short and there are objective means to characterize the average length and the variation and uniformity within a sample. With synthetic materials, staple length is controlled accurately through filament cutting and can be easily optimized to work well with existing processing equipment.

2.2.3 Shape and Cross Section

Not all fibers could be considered to be circular and the shape of a given fiber will affect bending and rigidity, handle, and luster. Nonround fibers are common in natural fibers and can be engineered into man-made fibers. With manufactured fibers, these shapes can be altered using new spinneret designs and processing techniques. Smooth circular fibers are typical for man-made fibers, but they can be irregular and include features such as crenellation, which can have a marked impact on optical properties.

2.2.4 Mechanical Properties

The response of fibers to mechanical stimulus is arguably the most important property, as without sufficient strength or flexibility there can be no yarn and no fabric. The mechanical properties, such as strength, stiffness, elasticity, and flexibility, will determine the behavior during processing and the resulting fabric properties. The strength and properties of a fabric and yarn is a complex combination of fiber and interfiber frictional properties; however, it should be realized that the yarn or fabric strength can never exceed the strength of the aggregate of textile fibers.

The elastic characteristics of a fiber determine how well it will recover from deformation and are of far more importance for many applications than the actual breaking point. Fibers with excellent elastic recovery will lend themselves to applications where tension and deformation is applied regularly and for prolonged periods. Spandex is an elastomer with fantastic elastic recovery, allowing nylon hosiery to recover very well from stretches and not sag or deform permanently. In comparison, a viscose sweater demonstrates less elastic behavior and is likely to deform at the areas of repeated strain such as elbows and necklines.

2.2.5 Optical and Aesthetic Properties of Fibers

The visual aspect of a fiber is determined by its size and shape, the internal microstructure, and the surface texture. Fibers can generate wildly different aesthetic effects through the

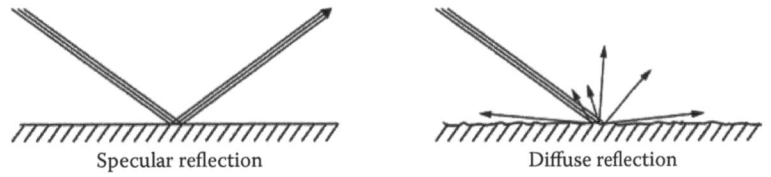

Specular reflection   Diffuse reflection

FIGURE 2.3
Specular and diffuse reflection on a smooth and rough surface.

variation of one or more of these values. This variation can be easily seen in the difference in light reflectance, luster, and gloss effects seen between polyester and woolen products. A smooth surface will generate a more specular reflection, creating a gloss effect, whereas a rough surface will generate more diffuse light refection and a matt effect (Figure 2.3).

Alongside surface reflection, there are a number of additional light interactions that will influence the luster and visual effect of a fiber. In broad terms, how light is reflected by a fiber is governed by the following physical properties:

- *Surface texture*: A rough fiber surface will generate reflections in multiple directions creating more diffuse reflections.
- *Refractive index*: This is determined by the relative velocity of light through a fiber in comparison to light in a vacuum. Changes in refractive index will skew how light is reflected and may also cause birefringence and dichroism within a fiber.
- *Adsorption*: Light that penetrates within a fiber can be absorbed by molecular agitation. Visible spectrum light is readily adsorbed by dyes, pigments, and additives to generate colored reflection light.
- *Shape*: The shape of the fiber will greatly influence the luster and shine of a fiber. Light is reflected depending on the angle of incidence and so complex shape variations generate different levels of shine and luster. Fibers with circular cross sections typically generate more specular reflection and so appear more lustrous.

2.3 Natural Fibers

Natural fibers are formed through a biological process, which determines the form and structure of the fiber. The abundance and variability of nature have given a range of fibrous materials that are readily formed into textiles; the major fiber forms and their groupings are listed in Figure 2.4. The key natural fibers by use and historic importance are cotton, wool, flax, and silk. These fibers are radically different in form and performance and generate drastically different textile designs and function.

In natural products, there is also the inevitability of variability that impacts greatly on the cost, appearance, and processing of these fibers. This inconsistency is often severe and most natural products undergo several stages of classification and sorting to increase uniformity and redirect poor quality fiber to an appropriate product stream. Although the fiber spinner and weaver often see material variability as a problem to be eliminated, the presence of stained or dead fibers can create a *natural* appearance effect that can add significant value to the cost of a garment or carpet.

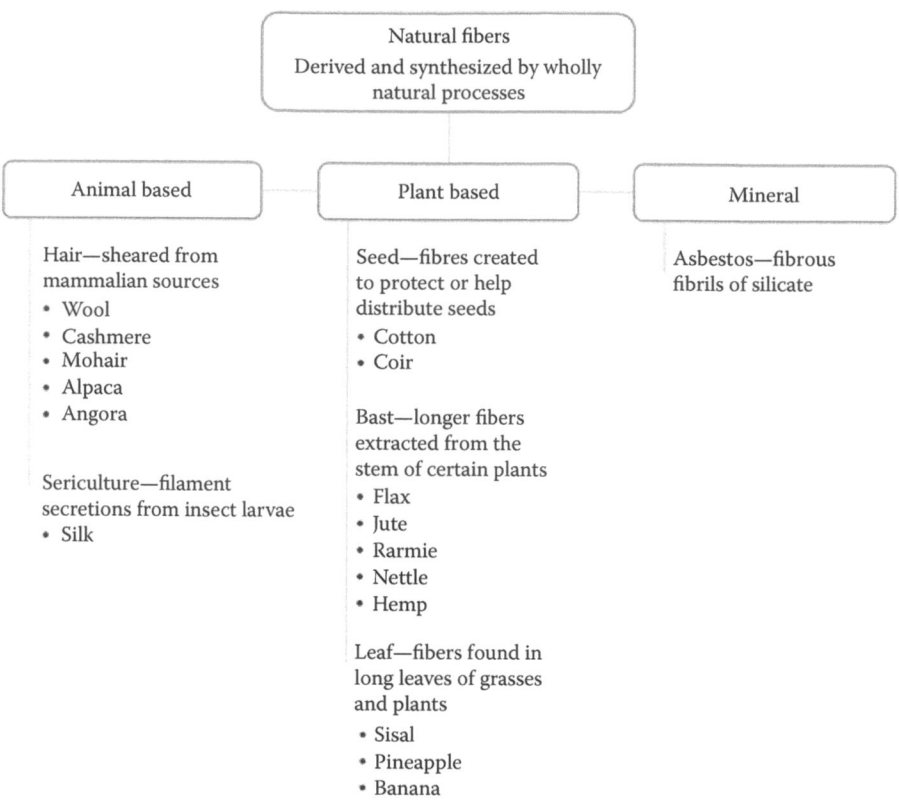

FIGURE 2.4
Diversity and classification of natural textile fibers.

2.3.1 Cotton

Cotton was used extensively in antiquity, with evidence of cotton utilization within India and China stretching back millennia. Cotton truly became westernized with the onset of the power loom and the explosion and dominance of cotton fabrics led to the fiber being known as *king cotton*. This domination faded with the emergence of synthetic materials, but cotton continues to hold more than 50% share for apparel and textile goods. For many apparel products, cotton is an indicator of quality and 100% cotton shirts carry a significant premium more than comparable synthetics and blended materials.

Cotton fibers are single-cell filaments that are harvested from plants belonging to the genus Gossypium. The cotton fibers are formed around a seed boll that has reached maturity to aid distribution and to protect the delicate seed. An individual cotton fiber appears as a contorted tube with a *kidney bean* shape with convolutions along the length. Typical mature cotton has a hollow-ribbon cross section as shown in Figure 2.5. This unique twisted shape creates a unique handle and appearance, allowing it to bend with freedom due to this shape structure. The high stiffness of cotton fiber means it can be readily processed in to high count yarns using the ring spinning process. Cotton can vary in length, diameter, and maturity and these factors will determine the quality of a cotton and which process and yarn type is most likely.

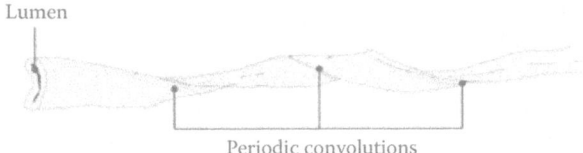

FIGURE 2.5
Ribbon-like structure of the cotton fiber with *kidney bean* cross section.

The cotton fiber is almost exclusively constructed from cellulose, which comprises around 90% of the total mass of the fiber. The noncellulosic material includes proteins, inorganics, pectins, and waxes. Cellulose appears in all plants to various extents, but it is the structure and organization of these cellulose chains that give cotton excellent mechanical properties and chemical resilience.

The cotton bolls are harvested and processed to remove trash and plant matter followed by a series of homogenizing steps prior to spinning. The processing route for cotton ring spinning, as shown in Figure 2.6, is extensive, comprising multiple steps all designed to improve the uniformity and quality of a cotton staple (see Chapter 3).

The *quality* of a cotton fiber is intrinsically linked to the fiber price. The factors affecting quality are length, fineness, color, strength, and uniformity, which were traditionally benchmarked against reference samples by human appraisers to determine the premium or discount a particular grade would command. However, the use of objective appraisal techniques is becoming the norm and most national trading bodies have developed a set of objective standards for the aforementioned properties along with trash and nep (tangled fiber balls) content. The development of the high-volume instrument (HVI) in the 1980s created an objective means of trading cotton, improving quality in the mill, and feeding valuable information back to cotton growers, leading to an overall increase in the average length (and quality) of Australian upland cotton since the introduction of the HVI (Gordon, 2007).

For the designer, there are additional aspects to consider in regards to processing. Chemical modification of cotton has been well established and can create different design features. Mercerization is a process of converting cellulose I to cellulose II via the treatment

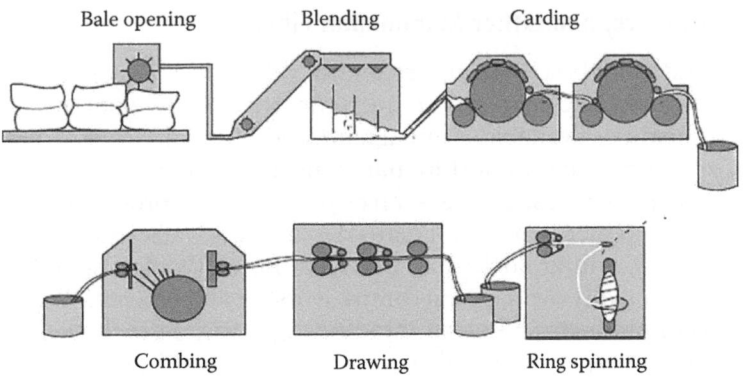

FIGURE 2.6
Processing route for cotton ring spinning.

with sodium hydroxide. This changes the appearance of the cotton fibers, the subsequent yarn, and the final fabric. Cotton also has a fine sheen of surface wax (0.4%–1.2%), which acts as a natural lubricant to limit the level of fiber breakage during the intensive opening, gilling, and carding phases. This wax is a fatty hydrophobic compound that also renders the cotton fiber impermeable to water and imparts a greasy handle on the finished fabric. For a crisper handle and improved dyeing, the wax must be removed in a scouring process. This will often be followed by a bleaching and mercerization step to improve the whiteness and luster of the fiber. These steps are typically done after fabric formation. Along with dyeing, it is these finishing processes that are water and energy intensive, generating waste products that must be handled accordingly.

2.3.2 Bast Fibers

Flax, hemp, ramie, stinging nettle, and jute are the main constituents in the classification of bast fibers. These fibers are derived from the inner bark (bast) of stems of these plants. Bast fibers are cellulosic but are completely different in appearance to cotton and share few common properties. These stems are chemically degraded to reveal bundles of bast fibers, which can be mechanically processed to reveal the number of fiber elements. The fineness can range from 19 µm for an individual fiber to 60 µm for a mechanically integral bundle. These bundles can typically have a length of up to 1200 mm with the individual fibers being of length typically in the order of 50 mm. Many bast products will contain a mixture of fibers and fiber bundles. This broad variation in diameter and length within a sample changes the handle of bast fibers, making them less suitable for next to skin applications and giving them a *woody appearance*. However, with the right processing steps, luxurious products such as linen can be obtained, which contain very few fiber bundles, ensuring softness. Bast fibers typically need an extensive preprocessing stage to disaggregate the fibrous bundles from the woody stem. This may involve retting, breaking, and scutching to weaken and eventually remove the bark element. The quality of this preprocessing route will determine how well the fiber cards and the eventual spinning quality.

The umbrella term of *bast* covers fibers from different plants, each providing fibers of differing length, diameter, color, and quality. High-quality flax fibers are typically formed into fine white linens, whereas jute and hemp fibers are a much darker brown and used extensively in sacks and low-cost ropes and yarns.

2.3.3 Wool, Cashmere, and Other Mammalian Fiber

Woolen fabrics and wool-based blends are highly valued in a range of applications. This range can be attributed in part to the unique properties of the wool fiber and to the different types and qualities available. Animal hairs and fibers are designed to help regulate the temperature and provide comfort to mammals in temperature and extreme environments, so it is no surprise that such fibers can be used to create products that are inherently comfortable. Wool, and by extension all natural hair fibers, varies significantly within the sheep, between sheep, within flocks, and between flocks. Wool is broadly classified into two broad and occasionally overlapping camps: worsted and woolen. Worsted wools typically cover all wool products processed into fine, tight yarns via the worsted processing route. These fibers are typically heavily combed and aligned in the sliver and ring spun into a strong, high-count yarn suitable for wool suits. In contrast, the woolen route uses a much shorter processing route, where slivers are ring spun immediately after carding and condensing. These yarns are much softer, less condensed, and suitable for carpet and upholstery manufacture and hand knitting. For worsted yarns, the fibers must be finer and uniform in length, with Australian merino wool being well suited for it. In contrast, the wools processed via the woolen route are often coarser, of poorer quality, and cost significantly less. Many British and Irish wools are directed via the latter route.

The finer, more expensive merino wool is one of the key elements that forms suiting. An inspection of such material will indicate that the wool yarns used are different to those formed for the use in hand knitting, broadknit sweatershirts, and scarves and in carpets and upholstery. Merino wool is now being formed into high-performance wicking base layers for a range of sporting and outdoor pursuits.

Throughout the twentieth century, significant effort was directed at understanding the internal structure of wool and hair for reasons of curiosity, necessity, and profit. A wool fiber is a complex natural product consisting of several layers of proteinous material assembled in such a way that create the unique mechanical and physiological properties of woolen fabrics. The ridged surface of the wool fiber also makes woolens susceptible to shrinkage and felting, particularly during washing. Hot temperatures and mechanical agitation encourage local sliding of fibers. However, once removed from the wash, the barbed shape of the fibers prevents the wool from recovering to its original shape, a phenomenon known as the directional frictional effect. The addition of easy-care finishes for wool is well established, but it is a chemical treatment that may influence the handle and appearance of a fabric. The well-known chlorine hercosett treatment is proven to resist felting but uses potentially undesirable chemical agents to achieve this.

As natural materials, wool and the other mammalian fibers also contain kempy fibers. Kemp is a highly medullated white fiber that is much more brittle than conventional wool. The medullation means that kemp does not take up dye readily and remains much lighter. This type of fiber is undesirable from a processing perspective but can add a dramatic visual effect that is often desirable in woolen coats and felted products.

Although wool is the dominant mammalian fiber, there are also fabrics sourced from other animals that typically find application in luxury markets or in local craft and artisan markets. Cashmere is considered a highly luxurious material due in part to the fibers being significantly finer than wool. Cashmere is sourced from the undercoat of a goat typical to China and Mongolia but which has also been exported for domestic breeding programs globally. The fibers of the undercoat, when separated from the coarser overcoat have a typical diameter of 15.3–17.3 μm (Couchman and McGregor, 1983). This separation process is known as dehairing and is essential to achieve the correct processing and comfort properties of cashmere. Mohair is sourced from the Angora goat historically bred around Ankara, Turkey. These fibers can be used to produce a soft and lustrous fabric from the finest mohair fibers. The age of the goat is a significant factor in fiber quality, with young kid goats producing the finest fibers (24–26 μm) and adult goats forming fibers significantly less valuable (40–44 μm) (Hunter, 1993). Along with fiber fineness, the cuticle height of cashmere and mohair fibers is significantly lower than for merino fibers, resulting in a smoother and more lustrous fiber (Wortmann et al., 1986). There are additional mammalian fibers available on the market in sufficient quantity to merit discussion:

- Alpaca, llama, and other camelid fibers are typically sourced from domesticated animals native to South America. Alpaca fibers are typically very fine. The nineteenth century industrialist and philanthropist Sir Titus Salt made his fortune via alpaca fibers. Interest in alpaca dipped in the twentieth century and quality and consistency suffered. It is only recently that prices have risen and interest in such fibers has returned. Allied to alpaca fibers are those sourced from the llama, vicuna, and guanaco. Llama fiber is less popular than alpaca but represents a growing market. Vicuna is a feral animal that produces very fine fibers (12–14 μm) with a price tag 10–40 times that of alpaca/llama. For now, vicuna remains a niche product with only 5 tons of greasy fiber produced in 2007 (McGregor, 2012). However, given its high cost, the vicuna is highly vulnerable to poaching, trapping, and cross-breeding. Such breeding programs do little to support the genetic diversity and wild population of the vicuna. The rarity of such fibers is likely to be a vicious cycle.

- Yak wool is collected from the undercoat of the Asiatic Yak and once dehaired has a soft feel and fiber diameters in the region of 18–21 μm. Fabrics made from yak rarely reach Western markets. However, the fineness, combined with a relatively low cost and geographical location means that yak wool can often find its way into 100% cashmere products, creating sourcing problems. Yak fibers can typically only be distinguished from cashmere via cuticle scale analysis by a skilled textile fiber microscopist.

- The Angora rabbit lends it name to Angora fibers, which are incredibly fine, with mean diameters in the range of 11–15 μm. Angora rabbit fibers are relatively smooth, with a shallow cuticle profile, creating a lustrous surface. However, a short staple length (25–40 mm) can make it difficult to process at economic speeds. They are typically blended with the longer lambswools or merino wools to facilitate spinning, weaving, and machine knitting (McGregor, 2012). Angora fibers are highly medullated (hollow) compared to wools, which changes the dye take-up and optical appearance (Mengüç et al., 2014).

- Analogous with the Angora rabbit in terms of processing and potential applications is the common possum. These so-called possum fiber yarns are being developed in New Zealand touted as a potentially sustainable source of luxury fiber. One challenge with wild possum fiber is quality control and coloration—possum fibers typically have a natural reddish brown hue and are difficult to bleach via conventional methods (Hassan, 2016).

2.3.4 Silk

Silk has been a highly valued fiber for centuries due to its soft handle and lustrous nature. It was a key material in East–West trade for millennia and the secrets of silk production was guarded closely by various Chinese dynasties before spreading to India in around AD 300. The secret and means of silk production would not reach Europe until the middle ages.

The production of silk involves the breeding and rearing of a large number of silkworms, specifically the genus *Bombyx mori*. This is a flightless moth that during the transition from larvae to moth produces a long continuous filament to form the cocoon. This continuous filament can then be uncoiled and formed in to a yarn of silk proteins. The as-spun silk is actually a natural bicomponent fiber consisting of two strands of fibroin embedded in a matrix of sericin. The majority of this sericin is dissolved in a bath of mildly acidic water to remove enough to separate the two strands of fibroin with just enough sericin coating to help bind the filaments together in the yarn. The individual silk filaments have a diameter around 10 μm, giving silk a very soft handle and luster. The silk filaments have a triangular shape, imparting unique light reflectance properties, which give the filament yarns a high sheen and clean appearance. Silk fibers are also incredibly strong with an inherent elasticity that allows silk fabrics to recover from stretch. The fineness and mechanical strength of silk impart excellent drape and handle characteristics, far superior than other natural materials.

The best silk is formed into continuous filament yarns combining silk strands of very long length to form a yarn known as *thrown silk*. These yarn types have a very smooth appearance and command the highest price. Silk is also available as staple yarn in the form of *spun silk*. These yarns are formed from silk that is not consistent enough for filament winding due to a flawed cocoon or from waste from silk throwing. The flawed cocoons are degummed in the same way, disentangled and then cut into staple lengths of various qualities. These staple silk fibers can be ring spun into yarns on conventional cotton and worsted processing lines. Spun silk yarns are hairier and thus less lustrous than continuous filament counterparts; the strength and yarn elasticity are also reduced and thus spun silk is available at a discount.

2.4 Synthetic and Regenerated Fibers

2.4.1 History and Background

Strictly speaking, the term synthetic fiber relates to fibers formed from polymers constructed from chains grown via a controlled chemical process. This category would include nylon, Kevlar, poly(ethylene terephthalate) (PET), and polyethylene, whereas fibers formed from so-called natural polymers are not considered to be true synthetics and are termed regenerated fibers. This latter category consists of viscose rayon and cellulose acetate along with some more recent developments such as chitosan, which is formed from the abundant chitin material found in sea crustaceans. Despite the now dominance of the true synthetics in the fiber market, it was the early development of regenerated cellulosics that would lay the ground work for many of the processes and techniques that are now used to make fibers from natural and synthetic feedstocks.

One of the key milestones in the story of synthetic fibers was the development and commercialization of nylon (Polyamide 6,6) in 1938 by DuPont and the group led by Carothers. This was closely followed by the development of an alternative form of nylon (Polyamide 6) at I G Farben in Germany. Nylon came to become the dominant early fiber due to the relative ease of manufacture and suitability for applications such as hosiery. The development of Dacron, strictly PET, more commonly known as polyester, is the next major milestone, with Dacron being commercialized in 1958. Polyester possesses excellent mechanical and aesthetic properties, rendering it highly suitable for textile applications. This fiber would grow to take up the largest market share of synthetics in 1972. Since then, there has been strides in using regenerated cellulose as a source material (lyocell, cellulose acetate) and in using biomaterials to derive the monomeric building blocks, for example, poly(lactic acid), which is synthesized entirely from renewable crop sources. In the technical sector, the development of the para-aramid Kevlar by Stephanie Kwolek and DuPont in the 1960s opened up a new field of research in liquid crystal polymers. This pioneering work laid the foundations for a constantly evolving environment, where established technologies are under constant challenge from new materials and processes.

2.4.2 Manufacturing Process

Synthetic fibers for textiles are all produced using the same fundamental processing techniques. In principle, a polymer fluid is forced through a series of fine holes that create the basic shape. The fluid is then encouraged to harden through cooling, chemical, or thermodynamic processes, which lead to a solid filament.

2.4.2.1 Melt Spinning

Melt spinning involves elevating thermoplastic polymers (PET, nylon, PP, PE) to a temperature at which they flow and can be passed through a spinneret to be then drawn into filaments. The generic design for a melt spinning line is given in Figure 2.7. Polymeric chips are fed to a heated screw extruder, where they are melted and homogenized to form a viscous fluid. A metering pump drives this fluid through to the spin pack, where it passes through a filter and distribution set to deliver polymer evenly to the fine holes of the spinneret typically 0.25–1 mm in diameter. The spin pack is typically at a temperature optimized for fiber production.

Reported processing temperatures for the most common polymers are given in Table 2.1. The pump drives the melt through these holes at a given rate to create many filaments. The filaments are pulled toward the winding unit and cooled in ambient or circulated air. They may pass along a set of rollers to add some initial draw and integrity to the filaments. Filaments at this stage are known as partially drawn yarn (PDY) or partially orientated yarn (POY).

Melt spinning has the advantage of requiring no solvents and uses the polymer *as is*. Polymer melts are typically highly viscous and generate a phenomenon known as exudate swell on leaving the spinneret. Exudate swell is caused by an essentially elastic material recovering from a temporary compression as it exits the orifice. The practical implications

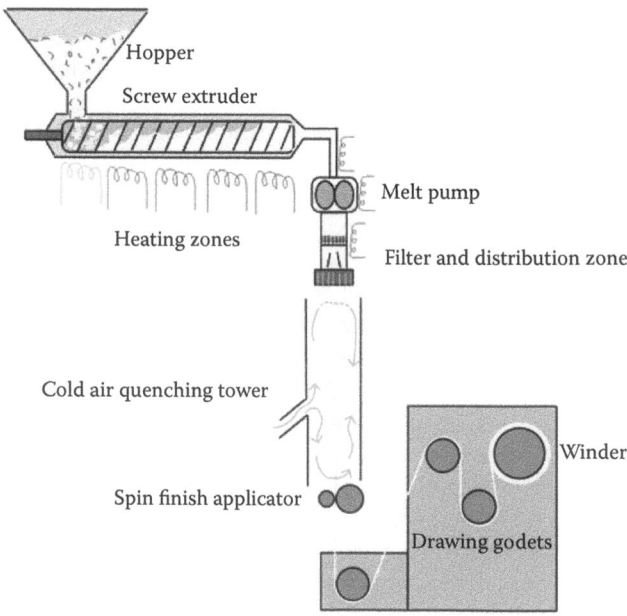

FIGURE 2.7
Schematic of melt spinning lines for partially orientated yarn filaments.

TABLE 2.1

Typical Extrusion Temperatures for Commodity
Thermoplastics

Polymer	Typical Spin Pack Temperatures (°C)
Polypropylene	200–230
High density polyethylene	210–250
Low density polyethylene	160–240
Poly(ethylene terephthalate)	290–320
Nylon 6	230–250
Nylon 6,6	260–300

of exudate swell are that the filaments will always deviate from the cross-sectional shape of a complex orifice, meaning that overly complex fiber geometries may not be possible with melt spinning. High melt viscosity also creates challenges with pumping and distribution; if excessive temperature and pressure are necessary to force polymer through the holes, then the polymer may become damaged during production and spinnerets may block. Each spinneret typically is constructed to work with a narrow range of materials and end fiber properties. Thermoplastic polymers with polar backbones (PET, nylon, polyurethanes) are usually hygroscopic and absorb moisture from the atmosphere. Water, with the addition of heat, can react with these polymers via hydrolysis causing havoc with regards to color, strength, and uniformity. These polymers will typically need to be dried for several hours to reduce water content to the region of 0.02% by weight.

A subclass of melt spinning is direct or reaction spinning. This configuration is connected directly to the reactor vessels and condenser units, with the initial feed material being monomer and catalyst. Heat must still be applied to bring the nascent polymer from the polymerization stage to the thermoplastic processing. This is less common in industry, but it is used to produce polyester yarns with very high throughputs. Direct spinning reduces the need for palletization and drying steps but increases the length of the *chain* in that a problem upstream or downstream can halt the entire processing line.

2.4.2.2 Wet Spinning

Solution spinning is typically used when melt spinning is not possible for non-thermoplastic and temperature-sensitive polymers. In this processing arrangement, the polymeric chains are dissolved in an appropriate solvent to form a viscous fluid. Typical solution concentrations can vary from 1%–25% depending on the polymer chain length, solvent system, and spin pack design. Once dissolved in solution, the chains are typically free to entangle and disentangle and move relative to each other. There are typically three variants of wet spinning, with the basic outline for each given in Figure 2.8. In all variants, a polymeric solution is pumped through a spinneret and filaments form through either evaporation or precipitation. In dry–wet spinning, the solvent is volatile enough to evaporate rapidly, leaving behind a gradually solidifying filament with only a small amount of residual solvent. In air-gap and coagulation spinning, the spinneret is submerged or suspended just above a spinning bath and the solvent is precipitated out of the filament using a coagulant or nonsolvent system. The filaments then harden and undergo several washing and drying steps before final winding. As the rate of diffusion of coagulant and solvent is critical, this variant of wet spinning is typically significantly slower than melt spinning.

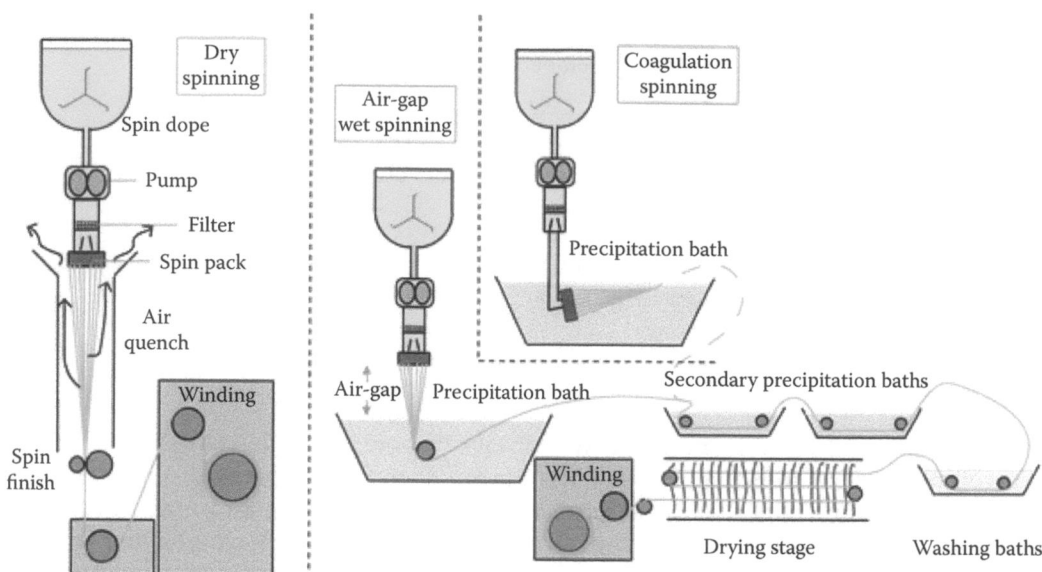

FIGURE 2.8
Variations of the wet spinning arrangement.

The spinning of continuous filaments also allows for additives to be incorporated into the fibers at the time of formation. For example, spun-dyed fibers are created via the batch-wise addition of colorants. This saves on any requirement to dye the fibers and is necessary for difficult to dye fibers such as polypropylene. However, the range of colors and minimum run quantity is often more limited.

An additional point on synthetic fibers is that the manufacturing process often locks in tension and strains within the fibers on a molecular level. When these fibers are subsequently heated for dyeing, bonding, or finishing, the fibers can contract as the strain is relaxed, which causes the yarn to shrink and the fabric to shrink in one if not two directions. This residual shrinkage is often removed through a finishing process known as heat setting. Here, fabrics are washed and allowed to shrink in a controlled manner through high-temperature ovens to remove as much as 20% shrinkage. This is a costly process, but the resulting fabric should be thermally stable in subsequent steps.

2.4.3 Commodity Polymers

The majority of synthetic fibers are derived from petrochemical sources, typically low cost and relatively easy to process. PET is the most common form of polyester and dominates the fiber market. This material is formed through the polycondensation of two petrochemical precursors. PET is melt extruded to form a strong, resilient fiber. Fully drawn PET yarn has a high tenacity and high modulus, rendering it suitable for demanding textile applications such as sewing threads. Nylon is also typically melt spun to form filaments. Nylon is more flexible than PET and can be drawn down to form fine denier filaments. This combination of flexibility and fineness makes nylon an excellent material for next to skin hosiery and stockings. Polypropylene and polyethylene comprise a smaller market segment but are important for carpet, automotive, and sportswear applications. These fibers typically demonstrate high abrasion resistance and are soil and stain resistant. However, these fibers lack the chemical functionality to facilitate dyeing and are typically colored at the fiber formation stage using thermally stable pigmentation. Acrylic fibers are typically wet spun, as the degradation temperature of this polymer is very close to the processing temperature. Wet spinning of acrylic involves several baths of coagulant and washing baths. Acrylic is frequently blended with wool to reduce the cost of knitted fabrics.

2.4.4 Regenerated Cellulose

The second classification of man-made fibers is those that are formed from existing natural polymers. These materials are spun into fibers using the filament processing discussed previously but with naturally derived building blocks. Cellulose is abundant in the environment with excellent mechanical properties, making it an ideal starting point. By dissolving cellulose into a dope and extruding into a coagulant, cellulose can be converted into a synthetic filament. The lyocell process uses N-Methylmorpholine N-oxide (NMMO) to dissolve wood pulp, whereas the older viscose rayon process converts cellulose into cellulose xanthate, dissolving this in caustic soda and coagulating into a coagulation bath containing sulfuric acid. Viscose rayon fibers, by the means of production, show skin-core effects that lead to crenellation of the fiber surface impacting light reflection and luster. Fibers produced via the lyocell process show no skin-core effects and do not have the crenelated surface texture.

Cellulose can also be modified by acetylation to produce cellulose acetate and cellulose triacetate. In these products, the hydroxyl (–OH) groups are replaced partially or fully by acetyl groups to aid dissolution in solvents such as chloroform or dichloromethane. The fiber is then dry–wet spun in ambient air, needing no precipitant. The acetylation of the –OH groups severely inhibits the formation of hydrogen bonds in cellulose acetate as the new chemical groups do not form such bonds with water. This means that cellulose acetate and triacetate holds crease and pleat very well, often being found in pleated skirts and in blends with silk, nylon, and PET.

2.4.5 Other Regenerated Materials

Besides cellulose, there are other natural, renewable materials that are used to form fibers. One such group is the regenerated proteins from sources such as milk. The use of milk as a material for wet spun fibers has been explored previously, with research dating back to the 1940s and produced as Fibrolane in the United Kingdom by Courtaulds (Peterson et al., 1948; Hearle, 2006). Regenerated casein fibers were found to have a smooth fiber surface giving a silky handle and were found to readily accept dye (Cheetham, 1957). These fibers were found to have similar tensile and elastic properties to wool and were originally conceived as artificial wool, which benefitted from shrink resistance and easy-care properties (Peterson et al., 1948). Research in such casein fibers has seen a renaissance in the last decade, with new companies looking to capitalize on the properties of such fibers. Researchers are also looking to mimic the performance of spider silk and companies have already demonstrated the use of wet spun regenerated proteins with high tensile strengths and additional functional properties (AMsilk, 2016).

Chitosan is a regenerated protein that is formed from the acetylation and processing of chitin, a material available in abundance as a by-product of the seafood industry. This fiber has attracted the attention of academia and has shown promise as a blend to improve comfort and crease recovery. However, volumes remain low and it is some way from being considered a major fiber type.

2.4.6 Elastomeric Fibers

One of the most interesting material developments of the twentieth century was the development of elastomeric fibers. These fibers are inherently elastic and demonstrate a remarkable capacity to stretch and recover from extreme deformations.

Rubber is an example of an elastomer and has found extensive use in waist elastic and hems where fiber denier is less critical. Rubber fibers are formed from a polymeric material harvested from the core of the *hevea* tree. Natural rubber has poor mechanical properties, which can be improved with vulcanization through the incorporation of cross-links between chains.

Thermoplastic elastomers are a subclass where the polymer backbone can be processed via thermoplastic routes such as melt spinning. These materials are typically a co-polymer containing a combination of so-called hard and soft phase materials. In 1958, a material from segmented polyurethane was introduced by DuPont with the trade name Spandex (United States) and Lycra (United Kingdom and others). Although not the first thermoplastic elastomer, it was certainly one of the most successful developments in this area.

Natural and synthetic rubbers are highly amorphous materials with a high amount of interchain cross-linking. These cross-links prevent irreversible chain slippage and pull a chain back into the original conformation with the removal of stress (Figure 2.9).

Elastomeric fibers are manufactured using a variety of spinning systems and as a result can have a range of cross section and shape with circular and irregular fibers available. Elastomeric materials are often hidden within garments to increase the elastic recovery in specific regions.

2.4.7 High Modulus Fibers

Some materials such as the high-performance aramids and carbon fibers offer little opportunity for coloration. However, the application for such materials is often orientated toward performance than aesthetic consideration. Kevlar falls into the class of materials known as liquid crystal or rigid rod polymers. Here, the polymeric chains have no flexing points and behave as linear rods. This makes processing more difficult but imparts fantastic properties to Kevlar fibers and fabrics. These fibers and fabric are highly resistant to heat with a melting point in excess of 625°C with a high specific

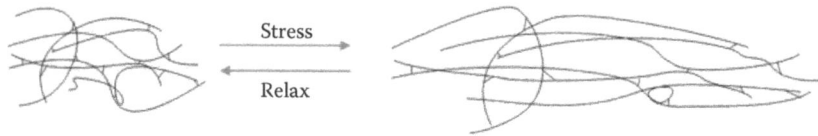

FIGURE 2.9
Rubber cross-links and application of stress.

modulus and high specific strength compared to steel due to a low density of 1.45 g cm^{-3}. The color of Kevlar was originally limited to yellow, but dyed and pigmented Kevlar has recently been developed to offer a broad range of colors (Du Pont, 2013). This reflects the growing use of Kevlar in direct-to-consumer products such as protective sports clothing and accessories.

2.4.8 Bicomponent and Ultrafine Fibers

Through spinneret design, it is possible to form fibers containing more than one polymeric component. In this arrangement, the polymers are not blended together but remain as discrete regions within the fiber. A range of bicomponent designs exist, but the principle types are side by side, core-sheath, segmented pie, and islands-in-the-sea (Figure 2.10). The segmented pie and islands-in-the-sea styles are often fibrillated to produce microfibers with excellent flexibility and softness. The 4DG fiber is a high surface area fiber designed for particle capture and wicking applications (FIT fiber, 2001).

Bicomponent fibers can have unusual physical and aesthetic properties, which make them a high-value product compared to conventional fibers. It is often done to combine the characteristics of polymers or to exploit the differences in a property such as melting point. For example, bicomponent astroturf produced from a core-sheath with a polyamide core and polyethylene sheath retains the resilience of the core but reduces incidence of friction burn through the polyethylene. Formation of such bicomponent material is not widespread but is relatively established as a technique.

Commercial microfibers produced using component melt spinning have diameters typically 0.5–5 μm and are often found in toweling and cleaning cloths, as they can pick up fine particles of dirt. Fibers smaller than 0.5 μm are intrinsically difficult to produce but have unique properties that make them attractive as functional units in membranes and scaffolds. Fibers less than 1 μm in diameter are termed submicron fibers and true *nanofibers* are fibers that have diameters less than 100 nm. Nanofibers are very fine and this influences the mechanical and reflective properties of the fabric. These fabrics can have a very fine pore size per weight of fabric, so have potential uses as selective filters and as breathable membranes. Ultrafine fibers are extremely difficult to handle and at present are not suitable for processing into yarns or fabrics due to the limitations of manufacture. Electrospinning is the principle means of producing submicron fibers. However, this method is mostly used to produce flat nonwovens and fibrous membranes for technical applications. As this technology grows, it is likely that electrospun products will find increasing use in apparel and nontechnical applications. However, at present, it is too limited with regards to production speed and material format to be considered as an apparel design element.

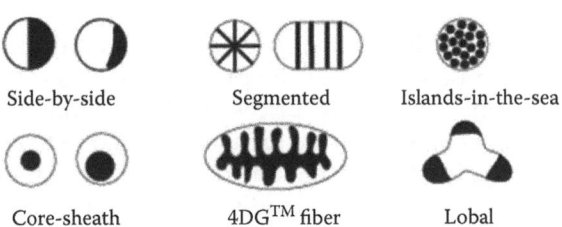

Side-by-side   Segmented   Islands-in-the-sea

Core-sheath   4DG^{TM} fiber   Lobal

FIGURE 2.10
Overview of common bicomponent fiber types.

2.5 Fiber Quality

Yarn spinning from staple and continuous filament is considered in much greater detail in Chapter 3. However, concepts of yarn quality are often viewed through the prism of fiber quality. This is especially true for natural materials. The principal function of the spinning process is to consolidate relatively short and loose fibers into a very long length of strong yarn. For staple fibers, this can be a multistep process of fiber opening, aligning, drawing, and then twisting into fine yarns. With continuous filament, this process may simply involve some light twisting or texturizing to improve cohesion and handle.

Yarn staple spinning is an intensive and high-speed process and its success depends on the processing parameters and quality of the raw material. Poor quality cotton, in the sense of high short fiber content, poor fiber maturity, and poor overall length will be very difficult to process into fine yarn beyond 40 s in the cotton count system. The tension and frictional abrasion encountered during ring spinning will create an excessive number of end breaks within the yarn, as thin and flawed sections will fail. Cotton not suitable for ring spinning may be processed into yarns using open-end techniques, which can tolerate shorter and less uniform feed but produce yarns coarser and weaker than ring spinning.

2.6 Economics and Sustainability

As much as there is variation in properties within a group of fibers, there is also significant variation with cost, both economic and environmental. In describing these elements, there is often a desire to be able to simply arrange the main apparel fibers with regards to total consumption, cost per kg, energy use, and water use. However, the reality is that there is no fixed order for fibers in any category except for consumption. Man-made fibers typically cost less than natural materials, but within the latter category, there is huge variation on price and low-quality natural materials are often available at significant discount. As of late 2016, the relative costs of fibers were

$$\text{polyester} \left(0.80 \text{ £ kg}^{-1}\right) < \text{nylon 6} \left(1.70 \text{ £ kg}^{-1}\right) < \text{rayon} \left(2.20 \text{ £ kg}^{-1}\right)$$

$$< \text{cotton} \left(2.70 \text{ £ kg}^{-1}\right) < \text{spandex} \left(3.50 \text{ £ kg}^{-1}\right) < \text{merino wool} \left(5.40 \text{ £ kg}^{-1}\right) \quad (2.1)$$

$$< \text{thrown silk} \left(34.00 \text{ £ kg}^{-1}\right)$$

The order and prices listed earlier should be regarded as only a snapshot of the 2016 market. However, it should be noted that the price of fibers is often linked to seasonal changes (cotton and wool), commodity prices (PET, PE, and PP), and almost never occur in isolation, such as a collapse in the price of cotton due to oversupply is likely to impact the price of staple polyester, which in turn cascades to other synthetic materials (Baffes and Gohou, 2005). As with all commodity materials the underlying cost of fibers linked to underlying cost of production, the risk of trading such materials, and general economic outlook.

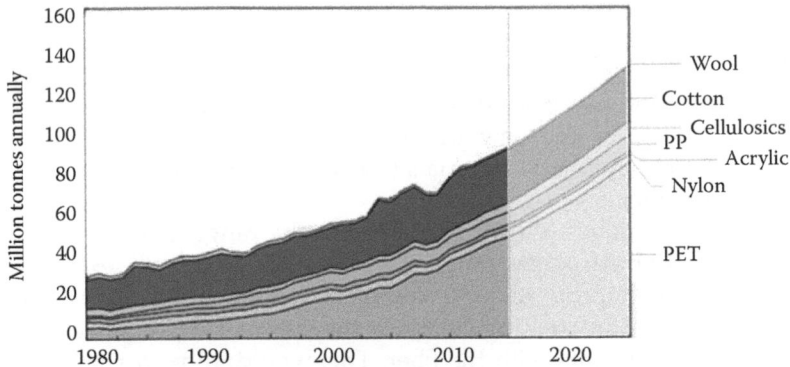

FIGURE 2.11
Recorded and predicted tonnage of commodity textile fibers. (Adapted from Qin, Y., Global fibres overview, *Synthetic Fibres Raw Materials Committee Meeting at APIC 2014*, Pattaya, Thailand, May 16, 2014.)

Reporting the water and energy usage for each textile fiber is intrinsically difficult due to the multitude of processing streams. This is made more complex by the contemporary tendency to perform life cycle analysis and to take a holistic view. If a fiber is proposed to have a significant chance of reuse, the environmental cost becomes more difficult to compare with rival fibers. Fiber volumes are easier to estimate, as a lot can be drawn from historical data. Figure 2.11 lists the historic and predicted trends for fiber consumption worldwide toward 2025.

It should be noted that blending staple fibers is a widely used means of reducing the cost of fabrics. Blends of polyester/cotton and wool/nylon are well established and the addition of uniform synthetic staple can also ease processing of natural materials, reducing costs again. Blending is also done for functional purposes, for instance to improve stretch or to impart crease resistance to a garment. However, it should be acknowledged that some blends may be more difficult to dye uniformly and may result in a mélange effect.

2.6.1 Recycling

The recycling of textiles and fibers has been practiced long before consumers and corporations adopted an agenda of sustainability. Garments from natural materials could be processed down into fibers via the shoddy recycling process. This involves shredding the fabrics to break up the structure with minimal damage to the yarns. Such materials were then used as mattress stuffing and formed into nonwovens for domestic insulation. The manufacture of fibers from sustainable and recycled sources is also worth considering. Polyester fibers produced from recycled bottles (rPET) have acceptable mechanical performance to be used in blended apparel (Telli and Özdil, 2015). Textile products, at end of life are typically downcycled into low-grade shoddy material. There has also been a recent drive to be able to recycle textile materials into products of comparable quality. This has been made increasingly difficult by the high number of different materials used to construct products. Manufacturers are recognizing that future products may need to be produced from materials designed to be compatible with thermoplastic or solvent recycling.

2.7 Future Trends

It is expected that as the twenty-first century progresses there will be further developments in fiber technologies and new and exciting materials will be on hand. Likewise, some of the natural and synthetic materials featured in this chapter may fall out of favor to the point of irrelevance.

New processing techniques will further develop the range of fiber shapes, sizes, and surface textures on offer. Also, new polymeric materials will create textile materials with mechanical and aesthetic properties previously unseen in fibers. One possibility being explored is the potential of creating color via the use of surface structural effects and crystallographic arrangements with the fiber. This would allow for iridescent elements, which would respond differently to light, creating unique effects (Eadie and Ghosh, 2011). Phenomena such as light transmittance and leakage for textile are only recently being explored and Bragg fibers, and materials with a unique photonic response are currently being demonstrated (Gauvreau et al., 2008).

References

AMSilk (2016) *Biosteel fibers.* Online resource, available at https://www.amsilk.com/industries/biosteel-fibers (Accessed February 11, 2017).

Baffes, J. and Gohou, G. (2005) The co-movement between cotton and polyester prices, World Bank Policy Research Working Paper No. 3534, World Bank, Washington, DC.

Cheetham, R. C. (1957) Fibro and Fibrolane in the carpet trade with particular reference to dyeing. *Journal of the Society of Dyers and Colourists*, 73(5), 189–198.

Constable, G. A. (2006) *Fiber quality data (1976–2006).* Australian Cotton Research Institute.

Couchman, R. C. and McGregor, B. A. (1983). A note on the cashmere down production of Australian goats. *Animal Production*, 36, 317–320.

Du Pont (2013). *DuPont™ Kevlar® is not just about yellow.* Du Pont press release: June 24.

Eadie, L. and Ghosh, T. K. (2011) Biomimicry in textiles: Past, present and potential. An overview. *Journal of the Royal Society Interface*, 8(59), 761–775.

Eichhorn, S. J., Hearle, J. W. S., Jaffe M. and Kikutani T. (2009) *Handbook of Textile Fibre Structure.* Oxford, UK: Woodhead Publishing.

FIT Fibre (2001). http://www.fitfibers.com/

Gauvreau, B., Guo, N., Schicker, K., Stoeffler, K., Boismenu, F., Ajji, A., Wingfield, R., Dubois, C. and Skorobogatiy, M. (2008) Color-changing and color-tunable photonic bandgap fiber textiles. *Optics Express*, 16(20), 15677–15693.

Gordon, S. (2007) Cotton fibre quality. In: Gordon, S. and Hseih Y.-L. (Eds.) *Cotton: Science and Technology.* Manchester, UK: The Textile Institute.

Hassan, M. M. (2016) Possum fiber—A wonderful creation of nature. In: Muthu, S. S. and Gardetti, M. A. (Eds.) *Sustainable Fibres for Fashion Industry.* Singapore: Springer.

Hearle, J. W. S. (2006) Protein fibers: Structural mechanics and future opportunities. *Journal of Materials Science*, 42, 8010–8019.

Hufenus, R., Affolter, C., Camenzind, M. A., Halbeisen, M., Spierings, A. B., Tischhauser, A., Zogg, K. and Schramm, G. (2010) Bicomponent artificial turf fibers for future sports flooring. *Fiber Society Spring Conference 2010*, Bursa, Turkey, May 12–14.

Hunter, L. (1993) Mohair: A Review of its Properties, Processing and Applications. The CSIR Division of Textile Technology, Port Elizabeth, International Mohair Association and The Textile Institute.

McGregor, B. A. (2012) *Properties, Processing and Performance of Rare and Natural Fibres A Review and Interpretation of Existing Research Results.* Canberra, Australia: RIRDC.

Mengüç, G. S., Özdil, N. and Kayseri, G. O. (2014) Physical properties of textile fibres. *American Journal of Materials Engineering and Technology*, 2, 11–13.

Morton, W. E. and Hearle, J. W. S. (1993) *Physical Properties of Textile Fibres*, 3rd ed. Manchester, UK: The Textile Institute.

Naylor, G. R. S., Phillips, D. G., Veitch, C. J., Dolling, M. and Marland, D. J. (1997) Fabric-Evoked Prickle in Worsted Spun Single Jersey Fabrics Part I: The Role of Fiber End Diameter Characteristics. *Textile Research Journal*, 67, 288–295.

Peterson, R. F., McDowell, R. L. and Hoover S. R. (1948) Continuous filament casein yarn. *Textile Research Journal*, 18(12), 744–748.

Qin, Y. (2014) Global fibres overview. *Synthetic Fibres Raw Materials Committee Meeting at APIC 2014*, Pattaya, Thailand, May 16.

Sommerville, P. (2009) The objective measurement of wool fibre quality. In: Johnson, N. and Russell, I. (Eds.) *Advances in Wool Technology*. Manchester, UK: The Textiles Institute.

Telli, A. and Özdil, N. (2015) Effect of recycled PET fibers on the performance properties of knitted fabrics. *Journal of Engineered Fibres and Fabrics*, 10(2), 47–60.

Ureyen, M. E. and Kadoglu, H. (2006) Regressional estimation of ring cotton yarn properties from HVI fiber properties. *Textile Research Journal*, 76(5), 360–366.

Wortmann, F. J. and Ams, W. (1986) Quantitative Fiber Mixture Analysis by Scanning Electron Microscopy, Part I: Blends of Mohair and Cashmere with Sheep's Wool. *Textile Research Journal*, 56, 442–446.

Wynne, A. (1997) *Textiles*. Oxford, UK: Macmillian Publishing.

3

Staple Yarns

Tom Cassidy and Dian Li

CONTENTS

3.1 Introduction

This chapter will describe the characteristics of the conventional staple yarns and how these are influenced by the technology used to produce them and how the characteristics influence their use in fabric manufacture and the design possibilities they offer. This chapter will go on to discuss some of the more important nonconventional yarns and important developments in conventional ring spinning machines.

The design technology elements that can be manipulated in staple yarns are as follows:

- Fiber length
- Fiber fineness
- Yarn structure (effected by fiber preparation and the spinning process)
- Color (including effects such as mixtures/melanges; roved unions/spun marls; twisted marls; color injections such as Knop yarns/Knickerbocker/Donegal)
- Linear density
- Single strand twist content and direction
- Folding twist (multi ply) content and direction

The design technology principles available are as follows:

- Symmetry/asymmetry
- Balance
- Evenness/regularity
- Strength and elongation
- Liveliness
- Bulkiness
- Texture

The authors' intention is to discuss these elements and principles in individual sections; however, there are many overlapping causes and effects that will be given attention at the appropriate points. First, a design technology overview of the conventional/traditional staple spinning systems is provided.

3.2 The Conventional/Traditional Staple Spinning Systems

The first industrial revolution began in the United Kingdom around 1750 and then spread to mainland Europe and the United States in progressive stages. If we go back to the early days, it can be observed that the main fibers in use in the United Kingdom were cotton and wool, so it was natural that the spinning systems that were developed had these fibers as their targets (Figure 3.1). The length of cotton fibers being spun into yarns was from 30 to 45 mm (nowadays man-made fibers can be broken into any lengths and typically 30–60 mm for short-staple spinning). The system used to spin these yarns was collectively known as cotton spinning and later as short-staple spinning and was concentrated around the county of Lancashire, UK. On the other hand, wool fibers vary from 1.5 in. to around 7 in. and so two systems were developed to process this range: the woolen system, later known as condenser spinning, and the worsted system, later known as roller drafted. The former used shorter fibers and produced a yarn with a disoriented fiber arrangement and having a wide range of fiber lengths in its cross section.

3.2.1 Cotton/Short-Staple System

This is a long process and was developed originally to deal with the processing of cotton fibers into yarns. The system begins with the delivery of the cotton fibers in bales from the Ginning plants, wherein the hand or mechanically picked bolls of cotton are separated from the seed fragments and any other trash that is attached. Ginning is, however, a relatively coarse-level operation and does not clean and open the fibers to a sufficient extent for spinning. Therefore, the first stage of fiber preparation in a spinning mill is known as the *Blowroom*. This consists of a series of machines containing beater bars, blades, and grids along with the clever use of airflow through a number of tubes and pipes to effect thorough cleaning and opening of the fibers into a form required for feeding to the very important stage of carding. As with every stage of the processing in a cotton spinning mill, it is vital to keep the ambient temperature and relative humidity at the optimum level for each stage. If this is neglected, many problems will be caused by the formation of static and changing fiber properties, which will make the fibrous assembly very difficult to control.

The next, and extremely important, stage is *Carding*. In contrast to the Woolen and Worsted systems, which use Roller and Clearer carding machines, cotton/short-staple cards use revolving or stationary cards covered with saw-like metal teeth, which work with the similar teeth of the main roller to open the fiber clumps into individual fibers, which are delivered as a fibrous web and then consolidated into sliver (Figure 3.2).

From now on in this text, the term short-staple spinning will be used to describe this system.

At this stage, the card slivers can take two processing routes. One is through Drawing machines (one or two times) and then straight to roving/speed frame for final preparation for spinning. The other also uses one or two drawing processes and then a machine is used to form laps, which are the feed for the combing process. After combing, the fibers are once again assembled as slivers and then proceed to one or two more drawing machines

| Short staple (cotton) | Worsted (roller drafted) | Condenser (woolen) |

FIGURE 3.1
The three conventional yarn structures.

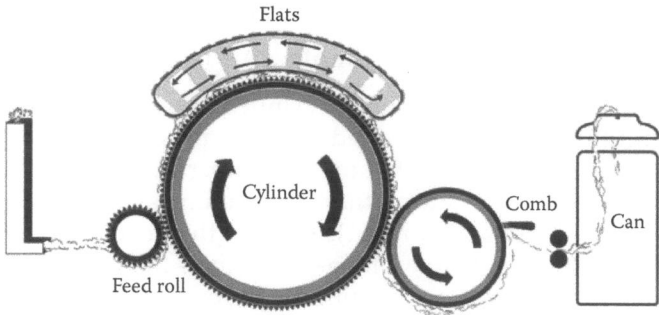

FIGURE 3.2
A cotton card.

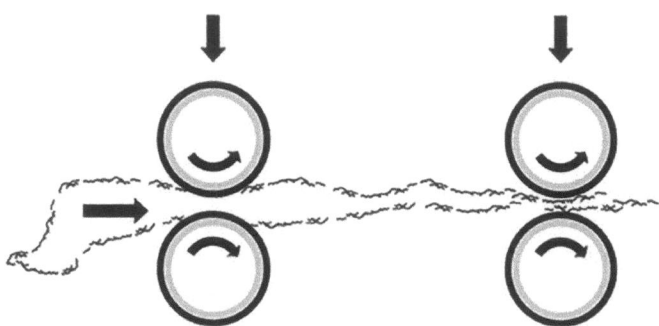

FIGURE 3.3
Roller drafting.

before being processed at the roving/speed frame stage. So, whether carded cotton yarns or combed cotton yarns (the former tend to be thicker and coarser, whereas the latter are finer and smooth surfaced), the Drawing process is next.

It is now timely for an introduction to a technological method that is commonly used in many yarn spinning processes to attenuate/draw the fibrous assembly to reduce its thickness. At this stage, the term thickness is used as it is probably the easiest term for the beginner to understand, but later in Section 3.2.1.5, linear density will be introduced as the much more accurate term for this yarn/fibrous assembly parameter. The technological term is called *Roller Drafting* (Figure 3.3). This is how the drawing machines reduce the thickness of the slivers and begin the alignment of the fibers in the yarn to be as parallel as possible along the yarn.

The amount of drafting achieved is therefore given by the simple equation:

$$\text{Draft} = \frac{\text{Speed of front rollers}}{\text{Speed of back rollers}} \tag{3.1}$$

For example, if the back rollers are revolving at 5 rpm and the front rollers at 100 rpm, then the draft would be 20 (assuming the diameter of both sets of rollers are the same). There are two points to note here:

1. If the rollers are of different diameters, then this will affect the relative speeds. For example, in the aforementioned example, if the back rollers are 4 cm in diameter and the front rollers are 2 cm in diameter, then the relative speeds become 5×4 (20) cm min^{-1} for the back rollers and 2×100 (200) cm min^{-1} for the front rollers and therefore the draft would be 200 divided by 20 equals 10.

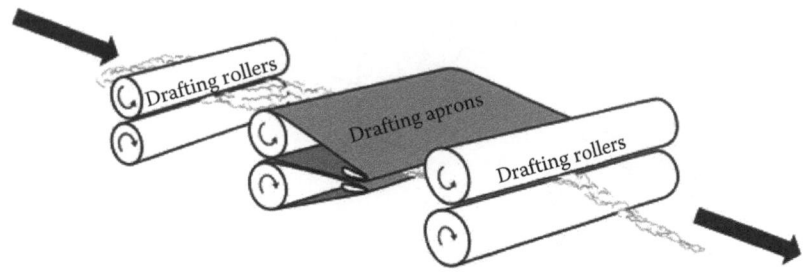

FIGURE 3.4
Apron control in drafting.

2. Some form of fiber support can be used to help control the fiber movement during drafting. In the early stages of drawing, the machines will have a set or sets of intermediate rollers between the back and front rollers, which runs at intermediate speeds. At the later stages of roving and spinning a pair of aprons, again running at intermediate speeds are positioned between the back and front rollers (Figure 3.4). The aprons give better fiber support and control as the drafts get higher.

In the drawing process, multiple number of slivers are fed into the back rollers of the drafting system so that the process will allow blending of the slivers to reduce unevenness between individual slivers and the drafting will reduce the output linear density/thickness to the same or almost the same as the individual slivers being fed in. So, the main purpose of the drafting or drawing is to improve the fiber alignment. After drawing (one or two times), the slivers then go straight to roving/speed frame (if they are going to be spun into carded cotton yarns) or a number of slivers are arranged together side by side and consolidated to form one coherent band known as a lap. This is the form of fibrous assembly fed to the *Combing* machine. Think of a lap as similar to the hair on your head before the hairdresser is able to comb through it and cut your hair to the same or similar lengths and remove what has been cut off and any damaged hair that he may feel should go.

The combing machine is mechanically complex, but the important idea for the reader to understand is that the fibers are combed through by pins to produce a very high degree of fiber alignment and orientation and that short fiber and any residual trash content is removed. At the output end of the comb, the fibers are once again formed into a sliver. These combed slivers are then fed through one or two more drawing machines and then the final slivers are fed to the *Roving* machine often called a speed frame in short-staple spinning (Figure 3.5).

Each spindle of the roving machine/speed frame will receive a sliver from the final drawing machine and this will be fed into the back rollers of a drafting system as described earlier but in a smaller scale. The rollers are smaller, the space between them shorter and now a set of slowly revolving aprons are used to control the fiber flow. Now the sliver is reduced in thickness/linear density to a value that is suitable for spinning (Ring Frame). The draft will be in the region of 10 and because the sliver will be quite thin and now called a roving, a small amount of twist will be added using a revolving flyer at the delivery end of the roving machine/speed frame and the flyer will also cause the roving to be wound onto a bobbin, which will be used as feed for the spinning machine.

Finally, we have reached the Spinning process (Figure 3.6), and once again, this uses roller drafting to reduce the thickness/linear density of the roving down to the required value. Also, at this stage, the amount of twist inserted in the yarn as it emerges from the nip of the front rollers is very important for future use of the yarn in fabric. The nip is the contact point

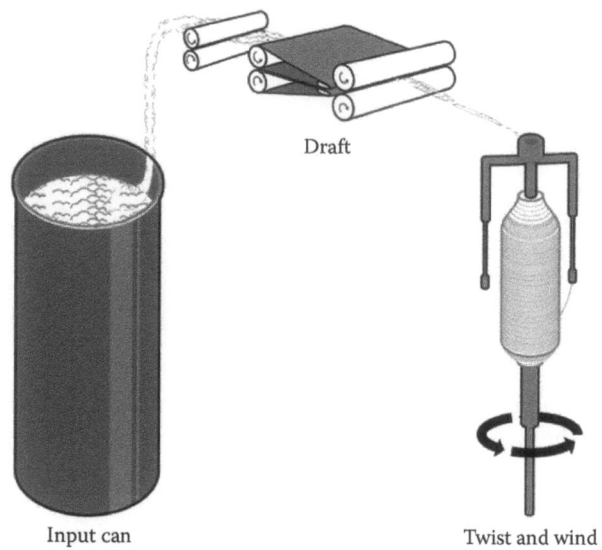

FIGURE 3.5
Roving frame often called a speed frame in short-staple spinning.

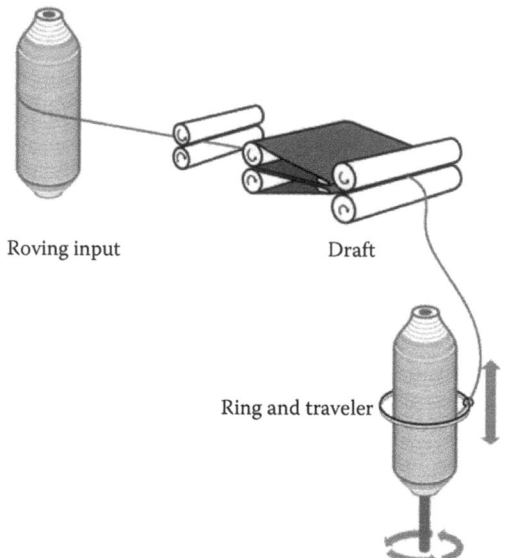

FIGURE 3.6
Short-staple ring frame.

between the top and bottom rollers at which the fiber stream comes under the control of the rollers. The twist is inserted by a small metal C-shaped element called a traveler. The traveler is caused to rotate around a ring at high speed by the pull exerted by the spindle rotating inside the ring. The pull on the traveler also causes the yarn to be wound onto a tube, which has been mounted on the spindle, and the upward and downwards movement of the ring and travelers means that the tube is slowly filled with yarn over most of its surface.

Finally, the single yarn strands will be wound from the tubes (sometimes also referred to as bobbins) onto cones (Figure 3.7) by a *Winding* machine. This allows the yarn to be

FIGURE 3.7
Yarn wound on a cone package.

passed through a clearing device to remove any thick places or knots that have been caused during fiber preparation and spinning and for a much larger package (called a cone) containing more yarn to be produced saving time at later processing. The later processing would be weaving or knitting if the weavers or knitters want to use single yarns, or twisting/folding if they want plied yarn using two, three, or more single yarns plied together.

Now the reader can start to consider the application of design elements and principles to the technology of short-staple spinning.

3.2.1.1 Fiber Length

In short-staple spinning, the length of fiber is limited from around 20–60 mm and short fibers are removed at combing for the spinning of fine yarns. Short fibers and other factors can cause the formation of neps in the yarn and fabric, which are mostly considered a detrimental factor to the quality of fabrics. Fashion designers, however, have been known to make use of neps to produce effects and exclusivity in their fabrics and apparel (Peter and Bowes, 1988).

LEARNING ACTIVITY

When using man-made fibers in the short-staple system, cutting or breaking systems can be used to engineer the fiber length so that the fibers can be staple spun in blends with cotton most common would be polyester/cotton, lyocell/cotton (Figure 3.8). Cotton fibers can also be blended with wool and other protein fibers, but that would normally be carried out using the Worsted/Roller Drafted spinning system or the Woolen/Condenser system as the wool fibers would be too long and coarse for the short-staple system. Of course, there are always new mixtures of fiber types using variations of the spinning systems and combinations of their stages being tried, but the basics are as have been described here.

FIGURE 3.8
Neps incorporated into fabric as a design feature.

3.2.1.2 Fiber Fineness

Using the short-staple spinning system, the minimum fibers required in the cross section is 30, although most spinners are more comfortable if they have 100 or more fibers in the cross section of the spun yarn. Fiber fineness is evaluated using a micronaire instrument, which measures the airflow through a specially prepared clump/tuft of fibers. The micronaire value is therefore a measure of fineness and fiber maturity. Basically, it can be stated that values of 3.1 and below indicate that the cotton is very fine, 3.1–3.9 fine, 4.0–4.9 medium, 5.0–5.9 slightly coarse, and above 6.0 coarse. Fiber fineness is measured in *decitex* (see linear density explanation next), which is obtained by micronaire value multiplied by 0.394. This measure allows the comparison of fiber fineness with man-made fibers, which are spun to particular decitex values. Fiber finesses can also be given as a micron value obtained by measuring individual fiber cross sections through a microscope, but this is a lengthy and tedious process and therefore micron value is rarely, if ever, used commercially. Moreover, natural cotton fibers are not circular in cross section, so using fiber diameter could be misleading. Fiber fineness affects properties such as fabric drape, yarn strength, evenness, and luster. These properties can be considered as design principles. Fiber fineness influences the yarn strength; the finer the fibers, the higher the strength. In a conventional/traditional spinning process, the fine fibers accumulate to the core and coarser fibers move to the outside of the yarn structure, causing hairiness. Therefore, the less coarse fibers allowed through to the spinning stage, the better. In particular, the handle of the yarn and resultant fabric will be affected by fiber fineness and, therefore, this property is of major interest to the design technologist though not at the neglect of other parameters.

3.2.1.3 Yarn Structure

For all conventional/traditional spinning systems, the final twist insertion is carried out on a continuous strand of fibers (Figure 3.9), thus the structure of the yarn means that it

FIGURE 3.9
Continuous fiber strand being twisted.

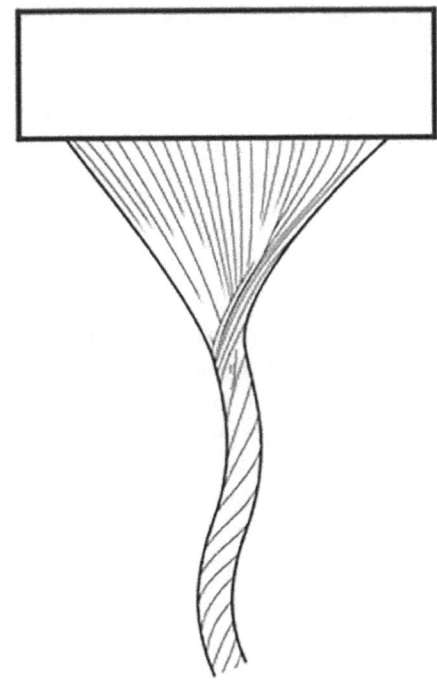

FIGURE 3.10
The twist V.

can be easily untwisted. The twist content can be easily measured and the twist direction is easily discernible. These are all important factors, as we will see soon when twist content and direction are discussed. The structure of conventional/traditional yarns also means that twist is more easily chosen to achieve certain properties and effects.

Of course, in the case of conventional short-staple spinning, the strand is twisted as it emerges from the nip of the front rollers and forms a *Twist V* (Figure 3.10). This results in the fibers on the outside of the V being less controlled which may result in a degree of hairiness on the surface of the yarn, which the spinners normally want to reduce as much as possible for most yarns though the hairiness can be used as a design feature, if desired. Also, fiber strands that have followed the carded yarn processing rather than the Combed yarn route will definitely have hairiness as a feature of their structure.

3.2.1.4 Color

It is very unusual to find cotton fibers or other fibers used in short-staple spinning being dyed in the fiber form. They are almost always spun in *ecru* (undyed) form. The dyeing is mostly carried out in fabric form for woven fabrics, but yarns for knitting are often dyed in the yarn form. There is therefore not the plethora of fiber color design techniques found in short-staple spinning as can be found in woolen/condenser spinning and to a lesser extent in worsted/roller-drafted yarns. Having said that, interest has been shown in recent times about the possibility of dope dyeing man-made filaments to produce mixture/mélange staple yarns to reduce the costs of holding large stocks of different colors in the yarn or fabric stage. Mixture and mélange yarns will be discussed in the sections on condenser spinning and roller-drafted spinning.

One color effect that is often found in short-staple yarns is known as marl or sometimes twisted marl. This is a simple but effective technique whereby two or more yarns of different colors are folded/twisted together. Careful choice of colors can offer exciting design possibilities. The reader may find that people from different parts of the textile community use the term *marl* in different ways and all will claim that their definition is the correct one. Just be prepared to be flexible or stand your ground, it depends on the situation.

3.2.1.5 Linear Density

With regard to interest and fun, very few areas of design technology offer more than linear density. Earlier in this chapter, the term thickness was used to describe yarns as this is a term most of us are familiar with. The term linear density, however, was introduced and it was explained that this is the more useful/accurate term when describing yarns. Thickness cannot be accurately controlled in a yarn due to the structure of fibers of different fineness, length, and numbers in the cross section held together by twist. The cross section of a worsted yarn is shown in Figure 3.11 and you can see the amount of air between the fibers that would be altered depending on the fiber length, spinning process, and fiber fineness among many other factors.

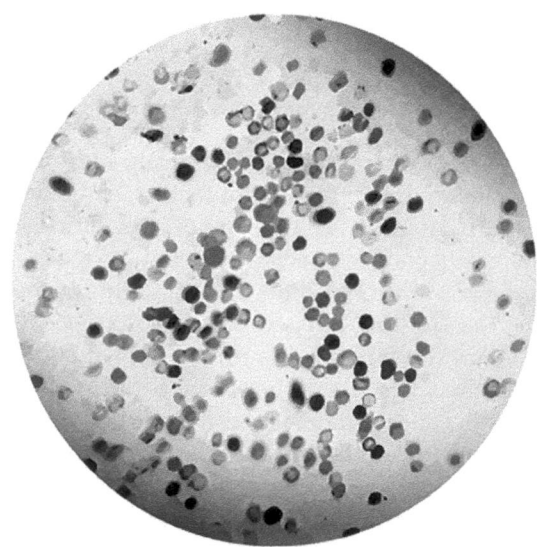

FIGURE 3.11
Cross section of a worsted yarn.

It was decided long ago, by spinners and fabric manufacturers, that the easiest and most accurate way to describe yarns for their different end use types was weight per unit length. This was probably because the spinners would use hanks/skeins as the form of yarn package in early times and even well into the establishment of the Industrial Revolution in the United Kingdom. Now these hanks would be formed in a multitude of different lengths of yarn, but at least each spinner knew the length of their own product and could place a number of them on one side of a balance with a known weight on the other side. The hanks/skeins would be measured in yards and the weights would be in pounds and ounces in some cases. Therefore, if the spinner used a pound weight, then the number of hanks of a certain yardage used to balance that weight would be called the yarn count. The more hanks, the finer the yarn, the higher the count; therefore, these count systems were later called *Indirect Counts* (remember, the higher the number, the finer the yarn). Fabric manufacturers would soon get to know which count suited their fabric-manufacturing process and would produce the correct fabric type and weight.

Once the industrial revolution arrived and yarns were being produced in volume, it was necessary to have some standardization of the hank/skein lengths and the weights being balanced against. In most cases, the easiest weight to use would be a pound, but what about the hank/skein lengths. Two scenarios are suggested: either the spinning mill owners in various localities would get together and decide on a particular length or the biggest mill owners in a locality would impose the length that they used due to their financial power. In the case of short-staple yarns, the length that finally emerged as the standard was 860 yards. So, the number of hanks/skeins that weighed a pound was known as cotton count (New English cotton). This would also apply to short-staple yarns produced from other natural fibers or man-made staple fibers when they came along.

At this stage (before the 1960s), linear density, therefore, meant the length per unit weight. The description of how this type of count system evolved will apply to the two other forms of conventional/traditional spinning system that will be discussed in Sections 3.2.2 and 3.2.3.

The reader might say, but they did not use grams in general society much before the 1960s, and that would be true, but when metrication came into general use, a new count system was developed, which was call metric count Nm (New metric). This was the number of hanks/skeins of 1000 m required to weigh 1 kg.

What then happens if the single yarn is folded/twisted with another to form a two-ply yarn, which is the most common form of yarn used? The resultant yarn is then designated as two times the single count. For example, if two single 30 Ne yarns were folded together, the resultant count would be 2/30 Ne, which would mean that the count of the folded yarn (two strands) would be 15 Ne. If we wanted to convert from Ne to Nm, the 30 Ne single yarn would be 50 Nm and the two-ply form would be 2/50 Nm, which would result in 25 Nm.

Are you with it so far? This topic can be confusing, so now is a good time to take a rest and thinks about linear density. It is a very important parameter for the design technologist to understand. We will come upon it again for the other staple yarn types, but the basic story you have learned here will apply for all.

By the late 1950s and early 1960s, the situation with the number of different count systems was becoming hard to handle and often caused confusion and sometimes even litigation, when yarn suppliers and fabric manufacturers found themselves using and communicating in different count systems/languages. This will become more evident to the reader as we learn about the other conventional/traditional spinning systems. Therefore, the industry, mainly through the auspices of the Textile Institute (the main professional body representing the industry through Royal charter), decided to develop a new count system that would be both metric and direct. The latter term meaning that it would use the weight per unit length, and thus the higher the number, the coarser the yarn would be and the lower the number, the finer the yarn. More sensible, I hope you agree. The weight decided on was grams and the length for yarns was a kilometer (1000 m). Therefore, a 30 Tex yarn means that a 1000 m of the yarn weighs 30 g. Also, in a common sense manner, when we fold/twist two such yarns together, we get a resultant 60 Tex yarn (designated as R60 Tex/2). If we fold/twist three such yarns together, we would get R90 TEX/3. Much simpler, and it was hoped that all sectors of the textile industry would adopt this system. However, till 2016 the Tex system was nowhere near being adopted universally; for example, knitters still prefer the metric count Nm.

A last note here is to be aware that fiber/filament linear densities are given in decitex (grams/10,000 m) and thicker assemblies like slivers and rovings are given in kilotex (kilograms/1,000 m).

3.2.1.6 Single Yarn Twist Content and Direction

The amount of twist inserted into a yarn is very important and affects many characteristics such as strength, elongation, liveliness, fabric handle, appearance, durability, pilling propensity/tendency, and luster. We are therefore discussing a very important design technology parameter. The idea is to try to keep the angle between the diagonal formed by the turns of twist in the yarn and the yarn axis the same no matter what the count is (Figure 3.12).

A useful way of doing this is to make use of a twist factor and here are three examples of how this would work for the three count systems you have met so far.

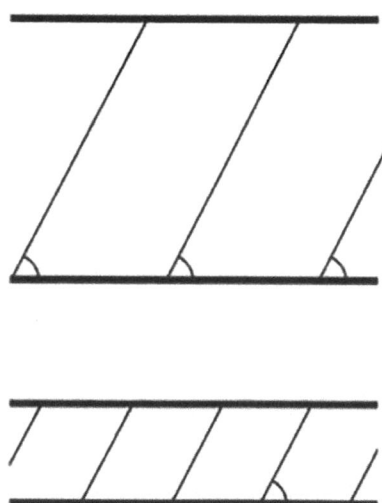

FIGURE 3.12
Two yarns of different counts with the same twist angle/factor maintained (Z twist).

$$\text{Turns per meter } (t/m) = \frac{\text{twist factor (called K factor for short-staple yarns)}}{\text{square route of Tex}} \quad (3.2)$$

$$\text{Turns per meter = twist factor} \times \text{square route of Ne} \quad (3.3)$$

$$\text{Turns per meter = twist factor} \times \text{square route of Nm} \quad (3.4)$$

Figure 3.12 shows two yarns of different count but using the same twist factor to determine the amount of twist to be inserted. Many design technologists use twist factors for different customers because they understand the differences in target consumers. For example, if a spinning company is supplying two knitwear companies with a high value soft handle yarn, they might know from experience and knowledge of their customer that one set of consumers are relatively well off and desire maximum softness of handle with little regard for durability and pilling performance. This yarn will therefore use a low twist factor (e.g., 3000 divided by square route of 50 = 424 t/m). The other company's consumers may desire a nice soft handle but may also desire the knitted garment to last longer and pill less, and so a higher twist factor may be used (e.g., 3200 divided by square route of 50 = 453 t/m).

In terms of twist direction, we have two options S or Z (Figure 3.12; Z twist). Perhaps putting it a bit simply but not without foundation, mechanized weaving came first and turning spinning machine spindles in a clockwise direction (S twist) was a natural mechanical choice, so this was called normal twist and used mainly to supply weavers. Turning the spindles in the opposite direction (Z twist) was then used to supply the knitting industry and was called reverse twist. It is also important to realize that using twist direction can give light reflection (luster) and fabric handle effects and is therefore an effective tool for the design technologist.

3.2.1.7 Folding Twist (Multi Ply) Content and Direction

When plying two single yarns together, it is most common to use *twist-against-twist*; in other words, if the singles twist direction is Z, then we use S for the folding twist and vice versa, if the singles twist is S, then we use Z in the folding twist. It is difficult to give definitive values for the amount of plying twist to use, but it will often be approximately half as much as the singles twist, but this value will decrease as the number of plies increases. For example, in commercial handknitting four ply yarns, the value tends toward 1.69 times the singles twist, but again will differ depending on the yarn design technologists' subjective assessment of optimum appearance and handle.

Let us now turn our attention to the design principles that can be achieved through the use of design elements that are parameters affected or manipulated through the component fiber properties and machine settings used in the spinning system.

3.2.1.8 Symmetry/Asymmetry

As most yarns are used to create knitted and woven fabrics, it is important to maintain symmetry in the yarns to not adversely influence the fabric appearance (e.g., through a skewed effect on knitted stitches called *spirality*). However, there are a number of options that can be used to create asymmetry in yarns and fabrics if such an effect is desired to enhance design aspects of the fabric. The following are the main ones, but more detail will be given on this principle in Chapter 4 on fancy yarns.

- A spiral yarn can be achieved by mixing two or more yarns having two opposite twist directions. If a Z and an S twist yarns are folded using S twist, the Z singles twist yarn will have as much singles Z twist removed as the amount of folded twist used and the S singles yarn will gain the same amount. This will make the S singles yarn contract and become thinner, whereas the Z singles yarn will extend and become thicker and so this yarn will spiral around the thinner yarn.

- Another way to produce a spiral effect is to fold together two (or more) yarns of different linear density. In this and the previous example, there are no precise twist or linear density values that can be prescribed. Experimentation will lead to different effects and experience will be gained along with a lot of fun. Both these methods use conventional/traditional spinning equipment. Another method is to feed two or more yarns at different delivery rates, but that is moving into the realms of fancy yarns and will be dealt with in the respective chapter. Asymmetrical yarns are also produced by hollow spindle spinning. Again that will be covered in Chapter 5.

- Yarns with an elastic nature can be produced by folding them together using the same twist direction as was used in the single yarn spinning process. The choice of twist contents, however, is very critical and the yarn would only be sparingly used in the fabric or skewness will be inevitable. This is how elasticity was achieved in yarns before elastane filaments and fibers became available.

3.2.1.9 Evenness/Regularity

In most cases, the aim of the spinner is to produce as even/regular a yarn as possible. In other words, they do not want the yarn count to vary along the yarn length, as this will cause many different types of faults in the knitted or woven fabric. There are three different wavelengths of fault that are commonly found as a result of poor fiber qualities and/or poor fiber preparation and spinning equipment settings and/or condition. The most common parameter used to describe the degree of evenness/unevenness in a yarn is the U% or CV%. For most of us, the latter value is easiest to understand: it is the amount of variation from the average count expressed as a percentage of that average count. The U% is more or less the same but was used in the days before the readings of the measuring equipment were dealt with by digital computers and an analogue value was produced by less sophisticated equipment. As people can take a long time to change, the reader might still find individuals who prefer to use U% rather than CV%. In both cases, the higher the value, the poorer the yarn in terms of evenness and you can use the expression $CV\% = U\% \times 1.25$ to convert, if required.

3.2.2 Worsted/Roller-Drafted Spinning

Like the short-staple system, this form of conventional/traditional spinning takes a long time to go from fiber to yarn and combing is again used to remove short fibers and to achieve the optimum degree of fiber alignment/orientation along the yarn axis. A major difference from short-staple spinning is the use of *scouring* (washing) at the beginning of the process when using natural protein fibers such as wool for which the system was first

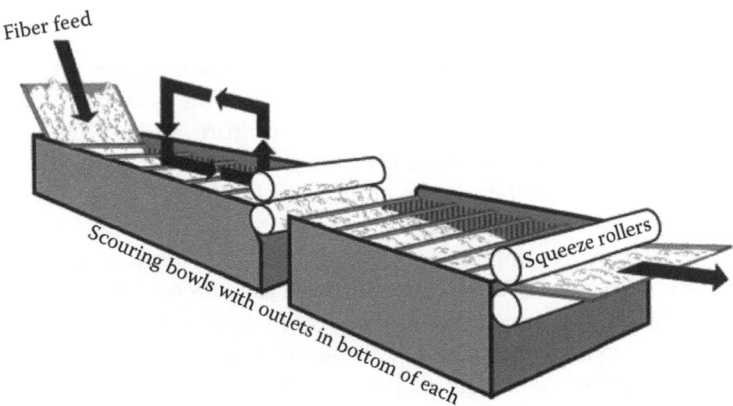

FIGURE 3.13
Scouring bowls.

intended. When the wool arrives at the Scouring plant, it is likely to have been *classed* into the general quality it belongs to. The classing system most commonly used was promulgated by the U.S. Department of Agriculture, and the higher the number, the finer the quality. For example, fine quality merino wool would be classed from 60s up to perhaps 76s and sometimes a little higher and the diameter of these fibers would be between 22 and 18 μm. The wool at this stage is full of impurities, which must be removed at the early stages of fiber preparation for spinning. Natural secretions are suint (sweat) and grease. The fine the fiber, the more of these natural secretions will occur (e.g., a fine merino may hold as much as 8%–10% suint and 30% grease). These will be removed at scouring, but other acretions such as kemp (dead fibers) bugs, dirt, and vegetable matter must be removed at blending or at carding.

The scouring process consists of a series of tanks/bowls (Figure 3.13) holding warm water with soap and alkali and a series of forks or rakes to push the wool through the tanks. The level of soap and alkali is reduced at each subsequent tank and the last tank generally contains only water to rinse the wool before it is fed into a dryer. The suint, which has been removed in the scouring process, is a natural form of lanolin and often sold to cosmetic companies by the Scorers. Radio frequency dryers were popular in the 1970/1980s, but most plants today use conventional hot air dryers and the main technological issue is to avoid over-drying.

3.2.2.1 Blending

Blending is carried out at this stage to ensure uniform properties of the yarn; to reduce variation in fabrics; to exploit the peculiar characteristics of each component of the blend to the full; and to reduce cost. It is rare to dye at the fiber stage in the roller-drafted system and so blending is carried out in large bins into which the fibers are fed through pneumatic pipes and a rotary spreader, which horizontally layers the fibers down the full length of the bin. They can therefore be removed to the carding machine in vertical layers giving more blending power. This is a large-scale process, but not as interesting to the design technologist as blending in the woolen/condenser system, so we will not spend a lot of time on this. The important issue is that the blender ensures as homogenous a mix/blend as possible.

3.2.2.2 Carding

The Worsted carding system uses roller and clearer and is a bigger and longer machine than was described for short-staple spinning but not as long as in the woolen/condenser system, which also uses roller and clearer carding (Figure 3.14). This is an appropriate place to describe the fundamental technology used in roller and clearer carding. The objectives of this process are to disentangle the clumps of wool fiber, to remove any burrs and other vegetable matter, and to continue the blending process. The main disentangling takes place between the teeth (wires) of the cylinder or swift and the workers using a point-to-point action (Figure 3.15). To move the fibers through the process, stripping rollers are used to remove or strip the fibers from the workers through a point-to-back action and these are in turn stripped by the cylinder or swift by the same action. Finally, a fancy roller is used to raise the fibers out of the swift using a back-to-back action, so that they can be formed

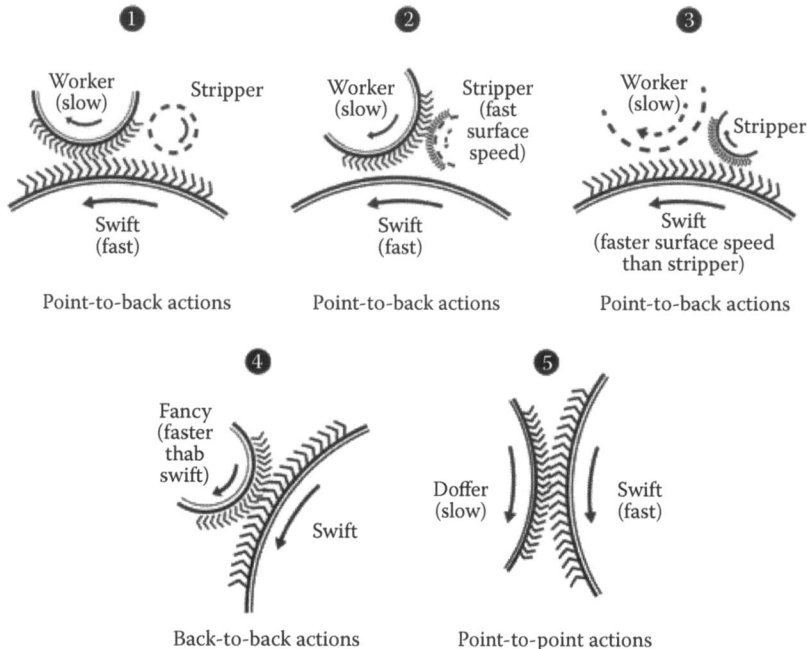

FIGURE 3.14
Carding actions between the rollers.

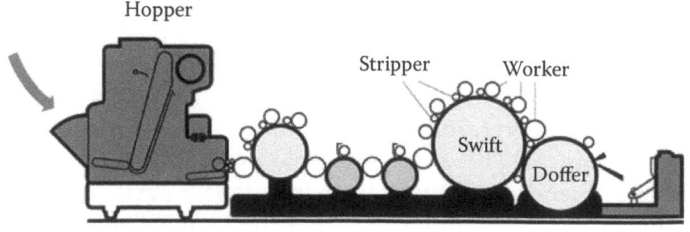

FIGURE 3.15
A worsted card.

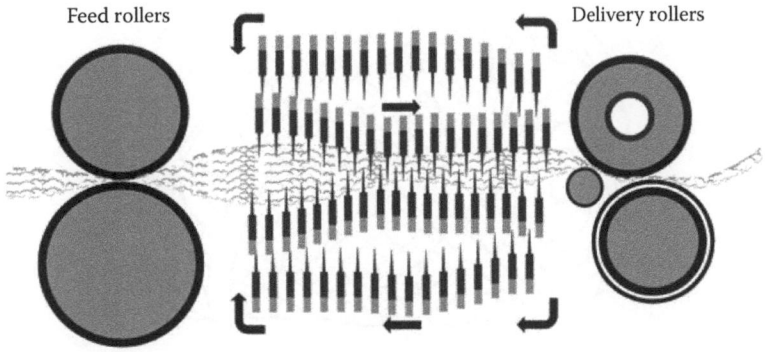

Feed rollers   Delivery rollers

Typical draft

FIGURE 3.16
Gill box.

into a wide fibrous web. The actions are brought about due to the direction of the teeth or wires on the various roller types and their relative surface speeds.

At the end of the Worsted card, the web of fibers is condensed into a rope-like structure called a sliver (as in the short-staple system), which is then coiled into a can for delivery to the drawing stage.

3.2.2.3 Drawing

Drawing on the Worsted system uses roller drafting as discussed for short-staple and a number of slivers (often called doublings) are fed in to continue the blending process. The main difference between a short-staple drawbox and a Roller Drafted spinning gill box is the steel pinned combs that act exactly like a comb being drawn through your hair (Figure 3.16). These control the fiber movements and are very effective in orienting the fibers all in the same direction (parallelizing).

3.2.2.4 Combing

The carded slivers are fed through two or three of the gill boxes and are then fed into a Worsted combing machine. As with short-staple spinning, the combing machine uses a technologically complicated process. As design technologists, we do not need to study this process, but rather it is important that we understand that it is used to remove short fibers (known as noil) from the slivers to allow the processing of the highest quality worsted/roller-drafted yarns with as little evidence of hairiness as possible.

3.2.2.5 Top Making

This stage consists, once again, of a gill box at the delivery end of which the sliver is wound/coiled into a self-contained and easy-to-transport ball top (Figure 3.17) of around 3–5 kg or packed into a large 25–30 kg bump top. Both forms are then transported to the fiber preparation for spinning stages. Nowadays, the conversion of raw material to a combed top form is mostly carried out in a separate mill to the fiber preparation and spinning mill.

Sometimes the tops will be dyed before delivery to or at the spinning mill.

FIGURE 3.17
A worsted ball top.

3.2.2.6 Drawing

This process also uses gill boxes (though at this stage they are usually called drawboxes). There are normally three drawing/gilling stages used at this stage with progressively finer toothed combs and increasing drafts. The main difference is that an auto-leveling system is used on the second and third box to continuously alter the draft through the speed of the back rollers. If the slivers get heavier, the back rollers are slowed down giving more draft and if the slivers are lighter, the back rollers are speed up giving more draft. By this means, the linear density of the emerging sliver is kept as even as possible.

3.2.2.7 Roving

This is very similar to the speed frame/roving process described for short-staple spinning. Again, a sliver is fed in to the back rollers of a roller drafting system and aprons are used to control the movement of fibers in the drafting zone. The linear density (count) of the sliver fed to the roving machine will be in the region of 10 kilotex and a draft of around 10 will be used to reduce the count of the sliver to 1 kilotex. At this linear density, a small amount of twist is required to provide cohesion to the roving, which is what the sliver is now called and this twist is also of importance to help in the drafting of the roving down to the required linear density at the spinning process.

3.2.2.8 Spinning

Worsted spinning frames are very similar to short-staple frames, using roller drafting, aprons to control the fibers during drafting and a ring and traveler system to insert the required amount of twist and to wind the yarn onto a tube. Winding onto cones and folding or twisting into two ply yarns is carried out as for all conventional yarns.

3.2.2.9 Introduction of Man-Made and Synthetic Fibers into Worsted/Roller Drafted Yarns

To use man-made and synthetic fibers in any yarn, the spun continuous form in which they are extruded must be cut or broken to form staple fibers of the correct/required length.

Probably the most common yarn type using stretch-breaking technology would be high bulk acrylic yarns, which can be found in cheaper knitwear garments and is particularly well used for the cheaper end of the handknitting yarn market. The continuous filament tow of acrylic filaments is passed through a set of heating plates and then a series of rollers, which are increasing in speed from start to finish therefore stretching the filaments until they break.

3.2.2.10 Design Elements and Principles of Worsted/Roller Drafted Yarns

Most of the design elements and principles discussed for short-staple yarns apply for Worsted yarns; however, the range is wider. In other words, the length and fineness of fibers is much more varied and can be used to produce a wider range of effects. The wool fibers are naturally coarser than cotton, but Worsted spinners will rarely use fibers more than 26 μm and often prefer the 18–24 range. That does not mean they cannot spin coarser fibers if required. Worsted spinners also prefer longer length fibers and will remove shorter fibers (noil) at combing and often sell these on to the woolen yarn spinners.

Worsted tops are often dyed allowing mixtures/mélange colors to be produced by blending different colored tops at the drawbox stages. A *Roved Union/Spun Marl* (depending on the terminology used by the people in the mill location) is spun by feeding two different colored rovings into the drafting zone of ring spinning machine. This type of yarn can be found in the tartan stockings/socks worn by Scottish country dancers or the Scottish military regiments. It is also found in many effect handknitting yarns. A *Marl/Twisted Marl* (again depending on local terminology) is where two or more spun yarns of different colors are plied together.

Worsted yarns are finer and smoother than their woolen spun counterparts. Worsted/roller-drafted yarns are used for by far the majority of handknitting yarns, although cotton/short-staple yarns are also used. Woolen spun yarns, which will be discussed next, are rarely used in the market. If the reader wants to observe a very effective comparison, then he or she can locate a woolen spun lambswool sweater and a botany worsted sweater. Both will have been knitted with yarns of similar quality yarns, in terms of fineness, but the clarity of the knitted loops in the botany will be much higher, the botany will also tend to look smarter and will be less prone to abrasion. Fine suiting is made from worsted spun yarns and tweed jackets are made from woolen spun yarns.

3.2.3 Woolen/Condenser Spinning

This is by far the shortest of the conventional spinning systems and a yarn can be produced from scoured and dyed fibers. The yarn of this type will of course have a hairier, coarser, and less sophisticated appearance than either short-staple or worsted yarns. The fibers are arranged in a disoriented, haphazard manner (sometimes termed *higgledy piggledy*) and the fiber lengths will vary much more than in the other conventional yarn forms. Dyeing and/or bleaching will be carried out at the fiber stage, which is also different from the other systems and so the first stage of fiber preparation in the spinning mill is that of blending.

3.2.3.1 Blending

Much of the description given for blending in the Worsted system applies also to this form of blending. The main difference, however, is that the blending equipment used in this system has to ensure that the mixing of the color effects is carried out effectively to produce as homogenous a mix as possible. The color effects can be simply considered in two categories: (1) solids, where the yarn will only contain one color of fibers and (2) mixtures/melanges, where a number of different colors are blended together. The latter produce very interesting color effects, which can be used valuably by fabric designers. Often 5–8 colors are used, but use of as many as 12–14 colors is also possible. A small percentage of oil, normally as an emulsion with water, is added at this stage. The oil is used to help the fibers through the subsequent carding and spinning processes, but it also has an important role in fabric finishing.

For example, when a fine lambswool or cashmere woolen spun yarn is knitted, the space between the loops is quite wide, but the oil acts with the soap and water during the finishing or washing processes to close up the loops to give an appearance that makes it quite difficult to discern the stitches visually, this provides a more casual appearance than for worsted or cotton yarns where the fabric stitches can be easily seen and the appearance tends to be smarter and more formal.

3.2.3.2 Carding

The woolen/condenser spinning system uses the longest, by far, carding machines of the conventional spinning systems and it can be well argued that it is at this stage that the woolen yarn's quality is decided; some would say this is where the woolen yarn is made. Of course, as we will see, the spinners have their say and their input, but it is fair to say that carding is far more important in woolen yarn production than in worsted or cotton spinning. The machine uses the same roller types as in worsted carding and the same actions, so we do not have to go over those again, though it may be good for the reader to recap.

The woolen card consists of two or three sections depending on the quality of yarn being processed. The fibers are fed by a hopper to the first part of the card called the scribbler. Nowadays, the hopper will almost always use a system to make sure that the fibers are fed in evenly weighed amounts to ensure that the feed is even along the length of the machine, resulting in an even final web (just like the autoleveller in worsted and cotton carding). The scribbler consists of two cylinders/swifts and the opening teeth or wire are quite coarse/severe to give a strong working action.

At the end of the scribbler, the web is removed from the doffer and brought together as a continuous sliver-like assembly and fed to the next section, which is an intermediate, and the sliver is cross fed onto this to ensure that the fibers do not stay on the same part of the carding rollers' surface throughout the machine; this gives improved mixing. The opening teeth/wire at this stage is finer than at the scribbler. At the end of the intermediate, the web is taken from the doffer and again brought together as a form of sliver and again cross fed to the final part of the machine called the carder. This is normally a two swift section and at the end of this is a particularly important section called the tape condenser (this is the reason why woolen spinning is also known as condenser spinning). The web is removed from the doffer by the doffer comb and is then fed into a set of tapes moving continuously and splitting the web onto the surface of the tapes as a ribbon of fibers (Figure 3.18). These fibrous ribbons are then fed through a set of rubbing aprons, which condense the ribbons into a cohesive circular fiber assembly called a slubbing.

After this, the slubbings are wound onto bobbins each holding 15–30 slubbings depending on the linear density of the slubbings, and there are normally between 4 and 12 bobbins depending on the design of the carding machine.

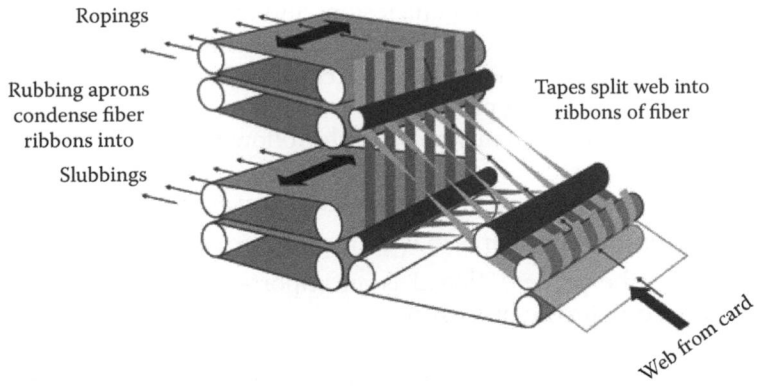

FIGURE 3.18
A woolen card tape condenser.

3.2.3.3 Spinning

There are two types of drafting that can be used for woolen spinning. Spindle drafting or ring frames. The drafts used are rarely more than 1.5 and are usually much less that (1.3–1.35). Spindle drafting is now only used for the finest yarns such as cashmere, lambswool, or vicuna. The machine using this type of drafting is called a mule or Jenny (Figure 3.19).

The slubbings are fed from the bobbins to a pair of rollers and from there to a tube set upon a spindle, which revolves to insert twist and to wind the yarn onto the tube. The spindles are set in a carriage, which moves away from the delivery rollers for a distance of about 2 m and the rollers stop after the spindle carriage has moved about 150 cm. Therefore, the draft is 200 divided by 150 equals 1.33. So, if the slubbing is 70 Tex, then the yarn will be 70 divided by 1.33 equals 53 Tex. The draft can be altered by delivering more or less slubbing before the rollers stop. The twist that is continually inserted as the carriage moves outward will run into any thin places in the slubbing making these tight and so that the thick places are given more draft and thus the yarn will be evened out. At the fullest

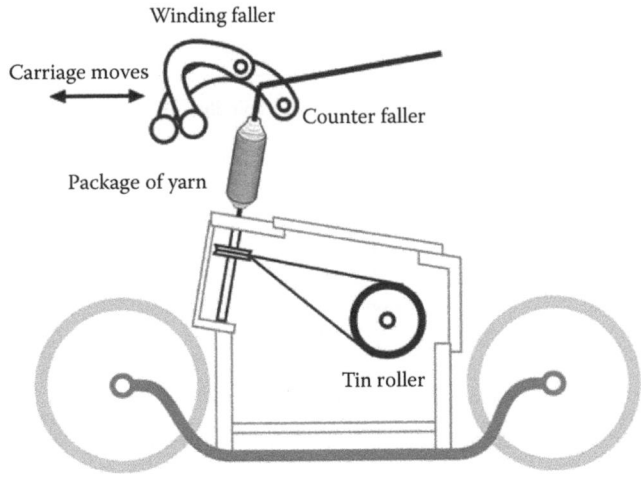

FIGURE 3.19
Woolen mule carriage.

extent of the carriage's outward movement, it stops and the spindles revolve at high speed to insert the final twist content to be inserted. Then, the faller bars engage to provide a means of winding the yarn onto the tubes as the carriage moves back in toward the delivery rollers to begin a new drafting phase.

Mule spinning gives the best quality yarn for the following reasons:

- The long drafting zone, which allows an excellent evening effect.
- The gentleness of the faller wires in winding the yarn onto the tubes.
- The slowness of the process. Speed is an important parameter in any type of machine involved in textile processing.

Nowadays, a form of mule in which the bobbins move backward and forward instead of the spindles has been adopted. This makes sense, as it is much less complex than a moving spindle carriage. Also, digitally controlled electric motors are used for the various moving parts. Is it assumed that these give better quality than ring frames, but it is difficult to find research to back this up.

3.2.3.4 Woolen Ring Frame

The woolen ring spinning frame uses roller drafting, but with a false twist tube just before the nip of the front rollers (Figure 3.20).

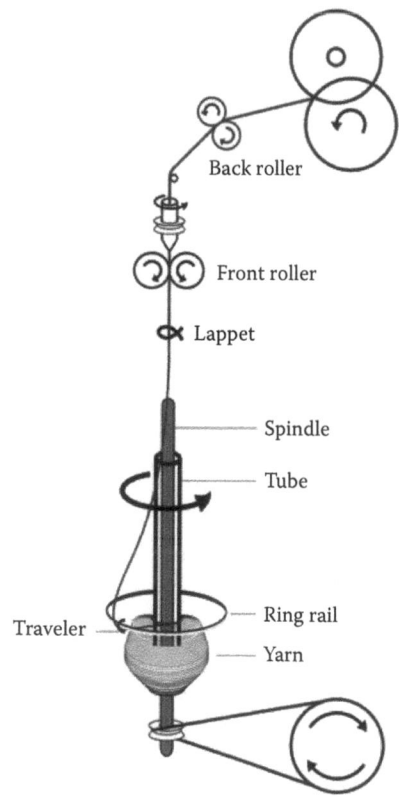

FIGURE 3.20
Woolen ring frame.

LEARNING ACTIVITY

Try getting two friends to hold each end of a ribbon of material and you start to twist/turn it in the middle. You will find that the twist direction is different on either side of your twisting hand(s).

If the ribbon was moving through your twisting hand, then the twists on either side would cancel each other out. This is what happens when a false twist tube is used. The twist inserted from the top of the revolving tube to the back rollers is real, but is cancelled out by the twist inserted as the slubbing emerges from the tube and so the slubbing is delivered by the front rollers with zero twist, to a ring and traveler arrangement the same as is used in short-staple/cotton spinning and worsted spinning. The twist inserted into the resultant yarn is therefore determined by the rotational speed of the traveler relative to the delivery speed of the front rollers.

The twist inserted in the drafting zone has the same effect as it does on the mule (spindle drafting), but the drafting zone is much shorter (about 5 m) and the machines are faster. These types of woolen ring frames are used for all yarns apart from the very finest and even sometimes for these as long as the quality produced is adequate for the requirements of the fabric producers. The yarn is then cleared and wound onto cones and can be used as a single yarn, but most commonly folded into two or more ply.

3.2.3.5 Design Elements and Principles

The elements that apply to woolen/condenser spinning will mostly be the same as for short-staple and roller-drafted yarns. The fiber length and fineness, however, can often be more variable, as the nature of the disoriented structure of woolen spun yarns is more forgiving particularly because there will be many more fibers (minimum is normally around 100 as opposed to around 40 in the other yarn types) in the cross section of the yarn. Of course, the finer the yarn, the less is the variability that can be accommodated. For example, fine cashmere and lambswool woolen spun yarns will demand a very low level of variability in fiber length and fineness, whereas carpet yarns can allow a much greater level of variability. Color is applied at the fiber stage of woolen spinning and so it is possible to create wonderful mixtures (often called melanges) using a wide range of fibers of different colors mixed together at the blending process. Also, it is possible to drop small multicolored knops onto the web during the carding process, which gives yarns known as KB (Knickerbocker), Knop, or Donegal yarns. The Knickerbocker name comes from the use of these yarns in the tweed knickerbockers used by golfers in the early part of the twentieth century.

In terms of principles, symmetry, asymmetry, and balance can be achieved by the manipulation of singles and folded twist contents and direction. The evenness (U% or CV%) of woolen yarns will be not as good as short-staple and roller-drafted but still has to be carefully checked and controlled to give good-quality fabrics and garments without short-, medium- or long-term irregularities, which can cause thick and thin places, striped, and bars in the fabric. Liveliness will be affected by twist content and if the yarn is too lively, it can be settled down with exposure to steam. woolen/condenser spun yarns will be bulkier than the other two conventional systems and tend to be coarser. The relative bulkiness and coarseness of texture of woolen yarns give a denser appearance to the resultant fabrics and a more casual appearance in garments, better cover in upholstery fabrics, and carpets.

3.2.3.6 *What Have You Learned about Conventional Yarn Production?*

Try the following questions:

- Which is the shortest of the conventional yarn spinning systems?
- What are the carding actions in roller and clearer carding?
- What is the combing process used for?
- What is the difference between roving and slubbing?
- What is the definition of Tex?
- Name three other count systems that you know?
- How are fibers controlled in the drafting zone of a Worsted Gill Box?
- How are fibers controlled in the drafting zones of roving/speed frames and cotton or worsted spinning machines?
- If the back rollers of a spinning frame are revolving at 100 cm min^{-1} and the front rollers are revolving at 2000 cm min^{-1}, what will be the draft?
- If a roving of 1 kiloTex is fed into the spinning frame, what will be the yarn count in Tex?
- What twist direction will be used to produce a single yarn for knitted fabric production?

If you answered eight of the aforementioned questions correctly, you can proceed to the next section, but still read up on the three wrong answers. If you answered less than eight correctly, then read this chapter from the beginning again.

3.3 Unconventional and New Spinning Technologies

In the latter part of the last century, most spinning innovations were aimed at improving production rates. The thinking that drove many of these developments was that if the twist insertion method could be separated from the yarn package formation, then larger packages could be achieved and at much higher speeds. The method that made the first and probably the biggest entry into spinning mills was that of rotor spinning also called open-end spinning.

3.3.1 Rotor/Open-End Spinning

Rotor spinning is used as an alternative to the ring spinning of short-staple yarns. A sliver from a drawing frame is fed into a very fast revolving opening roller, which breaks down the sliver and feeds it into a tube, which delivers the fibers individually into the groove around the inside edge of a high-speed revolving rotor. The revolving of the rotor causes a centrifugal airflow, which allows the fibers to be sucked off the opening roller and fed to the groove. An open end of preformed yarn is then sucked into the groove and the fibers join onto the end of this yarn and twist is inserted by the revolving rotor (Figure 3.21). The winding force (pull) of the package is sufficient to overcome the suction force of the rotor and so the formed yarn is delivered to the winding system, which produces a large cone package.

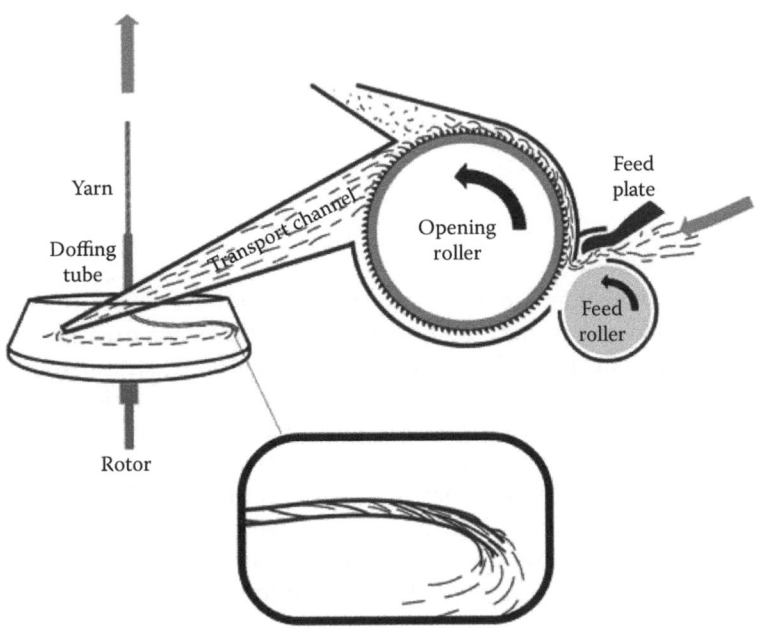

FIGURE 3.21
Rotor/open-end spinning.

The structure of rotor spun yarns is different from that of ring spun. The fibers closest to the core of the yarn have much more twist than those on layers closer to the surface of the yarn. An easy way to identify if a yarn has been rotor spun is to try to carry out a conventional twist test, it simply cannot be done. The strength and elongation properties are adequate for most purposes and the evenness is often better. The handle of fabrics made from rotor spun yarn, however, is significantly harsher than ring spun and the abrasion resistance is poorer. These last two properties/design principles limit the end uses of rotor spun yarns, although the volume of consumer demand has resulted in rotor spun yarns having around 35% of the textile and apparel market, much higher than any other alternative spinning system developed since the late 1960s. In recent years, a number of marketing ploys have been used by fashion retailers by branding products as *ring spun*, but the significance is rarely understood by consumers, unless of course they are design technologists.

3.3.2 Air Jet/Fasciated Yarn Spinning

Air jet is the next invention of the 1970/1980s that has made a significant impression on the textile market. The generic term is *fasciated* and air jet is the trade name of the most successful of this type of spinning system. The term *fasciated* comes from the structure of the yarn (Figure 3.22), which resembles the *fasces* held as a badge of office by the praetorian guard in ancient Rome. The fasces was a bundle of axes held together by leather straps.

LEARNING ACTIVITY

Look for images of a Roman fasces and compare it to the structure of an air jet yarn given in Figure 3.22.

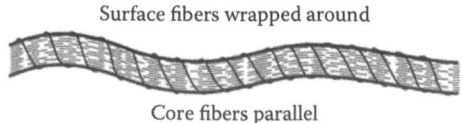

Surface fibers wrapped around

Core fibers parallel

FIGURE 3.22
Air jet/fasciated yarn structure.

   This spinning system makes use of false twist, which you were first introduced to in the description of a woolen ring frame. You will remember that if we turn a strand of fibers in the middle, we get opposite twist directions on both sides of the strand and if the strand is moving, we get a resultant zero twist. This is what happens with air jet/*fasciated* yarns, but the action of the fast-revolving air jets causes a few of the fibers on the extreme outside of the strand to escape the twist in one of the directions and therefore they receive real twist as they emerge from the air jet and so they wrap around the core. Nowadays, two air jets in series, revolving in opposite directions, are used to give more strength (Figure 3.23). Even so, air jet/*fasciated* yarns have low strength and elongation but sufficient for certain fabrics and end uses and their softness is good. The market for these yarns has been dwindling from around a maximum of 5%.

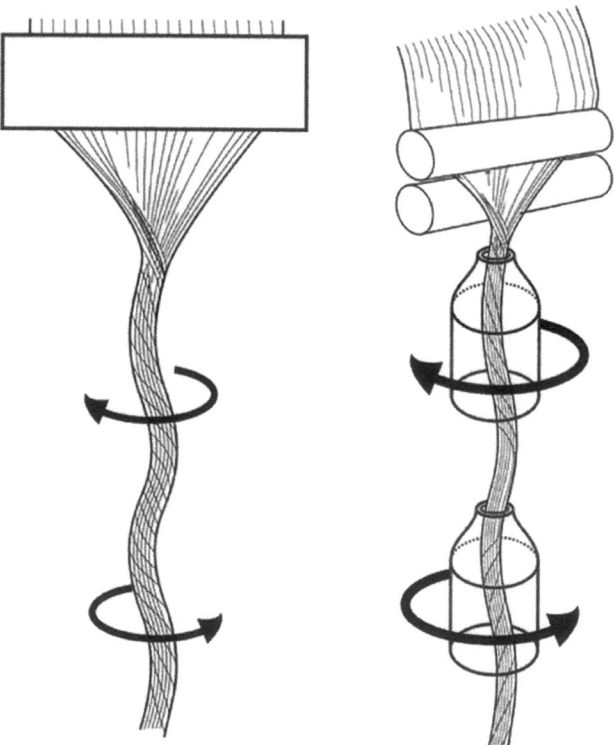

FIGURE 3.23
Air jet spinning.

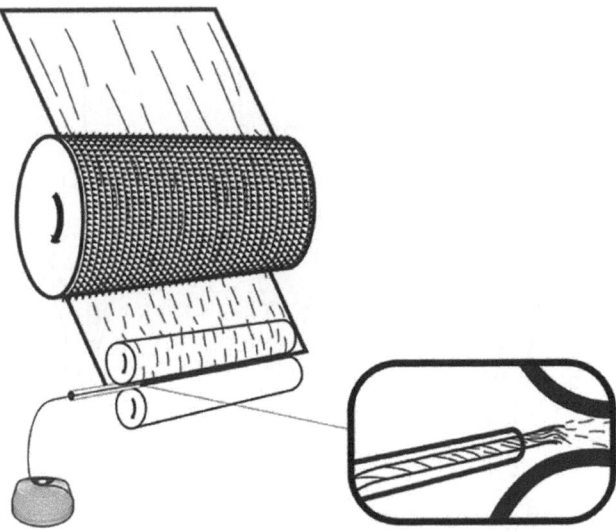

FIGURE 3.24
Friction spinning.

3.3.3 Friction Spinning

The best known of the friction spinning systems was and is Dref, an acronym formed from the name of its inventor Dr. Ernst Fehrer. The system employs a rotating carding drum for opening slivers into single fibers (Figure 3.24). The fibers are then stripped off the carding drum by centrifugal force and carried into the nip of two perforated spinning drums rotating in the same direction. The fibers are then twisted by mechanical friction on the surface of the drums and join an open end of yarn introduced to collect the fibers in the same way as in rotor spinning. The yarn is then wound onto a cone.

The main problem for friction spinning has been the low breaking strength and some similar systems, for example, the Platt friction spinner, failed commercially. To overcome this problem, a core continuous filament is introduced around which the staple fibers gather. This limits the end uses for which this type of yarn is used, but it can be successful for elasticated hosiery yarns and other speciality products; however, market share is and will remain very low.

3.3.4 Compact Spinning

This is not a nonconventional spinning system, rather an effective development and refinement of normal ring spinning that produces a quality of yarn that is a real improvement on conventional ring-spun short-staple/cotton yarns (Figure 3.25a). Basically, the bottom front drafting roller has a perforated center of around 1 cm through which air is drawn, which causes the drafted fiber strand to be drawn/compacted to form a narrow fiber stream (Figure 3.25b) as it emerges from the nip of the front roller to be met by twist. This means that there are very few fibers on the outside edges of the fiber stream and therefore an excellent reduction in the level of hairiness. This system has made great inroads into the fine cotton yarn system (i.e., 4-10 TEX). For coarser counts like the ubiquitous 2/30 Ne or R40 TEX × 2 (each single yarn being 20 Tex), it is probably an unnecessary expensive complication. The demand for yarn of this quality, however, may continue to grow.

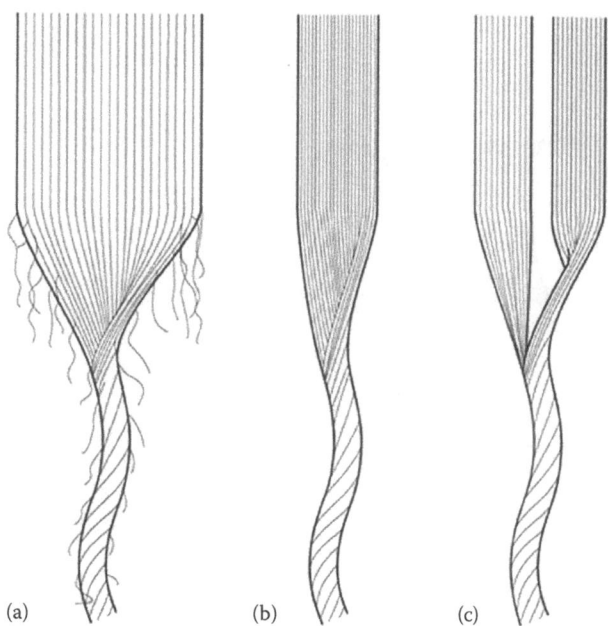

FIGURE 3.25
(a) Normal ring spinning, (b) compact spinning, and (c) Siro-spinning.

3.3.5 Siro-Spinning

This was developed in the 1970s and was aimed at Worsted spinning of wool. The idea was that, rather than spinning two single yarns and plying them into a two-fold yarn, two rovings could be fed side-by-side into the drafting zone of a worsted ring frame. The fiber strands are kept separated until they emerge from the nip of the front rollers and the strands are twisted together (Figure 3.25c). If the resultant yarn is untwisted, the two strands will be easily observed and therefore identification of this type of yarn is quite simple. The advantage of Siro-spinning is that hairiness is reduced, as the surface fibers are better trapped in the yarn structure and a process stage (twisting/plying) is removed. Disadvantages are that the yarn is not as well balanced, which may show in the fabric, and the load/elongation values are not as good as conventional plied yarns. Siro-spun yarns are not popular now among worsted spinners, but the technology has been, to a limited extent, adopted by the spinners of very fine cotton yarns in their efforts to reduce hairiness and the technology can also be combined with compact spinning technology.

Finally, it has to be noted that other nonconventional yarn spinning techniques have been developed, particularly in the second half of the twentieth century. Most of these enjoyed limited success and many have simply disappeared. It is unlikely that the design technologist will be called upon to use the yarns produced by those technologies, but he or she can use the understanding provided by this chapter to help them to observe, investigate, and understand new entrants to the market place, although, so far in this century, new spinning technology entrants are not much in evidence.

Bibliography

Brearley, A. (1963). *The Woolen Industry*. London, UK: Pitman.

Brearley, A. and Iredale, J.A. (1980). *The Worsted Industry*. Leeds, UK: WIRA.

Lawrence, C.A. (2003). *Fundamentals of Spun Yarn Technology*. Boca Raton, FL: CRC Press.

Peter, V. and Bowes, Q. (1988). Quality variations in cotton and their effects on bleaching, dyeing and finishing. International Cotton Conference, Bremen, Germany, p.5.

Shaw, C. and Eckersley, F. (1967). *Cotton*. London, UK: Pitman.

Soltani, P. and Johari, M.S. (2016). A study on siro-, solo-, compact-, and conventional ring-spun yarns. Part 1: Structural and migratory properties of the yarns. *The Journal of the Textile Institute*, 103(6), 622–628.

4

Continuous Filament and Texturized Synthetic Yarns

Arobindo Chatterjee

CONTENTS

4.1 Preamble

The production of artificial silk from natural cellulose polymer by De Chardonnet in France in 1892 marks the origin of the man-made fiber (MMF) industry followed by other cellulosics and acetates until the development of nylon by DuPont in the year 1939. Nylon was followed by the ICIs development of poly(ethylene terephthalate) (PET), discovered in the early 1940s.

From these early beginning, the MMF industry continues to develop and recorded a demand of 55.2 million tons of synthetic fiber in 2014. Out of all the different MMFs, the growth of polyester is significant and demonstrates a dominant role in the growth pattern. In 1980, polyester demand was only 5.2 million tons globally and in 2014 demand was put at 46.1 million tons, 73.4% of the total demand.

A very large part of the growth in polyester has come from China, with India and Southeast Asia also contributing. China accounts for 69% of all polyester fiber production globally, and if India and Southeast Asia are added, these three regions represent 86% of the global production.

TABLE 4.1

World Synthetic Production (1000 tons)

	Synthetics		
	Yarn	Staple	Total
2000	17,551	12,737	30,288
2001	17,719	12,596	30,315
2002	19,031	13,213	32,244
2003	20,016	13,753	33,769
2004	21,403	14,705	36,108
2005	22,706	15,431	38,137
2006	24,155	15,752	39,907
2007	26,562	16,596	43,158
2008	25,750	15,331	41,081
2009	26,551	15,964	42,515
2010	30,509	16,887	47,396
2011	33,106	17,662	50,768
2012	36,133	18,522	54,655
2013	38,817	18,653	57,470
2014	41,454	18,593	60,047

Source: European Manmade Fiber Association. http://www.cirfs.org/keystatistics/worldmanmadefibresproduction.aspx.

Polyester is dominant, but nylon, the oldest MMF, still plays an important role in the fiber business, with 4 million tons of global production in 2014. Nylon has developed into a niche fiber, in that it is focused on a limited number of end use, but some of these are quite large markets. Carpet, airbags, heavy-duty tires, cap ply for radial tires, intimate apparel, sheer hosiery, and swimwear are some of the important end use of nylon. There, however, has been remarkable growth in PET bulk continuous filament (BCF) yarn for carpet and also is now making inroads into the airbag market, particularly for the larger curtain air-bags. Table 4.1 shows the world's synthetic fiber production data for the last 15 years.

4.2 Introduction

A fiber or a filament is the basic building block or fundamental unit of any textile structure. It is a fine hair-like structure having the requisite properties specific to its application and should have the ability to be converted into a yarn or fabric form. A hair-like structure to be used as fiber must satisfy certain properties. Some of the basic property requirements are mechanical properties (strength, modulus or flexibility, extensibility, recovery from deformation, etc.); absorbency (ability to absorb moisture, dyeability, etc.); and stability (chemical, thermal, environmental, etc.). The dimension of the fiber and its diverse property requirements can only be met when fibers are made from long-chain molecules or polymers. Hence, all natural and MMFs are typically made of polymers. In a fiber, not only the type of molecules but their arrangement, that is, morphology is also unique. As fibers are very fine hair-like structures, most of the mechanical properties in a fiber are observed in one principal direction, that is the fiber axis. Accordingly, the long-chain molecules need to be arranged along the fiber axis, so that we have an oriented semicrystalline structure.

In most of the natural fibers, the oriented semicrystalline morphology is evolved with time during the whole growth cycle of the fiber or may be during the fiber formation stage, as in case of silk. Silkworms or spiders actually extrude liquid protein, which results in a highly stable filament on solidification. The stress acting on the polymeric fluid during extrusion and subsequent stretching leads to an oriented arrangement of molecules.

For producing MMFs, a process similar to the extrusion of silk is followed to achieve the unfolded or extended arrangements of polymer (either natural or synthetic) chains along the fiber axis and the process is known as fiber spinning. All MMFs are produced in the form of continuous filaments and subsequently converted into staple form as and when required.

The synthetic fibers, although they have many advantages compared with natural fibers, lacks in tactile and comfort properties. Most of the common synthetic fibers lack hydrophilicity due to the absence of polar functional groups and their smooth glass rod-like structures give them a typical plastic feel, which is not acceptable for their application in conventional forms of textiles.

Natural fibers have many advantages such as comfort, environment friendly, elegance, and safe to use (fire resistance), but they also have a number of disadvantages such as they are expensive, dimensionally unstable, prone to attack by mildew or insects, and require land or favorable conditions for the production. Similarly, MMFs also have a number of advantages and disadvantages. Synthetic fibers have high strength, durability, dimensional stability, resistance to bacteria or insects, wash and wear properties, no demand of fertile land, production independent of weather conditions, and most importantly the ability to have properties engineered. The main disadvantages of synthetic fibers are lack of tactile property and comfort, which are considered important for apparel applications.

For producing a fabric either by weaving or knitting, a continuous length of feed material in the form of yarn is necessary and from this point of view continuous filament yarns of MMFs can directly be used as feed material for fabric formation. There are certain problems associated with the material, process, properties, or production cost of the resulting filament yarns that makes them not always suitable to be used in their original form. In such cases, these filaments have to be converted into staple form and then further processed to convert theses staple fibers into continuous yarn form.

The staple yarn, due to its bulk and uneven and hairy surface, results in better absorbency and improved tactile property and hence many synthetic filaments are converted to staple to overcome these associated problems. Conversion of filament to staple is also done for blending them with natural fibers to produce a blended yarn with the object of balancing the attributes of both the natural and synthetic fibers to have an improved final product. The production of staple yarn involves additional process steps, such as crimping, cutting/breaking, and reprocessing through a series of machines/processes to convert them into the form of continuous staple yarn. To use synthetic fibers more advantageously in their continuous form, techniques have been developed for imparting bulk and surface texture and thereby imparting many of the characteristics associated with spun staple yarn bypassing all the process steps necessary for making a staple yarn. These processes of imparting bulk or modification of the characteristics of the continuous flat filament yarn are known as *texturing* or *texturizing* and are discussed later in Section 4.3.2. A brief classification of yarn is shown in Table 4.2.

TABLE 4.2

Yarn Classification by Physical Properties and Performance Characteristics

Yarn	General Yarn Properties
Staple yarns	Excellent hand, covering power, comfort, and textured appearance
Continuous filament yarn	Excellent strength, uniformity, fineness, fair hand, and poor covering power
Novelty yarn	Excellent decorative feature
Industrial yarns	Purely functional and designed to satisfy a specific end use
High bulk yarn	Great covering power, light weight, and fullness
Stretch yarn	Stretchability, good hand, and covering power

4.3 Fiber Formation Process

The process of MMF formation is known as fiber extrusion or fiber spinning. As has already been mentioned, the starting material for fiber formation is a polymer that may be available in nature or may be synthesized. According to the origin of the starting polymeric material, the MMFs are classified into two groups: regenerated fibers and synthetic fibers. Regenerated fibers are those for whom the starting material is a natural polymer but not in the form of fiber and they are converted into the form of a fiber or regenerated in the shape of a fiber through some physiochemical processes. Synthetic fibers, however, are completely synthesized by man, right from the starting polymer to the final fiber form.

Conversion of polymer into a long continuous hair-like structure involves three basic steps:

1. Spinning/extrusion
2. Drawing/stretching
3. Heat setting

These steps have their own significance: spinning/extrusion is the process of imparting a long continuous fine fiber form to the polymer; drawing/stretching involves large-scale deformation of the polymer molecules to impart orientation and crystallinity to the fiber; and heat setting is necessary to improve the dimensional stability by exposing the fibrous structure to a specific thermal environment.

4.3.1 Spinning/Extrusion

Spinning is the process of extrusion of polymer through a die, called a spinneret, containing large number of fine holes and subsequently solidifying those fine fluid polymer strands to obtain long continuous fine filaments. Most of the natural and synthetic polymers used for fiber formation are available in solid form and so they have to be converted into the fluid form for extrusion. Wherever the polymer is already available in fluid form, conversion is

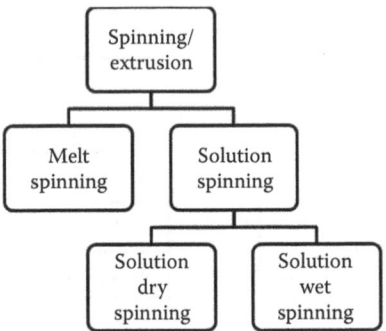

FIGURE 4.1
Different fiber extrusion processes.

not necessary and they can directly be used for spinning. Conversion of solid polymer into a liquid may be done by either melting the polymer or dissolving the polymer in a suitable solvent. On the basis of the technique used for the conversion of polymer into fluid form, the subsequent spinning process has been categorized into two classes: melt spinning and solution spinning. Melt spinning is preferred as a technique for conversion of polymer into fiber. Whenever melt spinning is not possible, solution spinning is the only viable alternative. Polymers, which do not melt or may be melted but the melt is not stable or requires a high melting temperature, cannot be processed through the melt spinning route and are dissolved in a suitable solvent to produce a sufficiently viscous solution for spinning. This solution spinning method is more complicated than melt spinning, owing on one hand to the necessity of dissolving the polymer in a proper solvent and on the other to the necessity of removing the solvent and recovering the polymer in a continuous filamentary form after extrusion. Solution spinning can further be classified into solution dry spinning and solution wet spinning (Figure 4.1). As this chapter deals with continuous filament yarns and texturized synthetic yarns, the discussion will mainly be focused on melt spinning and subsequent modification processes of the melt spun filaments.

4.3.1.1 Melt Spinning

Advances in the synthetic fiber industry have led to an improvement in melt spinning technology to increase productivity and control of the quality of the end products. The polymer melt is extruded and drawn at speeds of 1000–8000 m/min. Lower spinning speeds result in an undrawn yarn, which is then further drawn and heat treated to improve the mechanical properties. At higher spinning speeds, the fiber is partially or fully drawn, and can be texturized, drawn further, or used as spun yarn.

The basic filament formation process by melt spinning may be represented by the following flow diagram (Figure 4.2).

Irrespective of batch or continuous spinning, the basic spinning operation is more or less similar. The general features of a spinning plant are shown in Figure 4.3.

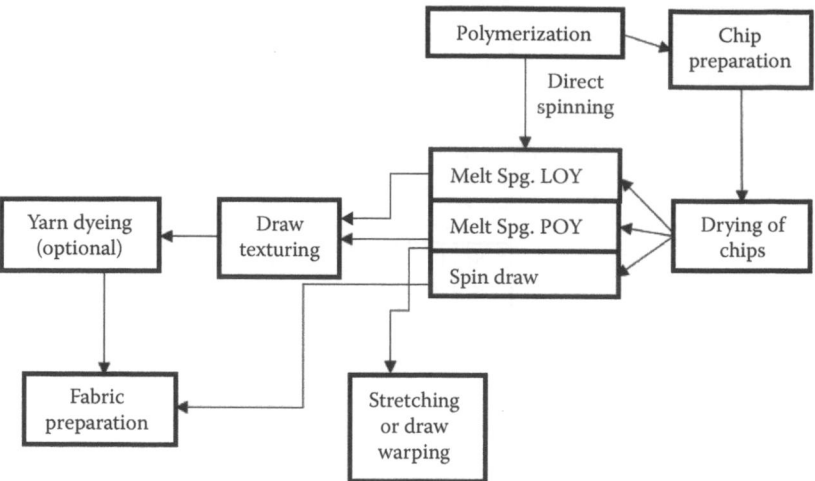

FIGURE 4.2
Flow diagram of basic filament formation process.

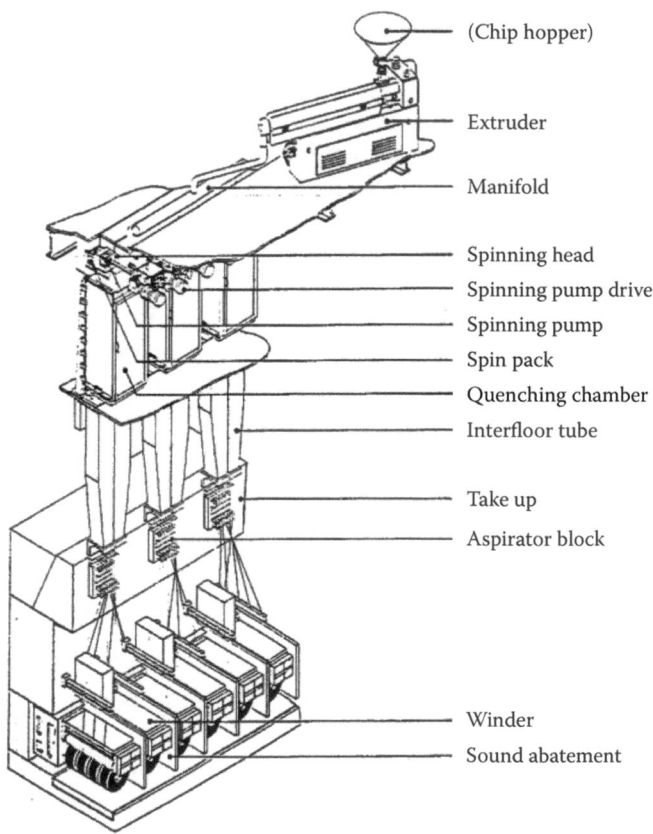

FIGURE 4.3
General features of melt spinning.

4.3.1.2 *Various Components of Melt Spinning Systems*

1. *Extruder*: The function of the extruder is to melt, mix, and provide a constant supply of molten polymer to the spinneret. Modern extruders are equipped with various mixer elements, which help in proper mixing of the melt and minimizing the temperature distribution (±1°C) in the melt.

2. *Prefilter*: The prefilter helps to remove any solid component from the melt, thereby increasing the efficiency of the spin pack filter and also protecting the gear pumps from wear.

3. *Manifold and static mixers*: From the extruder molten, polymer is carried to the individual spinning heads through a manifold. A manifold is a simple network of pipes fitted with a metering pump at the end. The manifold is designed in such a way that the residence time of the polymer from the extruder to individual spinning heads is the same irrespective of its distance from the extruder. This helps to maintain uniform thermal history, rheological properties, and pressure head at all positions of the spinning and thereby ensure more uniform product quality. When the distribution lines are long, at times, static mixers are also installed inside the distribution pipes. A static mixer helps to maintain the homogeneity and temperature distribution in the melt. With the development of extruder design and incorporation of various mixer elements, the use of a static mixer has become optional. The use of a static mixer is particularly recommended for the production of dope-dyed filaments. A static mixer also helps to compensate the parabolic velocity profile of the melt caused by the friction against the walls of the pipe during flow. Homogeneity and quality of melt is extremely important, as it determines the quality of the final product and is a function of temporal and local homogeneity (Figure 4.4).

4. *Spinning head*: Spinning heads are equipped with a spin pack assembly, which in turn consists of a filtration unit and a spinneret. Owing to the unavoidable chimney stack effect, *top loading* spinning heads cannot guarantee an even spinneret temperature, which is required for the production of fine denier filaments. To overcome the disadvantages of the top loading system, *bottom loading* spinning

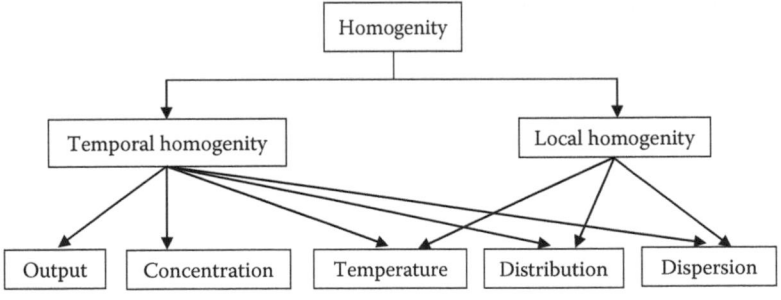

Local homogeneity   = f (ability of the extruder for distribution and dispersion)

Temporal homogeneity = f (transport-behaviour, temperature-treatment, dosing-equipment)

FIGURE 4.4
Temporal and local homogeneity.

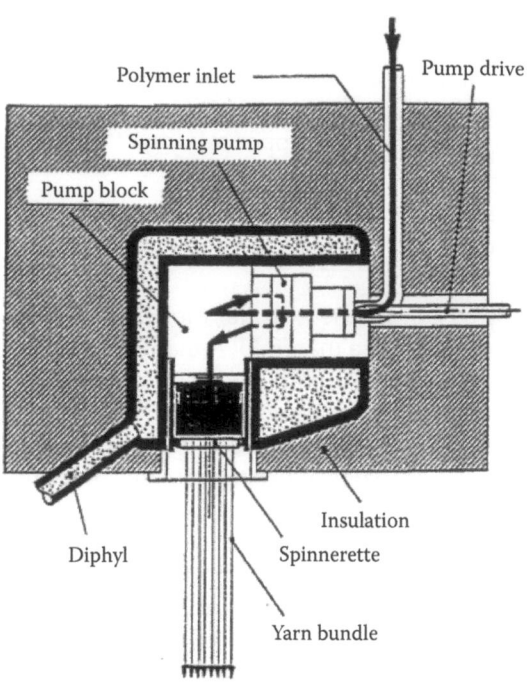

Polymer inlet — Pump drive

Spinning pump

Pump block

Insulation

Diphyl   Spinnerette

Yarn bundle

FIGURE 4.5
Schematic diagram of bottom loading spinning head.

heads have been developed (Figure 4.5). This system ensures that the temperature of the melt and the spinneret is extremely even and constant.

It is apparent that variation in the melt temperature of the polymer influences the viscosity of the melt, the pressure in the spin pack and the degree of preorientation of as-spun filaments. The actual changes in the properties in the as-spun filament are also dependent on other technical parameters. The effect of variation of the spinning head temperature on the pressure of the spinning head and the boiling water shrinkage during polyester spinning is shown in Figure 4.6.

An increase in the temperature of the spinning head leads to a considerable increase in the thread breakage rate.

5. *Quenching*: A stream of air is used to cool the filaments under the spinneret. It is extremely important to ensure both spatial and temporal uniformity of the air stream to have uniform filaments with good evenness. To have uniform preorientation, all filaments must be processed under conditions as uniform as possible.

The quenching system can be divided into three types:

   a. Cross-flow quenching

   b. Radial inflow quenching

   c. Radial outflow quenching

Cross-flow quenching is more commonly used for fine and coarse denier filaments and for both round and rectangular spinnerets. A quench distance between 1.2 and 1.5 m is suitable for textile filament yarns. The disadvantage of the cross-flow quenching system is that the air flow does not reach all filaments equally, but this is an unavoidable part of the system. In this regard, quenching

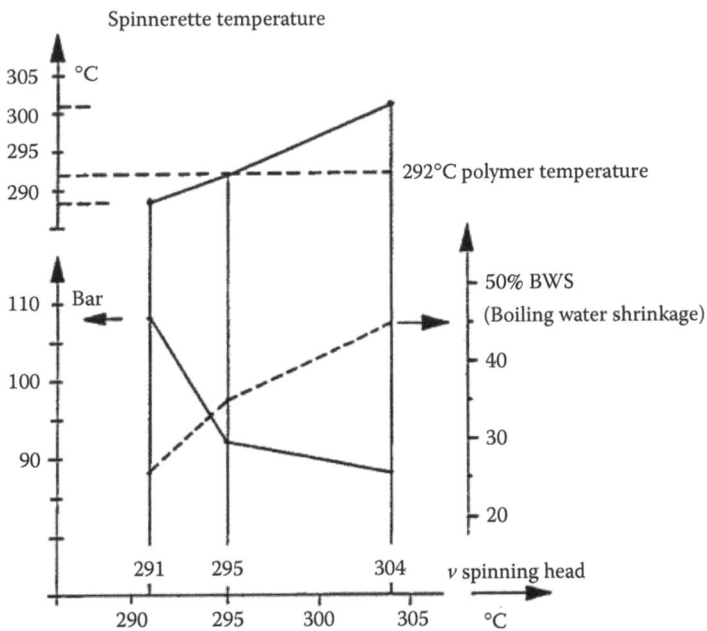

FIGURE 4.6
Effect of spinning head temperature.

with a radial air stream is better. In radial quenching, a quenching distance of about 40 cm or less is sufficient. Partially orientated yarns (POY), which are cooled by radial quenching, have better elongation properties than those cooled by cross-flow quenching. They are therefore better feeder yarns for the production of DTY. The influence of the quenching system on air consumption and yarn quality is shown in Figure 4.7.

6. *Spin finish*: Synthetic filaments are generally given a spin finish during spinning to make them suitable for further processing. Although spin finish is present as a thin layer, only a few molecules thick, on the surface of the filament, it is one of the most important variables in deciding the performance and quality of the process. There is no universal spin finish that can be used for all types of fiber and processes and hence spin finishes are specific based on the fiber and subsequent processes. The primary functions of a spin finish are to provide surface lubrication to the fiber/filament, antistatic action, and good fiber to fiber cohesion. To perform these functions, a spin finish should have certain desirable properties. Some of the important desirable properties of spin finish are as follows:

- It should help reduce fiber/filament to metal friction to minimize fiber abrasion and processing tension, but at the same time it must ensure good filament-to-filament cohesion to facilitate package formation and unwinding. The trend of increasing process speed has further complicated the situation. Products with too low dynamic friction are not always favorable, as sometimes it may lead to the problem of *dancing thread* or unstable yarn path. Therefore, it is wise to adjust the spin finish properties with respect to the godet materials such as, ceramic, chromed, or coated steel to get optimum result and a good yarn quality.
- It should help reduce static charge build up during processing.

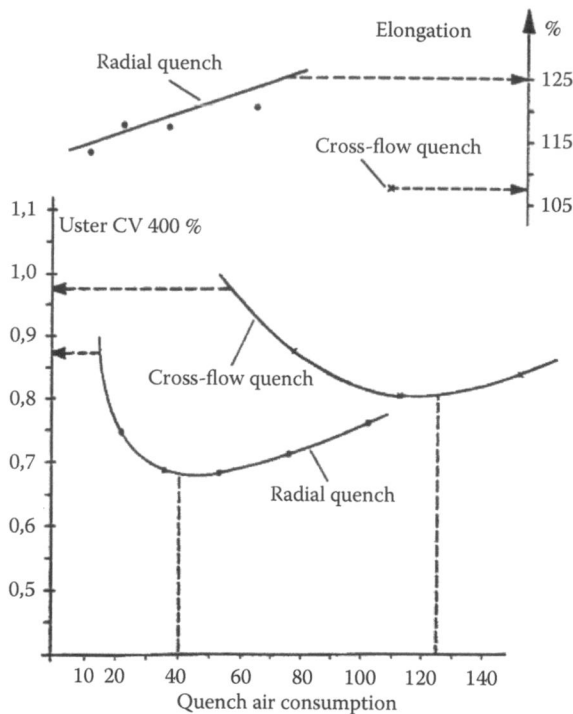

FIGURE 4.7
Influence of cross-flow and radial quenching system on air consumption, Uster CV%, and elongation potential of the filament.

- It should provide a balanced degree of cohesion, as too much lubrication may cause fiber slippage resulting in package distortion.
- It should have a controlled viscosity to facilitate the right amount of spin finish application.
- It should be easily removed before dyeing.
- It must not damage the various machine parts with which it comes into contact during processing.

7. *Winding*: The extruded filaments, after the application of spin finish, are wound on a suitable holder with the help of a winding device. The winding device is an important element in all spinning systems. With the increase in spinning speed, wide variation in denier of spun filaments, increased package size, and pack weight per spindle of more than 100 kg, the demand on the winding system has increased considerably. The basic considerations are to wind the filaments securely with minimum damage and to allow high-speed rewinding with low breakage rate during subsequent processing. Good storage stability and package stability are also important, as often the packages are to be transported over long distances. New and improved traverse mechanism has been developed by different machine manufacturers to ensure proper cylindrical form, uniform shore hardness, and extremely quick turns of the filaments at the package ends. Precision winders developed by various manufacturers, though costly, can take care of most of the requirements of high-speed winding and also can significantly influence the unwinding tension on the filaments during subsequent processes to minimize breakage rates.

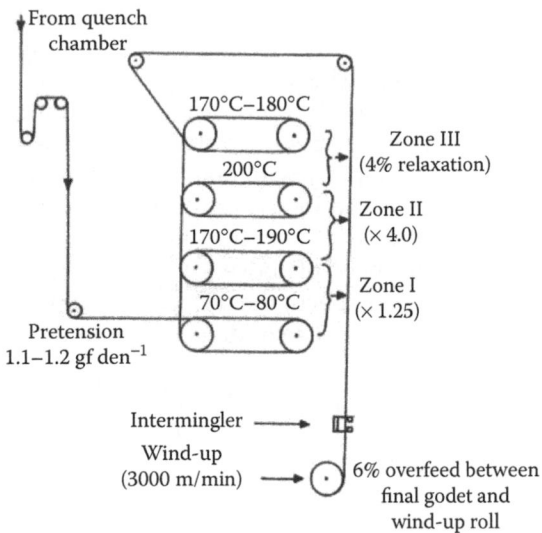

FIGURE 4.8
Outline of spin-draw process.

4.3.1.2.1 Integrated Spin-Draw Process

The draw-twisted yarn produced in the conventional two-step process has a problem of high level of stress generation during twisting and also variation in tension and orientation during winding, which leads to a variation in shrinkage potential of the yarn along its length. The cost of the yarn is also high due to additional process steps and material handling at each step. The package size is also relatively small, which is not favorable for subsequent processing. As a result, conventional draw-twisting processes are now being replaced by a one-step spin-draw process particularly for tire yarn production. In this process, a relatively high working speed of 3000–4000 m/min can be achieved with adequate mechanical properties, high dimensional stability, and low residual shrinkage. A typical spin-draw process outline for nylon tire yarn production is shown in Figure 4.8.

4.3.1.3 Structure Development During Spinning

Structure development in a filament or fiber refers to the development of molecular order or arrangement that is characterized by two structural parameters: orientation and crystallinity. Orientation in a fiber means the overall orientation which involves both the orientation of molecules in the amorphous region and the orientation of the crystals. The nature of elongational flow of the polymeric fluid during spinning primarily influences the development of orientation and subsequently crystallinity in as-spun filament. The spinning is done from a molten polymer and the extruded molten fluid filament is cooled down along the spin line, therefore crystallization occurs during the process. The time available to the molecules before solidification, temperature, and the stress or rate of deformation of the fluid filament influence the extent of crystallization. During spinning, molecules are oriented and extended, and when many such molecules come close to form a bundle of extended molecules, they act like a nucleus and promote the growth of crystals. Extended crystal structure or fibrillar structure has been observed in melt spun fibers.

The development of fine structure during spinning is influenced by different factors, such as type of polymer, molecular weight and its distribution, throughput rate, spinning speed, and quenching condition and system. The polymer type influences several important variables, such as die-swell, glass transition (T_g), and crystallization rate. Polypropylene (PP) has a relatively fast crystallization rate. It is assumed that crystallization starts almost immediately after extrusion. It increases rapidly with take-up speed or take-up tension. The orientation of molecules is also influenced by the crystallization and increases significantly as the crystallization is almost complete. As in the case of polyester, the spinning-induced crystallization sets in at a take-up speed of above 4,000 m/min and increases further with an increase in the speed. At a higher speed, the induced crystallinity results in an abrupt rise in velocity of the filament because of a sudden change in the filament structure from viscous to semicrystalline solid.

With increased speed, the necking point of the filament also moves upward, closer to spinneret as the spin line stress induced crystallization shifts the crystallization to a higher temperature level. Similar behavior has been observed for nylon also. In nylon, the crystallization during spinning does not reach completion at speeds below 3000 m/min. At higher winding speeds, the crystallization reaches completion. There exists an important difference between polyester and nylon and that is the T_g and moisture regain values of these two polymers are significantly different. The nylon due to its low glass transition temperature and high moisture regain permits crystallization in the winding room conditions. It follows that the structure present in conditioned as-spun yarn at lower speed is mainly due to crystallization after moisture pickup, whereas crystals are already generated in the spinning process for the yarns spun at higher speeds (Figure 4.9).

It has been proposed that during high-speed spinning, as the fiber diameter decreases, the molecules that were in a random state gradually tend to form an oriented mesophase.

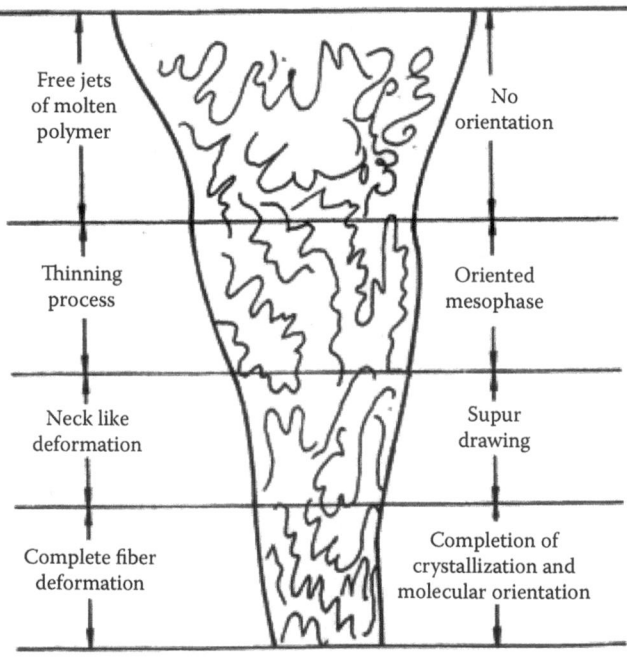

FIGURE 4.9
Proposed structure development during high-speed spinning.

At a certain point along the spin line, neck-like deformation appears, after which no further change in diameter takes place and a high degree of crystallinity and a high molecular orientation is achieved. For neck line to appear in a spin line, a minimum stress or deformation rate or spinning speed is imperative. Thus by controlling the rate of deformation or the stress on the spin line along with other spinning parameters, the level of fine structure development, that is, the development of orientation and crystallinity can be controlled.

4.3.2 Textured Yarns

The texturing of thermoplastic yarns has been one of the most exciting developments in the area of textiles. This process has radically changed the use of nylon and polyester yarns in the apparel, carpet, hosiery, and knitting industry. Texturing imparts many characteristics associated with spun staple yarn to continuous filament yarns. Other than bulk, texturing also influences other properties of the filament yarn, such as

1. texturizing improves pill and crease resistance, dimensional stability, durability, shape retention ability
2. textured yarns are soft, flexible, and have high extensibility
3. improved absorbency and comfort properties

4.3.2.1 Brief Overview of Texturizing Processes

Depending on the mechanism of imparting texture, the texturizing process may be categorized as follows:

1. *Mechanical*: air jet texturing
2. *Thermo-mechanical*: false twist process, stuffer box crimping, bulked continuous filament (BCF) process, knit-de-knit process, edge crimping process, gear crimping process
3. *Chemical/thermal*: bicomponent process

Among these three categories, mechanical and thermo-mechanical processes are commercially more important compared to the bicomponent process.

Among the different thermo-mechanical processes, knit-de-knit, gear crimping, and edge crimping are becoming less important for textile yarns. The mechanical stuffer box is also no longer used for textile yarns and has limited application for BCF yarns, but it is widely used for crimping staple fiber.

On the basis of the yarn characteristics, textured yarns may be classified into three major groups:

1. Stretch yarns
2. Modified stretch yarns or set yarns
3. Bulk yarns

Stretch yarns are characterized by their high stretch and recovery potential but possess moderate bulk. Stretch yarns are produced mainly by the false twist process and edge crimping process. Stretch yarns are used for the production of stretch-to-fit garments, where both extensibility and recovery from stretch are important.

Modified stretch yarns have intermediate stretch and bulk. They are produced by the same process as used for stretch yarns but with some modification. Modified stretch yarns have lower stretch potential and better hand and appearance. These yarns are generally used for knitted fabric, as they impart greater stitch clarity, softness, smoothness, and bulk and lesser extension under load.

Bulk yarns are characterized by their high bulk and moderate stretch potential. They are used where bulk and fullness of hand are of greater importance. Although they have moderate stretch, they generally possess adequate recovery characteristics. Bulk yarns are mostly used in upholstery, carpets, and winter garments. Bulked yarns can be produced by air jet texturing, stuffer box, knit-de-knit, gear crimping, and so on.

A brief description of the different approaches to the production of textured yarns is as follows:

- *Stuffer box texturing*: Stuffer box texturing can be of two types: mechanical stuffer box and gas dynamic stuffer box. In a mechanical stuffer box, fully drawn synthetic fiber yarns or tow is forced into a stuffer box chamber, which may be heated or cold or can have steam injection at the inlet.

  The gas dynamic stuffer box texturing can be a separate process independent of spinning or may be integrated specially with the spin-draw process also called BCF process.

- *Knit-de-knit process*: The yarn is produced by knitting a sleeve from drawn yarn on a circular knitting machine. The sleeve is then heat set and unraveled to give single yarns having a crimped structure. The process is relatively cumbersome, which gives a specific crimp structure. Polyester is not suitable for this process, as it does not produce enough crimp stability.

- *Edge crimping*: Stretch and modified stretch yarns can also be produced by the edge crimping process, which is also known as the Agilon® process. In this process, the thermoplastic yarn is heated, stretched, bent, and drawn around an edge and then cooled. In a multifilament yarn, all the filaments will not come into contact with the edge, but all the filaments will be bent. The line diagram of the process is shown in Figure 4.11a. This will generate an asymmetric stress distribution across the diameter of the filaments, there will be tensile stress on the outside of the bend and compressive stress on the inside of the bend. As the filaments are heated, there will be relaxation of this induced stress through molecular rearrangement and the structure will be set on cooling of the filaments. In this process, they form alternate helices as shown in Figure 4.10.

  The disadvantage of the process is its poor crimp stability and hence now the process is seldom used.

- *Gear crimping*: The concept of the development of texture by gear crimping is similar to that of edge crimping. Here, the preheated thermoplastic filament yarn is passed through a pair of hot gear wheels having specially profiled gear teeth and then cooled. While passing through the gear wheels, the yarn will be forced to

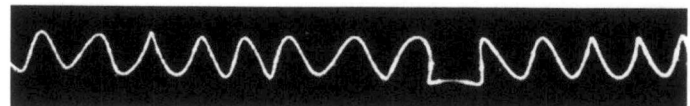

FIGURE 4.10
Alternate helices in edge crimp yarn.

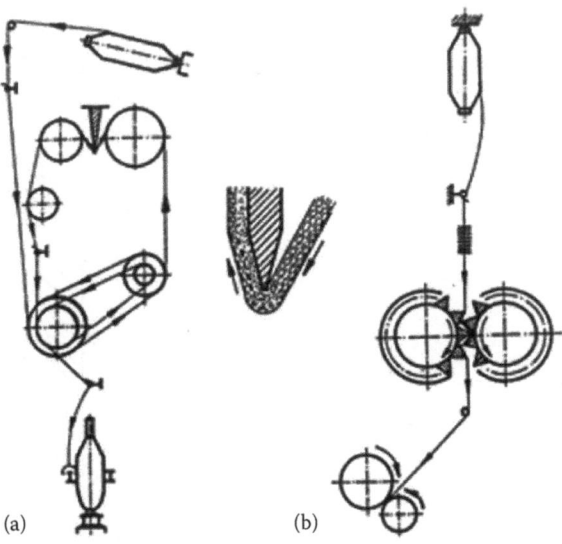

(a)   (b)

FIGURE 4.11
Outline of (a) edge crimping and (b) gear crimping process.

bend, leading to the development of asymmetric stress across its diameter and consequent development of crimps on subsequent cooling. Outline gear crimping process is shown in Figure 4.11b.

4.3.2.2 False Twist Texturing

False twist texturing is the most widely used method for the production of synthetic textured yarns. The evolution of false twist texturing process is shown in Figure 4.12. False twist texturing involves twisting, heat setting of the twisted filament yarn followed by untwisting in one continuous process. Different types of false twist insertion mechanisms or aggregates available are:

- *Magnetic pin spindle twister*: The spindle rotates in the gap between two pairs of rotating disks and is held in position by magnetic force. This type of system generates a positive real twist similar to Helenca® process and has some limitation in twist insertion rate and yarn speed. This type of system is used only when a very high or very low twist is required or may be for processing profiled special yarns.
- Three-spindle disk friction false twisting aggregate generates twist by frictional contact between the rotating disks and the yarn. This is the most widely used technique these days, but the problem of slippage resulting in surging cannot be avoided.
- The nip-controlled vector drive twisters achieve almost slip-free friction twisting at the point where the two belts or two rotating friction surfaces press against each other. The width of the belts varies with the denier of the yarn to be processed.

For effective heat setting of the twisted yarns, the minimization of stress developed in the yarn due to twisting is achieved through molecular rearrangement by exposing the synthetic filament yarns to a temperature above the glass transition temperature of the polymer from which the filaments are prepared. If the structure of the feed yarn has high

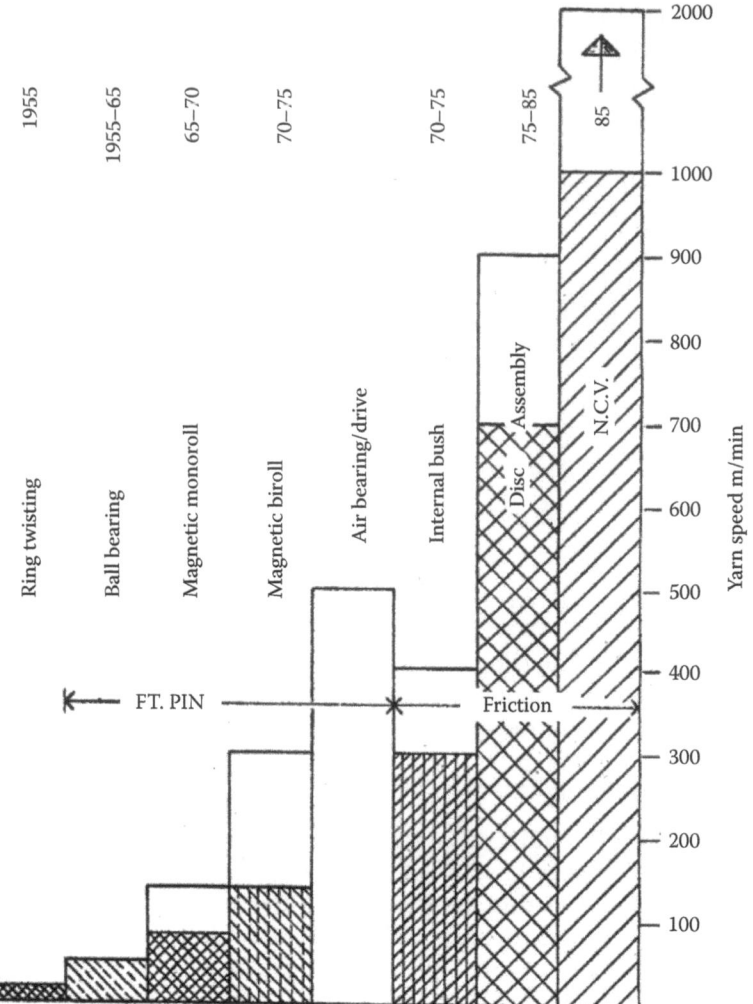

FIGURE 4.12
Evolution of false twist texturing process.

level of orientation and crystallinity, more time and energy will be required to bring about the necessary molecular rearrangement to achieve effective setting. So, the feed materials having relatively low level of orientation and crystallinity such as LOY or POY filament yarns are preferred for false twist texturing. Use of LOY or POY makes it feasible to achieve setting of the twisted structure in millisecond scale and operate the machine at high speed.

The use of LOY in false twist texturing, however, has some practical problems, such as

1. LOY is generally unstable, difficult to handle, and has poor shelf life
2. It is sensitive to moisture and temperature leading to variation in product quality

To take care of the above-mentioned problems, POY has been developed specially as a feed material to be used for texturing. As LOY or POY is used as feed material for false twist texturing, a drawing component is incorporated in the texturing process to achieve the desired mechanical property of the final yarn and hence the process is named as *draw texturing*.

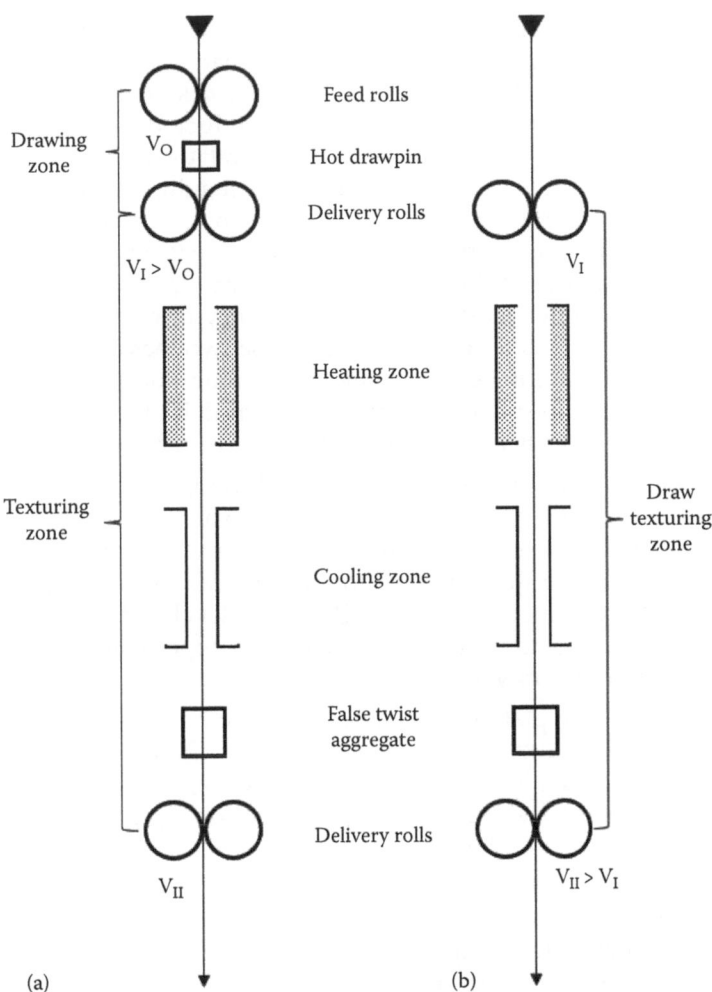

FIGURE 4.13
Schematic diagram of (a) sequential and (b) simultaneous draw texturing process.

On the basis of the drawing mode, the false twist texturing may be further distinguished as sequential draw texturing and simultaneous draw texturing. In case of LOY, as the structure is unstable and delicate, the yarn is first stabilized to some extent by drawing before exposing it to further texturing conditions. The yarn drawing is done in first stage and texturizing in the next stage. Both these stages are integral part of one continuous process called sequential draw texturing. Although POY can also be processed by sequential route, it is advantageous to draw and texturize POY in one step: simultaneous draw texturing.

Schematic diagram of sequential and simultaneous draw texturing is shown in Figure 4.13.

The variable in the false twist texturing process can be categorized as machine variable and process variables. Like any other process involving synthetic fiber processing, time, temperature, and tension are the three important parameters that influence the process. The important machine variables or elements that influence these three parameters in a false twist texturing machines are heating and cooling elements, false twisting elements, second heater in case of set yarn production, and machine speed.

- *Primary or first heater*: Primary or first heater may be contact type or noncontact type. Contact heaters have been used for many years, as they are reliable, cheap, and easy to operate. As the contact-type heaters work at a temperature of about 200°C and the synthetic fiber polymers are nonconductors, they have limited ability to transfer heat to the yarn and accordingly relatively long length of heater, 2–2.5 m depending on the linear speed of yarn, is required. Design of the machine becomes awkward with long heaters and the machine requires more floor space. Contact heaters also have a problem of increased tension build up in the yarn; in addition, the surfaces of the heaters become dirty from the deposition of spin finish, leading to further reduction in the efficiency of heat transfer and change in the surface characteristics of the heater. These two factors lead to an increase in both yarn breakage rate and machine down time due to frequent stoppages required for cleaning the heater. High twist insertion is also a problem due to the surface contact with the heater and hence limits bulk generation.

  Short noncontact-type heaters, which were developed later to avoid the problems related to contact-type heaters, where the yarn passes through a hollow heated tube at a high speed. Noncontact-type heaters can work at a relatively high temperature and hence the length of the heaters can be reduced to 1 m range, machine speed can be increased, and higher twist level and bulk can be achieved with minimum damage to the yarn. For noncontact-type heaters, however, threading is a problem in case of yarn breakage.

- *Cooling plates*: Cooling plate is placed between the heater and the twisting unit to cool down and stabilize the highly twisted yarn. Nonstabilized yarns have a tendency of ballooning, leading to high yarn breakage rate. The cooling plate is normally contact-type metal plate. Length of the cooling plate is proportional to the speed of the yarn/machine.

- *Twisting unit*: The heart of the draw texturing process is the twist insertion device or friction aggregate or false twist unit. As the twist is inserted by the frictional torque transfer between the yarn and the rotating disk surface (Figure 4.14), the amount of twist inserted depends on

  1. The angle of wrap of the yarn over the disk assembly
  2. The relative speed of the friction disks and the yarn speed (D/Y ratio)

The angle of wrap or the frictional contact between the yarn and the disk assembly can be varied by

  1. Changing the diameter of the friction disk
  2. Changing the horizontal and/or vertical distance between the disks
  3. Changing the number of disks

Different types of materials have been used for the manufacture of friction disks and they have their own influence on the process or the product. The basic differences between these disks are their surface characteristics such as their softness (rubber, polyurethane, etc.) or hardness (ceramic, nickel, etc.) and frictional characteristics. The most commonly used disks are either soft surface such as polyurethane type or hard surface type such as ceramic. If the surface is soft, it helps better torque transfer and hence a higher level of twist can be achieved. The possibility of damage to the filament due to abrasion between

FIGURE 4.14
Schematic diagram of friction disk assembly.

filaments and disk is also less in case the friction surface is soft. In case of ceramic surface, the chances of slippage between the yarn and the disk is more, resulting in less torque transfer and hence less build-up of twist. The chances of filament damage due to abrasion and snow generation is also more in case of hard ceramic disks. The life of ceramic disks, however, is much higher compared to polyurethane disks.

Second heater: Second heater is employed when a set yarn having moderate stretch and bulk characteristics are required. It is also normally a noncontact-type heater of about 1 m length through which the textured yarn passes under controlled relaxation. The treatment leads to a reduction in the amplitude of the crimp and shrinkage of the textured yarn, hence the structure become less twist lively, less bulky, and less stretchable. The amount of bulk, however, can be increased by increasing overfeed in the second heater zone. These set yarns find application in all types of fabric manufacturing due to their low stretch and stable character and exhibit a better drapeability and a crisper feel. Finally, the yarn is wound on a package after intermingling and application of coning oil.

Schematic diagram of stretch yarn and set yarn production process is shown in Figure 4.15.

4.3.2.3 Process Variables

The important process variables that influence the properties of the textured yarn may be summarized as follows:

1. Draw ratio
2. First heater temperature/contact time
3. D/Y ratio
4. Second heater temperature and overfeed

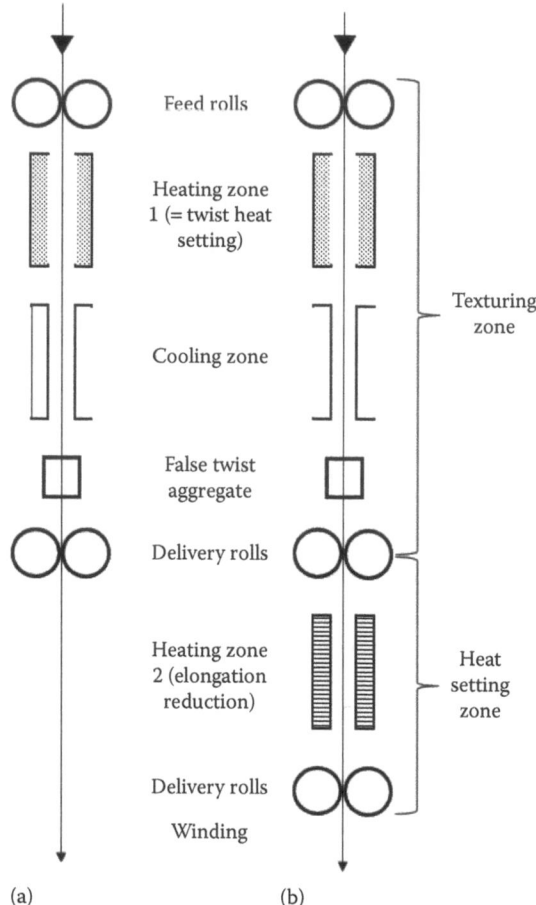

FIGURE 4.15
Schematic diagram of stretch yarn and set yarn production process: (a) HE-yarn single heater and (b) SET-yarn double heater.

Draw ratio typically influences the molecular order of the filaments, the final denier of the individual filaments, and the resultant textured yarn. As a result, dye uptake, yarn breakage rate (both in case of high and low draw ratio), stability of the yarn, and its stretch potential are influenced by the draw ratio.

    First heater temperature/contact time of the yarn with that of the heater has similar effect. As commercial processes operate at high speed to facilitate production rate, there is very little scope to adjust the time window. Temperature, on the other hand, is the most common tool to derive and manipulate the desired properties of the yarn. First heater temperature influences the degree of set of the twisted yarn structure through molecular rearrangement. As a result, it will influence the bulk, stretch potential, dyeability, and broken filament level or yarn breakage rate. The working temperature of the first heater is decided on the basis of the type of the feed yarn and optimization of all the other parameters influenced by heater temperature. Interdependency of texturing speed, yarn count, and heater length is shown in Figure 4.16.

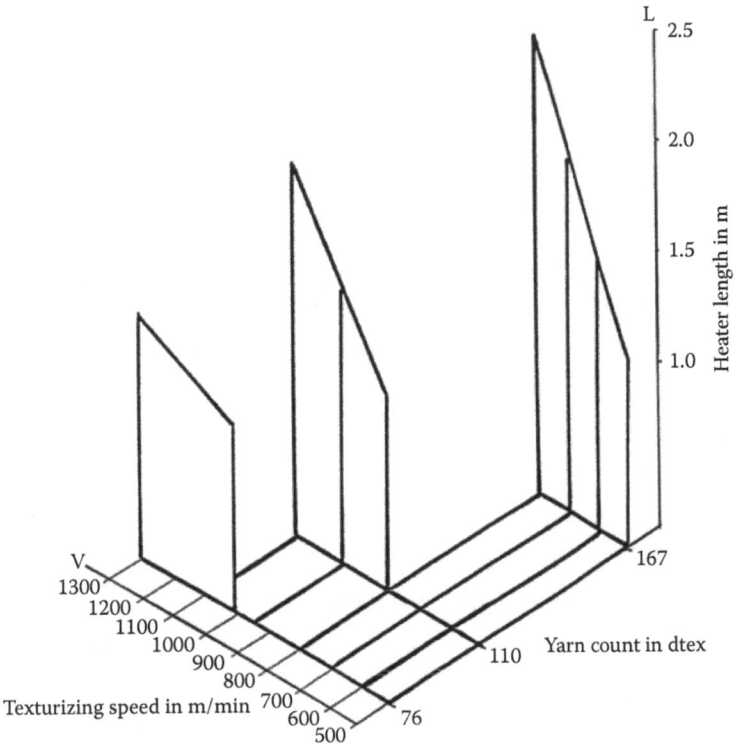

FIGURE 4.16
Texturing speed, yarn count, and heater length.

D/Y ratio influences the level of twist inserted in the yarn. It is the ratio of the circumferential speed of the disk to the linear speed of the yarn.

$$\frac{D}{Y}\ ratio = \frac{\text{circumferential speed of disk in meter per minute}}{\text{linear speed of the yarn in meter per minute}} \qquad (4.1)$$

The D/Y ratio also influences the tension of the yarn before (T_1) and after (T_2) the friction twisting unit. The D/Y ratio is adjusted to have a stable situation of $T_2/T_1 = 1$.

When D/Y ratio is high, the twist level will be high and the disk will store extra length of yarn between the entry and exit points of the disk. So, T_1 will be more than T_2 and T_2/T_1 will be less than 1. This situation will lead to an increase in surging, which leads to nonuniformity in the twist level along the length of the yarn and also occurrence of tight spots. Both of these will result in fabric fault.

In case of low D/Y ratio, that is, when the disk speed is less compared to the linear speed of the yarn, the yarn will be literally pulled through the disk. In this case, T_1 will be less than T_2 or $T_2/T_1 > 1$. Although the situation is stable, it will lead to mechanical damage to the yarn and more wear and tear of the friction unit. Influence of D/Y ratio on the yarn twist is shown in Figure 4.17.

Second heater temperature and overfeed primarily influence the stretchability and bulk of the final yarn. The interplay between the yarn tension in the second heater zone, the overfeed, and temperature of the second heater will decide the final property of the set yarn.

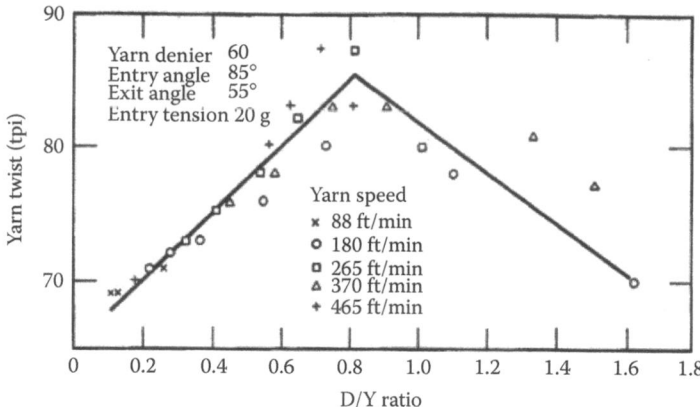

FIGURE 4.17
Influence of D/Y ratio on yarn twist.

4.3.2.4 Material Variables

Material variables may be summarized as follows:

1. Type and denier of individual filaments
2. Filament count and yarn denier
3. Spin finish
4. Mechanical properties of the parent yarn

The denier of the individual filaments, number of filaments, and denier of the yarn have profound influence on the process and properties of the final textured yarn. More is the denier of the yarn, more torque will be necessary to impart equivalent level of twist in the yarn. If more energy is required to impart twist, it is expected that the retained energy in the yarn will also be more, which will result in higher resistance to deformation, higher stretch potential, and quick recovery from stretch. The effect of the denier of the individual filaments will also be similar, that is, with coarser individual filaments, the resistance to deformation, stretch potential, and recovery from stretch will be higher. With a constant yarn denier, the denier of the individual filaments will influence the total number of filaments in the yarn. With reduction in the denier of individual filament or increase in the number of filaments in the yarn, the resultant fabric becomes soft, full, and more comfortable. Reduction in the filament denier will make the filaments more delicate and hence chances of damage to the individual filaments or yarn will be more. The selection of disk and other process parameters needs to be monitored accordingly. For a constant twist level, the variation in denier of the yarn will influence the helix angle and hence the stretch potential.

4.3.2.4.1 Spin Finish for Texturing

Continued demand for better yarn quality and higher productivity have also forced spin finish manufacturers to developed product with different chemical composition. A successful spin finish has to satisfy partially conflicting requirements. The spin finish, although applied during filament formation, it's effect only become evident during texturing.

Generally speaking, finishes that are too smooth, produce too little twist due to inadequate torque transfer. Excessive smoothness also increases the possibility of tight spot formation, which reduces the draw ratio. Conversely, if the spin finish is too rough, it increases polymer abrasion and breakage rate due to increased friction between the yarn and twisting unit. More friction will also lead to a higher snow generation. Moreover, the effect of spin finish increases with an increase in the texturing velocity. This contradictory requirement of spin finish during texturing makes life difficult for the spin finish manufacturers, as so many factors have to be taken into consideration. Nevertheless, there are now some very good finishes available, which largely satisfy the demand.

Some of the important requirements of spin finish for texturizing process are as follows:

- Yarn properties should not be altered during storage.
- In the drawing zone, important considerations are volatility, thermal degradation, heater plate residue, and so on.
- To minimize heat loss, the finish should have low specific heat and the heat of vaporization of the volatile component should also be low.
- The finish should have the required frictional characteristics depending upon the type of process and twisting unit.
- In case of draw texturing, high dynamic filament-to-filament friction leads to an increase in the filament break. High filament-to-filament friction or cohesion between the filaments, however, is necessary for POY package build-up. This contradictory requirement is met with by using a volatile cohesive component in the spin finish, which evaporates during the passage of the yarn through the first heater, thereby lowering the filament-to-filament friction.
- The finish should not soften or swell the polyurethane disks used.

4.3.2.4.2 Mechanical Property of the Parent Yarn

The physical property and the chemical nature of the filament would affect various parameters during draw texturing, such as thread line tension, temperature, twist level, dwell time on the heater, degree of set, and so on. To ensure good texturing, it is naturally important that the feeder yarn (POY) has a good Uster CV, and that the cross section of the filaments is fairly even. From the relationship between the yarn retractive force and the yarn parameters, the significance and role of yarn physical properties can be understood. The yarn retractive force equation is as follows:

$$\frac{F}{D} = \frac{(CEd)}{A^2} \tag{4.2}$$

where:
  F is the retractive force of the yarn
  D is the total yarn denier
  C is a constant
  E is the filament modulus
  d is the filament denier
  A is the diameter of the filament helix

It can be seen that the retractive force is directly proportional to the product of modulus and denier. Thus, a material with lower modulus will lead to lower retractive force when all other parameters are constant. As the modulus of PET > N66 > N6, accordingly the retractive force of the textured yarn made from these polymers will have similar trends. It is well known that draw textured nylon has very good stretch and recovery, texture, and covering power. Polyesters, on the other hand, have some difficulty in retaining their torque crimp memory if not textured properly. It must be pointed out that fabric made from properly textured and relaxed polyester yarns, however, has a better texture or hand, a better covering power, and a better crease resistance. Polypropylene yarns generally require more dwell time in the heating zone, which slows down the process to some extent. Textured polypropylene yarns, however, have very good torque memory, provide excellent covering power, and very pleasant apparent effect of lightness of hand due to low specific gravity.

Air jet texturing: It can be used for any type of feed yarns such as yarns produced from MMFs or from natural fibers. The yarn is overfed into the texturing zone, where texturing is carried out by venturi jet with an impact plate. The versatility of the technique makes it commercially significant.

The air jet process of producing bulk yarn was developed in the United States by E. I. Du Pont de Nemours & Co. Bulk in a multifilament yarn can be produced by blowing air in a twisted multifilament yarn, where some extra length of yarn is made available in the texturing zone, by overfeed, for the formation of random loops by the turbulence of the air stream. This leads to the development of bulk through mechanical interlocking of loops and reduction in yarn length. The yarn thus produced has an appearance of a staple yarn, but has higher bulk, greater covering power, reduced opacity, and a warmer hand.

When filament yarns are textured by air jet process, the most apparent change that occurs in the structure is the formation of large number of small loops distributed along the length of the yarn. The yarn has a well-defined core, there is a marked increase in the overall yarn diameter and bulk.

The line diagram of an air jet texturing machine is shown in Figure 4.18.

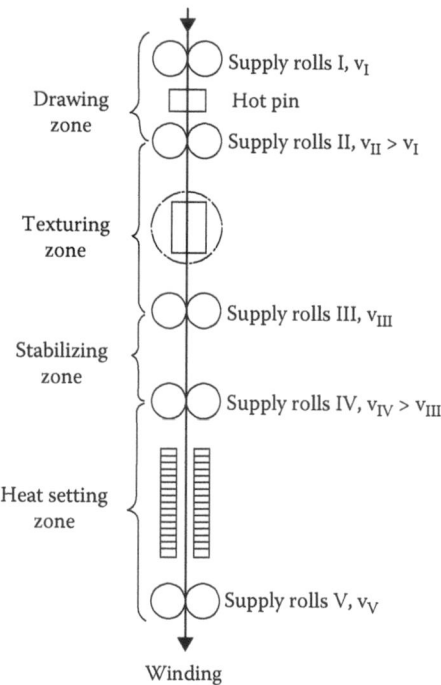

FIGURE 4.18
Flow diagram of air jet texturing machine.

The main components of the system are as follows:

1. *Drawing zone*: When processing thermoplastic yarns that are readily available in the form of POY, the yarn needs to be drawn to stabilize the structure. Drawing can be done by using hot pin or heated godets. This will help to produce yarns with low residual shrinkage.

2. *Yarn wetting system*: Most air jet texturing machines have provision of application of water on the yarn before it enters the into the air jet. Prewetting of the yarn helps to reduce interfilament friction in the yarn and facilitates loop formation through easy displacement under the influence of air jet. The introduction of yarn wetting facilitates the process of bulking and thereby helps to increase the operating speed or production of the machine.

3. *Air jet texturing assembly*: There are two basic types of air texturing jets: the radial and the axial. The jet is the heart of the process and the selection of correct jet is important. The selection of jet depends on the type of the yarn to be processed, end use characteristics, and the denier. In air jet texturing, a wide variety of denier range and yarns having different characteristic feature based on application can be produced. Different air jet textured yarns along with their properties and applications are shown in Table 4.3.

4. *Mechanical stabilization*: Mechanical stabilization basically involves stretching of textured yarn to improve the stability of the mechanically entangled loops. Higher is the overfeed, less will be stability of the loops and more stretching will be necessary to stabilize the yarn structure. In practice, the stretch ranges between 2% and 10%.

5. *Heat setting*: Heat setting is done in case the feed yarn is thermoplastic in nature. Unlike false twist textured yarns, air jet textured yarns are denser and the structure of the yarn is also different, so the effect of heat setting is apparently not as visible as in case of false twist textured yarns. The important changes that may occur after heat setting of air jet textured yarns are shrinkage and the shrinkage potential is related with the overfeed percentage. There may be some reduction in the loop size due to shrinkage of the individual filaments constituting the yarn and finally the overall shrinkage may increase the stability of the yarn by enhancing the mechanical interlocking.

6. *Coning oil application and winding or package building*: The coning oil application and winding are similar to that of false twist texturing system.

TABLE 4.3

Air Jet Textured Yarns Properties and Applications

Property	Application
Low friction yarn	Sewing thread
Spun yarn	Sports and leisure wear, outer wear, car seat cover, and furnishing
High friction	Table cloth, bed sheet, belts and straps, luggage, and rucksacks
Dimensional stability	Tarpaulin, coated fabric, printed circuit, and so on.
Blended yarn	Composite yarns
Structural effect	Curtains and wall coverings
Functional wear	Rain and sports wear

The structure and properties of the air jet textured yarns are influenced by both material and process variables. The important material or feed yarn variables are:

- Type of polymer
- Number of filaments and denier of the individual filament
- Total yarn denier
- Twist level in the yarn

The process variable includes:

- Amount of overfeed
- Speed of the machine
- Air pressure
- Type of jet

When the filament yarn is textured by air jet texturing, there is a resultant change in yarn linear density because of bulking. In general, variations in pretwist and air pressure do not have a significant effect on linear density, but the amount of overfeed has a direct influence on resultant yarn linear density. The overall yarn diameter and loop size increase with overfeed. Increasing the number filament for a given yarn denier will increase the number of loops per unit length. With an increase in twist, the overall yarn diameter and the loop size decrease due to more snarling tendency of the untwisted filaments. Level of pretwist significantly influences the stability of the yarn. Stability of the yarn is low both for very low and very high pretwist levels. At low pretwist level, there is little interlocking, leading to an unstable yarn; at high pretwist level, a large number of small loops are formed with less number of tensioned filament in between, which are primarily responsible for strain during mechanical deformation. Air textured yarn generally exhibits lower tenacity and elongation to break than that of the parent yarn. Processing condition such as twist level, overfeed, and air pressure significantly modifies the mechanical property of the air jet textured yarn.

4.4 Bulked Continuous Filament Yarn

The BCF process is a one-step integrated spinning and texturing process typically used for production of carpet yarns. Mainly nylon or polypropylene is used as raw material for the production of carpet yarns by BCF process, but polyester may also become important in near future as a BCF yarn. The BCF process works on the principle of stuffer box crimping, which is popularly used for imparting crimp in staple fiber. The mechanical process of crimp generation in a stuffer box suffers from an inherent problem of variability in crimp generation. As the staple fibers are blended during subsequent yarn formation process, the variability levels off to an acceptable level through blending. But in case of carpet yarn, there is no opportunity of blending and hence strict control over variability needs to be maintained. Development of dynamic hot fluid jet texturing has made it possible to produce carpet yarns by BCF process.

In a draw texturing machine, the drawing of the filament yarn is done prior to texturing. The process consists of spinning, drawing, texturing, intermingling, coning oil application, and packaging or winding. All other processes are almost similar to false twist draw texturing process with the exception of texturing unit. The texturing here is done by hot fluid jet. The design of the texturing unit is such that the hot fluid jet enters the unit at high speed through a narrow slot, whose size tends to increase as the fluid progresses through the unit, causing an expansion of the fluid jet. The speed of the fluid jet reduces after expansion and so does the speed of the yarn carried by the jet. The retardation of the yarn forms a plug against which yarns from the feed side carried by the hot jet tend to accumulate in a crimped configuration. The yarn is relaxed in the crimped state due to the presence of temperature leading to the formation of textured yarn, which then moves out of the jet and passes over a cold surface to set the developed texture. It is the hot fluid and the yarn plug that provide the bulk. Possibility of precise and accurate control of fluid temperature and flow helps to increase the uniformity in bulk generation and development of a three-dimensional bulk structure. Principle of hot fluid jet texturing is shown in Figure 4.19.

When the hot fluid enters the jet, due to the high pressure, it enters with high velocity and turbulence to carry the yarn forward. The major variable in the process is the air temperature and pressure. Increase in pressure increases turbulence, which in turn facilitates heat transfer from the air to the filament yarn and to some extent influences the yarn property. For nylon processing, the air temperature ranges from 200°C to 220°C, whereas for polypropylene, the temperature range used is 140°C–160°C. The air pressure ranges from 5 to 8 bar. In all thermoplastic yarns, the final bulk not only depends on the conditions inside the jet but also on the complete thermal history of the yarn. Outline of the BCF process is shown in Figure 4.20.

Schematic representation of different types of textured yarns are shown in Figure 4.21.

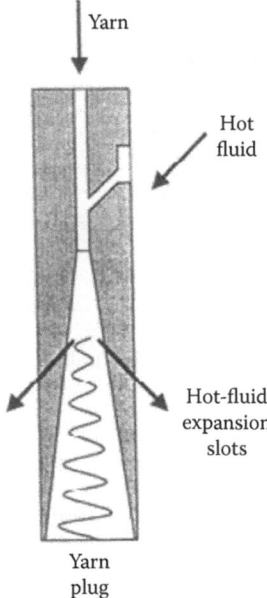

FIGURE 4.19
Principle of hot fluid texturing.

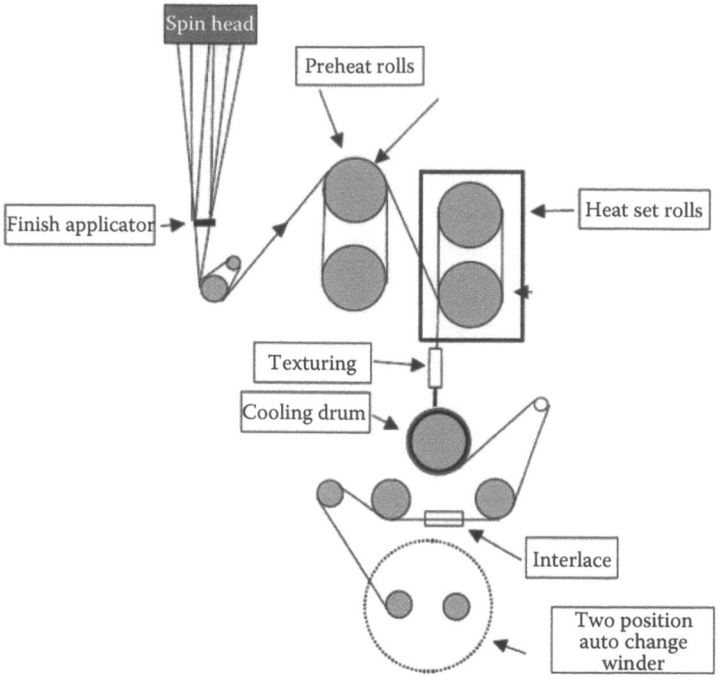

FIGURE 4.20
Outline of BCF process.

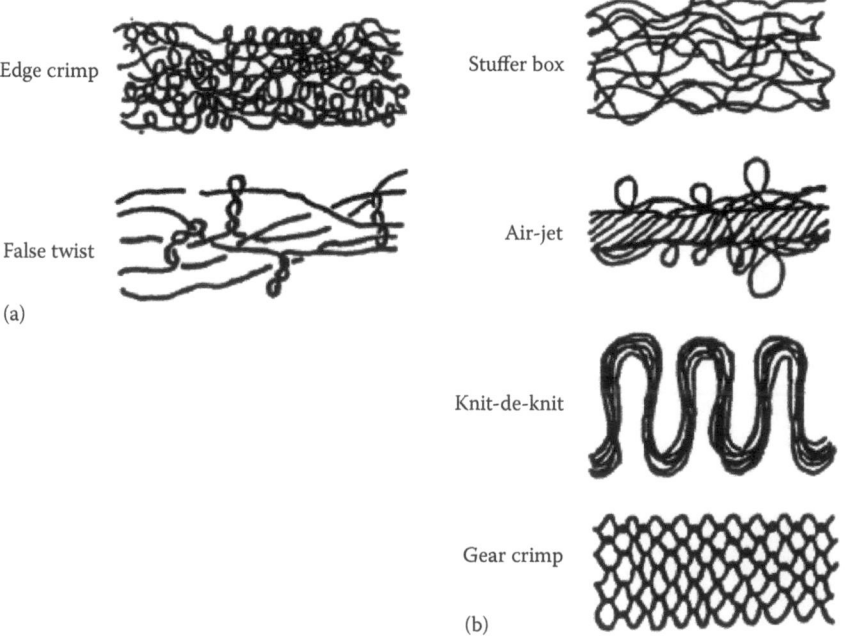

FIGURE 4.21
Two different types of yarns: (a) stretch yarns and (b) bulk yarns.

4.5 Influence of Yarn on Qualities of Apparel Fabric

Influence of yarn structure on the aesthetic and tactile qualities of apparel fabrics is a very complex subject and needs elaborate discussion. The discussion in this chapter is mainly focused to highlight the complexity of the subjects and a fundamental understanding about some important features of yarn structure contributing to the aesthetic and tactile qualities of apparel fabric. Among the more important factors in the design of apparel fabrics are the look, the feel, the performance, and the cost. Of course, other considerations may be important depending on the requirement, but visual and tactile qualities are always fundamental to fabric design. The look of a fabric depends mainly on the interaction of color and surface texture.

- *Visual aesthetics*: The contribution of yarn structure to the visual aesthetic of a fabric is mainly governed by the surface feature of the constituent yarns. This surface feature is a combined effect of luster, texture, softness, smoothness, bulk, and so on. The spun yarns provide a unique combination of structural features that contribute to unique natural textured appearance to fabrics that is not possible with any other yarn structure. Filament yarns also provide a unique combination of structural features. High filament orientation in yarn and high filament packing density, an almost flawless uniformity in yarn structure are responsible for a texture less or a very fine texture look. It also provides high luster, sheer, and smoothness of apparel fabric. The filament yarns, however, can be irregularly twisted to produce a fine subtle texture in apparel fabric. On the other hand, filament yarns do not contribute effectively to bulkiness, covering power and a soft look of the fabric. Textured filament yarn structural features are a combination of both the features of filament yarn and features of spun yarn, such as a high degree of fiber nonlinearity, a very uniform yarn structure in terms of fiber diameter, a very low fiber packing density, and so on. The combined effect of these features is a highly textured appearance, bulk and covering power, a toned down luster, and a soft look of fabrics.

- *Tactile aesthetics*: The appeal of apparel fabrics depends on the interactions of the visual effect of texture and its feel. It is not easy to separate the visual effects from the tactile or hand effects, as most of the time we see and feel the fabric simultaneously. The hand of a fabric is the total response of or combination of impression that arises when a fabric is touched, rubbed, squeezed, or handled by any other manner. Several descriptive parameters are used to characterize the fabric feel, such as soft, crisp, firm, hard, harsh, dead, live, cool, warm, dry, and so on. Several fabric properties such as stiffness, compressibility, resilience, stretchability, friction, drape, bulk retention, and so on are used in an effort to characterize different aspects of fabric hand. As fabric hand is an important parameter considered by the consumers, many techniques have been proposed for the objective assessment of fabric hand. No reliable and simple technique is available, however, which takes into consideration all the psychological and physiological parameters associated with fabric hand. In flat filament yarn, due to high packing density and linearity of filaments, the resultant fabric is smooth and has low bulk or compressibility. When textured, the same filament yarns, due to low packing density and nonlinearity of surface, produce a soft, supple, compressible, and resilient feel in fabric (Table 4.4).

TABLE 4.4

Relative Bulk and Surface Contact of Yarns under Relaxed and Compressed Conditions

Yarn Type	Relative Bulk		Relative Contact	
	Relaxed	Compressed	Relaxed	Compressed
Filament without twist	5	5	1	1
Twisted filament	4	3	2	3
High bulk filament	1	2	5	4
Stretch filament	1	4	5	2

4.6 Conclusion

Exceptional property, handle and aesthetics of natural silk, is one of the key factors in the development of MMF, which has led to the nomenclature of first MMF as artificial silk. After the discovery of nylon and polyester, MMFs have become an integral part of human life. Their use ranges from our day-to-day application in apparel, home furnishing, wipes, tooth brush, dental floss, and pillows to more demanding and specific applications such as filtration, insulation, geotextile, health and hygiene products, and wound management. Fiber-based scaffolds are increasingly becoming important for tissue engineering. Similarly, high-performance composite would not have been possible without high performance fiber. Much of modern life depends upon our understanding, production, and use of synthetic fibers.

Bibliography

Deopura BL, Chatterjee A and Padaki NV (2013), Process control in the manufacturing of synthetic textile fibres, in *Process Control in Textile Manufacturing*, Woodhead Publishing Series in Textiles. doi:10.1533/9780857095633.2.109.

Fourne F (1999), *Synthetic Fibres*, Carl Hanser Verlag, Munich, Germany.

Goswami BC, Martindle JG and Scardino FL (1977), *Textile Yarns Technology, Structure and Applications*, John Wiley & Sons, New York.

Gupta VB and Kothari VK (Ed.) (1997), *Manufactured Fibre Technology*, Chapman & Hall, London.

Hearle JWS, Hollick L and Wilson DK (2001), *Yarn Texturing Technology*, Woodhead Publishing, Cambridge, UK.

Vaidya AA (1988), *Production of Synthetic Fibres*, Prentice-Hall of India, New Delhi, India.

5

Fancy Yarns

Hugh Gong

CONTENTS

5.1 Introduction

This chapter describes a special category of yarns that are designed mainly for their aesthetic appearance rather than performance. These yarns are popularly called *fancy yarns*. A fancy yarn is defined by the Textile Institute (Textile Terms and Definitions) as "a yarn that differs from the normal construction of single and folded yarns by way of deliberately produced irregularities in its construction. These irregularities relate to an increased input of one or more of its components, or to the inclusion of periodic effects, such as knops, loop, curls, slubs, or the like." This definition is rather narrow, as it only includes yarns that vary deliberately in form. Many fancy yarns achieve deliberate variation in appearance by the way of color. To be more inclusive, a fancy yarn may be defined as any yarn that contains deliberate variation either in the form or in color, or both. The word *deliberate* is very important here because all yarns made from staple fibers are inherently variable due to the imperfection of the yarn spinning systems and the nonuniformity of fiber material. In the production of normal *regular* yarns, efforts are made to minimize these variations so that the eventual fabric uniformity and performance properties such as strength and abrasion resistance are maximized. In fancy yarns, however, the variations are introduced by design to enhance the aesthetic appearance. These yarns provide the fabric designer greater scope in achieving a more attractive and exclusive product, but also pose greater challenges as they usually suffer from poorer performance and higher costs.

This chapter will provide descriptions of the main types of fancy yarns first. These will be followed by the explanations of popular yarn production technologies that are capable of making fancy yarns. For a fancy yarn designer, knowledge of the production technologies is invaluable, as it enables him or her to exploit the capabilities of the technologies fully and to avoid costly design mistakes.

5.2 Fancy Yarn Types

It would be a futile attempt to try to describe all the varieties of fancy yarns, as these are designed to differ by definition. From the point of how the variation is introduced in the yarn, fancy yarn effects may be broadly divided into two categories: fiber effect and yarn effect. Fiber effects are introduced prior to the formation of yarn; yarn effects are introduced by combining two or more yarns after the individual yarns have already been made. The two categories can obviously be combined to make more complex effects.

5.2.1 Fiber Effects

Fiber effect fancy yarns are created during the spinning processes prior to the formation of the final yarn. These yarns are characterized by varying sizes of fiber lumps along the yarn length. Depending on the size of these fiber lumps, these fiber effects are often further divided into three subcategories: nepp, slub, and flake.

5.2.1.1 Nepp Yarns

Nepp yarn, or nupp yarn, has a compact yarn structure with specks of fiber clusters distributed along the base yarn structure. Figure 5.1 shows an example. These yarns are most effective when the effect fibers have contrasting colors to the base yarn fiber, whereas more subtle effects can be created when the effect fiber and the base yarn have similar colors. The distribution of the effects should generally be random to avoid the moiré effect as shown in Figure 5.2. The left side of Figure 5.2 shows a nepp yarn with a periodical effect. Once used in a fabric, the periodical effect results in a moiré pattern, which is usually undesirable. The right side of Figure 5.2 shows a nepp yarn with a more random effect and the resulting fabric has a more even appearance. It is, however, possible to deliberately create periodical effects if the moiré pattern is a designed outcome.

FIGURE 5.1
Nepp yarn.

 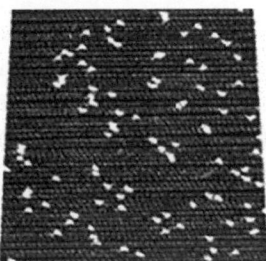

FIGURE 5.2
Influence of effect distribution on fabric appearance.

FIGURE 5.3
Wool nepps.

The production of nepp yarns are usually achieved during the preparation of fibers. The most widely used method is mixing prepared fiber balls such as wool nepps (Figure 5.3) into the main fiber stock before carding. The fiber effect can be varied by the mixing ratio and the card setting, in addition to the usual yarn parameters of linear density and twist. If the card settings are closer, these fiber balls will be opened up into smaller clusters and the effects will be smaller but more frequent; conversely, if the card settings are more open, these fiber balls will have larger and looser but less frequent effects. Due to the more vigorous actions of cotton cards, nepp yarn material is not normally prepared using

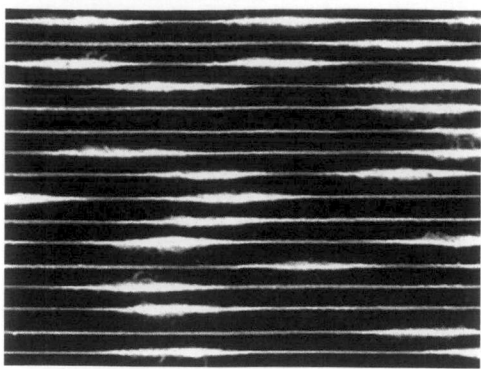

FIGURE 5.4
Slub yarn.

the cotton card; otherwise, the fiber balls will mostly be opened up and the effects will be lost. From a performance point of view, larger effects, relative to the base yarn linear density, and effects in a looser and softer yarn structure with low twist level will have lower resistance to abrasion and the effects can be more easily rubbed off during fabric production and subsequent end use.

5.2.1.2 Slub Yarns

Slub yarns are more pronounced effects compared with nepp yarns. Figure 5.4 shows an example of slub yarn. These are more commonly produced during processes after carding but prior to the formation of the final yarn. These effects can be formed by introducing additional fibers into the fiber stream in a controlled manner or by introducing deliberate unevenness into the fiber flow. The introduction of additional fibers can be carried out during drafting in the spinning process or during condensation in the woolen process. Unevenness can be introduced by varying the drafting roller speed in traditional ring spinning. It can also be introduced by deliberately creating local mechanical faults in machine components in a spinning system, such as the opening roller of a rotor spinning system. It is also possible to exploit the fact that fibers of dramatically different lengths tend to cluster during roller drafting, so mixing these fibers together before drafting and during drafting, the shorter fibers tend to cluster together to form slubs.

5.2.1.3 Flake Yarns

Flake yarns contain larger and usually looser fiber clusters than slub yarns. These yarns are sometimes also called flamme yarns. Figure 5.5 shows an example of flake yarn. In many ways, flake yarns may be considered as more pronounced slub yarns, but these effects can only be created by controlled introduction or injection of additional fibers into the fiber stream before spinning.

5.2.2 Yarn Effects

Yarn effects are also sometimes called ply effects, as these effects are created by plying two or more yarns together subsequent to the production of the single yarns. These yarns can mostly be created using the traditional ring spinning system but with additional feeding

FIGURE 5.5
Flake yarn.

and control devices, and more recently with the hollow spindle system. These fancy yarns always contain at least two basic component yarns: the ground or core and the effect. In the majority of cases, an additional component, binder yarn, is also required to fix the effect yarn on to the ground yarn. In reality, the variety of fancy yarns is unlimited, but based on the fundamental yarn structure, they can be classified into a few basic types. It is also possible to use knitting, braiding, or other techniques to make thin strands and use them as *yarns*. These types of fancy yarns are excluded here.

5.2.2.1 Marl Yarn

Marl yarns are probably the most simple plied yarn structures. These are effectively straight folded yarns made by plying two, and sometimes more, yarns together. The individual component yarns are usually exactly the same in linear density and twist but often differ in color or texture or both. The folding of these yarns creates a final yarn with subtle color or texture variation. A typical use of marl yarns is men's suiting fabric to create a pinstripe effect. Figure 5.6 shows an example of a marl yarn with three component yarns. If any of the component yarns differ in linear density or twist, yarn structure variation will occur and the final yarn will be unbalanced, although it may be a designed outcome.

5.2.2.2 Spiral Yarn

The basic spiral structure is formed by plying two yarns together so that one yarn spirals around the other as shown in Figure 5.7. Spiral yarns are also called corkscrew yarns due to their appearance. This structure differs from the marl yarn in that one component yarn has a longer length than the other, and the shorter component yarn remains substantially straight.

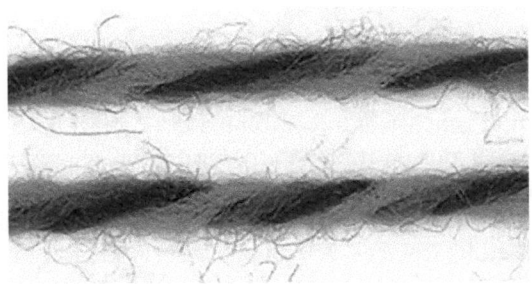

FIGURE 5.6
Marl yarns.

FIGURE 5.7
Spiral yarn structure.

FIGURE 5.8
Spiral yarn with thick effect yarn.

FIGURE 5.9
Spiral yarn with fine effect yarn.

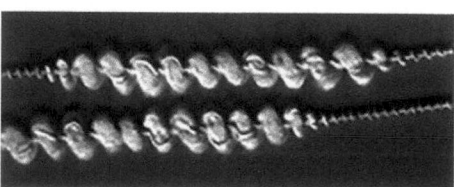

FIGURE 5.10
Spiral yarn combined with flake effect.

In most cases, two yarns of contrasting linear densities are used. Figure 5.8 shows one yarn with a thick yarn spiraling around a thin yarn, whereas Figure 5.9 shows a thin yarn spiraling around a thick yarn. It is possible to combine a fiber effect such as a slub or flake yarn with spiral effect to create a more complex yarn. Figure 5.10 shows an example of the spiral effect produced by a flake yarn.

Spiral yarns can be produced with one component yarn being fed faster than the other, requiring special overfeeding devices, which are described later in this chapter; they may also be made by simply plying yarns with differing linear densities but with the same feeding speed. In this case, no special feeding devices are required. Although the two component yarns have the same length during feeding, once plied, the two yarns will have

FIGURE 5.11
Gimp yarn structure.

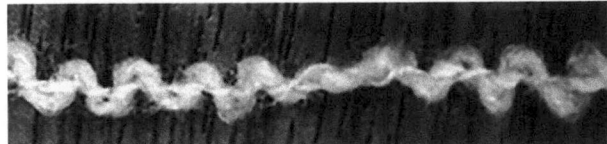

FIGURE 5.12
Gimp yarn.

different lengths depending on the ply twist direction. If the ply twist is in the same direction of the thick yarn twist, the thick yarn will contract, leading to a final yarn in which the thin yarn spirals around the thick yarn; if the ply twist is in the opposite direction of the thick yarn twist, the thick yarn will lengthen, leading to a final yarn in which the thick yarn spirals around the thin yarn.

5.2.2.3 Gimp Yarn

A gimp yarn consists of at least three component yarns—the core, the effect, and the binder—and is produced in two stages. In the first stage, the core and the effect, which is usually overfed, are twisted together, producing an intermediate yarn similar to a spiral. In the second stage, the intermediate yarn is twisted together with the binder yarn with a twist that is opposite in direction to the twist used in the first stage. This reverse binding process removes most of the first stage twist. This leads to the effect yarn forming wavy projections on the yarn surface, and these projections are secured onto the core yarn by the binder yarn. The basic structure of a gimp yarn is illustrated in Figure 5.11 and a yarn example is shown in Figure 5.12.

Although it is common to overfeed the effect in the first twisting stage, this is not essential. When the effect yarn is much thicker than the core yarn, which is often the case to emphasize the wavy effect, it is possible to produce a spiral effect without the specialist overfeeding device as described in Section 5.2.2.2. Reverse binding this spiral yarn can also produce a gimp yarn with more subtle wary projections.

5.2.2.4 Boucle Yarn

The boucle yarn is very similar in construction to the gimp yarn. It requires a minimum of three component yarns: core, effect, and binder; and it is produced in two stages. The yarn construction is illustrated in Figure 5.13. The main difference between a boucle yarn and a gimp yarn is that the wavy projections on the boucle yarn surface are further away from the yarn body, a result of greater overfeeding of the effect yarn during the first twisting stage. On the account of greater overfeed, the effect spirals very loosely around the core following the first twisting stage. The wavy projections can be more easily distorted during the second twisting stage, leading to a more variable yarn appearance, as shown by the example in Figure 5.14.

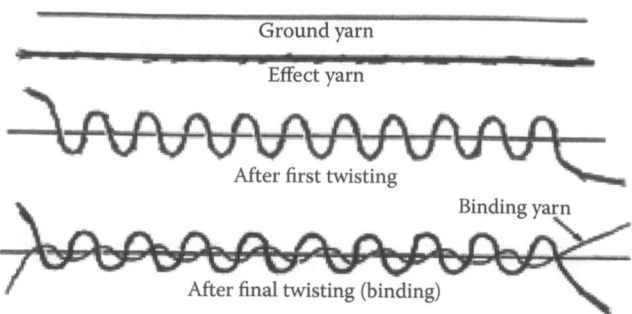

FIGURE 5.13
Boucle yarn construction.

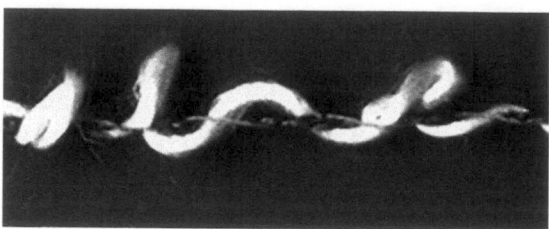

FIGURE 5.14
Boucle yarn.

5.2.2.5 Loop Yarn

Loop yarns are characterized by circular projections formed by the effect yarn. They are typically formed by at least four component yarns: two cores, the effect, and the binder. Two cores are required to form a stable triangular space in which the overfed effect yarn can accumulate to produce the loops during the first twisting stage, shown in Figure 5.15. The first stage yarn must be further processed by a reverse binding process to fix the loops

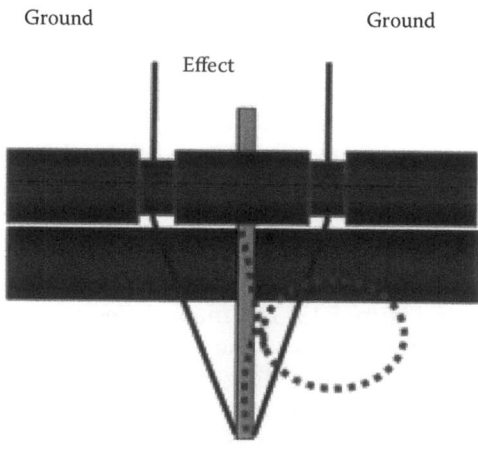

FIGURE 5.15
Loop yarn formation.

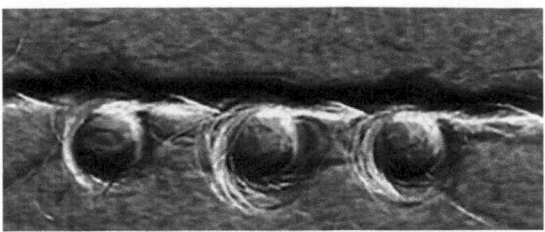

FIGURE 5.16
Mohair loop yarn.

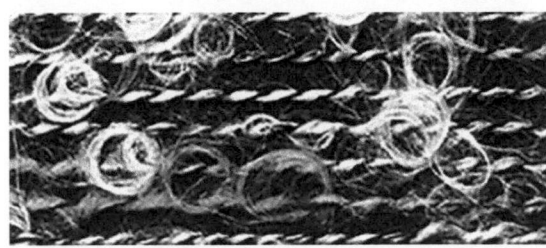

FIGURE 5.17
Loop yarn with untwisted effect fibres.

onto the core yarns because in the first twisting stage, the core and effect yarns simply twist around each other and the loops are not trapped by the core yarns. The effect yarn is usually overfed by 200% or more relative to the core yarns. To produce uniform and stable loops, it is important that the effect yarn is made from elastic and pliable fibers such as mohair, and is not twist lively. Rovings of long staple fibers may be used as effects to produce loose fiber loops. Figure 5.16 shows the uniform loops formed by a mohair effect yarn and Figure 5.17 shows the loose fiber loops using a roving as the effect. Although the loop size is obviously dependent on the effect overfeed ratio, it can also be controlled by the spacing of the two core yarns, the twist level used during the first twisting stage, and the yarn tension. Controlling the twist level is the easiest way to alter the loop size, as it does not require the change of any of the spinning machine parts. The reverse binding process also affords the opportunity to readjust the final yarn twist so that the yarn is not overly hard.

5.2.2.6 Snarl Yarn

Snarl yarns are made in exactly the same way as loop yarns, except that the effect yarn is twist lively instead of stable. Due to the twist liveliness, the loops formed by the effect yarn collapse under the influence of the untwisting stress in the yarn and form kinks. An example is shown in Figure 5.18. To enhance the formation of the snarls, the effect yarn overfeed is usually higher than that used for loop yarns.

5.2.2.7 Knop Yarn

A knop yarn contains sections, with only the effect yarn being visible as shown in Figure 5.19. These sections on knop are formed when the core yarn is stopped momentarily

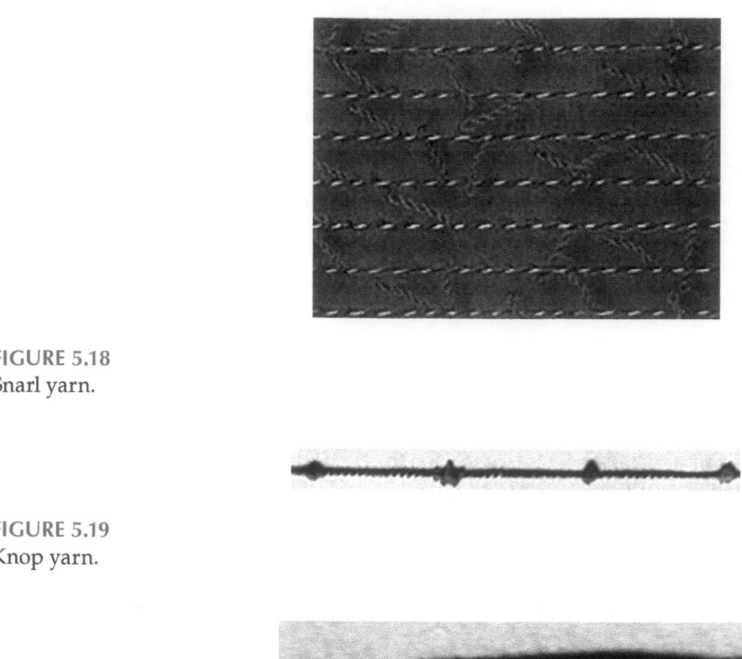

FIGURE 5.18
Snarl yarn.

FIGURE 5.19
Knop yarn.

FIGURE 5.20
Elongated knop.

while the effect yarn feeding continues. The excess effect yarn wraps around the core yarn at the same spot, forming these bunches. Between the knop sections, the yarn resembles a normal plied yarn. For all the other yarns described previously, the effect yarn overfeed, when used, is constant during production; for knop yarns, the overfeed is controlled. The feeding device for the core yarn must therefore be able to change speed during production. The formation of the knop can be controlled by a device called knopping bar or a control bar, shown in Figure 5.20. The extent of the knop can be spread by the movement of the knopping bar, producing elongated knops or stripes as shown in Figure 5.20. The core yarn and the effect yarn may be stopped at alternating intervals, leading to a yarn showing alternating sections of one of the component yarns. This type of yarn is also called cloud yarns.

5.2.2.8 Chenille Yarn

A chenille yarn consists of a cut pile that is trapped by the core yarns. The basic structure of a chenille yarn is shown in Figure 5.21 and some examples of yarn are shown in Figure 5.22. The production of chenille yarns can be accomplished on a dedicated chenille machine, which will be described later in this chapter. These yarns can also be produced by other methods such as weaving or flocking. Figure 5.23 shows a leno woven fabric, which can be cut along the warp at the points indicated by the arrows to form chenille yarns. Further twisting may be necessary to enhance the binding of the piles formed from the cut weft yarns. It should be obvious that chenille yarns made in this way are expensive. A faster and more economical process for making chenille yarns is flocking. In this process, shown in Figure 5.24, a continuous core strand, usually of filaments, is coated

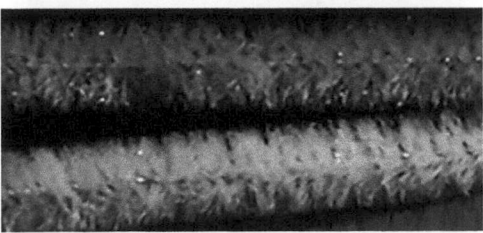

FIGURE 5.21
Chenille yarn structure.

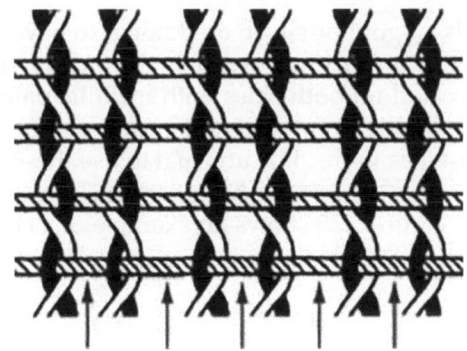

FIGURE 5.22
Chenille yarn.

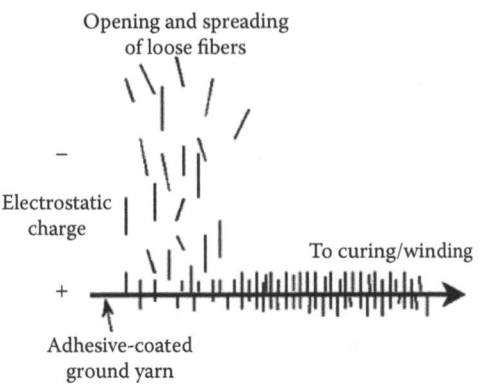

FIGURE 5.23
Chenille yarn production using leno weaving.

Opening and spreading
of loose fibers

Electrostatic
charge

To curing/winding

Adhesive-coated
ground yarn

FIGURE 5.24
The yarn flocking process.

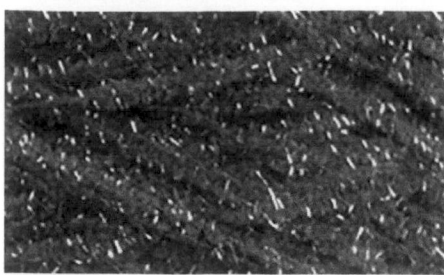

FIGURE 5.25
Metallic yarn.

with adhesive and flocked with loose fibers. To ensure the fibers form standing piles on the core strand, flocking is usually carried out in an electrical field. The loose fibers are charged with an opposite electrostatic charge to the core strand. Chenille yarns produced by flocking tend to have lower abrasion resistance, as the fiber pile is only stuck to the core by the adhesive, whereas the pile in other chenille yarns is trapped by intertwining yarns.

5.2.3 Metallic Yarns

Clothes of metallic threads, of gold or silver, are known to have been made thousands of years ago. Modern metallic yarns are, however, usually only made from laminated plastic films. The films are first coated on both sides with metallic paint of suitable color, a further transparent coating is applied on top to enhance the wear resistance. The films are cut into thin strips to be used as yarns. It is unusual to use these slit film yarns directly in fabrics because of poor abrasion and strength properties. They are often used as a component in a compound yarn. Figure 5.25 shows an example of a chenille yarn with metallic components.

5.3 Spinning Systems

Fancy yarns can be made using a wide range of spinning systems. To increase the machine flexibility, most manufacturers of modern spinning systems including filament yarn texturing systems offer special facilities to enable the production of fancy yarns. The most widely used systems are ring and hollow spindle, however. The chenille machine, as the name suggests, is a system that can only be used for making chenille yarns. The following sections describe the basic principles of the various spinning systems and how they can be used for making fancy yarns.

5.3.1 Ring Spinning

The principle of the ring spinning system is shown in Figures 5.26 and 5.27. Rovings are drafted by the drafting rollers to the required linear density before they are fed to the ring spindle. The drafted fibers are twisted by the rotation of the traveler on the ring. The rotation of the traveler is controlled by the rotation of the spindle on which the yarn bobbin is mounted and the speed of the roving delivery by the drafting system. The relationship

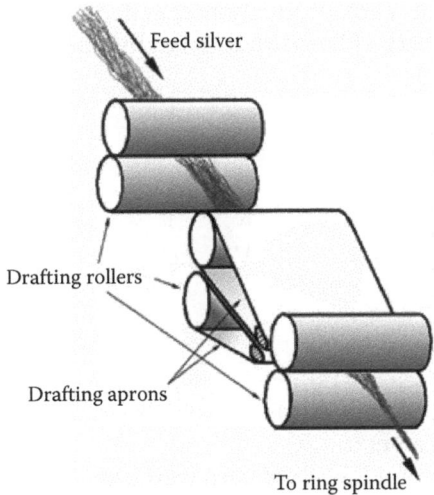

FIGURE 5.26
Roller drafting.

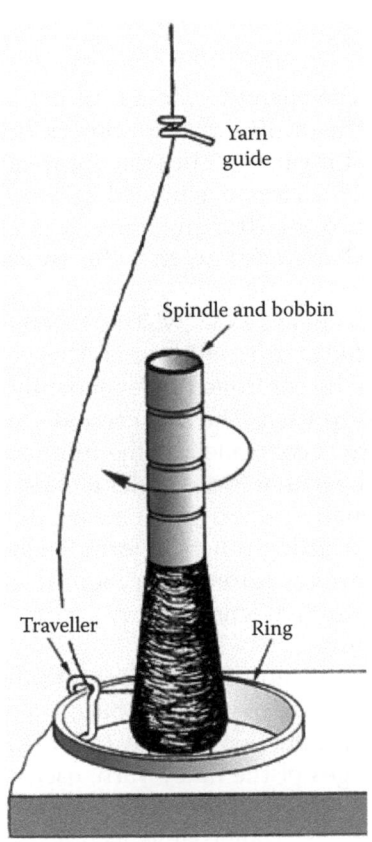

FIGURE 5.27
Ring spindle.

between traveler speed, the yarn production speed, and the spindle speed is given by Equation 5.1. The twist density of the yarn is given by Equation 5.2.

$$N_t = N_s - \frac{V_d}{\pi D_p} \tag{5.1}$$

$$t = \frac{N_s}{V_d} - \frac{1}{\pi D_p} \tag{5.2}$$

where:
  N_t is the traveler speed (turns/min)
  N_s is the spindle speed (turns/min)
  V_d is the yarn delivery speed (m/min)
  D_p is the yarn package diameter
  t is the turns of twist per unit length of yarn (turns/m)

The second term of the equation is a small percentage of the overall twist value and is usually ignored to simplify the calculation as follows:

$$t = \frac{N_s}{V_d} \tag{5.3}$$

In normal yarn production, the yarn evenness is of primary importance and the roving should be drafted with the minimum variation in linear density. For the production of fancy yarns such as slub yarns, deliberate local variations in the linear density of the yarn are introduced. This can be achieved by varying the draft roller speed or injecting additional fibers into the drafting zone. It is also possible to mix fibers of different lengths in the feed material so that the shorter fibers form slubs during drafting.

Fancy yarns with yarn effects require the twisting together of two or more yarns. These component yarns are usually fed at different speeds. This necessitates the provision of two or more sets of rollers that can be controlled independently. Figure 5.28 illustrates a feeding system for producing a loop yarn. The two ground yarns are controlled by the back rollers, whereas the effect yarn is controlled by the frontfeed rollers. As the effect yarn is overfed relative to the ground yarns, it is not under tension and must be controlled by the rollers closest to the yarn formation zone, which means the front rollers. The three yarns converge just below the front drafting rollers to form the loop effect. As described earlier in this chapter, many fancy yarns require two or more stages of production. For the loop yarn shown in Figure 5.27, a second twisting process is required to bind the loops on to the ground yarns with a binder yarn.

The ring spinning system is the most flexible yarn production system in terms of both raw material handling and the range of yarns produced. It is, however, slow in speed and requires multiple stages in the production of many fancy yarns. One has to remember that in addition to the twisting stages of the fancy yarn, each component yarn must be produced separately. The cost of producing fancy yarns on the ring spinning system is therefore very high. In recent years, faster alternative techniques for making fancy yarns in a single-stage process have been developed. The most widely used of these new techniques is the hollow spindle system.

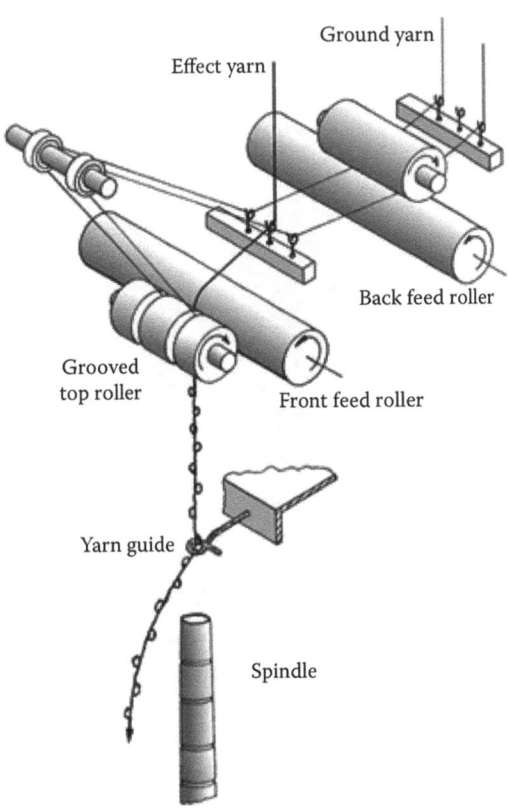

FIGURE 5.28
Feed system for loop yarn formation.

5.3.2 Hollow Spindle System

The hollow spindle spinning method was developed for making normal yarns. In this system, the short staple fiber strand is strengthened by wrapping it with a continuous binder instead of twisting. This resulted in a fasciated yarn structure with the core fibers lying mostly parallel to each another. It has not been very popular in the production of normal yarns because of the necessity of using at least two components in the yarn. Although most fancy yarns contain multiple components, hollow spindle system has become very widely used for their manufacture. One biggest advantage of hollow spindle spinning over ring spinning in the manufacture of fancy yarn is that most yarns can be made using a single passage of the machine. Figure 5.29 shows an example of a hollow spindle fancy yarn system. In this particular case, four independent feeding devices are used. This enables the production of a wide range of fancy effects. All the feeds are passed through the rotating hollow spindle. A bobbin containing the binder yarn is mounted on the hollow spindle and rotates with it. The binder is pulled into the hollow spindle from the top. The rotation of the hollow spindle wraps the binder around all the materials passing through the hollow spindle and binds them together.

Drafted staple fiber strands do not have cohesion prior to being wrapped by the binder. To prevent these fiber strands from breaking before being wrapped, false twist is usually inserted into the strands. This is achieved by wrapping the strand around a twist distributor usually located at the bottom of the spindle instead of passing the strand straight

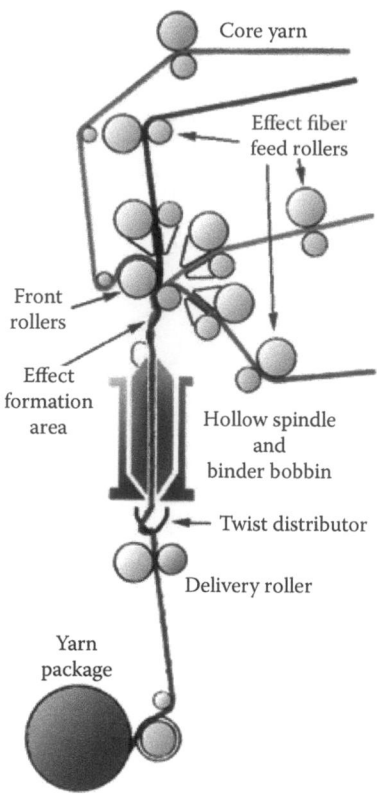

FIGURE 5.29
Hollow spindle fancy yarn system.

through the hollow spindle. The rotation of the twist distributor imparts false twist into the strand.

Fancy yarns made using the hollow spindle system have different characteristics to those made using the ring spinning system even when the yarns have similar structures. This is because in hollow spindle spinning, no real twist is imparted to the core strand, while in ring spinning all the strands are twisted together. Due to the lack of twist, hollow spindle yarns have lower twist liveliness, but they will disintegrate when the binder yarn is broken.

Fancy yarns can be made in a single process using the hollow spindle system. This makes the fine tuning of the effects much easier as the effects of any adjustment of the machine settings are seen immediately.

A wide range of fancy effects can be produced using the hollow spindle machine. In addition to the independent control of the feeding devices, the delivery roller can also be controlled to produce even more effects. More recent hollow spindle machines are equipped with computer control systems, which allow the programmed control of the machine for producing fancy effects.

To overcome the lack of twist in the yarns produced on the hollow spindle machine, it is possible to pass the yarn through a ring twisting stage afterward. This is more conveniently achieved by a system combining hollow spindle spinning with ring spinning, as shown in Figure 5.30.

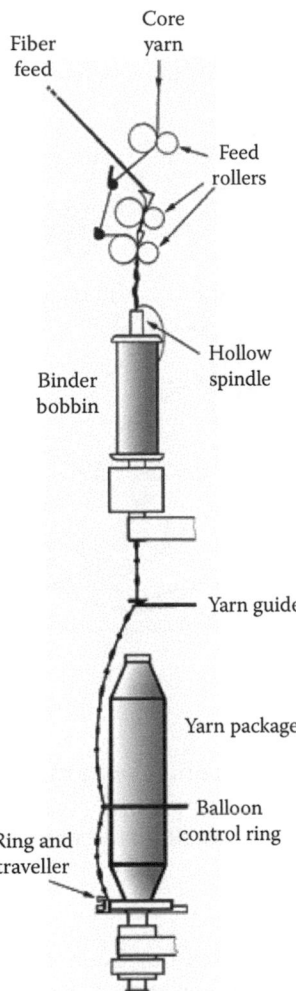

FIGURE 5.30
Combination of hollow and ring spindles.

5.3.3 Open End Spinning

There are two commercial open-end systems: rotor and friction. The rotor system is by far the more widely used system. The principle of the rotor system is illustrated in Figure 5.31.

The fiber material is fed into an opening unit by a feed roller in conjunction with a feed shoe. This feed material is usually a drawn sliver. An opening roller is located inside the opening unit. The surface speed of this opening roller is much faster than the feed roller surface speed, opening up the fibers to create a very thin and open fiber flow. The fibers are taken off the opening roller by an air stream and carried into the rotor. The centrifugal force generated by the rotor forces the fibers into the rotor groove. Only a very thin layer of fibers is deposited in the rotor with each rotation. Many such layers of fibers are needed to make up the yarn. This doubling up of the fibers in the rotor is called back-doubling. On account of this back doubling, any local variations in the feed material tend to be spread out, making the yarn more even.

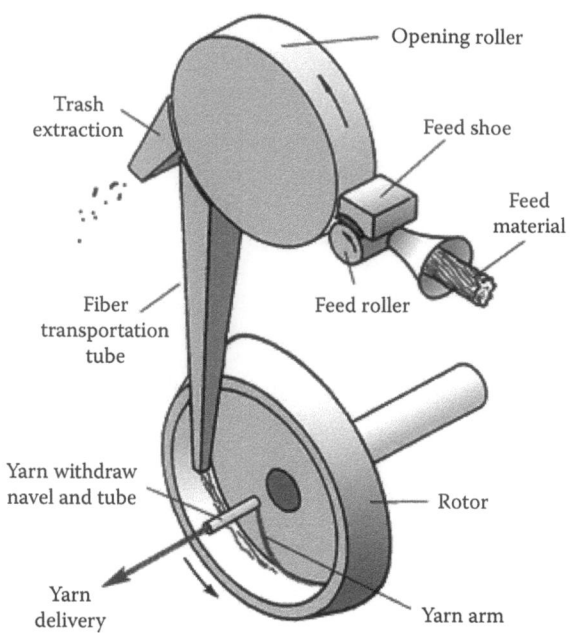

FIGURE 5.31
The rotor spinning system.

Although rotor spinning is used mainly for spinning plain short staple fiber yarns, machine manufacturers have started to provide electronic control devices or modified machine components to allow the production of fiber effect fancy yarns such as slub yarns.

5.3.4 Chenille Machine

A method of producing a chenille yarn has been developed that produces two yarns simultaneously at each unit. This is illustrated in Figures 5.32 and 5.33. The effect yarns are wrapped around a gauge, which is triangularly shaped at the top, narrowing toward the base to allow the effect yarn coils to slide downward onto the cutting knife. The width at the bottom of the gauge determines the effect length, which creates the depth of the pile in the final yarn. Although for simplicity, the cutting knife is shown here as a linear knife edge, actual machines all use a circular cutting knife. On each side of the knife, there are two ground yarns. One ground yarn is guided by the take-up roller, whereas the other is guided by the companion roller. The take-up roller is pressed against the profiled guide and intermeshes with the companion roller, allowing the two ground yarns to trap the pile created by the effect yarn in between them, at right angles to the ground yarn axis. The two ground yarns are twisted together, usually by a ring spindle at the lower part of the machine, to produce the final yarn. This machine can produce chenille yarns very conveniently and at high speed; however, it lacks flexibility, as it can produce nothing but chenille yarns. It is therefore only popular with dedicated fancy yarn producers.

5.3.5 Texturing

Although fancy yarns are mostly made from staple fibers, it is possible to produce fancy filament yarns during the texturing process. The most widely used filament texturing

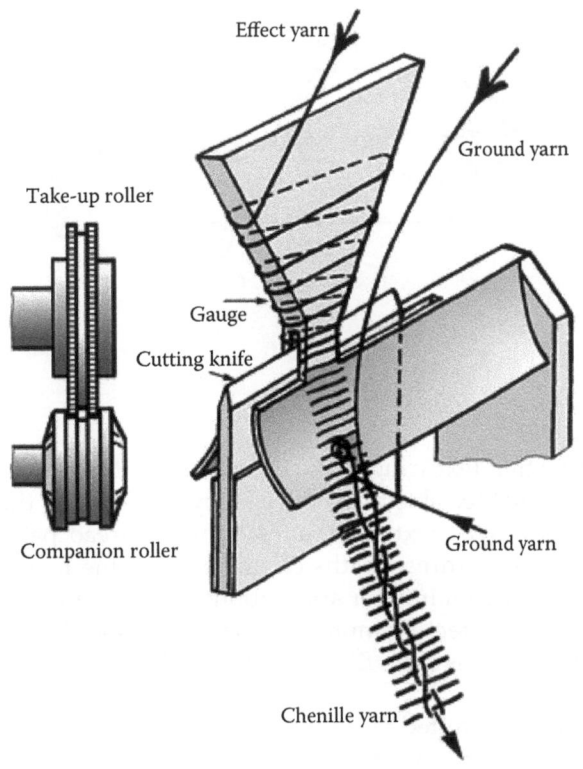

FIGURE 5.32
Principle of chenille yarn production system.

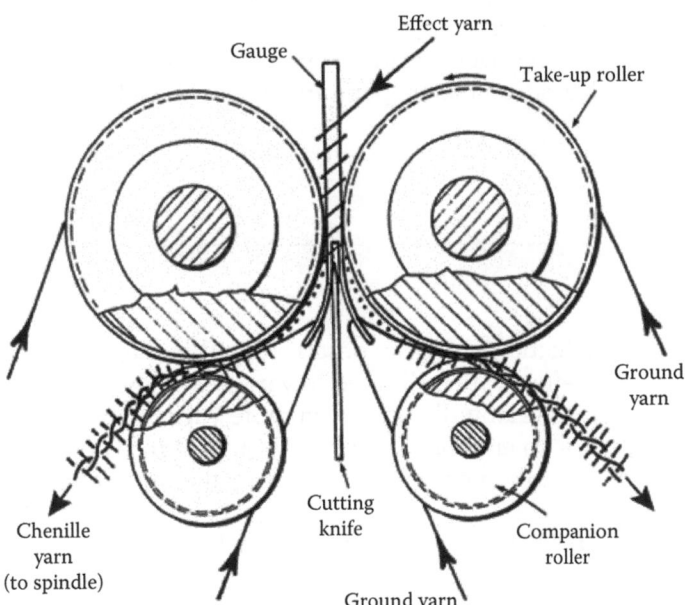

FIGURE 5.33
Side view of chenille yarn production system.

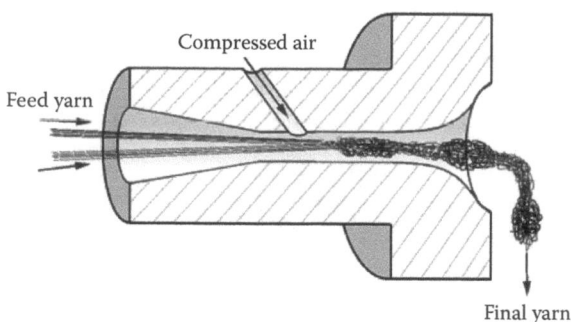

FIGURE 5.34
Air jet texturing.

processes are false twisting texturing, bulked continuous filament texturing, and air jet texturing. It is normally not feasible to produce fancy yarns using the bulked continuous filament process. In false twist texturing, alterations of thread path or combinations of filaments following different thread paths can result in some fancy effects. False twist-textured yarns, however, are mainly for applications that demand very high stretchiness, so the impact of any fancy effect is limited. Air jet texturing offers the most convenient route for producing fancy yarns from filaments. Figure 5.34 illustrates the basic principles of air jet texturing.

The flat filaments are first wetted to improve process stability by reducing filament-to-filament friction and are then fed into the texturing jet. The filaments are blasted with high pressure air or steam inside the jet, and the filaments buckle and entangle with each other in the turbulent air stream. By combining two or more feeds, which may differ in several ways such as linear density and speed, fancy effects including slubs and boucles can be made. In contrast to the high stretch yarn made from the other two main texturing processes, air textured yarns resemble more closely yarns from staple fibers. One clear advantage of air jet texturing over traditional processes for fancy yarn production is its speed, reaching several hundred meters per minute in contrast to a few tens of meters per minute.

5.4 Use of Fancy Yarns

The fancy yarns and production processes described in this chapter are by far not exhaustive. By definition, fancy effects are unlimited in variety and the continuous development of new technology and material will allow even more yarns to be made. As the primary aim of using fancy yarn is to enhance the aesthetic appearance of fabrics, there is a continued effort by producers to come up with new yarns that differentiate themselves from competitors.

However, one main drawback of using fancy yarns lies in the increased costs of yarn, fabric, and garment manufacture. The production speeds for fancy yarns are generally slower than for plain commodity yarns because of the inherent unevenness of fancy yarn. In addition, most fancy yarns require multiple components and several twisting stages, so the cost of fancy yarn production is typically several times that of commodity yarns.

Fancy yarns are also far more sensitive to fashion trends, so it is unwise to produce large quantities without order confirmations. Any redundant stock will be highly costly to the yarn producer. In fabric production, the uneven nature of fancy yarns will demand more careful handling and slower production speeds, leading to increased fabric costs. The interaction with the fabric structure also means that the final appearance of the fancy effects can be unpredictable, so in most cases, samples must be produced prior to normal production.

With some exceptions, the use of fancy yarns will lead to lower fabric performance in terms of strength, wear resistance, and aftercare. Careful consideration must therefore be given to the performance requirement of the end use and to the quantities of fancy in the fabric when fancy yarns are to be used. Given the higher costs and lower performances, the main use of fancy yarn is usually more for high value and high margin applications.

6

Fiber and Filament Dyeing

Asim Kumar Roy Choudhury

CONTENTS

6.1 Introduction

Dyeing is a process of applying color to the whole body of a textile material, and some degree of color fastness is desirable. Textiles are dyed using continuous and batch processes, and dyeing may take place at any of the several stages during the manufacturing process (i.e., prior to fiber extrusion, fiber in staple form, yarn, fabric, and garment).

6.2 Stages of Coloration

The stage at which coloration or dyeing is carried out depends on several factors, fashion and economics being the most important considerations. The decision of coloration is postponed as long as possible to cater to the dictates of fashion and dyeing in a fabric/garment form gives maximum flexibility.

The coloration of textiles may be carried out at the following stages (Clarke, 1982):

1. Mass pigmentation of polymer melt or solution
2. Dyeing of polymer gel
3. Fiber tow-dyeing
4. Loose-stock or fiber dyeing
5. Dyeing of fiber top or sliver
6. Yarn dyeing
7. Fabric dyeing
8. Garment dyeing

Man-made fibers can be dyed at all the above-mentioned stages, whereas only stages (4)–(8) are suitable for natural fibers.

In terms of volume, the major routes for synthetic fibers are gel, tow, and loose-stock dyeing, with yarn and fabric dyeing forming the next most important segments of the industry. Dyeing textiles in the form of fabric or garment offers the processor maximum flexibility in responding quickly to changes in fashion and market.

The selection of the coloration stage is decided by several factors such as

1. Economy
2. Availability of equipment
3. Suitability of the stage for the particular fiber and end-product

The relative importance of different methods of coloration also depends on the market demand. The progress in gel-dyeing techniques for synthetics has resulted in a marked decline in yarn-dyeing capacity and some decline in loose-stock dyeing. Continuous top dyeing has declined dramatically, mainly because of cost. Batchwise tow-dyeing processes have at least held their share of the market, with modest increase in certain European countries (Clarke, 1982).

Dyeing in fiber and partly in yarn state allows the use of dyes of high fastness but of poor leveling properties, as any unlevelness created during dyeing will be minimized by blending during subsequent spinning processes. Dyeing in the later stages of the manu- facturing sequence gives economic benefits due to the elimination of colored waste and a reduction in the stock holding of colored goods. The gray fabrics can be dyed at any time to suit the dictates of fashion; however, dye selection in this case is more restricted. Unevenness of the dyed fabric cannot be rectified easily—leveling types of dyes are to be selected even at the cost of the fastness properties.

Fabric dyeing is the most popular choice, as this is the final stage of coloration in the textile mill after which the material is sent for finishing. Fabric is the most stable form of textile material. It is easier to handle and available in long lengths so that continuous processing is possible.

6.3 Yarn Dyeing

The use of dyes of high fastness is required to achieve a high degree of reproducibility.

Colored yarns are required for some end-uses namely:

1. Ready-made garments
2. Hosiery industries
3. Bag closing industry
4. Sports goods
5. Umbrella and carpet making
6. Book binding
7. Handloom textiles
8. Furnishing textiles

Dyed yarns are also extensively used for making stripe, check, or designed knitted/woven fabrics or in sweater manufacturing. Yarns are broadly dyed in two forms—hank and package dyeing.

In the textile industry, a hank refers to a unit of yarn or twine that is in a coiled form. This is often the best form for use in hand looms, compared to the cone form needed for power looms. Hanks come in varying lengths depending on the type of the material. For instance, a hank of linen is 300 yards (270 m) and that of cotton or silk is 840 yards (768 m). In worsted yarn, made from combed fiber, there are 560 yards (510 m) to a hank.

Acrylic or wool yarns are mostly dyed in hank form. In the continuous filament industry, polyester, or polyamide yarns are always dyed in package form, whereas viscose rayon yarns are partly dyed in hank form.

The first patent was issued to Obermaier for package dyeing around 1882, whereas hank dyeing has a much longer history and probably dates back to antiquity when yarn was simply suspended on wooden poles and turned by hand. This process is still practiced in the cottage sector. Hank form is very convenient to handle and transport. Many spinners are still equipped with hank-reeling machines as the last operation in the spinning process. It is often claimed that with certain products, a better quality is obtained by the hank dyeing route. The products are cotton doubled-yarn, hand-knitting yarns in wool, nylon, acrylic and their blends, carpet yarns, and acrylic yarn for machine knitting.

The characteristic features of hank dyeing are (Park, 1981) as follows:

1. Hank-dyed yarn has a fuller bulk and softer handle.
2. Yarn may become entangled.
3. Hank-reeling and back-winding are costly and may generate waste.
4. Levelness may be inferior to package-dyed yarn.
5. Dyeing cost is less.
6. Manual dyeing in the absence of suitable machine is possible for small lots.

The advantages claimed for package dyeing over hank dyeing include the following:

1. The yarn is leaner but gives better fabric definition.
2. Back-winding is faster and lesser waste is generated.
3. There are savings in space, energy, water, chemicals, among others.
4. There are higher levelness and reproducibility.
5. Hank-reeling is eliminated.
6. Dyeing is more controlled with better levelness and fastness.
7. Lower liquor ratio causes savings in water, effluent, energy, dyes, and chemicals.
8. Less number of attendants are required.
9. High-temperature dyeing and rapid drying are possible.
10. Universal package dyeing may be used for multiple purposes—yarn package dyeing (cone or cheese), top or sliver dyeing, loose fiber dyeing, and so on.
11. Larger control on the process makes automation easier.

Hank-reeling from ring bobbin is a comparatively expensive process, the plant requires much space, and the yarn must be wound onto a bobbin, cone, or pirn at a later stage. The second winding from hank after dyeing is also slow and labor-intensive. A single yarn, on the other hand, can be wound directly from a cop or ring tube onto a cheese, dyed in this form, and then rewound with high-speed machines onto a cone. The spinner, before delivering the materials to the dyer, conveniently cones a doubled yarn. It is more economical for the spinner to cone the yarn than to reel it into hanks.

Disadvantages of package dyeing are the increased cost to the production plant and the possibility of producing less bulky yarns. The latter can be minimized by modifying the process or changing the yarn specification. Some end-users prefer package dyeing because of the following reasons:

1. Texturized yarns for both weaving and knitting could be produced.
2. Sewing threads could be produced.
3. Singles cotton yarns are compatible with the process.
4. Continuous filament yarns are compatible with the process.
5. Wool or acrylic yarn for apparel and furnishing could be produced using this process.

Scouring—removal of impurities before dyeing—is not easily carried out in package dyeing machines. Self-scouring, water-soluble, or emulsifiable lubricants are added during spinning. These are removed by rinsing with water or are stable in the dyebath during dyeing. Heavy deposits on the inside of the package can cause serious problems. To minimize this, scouring and rinsing are normally carried out on a two-way flow.

6.4 Hank Dyeing Machines

For many centuries, hank dyeing has been carried out in primitive equipment in which the yarn hanks are suspended on wooden poles in a rectangular wooden vat. The sticks are moved from end to end of the vat to give some degree of agitation and the hanks are simultaneously turned manually to dye the portion of hank in contact with the sticks. The system has been gradually mechanized, and wood has been replaced by metal and stainless steel.

The traditional Hussong hank dyeing machine is shown in Figure 6.1. It consists of a frame (F) carrying sticks or poles on which hanks (H) are hung and which can be lowered by a hoist on a gantry into the rectangular vessel containing the dye liquor. The liquor is circulated by means of a reversible impeller (I), which is separated from the hanks by a

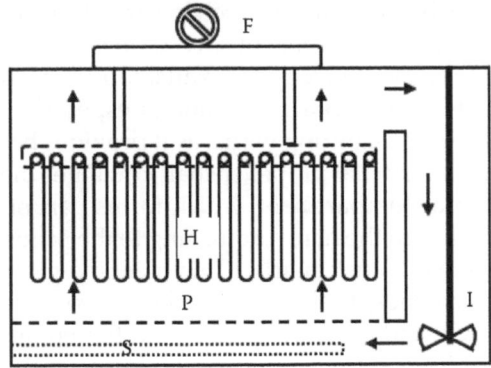

FIGURE 6.1
Hussong hank dyeing machine.

separator and is located at one side of the dye vessel. There is an adjustable weir at the bottom of the impeller compartment, which regulates the flow volume. A false bottom (grid) (P) is provided at the bottom. The holes in the grid are arranged in such a way that there are more holes at the remote end of the impeller than adjacent to it to distribute the liquor flow uniformly. The packing of the hanks is denser at the ends than in the middle, as the circulation is more powerful at the ends. The impeller can be rotated in opposite directions alternately so that the direction of liquor circulation can be reversed. The upward direction of circulation is maintained during majority of the dyeing cycle. During this period, the hanks are pressed against the perforated top of the pole frame, and the liquor, by penetrating through the hanks, is compelled to come into intimate contact with every part of the hank. The desired dyeing temperature is maintained by opening or closing the steam pipes (S); the steam pipes are situated below the perforated false bottom. In modern machines, the yarn is suspended from V-shaped sticks with perforations to prevent stick marking. With some qualities, when stick marking occurs even with V-shaped sticks, the yarn is turned on the sticks manually halfway through the dyeing process.

These types of machines are made by several manufactures and are still widely used for dyeing cotton hanks and in the carpet industry. A perpetual problem of dyeing in this type of machines is that the hanks at the outside of the carrier frame are often in direct contact with the sides of the dyebath. The inadequate liquor circulation at the sides of the vessel may result in poor levelness of the dyed fabrics.

The dye liquor inlets and outlets should always be beneath the false bottom so that the filling and emptying operations should cause minimum disturbance to the arrangement of hanks. The liquor ratios in these machines are very high—generally around 1:30. Power consumption is low. The electric motor driving the propellers usually operates at 1–2 kW. The dye vessel must be free from rough places or projections. The rectangular shape of the vessel does not normally facilitate pressurized high-temperature dyeing. A few machines, however, have been modified to operate at 105°C–108°C for level dyeing of acrylic fibers with basic dyes.

An improvement on the conventional Hussong machine is the Pegg pulsator hank dyeing machine. In pulsator mechanism, the flow of liquor is automatically interrupted for 5–10 s every 2.5 min, thereby giving a pulsating movement, which raises the hanks off the sticks and prevents stick marks (Park, 1981). There is no false bottom, the steam pipes are placed in a small end section, and the impeller shaft is horizontal. The penetration is assisted by an automatically controlled unidirectional pulsating flow. The dye liquor flows downward, so that the hanks remain stationary and stretched at full length throughout the dyeing cycle, resulting in better quality of yarn and less problem during winding. Due to its compact nature, the liquor ratio in such a machine is much lower (1:10 as opposed to 1:30 in conventional hank dyeing machine), with consequent savings in steam, water, and chemical costs. Machines ranging from 5 to 600 kg capacity are available and machine coupling is possible.

The improved Pegg GSH hank dyeing machine is especially designed for high-bulk acrylic and various types of hand-knitting yarns, as it provides better loftiness and acceptable uniformity of shade. A horizontal propeller shaft circulates the dye liquor in one direction only, up a narrow central compartment, which forms part of the frame, and through the top hank poles down through the hanks, returning to the propeller. The hanks are lifted slightly off the poles to prevent stick marks. This machine gives a gentle, but uniform, liquor flow through the yarn pack, with minimum entanglement, eliminates temperature differences through the pack, and prevents glazing and flattening of the yarn.

Mezzera s.p.a. of Milan, Italy, developed a special hank dyeing machine, model TMB (Figure 6.2), for dyeing of very delicate articles such as wool, viscose, silk, and finest cotton yarns including mercerized yarns. The machine incorporates features that ensure

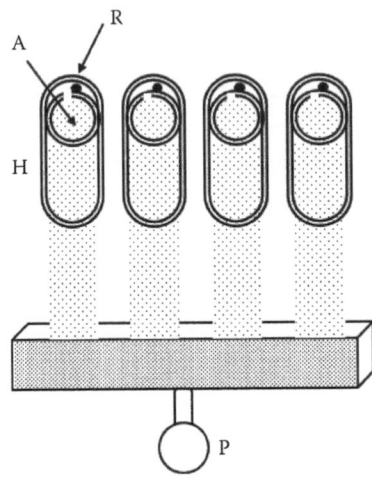

FIGURE 6.2
Mezzera model TMB hank dyeing machine.

preservation of yarn quality and economy of production. The machine may have 2–20 stainless steel supporting arms (A) on which the hanks (H) are placed. A unique feature is that the hanks are never immersed in the dyeing liquor. Each of the arms consists of a perforated tube from which the dye liquor flows. Plates at the end of tube prevent the hanks from falling off the arms. A hank displacing rod (R) rotates about the perforated tube to lift and move the hanks so that the point of suspension continuously changes, thereby ensuring uniform dyeing of the whole hank. A timer adjusts the interval between rotations of the rod according to the type of fiber and dye. Liquor stops flowing, whereas the rod rotates to avoid wool felting, silk fraying, or fiber entanglement. A three-speed pump sucks liquor from a specially shaped vat to minimize the liquor ratio and to distribute the dye liquor through the arms. The liquor then falls down into vat. In some other models, the rod is eliminated and the perforated tube is half-rotated on either side at some interval.

6.5 Package Dyeing Machines

In package dyeing machines, the yarn is wound into suitable packages (e.g., cones, cheeses), mounted onto perforated spindles, and then dyed by forcing the dye liquor through the package. The earlier machines were atmospheric machines with unidirectional flow (in-to-out) of liquor through the package. Subsequent machine modifications were made to improve the levelness by enclosing the vessel and frequently reversing the direction of the liquor flow. Virtually all modern package dyeing machines are now pressurized and capable of operating at temperatures up to 135°C.

For dyeing, the yarn is wound on perforated stainless steel cones or cheese formers. The dyers prefer cheese form, as they tend to give level dyeing because of their uniform diameter. The success of package dyeing, in terms of both levelness and yarn quality, is greatly influenced by the degree of care taken in the preparation of the yarn packages. A wide range of designs and materials are being used to support the packages, such as perforated cones, springs, rockets, plastic tubes, and nonwoven fabrics, to name a few. With most

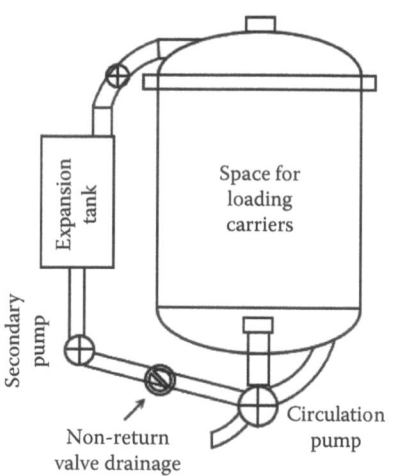

FIGURE 6.3
H.T.H.P package dyeing machine.

stainless steel formers, cost can be a serious problem. Plastic formers are less expensive, but they have a shorter life and are heavy stained by certain dyes causing contamination.

Nontextured nylon or polyester filaments are relaxed before winding on package formers, otherwise shrinkage during dyeing will cause damage to the yarn and crush the package former. The winding onto radially compressible formers is not very successful.

A vertical spindle machine, popularly known as H.T.H.P. package dyeing machine, is shown in Figure 6.3. The change from a rectangular to a circular shape allowed the vessels to be more readily pressurized for dyeing above 100°C.

The pressure vessels are cylindrical with domed ends, the upper being the lid or the cover, which is securely locked by a series of bolts in older machines or by a sliding ring or similar quick-locking device. The lids of the large-diameter machines are heavy and therefore raised by pneumatically or hydraulically operated cylinders. The machine capacity may vary from 100 to 1000 kg. Machine diameters go up to just more than 2 m and the vessel height is usually similar. The overall height of the machine is increased by the pump fitted immediately below the main vessel. The height of the machine and the need to load it by an overhead travelling crane necessitates either plenty of headroom or setting the machine in a pit below the shop-floor level. Where the machines are mounted at ground level, a raised platform with surrounding safety rail is required. A 5-ton hoist is usually sufficient for a machine of 2-m diameter.

The cheeses, cones, or other packages are loaded onto perforated vertical spindles, which are then carefully placed in the package dyeing machine. The spindles are up to 125 cm long and usually circular in cross section (7 cm diameter), although some have a Y cross section to permit easier longitudinal flow of dye liquor. Larger diameter spindles favor level dyeing. The number of packages (cones/cheeses) per spindle is usually around 8–10. The column is pressed down firmly by mechanical devices and the end-cap is screwed tightly to complete the seal. Packages of hydrophilic fibers should be allowed time to wet out and swell before pressure is applied. As some textured yarn packages tend to shrink when wetted for the first time, it may be necessary to wet out the loaded packages for 5 min, and then additional packages are put into each column to maintain uniform permeability. Compressing the packages lengthways not only provides a good seal but also

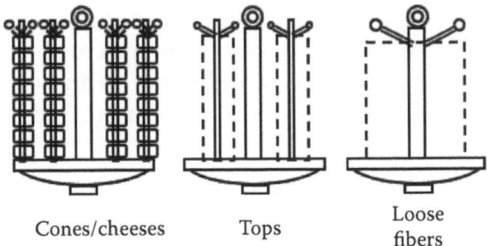

Cones/cheeses   Tops   Loose fibers

FIGURE 6.4
Carriers for package dyeing.

tends to compact some areas of low density and high permeability, thus providing more uniform liquor flow along the column. Nontextured continuous-filament yarns are sometimes wound on rigid formers, which either interlock or are fitted with spacers to provide a seal against excessive liquor flow. Where high rates of liquor flow are required, long spindles could create pressure drop along the column. The spindles are screwed firmly into holes set in a frame called carrier with a circular hollow base and a center pillar with an eye at the top for lifting. At the center bottom of the carrier frame, there is an inlet for dye liquor; this inlet fits onto the pump discharge when the frame is lowered to the bottom of the dye vessel. There may be separate carriers for cones/cheeses, tops, and loose-stock as shown in Figure 6.4.

The cylindrical dyeing vessel is provided with a hinged lid, which, when closed, makes a pressure-resisting seal. Before the lid is clamped down, the carrier carrying the cones, cheeses, tops, or loose fibers is lowered into position; it rests on a seating, which is connected to the circulating pump below. The liquor is forced through the perforated pipe on which the packages are mounted. Under working conditions, the pressurized vessel must be filled with dye liquor. An overflow pipe is connected from the side of the vessel to the expansion tank so that extra dye liquor can be taken out of the vessel to accommodate for the increase in volume, which accompanies the rise in temperature. At the beginning, the flow should always be from outside to inside, and the flow could be reversed after a few minutes.

The machine is connected to a circulating pump. When the vessel is open to atmosphere, circulation is possible only from inside to outside, meaning that the liquor moves from the interior of the carrier through yarn package and comes into the vessel from where it returns to the pump. The circulation from outside to inside needs suction, which cannot be done successfully by the circulation pump. The penetration will obviously be better if the flow is reversed periodically. To have a two-way flow, the vessel must be closed and capable of withstanding pressure of up to 20 lb in^{-2}. Developments in the package dyeing machines have resulted in improvements in pump design and spindle geometry, which benefits the circulation. The conventional machines have a flow rate of 30–45 L kg^{-1} min^{-1}, with complete circulation every 30 s. The machines with a flow rate of 50–150 L kg^{-1} min^{-1}, with a complete circulation every sixth second, are considered as rapid dyeing machines. These will allow the temperature to rise at a rate of 8°C–16°C min^{-1}, as compared to 2°C–3°C min^{-1} in conventional machines. The main pump is fitted to the bottom of the vessel, usually as close to the center as possible to minimize frictional losses in flow.

For heating the liquor, conventionally, high-pressure steam is passed through horizontal coils located in the lower part of the vessel. A large surface area is thus available for heat exchange. The system is compact and heat loss is minimized.

In conventional package dyeing machines, the material-to-liquor ratio is around 1:10 when fully loaded. In partially flooded state, the liquor covers all the packages but does not fill the dome of the machines, causing reduction in the effective liquor ratio. Only in-to-out flow through the packages can, however, be used, and the packages must be sufficiently rigid to avoid distortion.

6.6 Drying after Dyeing

One major problem of yarn dyeing is the removal of water from the dyed material irrespective of the physical form (hank or package) in which dyeing is carried out.

Following dyeing, the hanks are removed from the frames and tied into bunches by looping one hank through several hanks. Then these bunches are dried before further processing. The dyed packages are also dried in a similar way.

6.6.1 Hydro-Extraction

This is the cheapest method of removing the water from dyed hanks or dyed packages. This is a nonthermal process and, hence, there is no heating cost.

Centrifuge hydro-extraction (Figure 6.5) is the natural choice for drying materials in batch form, such as loose fibers, hanks, or packages of yarn, fabrics processed in small batches in rope form, and garments. The machine consists of two concentric cylindrical vessels (O and P), with the inside vessel (P) having a perforated periphery and capable of rotating at high speed. The wetted materials are placed inside the inner perforated vessel. When the inner vessel is set into motion, the water from the materials is forced to move away from the center due to centrifugal force and removed through the perforations to the drain (D). As the machine runs, the extracted water is accumulated at the bottom. If the goods are free to move, they will be compressed into a narrow band against the basket wall, but all the water removed must pass through the outer layer.

The significant variables in this process are the centrifugal acceleration and the duration. The acceleration is determined by the rate of rotation and proportional to the radius.

A basket having 120 cm diameter and rotating at 1440 r.p.m. will reduce the moisture content of acrylic yarns to 6%–8% and of wool to about 27%.

A centrifuge or hydro-extractor especially suitable for yarns in hank form may have a galvanized steel basket of diameter just above 1 m and a depth of about 735 mm. The top

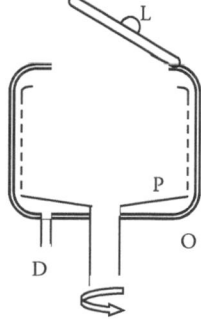

FIGURE 6.5
Centrifuge hydro-extractor.

of the machine has a hinged and a balanced safety cover (L) with a sight glass. The cover is operated pneumatically and interlocked so that it cannot be raised while the basket is moving. The basket is mounted on a central spindle carried by ball and roller bearings. Unlike conventional centrifuge machines that need belt drives or coupling, in modern machines, the electric motor (0.45 kW) is built round the spindle. The machine weighs about 2.5 tons and a strong foundation block is recommended.

Conventional hydro-extractors are labor-intensive, as the materials need to be loaded and unloaded. With high-speed versions, the packages can be loaded into bags directly; these bags are lifted by a small hoist in and out of the centrifuge.

6.6.2 Thermal Chambers

After hydro-extraction, there is still enough residual moisture in the material. Final drying and straitening of hanks, therefore, is necessary. This is usually carried out in a heated chamber. The hanks are hanged on the poles. Hot air is blown through perforated spindles of the packages. Air is circulated in the room by electric fans and moist air is driven out by exhaust fans. The method is labor-intensive and the hank takes a long time to dry. In modern hank drying machines, the hanks are placed on two moving endless belts, which, in turn, take hanks through a heated chamber for complete drying and straitening. Wet hanks may also be passed through a hot chamber on a conveyer belt, but they will not be straightened.

Hanks that do not require drying after centrifuging, such as acrylics, may be passed through a circulating cold air chamber for straightening.

6.6.3 Radio Frequency Drying

The radio frequency (RF) or dielectric energy and microwave have recently been used for drying of textiles industrially. Dielectric heating has a low frequency but high field strength, whereas microwave heating has a high frequency but low field strength. The advantages of these forms of heating are as follows:

1. The interior of the fabric is heated at the same rate as the surface, so that the migration of dyes or resins toward the surface during drying is lowest.
2. Drying times are considerably reduced.
3. Over-drying cannot occur. The yarn quality, therefore, is not compromised during drying.

Some molecules called dipoles, or molecules having dielectric moments (e.g., water), have structures in which the positive and negative charges are separated by a small finite distance and they exist in a position of minimum stress in an electrical field. A dielectric is the electrically nonconducting medium between the two electrodes in an electric capacitor. When the field is reversed, the minimum stress position is also reversed. In case of high frequency, these reversals take place several million times a second. Electromagnetic fields typically in use are of 13 and 27 MHz RF and 896 and 2450 MHz microwave frequency. This rapid dipole oscillation generates heat within the molecule itself, which in the case of water leads to evaporation. The textile materials have symmetric molecules; hence, there is generation of miniscule amount of heat in the substrate. The temperature, of course, will rise to the boiling point of water because of intimate contact. The amount of heat generated is related to the amount of water present, so that the wettest portions heat up more than those with low moisture content. The amount of heat generated decreases, as water is lost during drying. In conventional drying, water from the surface is dried and more water from inside is brought to the surface by capillary action.

Dielectric method generates heat irrespective of whether the water is bound (imbibed) or unbound. A limited number of frequencies are permitted and most dielectric heaters for textile drying uses 27.12 MHz. Dielectric machines have a conveyor belt moving slowly through a rectangular tunnel in which an electromagnetic field is applied.

In case of RF heater, the power can be drawn from the generator only when the material is present. In other words, the material itself is an essential electric component of the circuit and as the parameters that affect the electrical characteristics vary, so does the power being drawn from the generator.

For package dyeing, the majority of water is to be removed mechanically before applying the dielectric heater. Centrifuging or suction hydro-extraction based on carousel principle may be carried out. Radiation drying by RF or microwave heating is rarely employed for fabrics. An economic limitation is that the efficiency of conversion from electrical to heat energy is only about 60%.

6.7 Space Dyeing

Space dyeing is section-wise dyeing or printing of yarns and tows to achieve fancy effects after the fabric manufacturing. The effect obtained is localized application of dyes—effect similar to that obtained in printing, but by dyeing methods. The colors are applied at regular intervals during space dyeing, but the effects obtained after fabric manufacture are abstract and random. The style obtained, therefore, cannot be imitated by printing directly on the fabric. Space dyeing is the collective term for a variety of methods such as:

1. Yarn dyeing
   - Knit-deknit method (U.S.)
2. Space dyeing of warp
   - Laing method (Edger Pickering Ltd., UK)
   - Stalwart method (Stalwart Dyeing Co., UK)
   - Martin process (Martin Proc. Int. S.A., Belgium)
   - Zaza printing (Antonis Zanolo, Italy)
   - Eastern color dyeing method (Daito Sangyo Co. Ltd., Japan)
   - Superba method (Ets. Superba, France)
   - OPI method (Omnium de Prospective Industrielle S.A., France)
   - Unitika-Mixy method (Unitika Ltd., Japan)
3. Space dyeing of hanks
   - VH-Syn-O-Flow method (Vald Henriksen A/S, Denmark),
   - Multispace dyer machine (Callebaut-De Blicquy S.A., Belgium)
   - Hussong-Walker-Davis (H-W-D) method (Hussong-Walker-Davis Co., U.S.)

4. Space dyeing on wound packages
   - Astro-Dye method (Astra-Dye-Works, U.S.)
5. Space dyeing of tufted carpets, knitted, and woven fabrics
   - TAK space dyeing method (TAG, Germany)
   - Polychromatic method (James Farmer Norton, UK)

The methods have been discussed in details elsewhere (Bayer, 1976; Roy Choudhury, 2006).

6.8 Preparatory Processes

Textile fibers especially of natural origin contain considerable amounts of impurities and coloring matters. These unwanted substances are removed by various preparatory processes before dyeing. Preparatory processes may be broadly classified into two groups:

1. Cleaning processes, wherein bulk of the foreign matters or impurities is removed by physical or chemical means.
2. Whitening processes, in which trace coloring matters are destroyed chemically or the whiteness of the materials is improved optically.

6.9 Scouring

Scouring is done to remove unwanted oils, fats, waxes, soluble impurities, and any particulate or solid dirt adhering to the fibers, which otherwise hamper dyeing, printing, and finishing processes. The process essentially consists of treatment with soap or detergent with or without the addition of alkali. Depending on the fiber type, alkali may be weak (e.g., soda ash) or strong (caustic soda). When the impurities are removed, the cotton becomes absorbent; the success of scouring process is judged by the improvement in wettability of the scoured material.

6.9.1 Scouring of Cotton

Scouring processes for different textile materials vary widely. Among natural fibers, raw cotton is available in the purest form. The total amount of impurities to be removed is less than 10% of the total weight. Prolonged boiling, nevertheless, is necessary because cotton contains waxes of high molecular weight, which are difficult to remove. The proteins also lie in the central cavity of the fiber (lumen), which is relatively inaccessible to the chemicals used during scouring. Cellulose, fortunately, is unaffected by prolonged treatment with caustic solution up to the concentration of 2% in the absence of air. During scouring, it is, therefore, possible to convert all the impurities, except natural coloring matters, into a soluble form, which can then be washed away with water. For cotton, vigorous scouring may be helpful for removal of mechanically held particles of cotton seeds, husks, and leaves, which may have escaped the ginning process and are supposed to be removed during scouring. The recipe for scouring may vary widely depending on the impurities present and yarn parameters. A guideline recipe is as follows:

Caustic soda	1.0%–2.0%
Soda ash	2%
Detergent	0.5%
Wetting agent (T.R.Oil., Dedanol OTL etc.)	0.2%
Sodium silicate	0.5%
Sodium sulphite	0.5%
	(% on the weight of material)
Liquor:material ratio:: 10:1	

The treatment is done for about 2 h or more at boil till the yarn is absorbent.

For bio or enzyme scouring of cotton, pectinase is the only enzyme needed for wettability/dyeability, whereas other enzymes may have other functional benefits. The process is most efficient when run at slightly alkaline conditions. BioPrep 3000L (Novo Nordisk, Denmark) is an alkaline pectinase, free from cellulases. It works optimally at pH 7–9.5 and at temperatures up to 60°C in exhaust systems (Lange, 2000).

6.9.2 Scouring of Other Cellulosic Materials

Scouring of cellulosic fibers other than cotton is quite simple. Bast fibers like jute and flax cannot be severally scoured owing to the removal of several nonfibrous components, which consequently damage the material. These are generally scoured using soap or detergent along with soda ash. Regenerated cellulosic fibers like viscose are more sensitive to alkali. Viscose fibers are scoured with $2 \ \text{g dm}^{-3}$ anionic or nonionic detergent for about 1 h at 60°C (M:L ratio:: 1:10). No alkali (caustic soda or soda ash) is added. The man-made fibers (e.g., lyocell) are much cleaner, so scouring with nonionic or anionic detergents is sufficient.

6.9.3 Scouring of Wool

Raw wool contains about 40%–50% impurities. Wool is scoured in the fiber form, otherwise it is difficult to spin. Wool wax is difficult to saponify under conditions that would not damage the wool fibers. However, wool wax may be readily emulsified, particularly at a temperature slightly higher than its melting point (40°C–45°C). Emulsion scouring with soap is the most common method of cleaning loose wool. Scouring, most popularly known as wool washing, is done with 2%–4% soap and 2% sodium carbonate, calculated on the weight of the wool. At present, wool scouring is most economically done with nonionic detergents. Among nonionic detergents, octa/nona-ethoxylatednonylphenols are preferred for the efficiency of the treatment, although ethoxylated straight-chain alcohols are used where better biodegradability is required. The advantages of nonionic detergents over soaps include greater efficiency under neutral conditions, stability in hard water, lower cost, and more efficient removal of grease. These are easily desorbed in difficult rinsing conditions (e.g., in yarn cheeses), although they are not as efficient as soap for the suspension of dirt. Instead of sodium carbonate, sodium sulphate may be used as a detergent builder, preserving neutrality of the medium.

The oil content of wool is reduced to 0.4%–0.6% after scouring, but it is customary to add olive oil or specially prepared mineral oil during yarn spinning. Manufactured products may therefore contain about 5% oils in worsted yarns, to as high as 20% or more in carpet yarns, together with dirt picked up during the manufacturing processes. Yarn may be scoured in hank form or in packages such as cones and cheeses. An endless rope or chain of hanks, alternately, may be made and scouring could be carried out in Dolly scouring machine. The concentration of soap and soda is to be chosen very cautiously depending on the sensitivity of the wool under treatment.

6.9.4 Scouring of Silk

Raw mulberry silk is white, golden yellow, or yellow-green, whereas wild silk is light to deep brown. Raw silk does not possess the luster and softness for which this fiber is known. The gummy substance called sericin covering the fibrous material and fibroin imparts a harsh handle and must be removed to bring out the supple and lustrous qualities. Sericin and fibroin, the two components of raw silk, are both proteins, but they differ considerably in their relative compositions of various amino acids and accessibility. The process for the removal of sericin is known as degumming, boiling off, and, less commonly, scouring. Degumming also removes accompanying substances like fats, oils, natural pigments, and mineral components. The main degumming agent is soap. Boiling off or degumming of silk is carried out just below boil (90°C–95°C) in soap solution with or without alkali according to the quality of the fiber.

The concentration of soap is generally not less than 20%–30% of the weight of the material, which works out at 5–7 g dm^{-3} for a liquor ratio 30–40:1. The pH of the liquor should not fall outside the range of 9.2–10.5. The temperature of the degumming bath generally lies between 90°C and 95°C.

Enzyme degumming is regularly used in China. The recommended proteolytic enzymes are trypsin (of animal origin), pepsin, and papain (of vegetable origin). They hydrolyze peptide bonds formed by the carboxyl groups of lysine and arginine. Enzymatic degumming is not a single-step process. The gum must be swollen before the enzyme treatment. An additional treatment with mild alkali is necessary to remove natural waxes, soil, and lubricant oils.

6.9.5 Scouring of Synthetic Yarns

With polyamides and polyester, nonionic synthetic detergents (1.5–2 g dm^{-3}), with the addition of an alkali (0.5–1.5 g dm^{-3} sodium carbonate or trisodium phosphate), are suitable. For acrylics, usually a weak alkaline treatment with nonionic detergent (0.5%–1% o.w.m.) and ammonia (0.5%–1% o.w.m.) may be carried out at 80°C.

6.10 Bleaching

After scouring processes, the textile materials are in very absorbent form and can be dyed without much problem. But the materials are still yellowish or brownish, which may affect the tone and brightness of the shade obtained by dyeing, particularly for light shades.

The main objective of bleaching is to remove the natural and adventitious coloring matters with the aim to produce pure white materials. Bleaching may be carried out by

1. Reducing agents like sulphur dioxide or sodium hydrosulphite or sodium dithionite; however, the effect is temporary
2. Oxidizing agents like sodium or calcium hypochlorite, sodium chlorite, and hydrogen peroxide

However, sodium or calcium hypochlorite and sodium chlorite are not ecofriendly. They increase AOX value, which stands for *Adsorbable Organic Halogens*. AOX is a measured value for organically bound halogens in a given substance.

Hydrogen peroxide is the only ecofriendly and universal bleaching agent. Its advantages are as follows:

- It can bleach most of the textile fibers without damaging the materials.
- The whiteness achieved is permanent.
- As peroxide bleaching is done under alkaline conditions at/or near boil, both scouring and bleaching can be combined.
- It allows versatile processing such as batch or continuous, hot or cold, rapid or long dwell.

Hydrogen peroxide solution, however, is unstable at high alkalinity and in the presence of metallic impurities; hence, a stabilizer is necessary. Sodium silicate is the most widely used stabilizer. Sequestering agents, theoretically, should be the best stabilizers; stabilization may be affected by the elimination of heavy metal ions responsible for free-radical formation. Proteins also exert a stabilizing influence and can inactivate slight traces of copper in the bleaching liquor. Other stabilizers include magnesium silicate or hydroxide, trisodium phosphate, sodium pyrophosphate, tetrasodium pyrophosphate, and commercial nonsilicate stabilizers (e.g., a mixture of sodium oxalate and sodium pyrophosphate).

The residual peroxide may be harmful for dyes and, hence, it should be removed before dyeing (especially with reactive dyes). The advantage of catalase enzyme (Niels, 2000) is that it attacks only hydrogen peroxide and nothing else. The reaction is as follows:

$$2H_2O_2 + catalase \rightarrow 2H_2O + O_2 + catalase \qquad (6.1)$$

The reaction rate is extremely fast and, under optimum conditions, 1 mole of catalase is able to decompose 500 million moles of hydrogen peroxide in 1 min. The catalase is free to decompose more hydrogen peroxide as long as both desired pH (6.5–7.5) and temperature (70°C–80°C) are maintained. The need to neutralize before adding the dye is beneficial because catalase is most active in the pH range of 6–8.

When using catalase, the number of rinses can be reduced drastically. The catalase is applied in the dyebath prior to adding chemicals and dyes.

The steps to be followed are as follows:

- After bleaching, the bath is cooled to 70°C–80°C and, if needed, the machine is drained.
- It is then refilled and pH is neutralized with acetic acid.
- After running for 10–15 min, the pH is checked and adjusted to 6.5–7.5 before draining.
- The bath is refilled, catalase is added (amount depending on the concentration of the product), and run for 10 min.
- The dyes, electrolyte, chemicals, and alkali can be added to the machine without draining the liquor.

6.10.1 Bleaching of Cotton

A scouring and bleaching process for cotton yarn in package dyeing machine is as follows:

1. The yarn is scoured at boil for 30 min with soda ash and detergent.
2. The material is rinsed, cooled, and bleached below 40°C in liquor containing sodium hypochlorite solution with 1.5 g dm^{-3} of available chlorine.
3. The liquor is drained and the material is rinsed with cold water.
4. The machine is filled with bleaching liquor containing:

Sodium silicate	1 g dm^{-3}
Trisodium phosphate	1.6 g dm^{-3}
Caustic soda	1.7 g dm^{-3}
Soda ash	0.3 g dm^{-3}
Hydrogen peroxide (35%)	1–2 g dm^{-3}
Magnesium sulfate	0.03 g dm^{-3}

5. The temperature is raised to 90°C and the circulation is continued for 45 min.
6. The material is washed with soft water at 90°C.
7. It is finally cold washed with soft water.

Scouring and bleaching of cotton may be combined, as both processes are generally conducted under alkaline pH. The advantages of this process are increased production with reduction in labor cost and treatment time and lower consumption of water, steam, and electricity. The loss in weight and strength of material is less. The disadvantage is increased chemical cost, as a higher dose of hydrogen peroxide is required. In the presence of hydrogen peroxide, the scouring process is accelerated and less time is generally required to achieve good absorbency of the material.

A guide recipe for combined scouring and bleaching is as follows:

Sodium hydroxide	2%–3%
Hydrogen peroxide	1%–3%
Temperature	70°C
Potassium persulphate	0.5%
Organic stabilizer	0.4–1.0 g dm^{-3}
Time	2 h
M:L ratio	1:30

Attempts have also made to combine bio scouring with per acetic acid bleaching at low temperature. For example,

Tetra-acetyl ethylene diamine (TAED) (peroxide activator)	2 g dm^{-3}
Sodium perborate	0.15 mL dm^{-3}
Pectinase	2 g dm^{-3}
Nonionic surfactant (wetting agent)	5 g dm^{-3}
Treatment time	90 min
Temperature	60°C

6.10.2 Bleaching of Wool

Wool may be bleached with 0.3%–1.2% (w/w) hydrogen peroxide solution depending on the whiteness required, the discoloration of wool, and the time permissible to complete the bleaching. The bath may be prepared as follows:

Sodium silicate or organic stabilizer	5–10 g dm^{-3}
Soda ash	2 g dm^{-3}
Hydrogen peroxide (35%)	3–6 g dm^{-3}
Wetting agent	as required
pH is adjusted to	8.5–9
or	
Tetrasodium pyrophosphate	2 g dm^{-3}
EDTA (30%)	1 g dm^{-3}
Hydrogen peroxide (35%)	2–4 g dm^{-3}
Wetting agent	As required
M:L ratio	10

The material is allowed to stand for 4–16 h at 50°C. Recent optimization has been intended to reduce the bleaching time to 1 h. For overnight bleaching, the initial temperature should be 54°C and may be allowed to drop to 40°C by the next morning.

The studies showed that hydrogen peroxide bleaching in the presence of protease preparation, Bactosol SI (Clariant, now Archroma), considerably improved whiteness and hydrophilicity. A number of protease enzymes are commercially available in the market and these are regularly added to laundry detergents to aid in the removal of protein-based stains.

6.10.3 Bleaching of Silk

Bleaching of silk with hydrogen peroxide can be carried out in more severe conditions than wool, as silk is not easily degraded. The material is to be carefully checked for metal spots, which may cause catalytic degradation of hydrogen peroxide. Most varieties of silk are bleached by steeping, but simple machines may be used. The treatment is carried out with 0.55% w/w hydrogen peroxide solution made alkaline with sodium silicate and ammonia to give a pH of 10, or with the addition of tetrasodium pyrophosphate and EDTA at 60°C–75°C for 2–4 h. For overnight steeping, the concentration of hydrogen peroxide may be reduced. Over-bleaching should be avoided, as yellowing may occur. An alternate recipe for mulberry silk is as follows:

Hydrogen peroxide (35%)	15–20 mL dm^{-3}
Organic stabilizer	0.5 g dm^{-3}
Wetting agent	2 g dm^{-3}
M:L ratio	10
pH 9 with trisodium phosphate	

The material is immersed at 40°C and treated at 90°C for 1–2 h and then rinsed with hot and cold water.

6.10.4 Bleaching of Synthetic Fibers

Synthetic fibers, such as polyester, nylon, and acrylic fibers, cannot be satisfactorily bleached with hydrogen peroxide. They are usually bleached with sodium chlorite, which is harmful to health and should be used in a closed vessel. A guide recipe is as follows:

NaClO$_2$ (80%)	1–2 g dm^{-3}
Sodium nitrate	1–3 g dm^{-3}
Formic acid (85%)	2 mL dm^{-3} (to achieve pH 3.5–4)
Time	1 h
Temperature	80°C–85°C
M:L ratio	10

6.10.5 Whitening

For exceptional whiteness, the material after bleaching may be treated with about 0.5–1.0 g dm^{-3} solution of Ultramarine Blue (C.I. pigment blue 29/770077), Milling Violet 5B (C.I. Acid Violet 49/42640), or fluorescent brightening agents (FBAs) (Tinopal BV of Geigy or similar product) at room temperature.

6.11 Dyes

The dyes are coloring matters that are either available in soluble form or are made soluble before application. They should have some substantivity or affinity for one or more textile materials and be absorbed from the aqueous solution.

The four major characteristics of dyes are as follows:

1. The colors should be intense, so that very little quantity is required for dyeing.
2. It should be permanently or temporarily soluble in water. Different dye classes require different extent of solubility and, hence, different solubilizing groups. Such groups also decide the ionic nature of the dye, for example,

Permanent Groups	Dye Classes
–SO$_3$Na, –COONa	Direct, acid, mordant, 1:1 metal–complex, reactive dyes
–$^+$NH$_2$HCl$^-$, –NR$_3^+$Cl$^-$	Basic and modified basic dyes
–OH, –NH$_2$, –SO$_2$NH$_2$	Disperse and 1:2 metal–complex dyes
Temporary Groups	
–OH	Azoic (naphthol) colors
–C=O	Vat dyes
–OSO$_3^-$ Na$^+$	Solubilized vat dyes

3. The dye should have affinity or substantivity toward one or more type of textile fibers. The substantivity may be imparted to a dye by introducing various chemical groups into it, such as anionic, cationic, polar groups, a large number of benzene group, and phenolic groups. The dyestuff may have reactive groups (e.g., chlorotriazine), which can form covalent bond with the fiber.

4. The dyed materials have diversified use. They are, consequently, subjected to treatment with several external agents under diversified conditions. The resistance to such external agents is known as fastness properties.

The most important fastness properties are

a. Washing fastness
b. Light fastness
c. Rubbing fastness—wet and dry
d. Fastness to perspiration
e. Fastness to bleaching
f. Fastness to dry heat

The degrees of fastness are expressed numerically—"1" being the lowest and "5" the highest (8 for light fastness using blue wool method). The fastness grading may be intermediate between two full numbers and may be expressed as 2–3, 4–5 or 2.5, 4.5, and so on. In most of the tests, the grading is decided by the difference in color of the material before and after the test. The contrast is compared visually with the contrast represented by Grey Scale (ISO 105 A02 or BS1006 A02, 1978).

Detailed discussions can be found elsewhere (Roy Choudhury, 2006).

6.11.1 Classification of Dyes

Dyes are classified in two ways based on:

1. Chemical constitution
2. Dyeing properties

There is little correlation between the two methods. The members constitutionally classified as azo dyes are found among several of the classes based on application. The practical dyer will be concerned only with the classification according to methods of application. The chemical classes are more important for dye makers and researchers.

6.11.2 Classification According to Methods of Application

The dyes may be classified according to the methods of application as follows:

1. Direct dyes
2. Acid dyes (including metal–complex acid dyes)
3. Basic or cationic dyes
4. Mordant dyes
5. Azoic dyes

6. Vat dyes

7. Solubilized vat dyes

8. Sulphur dyes

9. Reactive dyes

10. Disperse dyes

11. Pigment colors

12. Mineral colors

13. Oxidation color

14. Ingrain dyes

6.11.3 Dye Classes and Textile Fibers

Suitability of various dye classes for textile fibers are summarized in Table 6.1.

6.11.4 Indexing of Dyes

Thousands of dyes are being produced by various dyestuff manufacturers around the world. The industry is never static; new products are continually being introduced and established ones are sometimes withdrawn. The situation becomes complicated with frequent changes in the names of the products.

These difficulties were felt long back, and the need for a systematic classification and recording resulting in a system called color index was developed by the Society of Dyers and Colorists, UK, in 1924 and supplement in 1928. In 1945, American Association of Textile Chemists and Colorists (AATCC) collaborated and the publication is now being jointly published in eight volumes. The first four are original volumes, whereas the others are revisions and supplements.

Color Index International is a reference database, which was first printed in 1925 but is now published exclusively on the web (http://colour-index.com/).

Two numbers, such as C.I. generic number and C.I. constitution number, are allotted for each dye. The C.I. designations of two dyes, for example, may be C.I. Basic Violet 7 (48020) and C.I. Disperse Blue 3 (61505). Whereas the first portion is a generic number, the bracketed portion is a constitution number. The generic name consists of dye class according to usage (e.g., Direct, Acid, Basic), hue name (whether yellow, orange, red, violet, blue, green,

TABLE 6.1

Different Textile Fibers and Applicable Dye Classes

Fiber to Be Dyed	Suitable Dye Class
Cellulosic (cotton, viscose, lyocell, etc.)	Direct, azoic, sulphur, vat, solubilized vat, reactive, ingrain, mineral, pigment, oxidation colors (limited use).
Cellulosic (jute, flux)	Most of the aforementioned classes, except those require caustic alkali, in addition, acid and basic dyes.
Protein (wool, silk, nylon)	Acid and selected direct dyes, basic dyes (limited use).
Polyester, cellulose acetate, and nylon	Disperse dyes.
Acrylic and modacrylic	Conventional and modified basic dyes.

brown, or black shade imparted by the dye on the substrate), and a unique number for each dye of a particular chemical structure.

6.11.5 Commercial Names of Dyes

The commercial name of a dye usually consists of the following:

1. A brand name, which denotes the specific dye class of a particular manufacturer, such as Sandolan, Cibacron, Dispersol, and so on.
2. Hue or color obtainable on the material after dyeing.
3. One or more suffix letters and figures indicating secondary hue or tone such as R, G, and B for red, yellow, and blue, respectively. The suffix G is for yellowness or greenness. These suffixes may be numbered to compare the tone among various similar dyes. One blue with suffix 2R will be redder than the blue dye with suffix R. The blue dye with suffix RR will be redder than R and less red than 2R.

Some special qualities such as F for fine, FF for superfine, L or LL for lightfast, N, W, or K for IN, IW, or IK class of vat dyes, respectively.

Strength, for example, 250%, 200%, or 150% for a brand of increased dye strength having less diluent as compared to normal brand of 100% strength. Nearly all dyes are standardized by mixing with colorless diluents, mostly common salt. The actual dye contained (purity) in a commercial dye is most often less than 40%.

As mentioned earlier, the dyes for yarn dyeing should have high fastness and stability to various chemical and physical treatments done during textile production processes. The most suitable dye classes for yarns of different fibers are listed in the following:

1. *Cotton and other cellulosics*: Vat, reactive dyes, and to a small extent azoic dyes in the decentralized (cottage) sector.
2. *Wool, silk, and nylon*: Acid dyes and metal–complex (1:1 and 1:2) dyes.
3. *Acrylic*: Conventional and modified basic dyes.
4. *Polyester*: Disperse dyes.

6.12 Dyeing of Cotton Yarn

Cotton yarns are mostly dyed with vat and reactive dyes; azoic dyes are also used, but to a lesser extent.

Process flow diagram for package dyeing of yarn

Yarn → Winding on soft package → Dyeing → Drying → Winding

or reeling → Quality control → Packing → Storage → Market

The dyeing methods are discussed briefly.

6.12.1 Vat Dyes

This is the most important dye-class for cellulosic materials and provides excellent all-round fastness properties. These dyes are water-insoluble and are made soluble (leuco form) by reacting with sodium hydrosulphite and caustic soda.

$$Na_2S_2O_4 + 4H_2O = 2NaHSO_4 + 3H_2 \qquad (6.2)$$

The reduced leuco vat dye has affinity for cellulosic material and after satisfactory exhaustion the fabric is kept in air or treated with oxidizing agent to bring the soluble form back to the original insoluble form. The dyed yarn is soaped, rinsed, and dried. Vat dyes are mainly divided into two main classes:

1. Indigoid vat dyes, which are usually derivatives of indigotin or thioindigo.
2. Anthraquinoid vat dyes, which are derived from anthraquinone.

The vat dyes may be applied to cellulosic fibers in each of the three forms:

1. As the leuco compound (water-soluble form)
2. As the pigment (water-insoluble form)
3. As the vat acid or acid leuco compound

Hanks of yarn, most often made of mercerized cotton, are dyed with leveling type of vat dyes by leuco vat method or by semi-pigmentation method. The dyeing may be carried out in hand becks, preferably provided with a hood, by hanging the wetted and squeezed hanks on bent sticks and manually rotating the hanks. The hanks are given several quick rotations followed by slow periodical rotations. The hanks are then squeezed, rinsed with 0.2–0.3 g dm^{-3} sodium hydrosulphite solution, subsequently rinsed twice with cold water, and soaped. In mechanical becks, the movement of the hanks is automatic.

Theoretically, every vat dye will have its own optimum temperature and require specific quantity of alkali and reducing agent. For simplicity, anthraquinoid dyes are classified from application standpoint into three or four classes, named as Method 1, 2, 3 (Ciba) or for Indanthren dyes (BASF) as:

1. IK (*K* for cold or Kält in German)
2. IW (*W* for warm)
3. IN (*N* for normal)
4. INS (*S* for special) dyes

The first step is vatting (i.e., reduction), followed by dyeing and oxidation (by air or oxidizing agent). The recipe is given in Table 6.2.

The dyeing in package form may be carried out by leuco vat process or semi-pigmentation process. In leuco vat process, the vatted dye is added using a filter in two or four portions during outside-to-inside flow. The direction of liquor flow is changed every 2–3 min during the first 15 min of dyeing. The flow cycle may be 6 min out-to-in and 4 min in-to-out, thereafter. After 30 min of dyeing, the salt is added in portions

TABLE 6.2

Dyeing Conditions of Various Vat Dye Classes

Methods →	IK	IW	IN	INS	Indigoid
Vatting (quantities on the weight of dye)					
Caustic soda	Equal	Equal	Double	Double	Double
Sodium dithionite	Equal	Equal	Equal	Equal	Equal
Temperature, °C	45	50	60	60	70
Time, min	15	15	15	15	15
Dyeing					
Caustic soda, g dm^{-3}	2	4	6	8	2
Sodium dithionite, g dm^{-3}	4	4	4	5	4
Common salt, g dm^{-3}	25	15	x	x	25
Leveling agent, Lyogen V, g dm^{-3}	x	x	1	1	1
Temperature, °C	30	50	50–60	60	60
Time, min	30	30	30	30	30

and dyeing continued for a further 15–30 min. The material is subsequently rinsed, oxidized, and soaped.

Semi-pigmentation method is suitable for automated package dyeing machines. The steps are as follows:

- Caustic soda is added to the dyebath and circulated for one complete cycle at 15°C–20°C. The required amount of caustic soda (66°Tw) is 15–30 mL dm^{-3} for IN dyes and 10–15 mL dm^{-3} for IW dyes, depending on the depth.
- The pigment dispersion is added using a filter and circulated for one complete cycle.
- The required amounts of sodium hydrosulphite (4–8 g dm^{-3}) and common salt (8–20 g dm^{-3}, IW dyes only) are added more than two to three cycles.
- Dyeing is continued for 20 min at 20°C.
- The temperature is raised by 1°C min^{-1} to the required dyeing temperature.
- Dyeing is carried out for 20–30 min for complete exhaustion and leveling.
- The product is rinsed, oxidized, soaped, and rinsed.

6.12.1.1 Denim Processing

Jeans made from blue denim have been popular over a longer period than any other items of apparel. They are inexpensive, durable, and versatile. The blue-dyed warps are washed down to an attractive blue without staining the white weft. For denim fabric, warp yarns are dyed with indigo, a vat dye with poor affinity for cotton, so that after weaving, the fabric dye can be removed from selected portions by rubbing or by other appropriate methods.

In the continuous indigo dyeing of warp yarns, the yarns are in the form of a full-width warp beam or in ball form, containing 300–400 individual threads. It passes through several (five to six) vats, each followed by an air passage for oxidation. The processing speed is 20–30 m min^{-1} with an immersion time of 20–30 s in each vat. On leaving the vat, the material is squeezed to a liquor pick-up of approximately 100%, after which the dyeing

TABLE 6.3

Salt (g dm⁻³), Soda Ash (g dm⁻³), and Temperature (°C) Requirement for Different Brands of Reactive Dyes

Brand →	Cold Brand		HE/H Brand		Vinyl Sulphone/ME	
% Depth of Shade	Salt	Soda	Salt	Soda	Salt	Soda
≤ 5.0	25	5	30	10	3–5	10–15
0.51–2.0	35	5	45–60	15	15–50	20
2.01–4.0	45	10	70	20	40	20
>4.0	55	10	90	20	50	20
Temp	Room		70–80		65	

requires around 2 min for oxidation before it passes into the next vat. Rinsing at ambient temperature in two or three rinsing baths after oxidation is normally adequate. The dip tanks are coupled to keep the dyebath composition exactly the same. A good volume of flow with minimum turbulence is essential to ensure uniformity.

6.12.2 Reactive Dyes

These dyes contain reactive groups, which react with hydroxyl (e.g., cellulosic) or amino groups (e.g., protein) of textile fibers forming covalent bonds. Depending on the reactive group, these dyes are classified into a number of subgroups: cold-brand (M-brand), Remazol (vinyl sulphone), high exhaust (HE-brand), medium exhaust (ME dyes), and so on. The salt–soda requirements and dyeing temperature of different brands of reactive dyes are shown in Table 6.3. Common salt (sodium chloride) is normally used, but Glauber's salt (sodium sulphate) may be used for brighter shades of vinyl sulphone and ME dyes.

The process flow chart for dyeing 100% Cotton Yarn (Dark/Medium/Light shade) yarn with HE/H dyes in package dyeing machine is as follows:

- *Pretreatment*: Batch loaded → Demineralization (50°C, 20 min; pH = 4.5) → Scouring and bleaching (100°C, 40 min) → Draining → Rinsing → Draining → Neutralization with acid (50°C, 20 min) → Hot washing with peroxide killer, catalase enzyme (60°C, 20 min) → Drain.

- *Dyeing*: Leveling agent and Salt (60°C × 20 min; pH = 6) → Color dosing (60°C × 20 min) → Run time = 10 min (60°C) → Color migration (80°C × 20 min) → Cooling (60°C) → Level check → Soda dosing (60°C × 30 min) → Dyeing run (Dark-60°C, 60 min; medium-60°C × 40 min; light-60°C, 30 min) → Dyeing sample check → (If Ok) → Drain.

- *Posttreatment*: Rinsing (with cold water) → Neutralization after dyeing (50°C, 20 min) → Draining → Soaping (Hot wash) → Draining → Rinsing → Add finishing chemical (60°C, 20 min) → Draining → Unloading.

6.12.3 Azoic Colors

While azo dyes are readymade dyes, azoic colors are formed inside the textile materials by the reaction of two colorless or faintly colored compounds called naphthols, or C.I. Azoic Coupling Component, and fast bases organic aromatic bases, or C.I. Azoic Diazo Components. The naphthols are first applied on textile materials by a process called naphtholation, as they have some affinity for the cellulosic materials. The treated material is subsequently treated with a solubilized diazotized form of the bases, when intense color is formed. The latter step is known as coupling or development. The solubilization of the base is carried out with sodium nitrite and hydrochloric acid at low temperature in the process called diazotization.

The four steps of application of azoic colors are as follows:

1. Dissolution of naphthols
2. Naphtholation
3. Conversion of bases into diazo compounds
4. Development or coupling

In hot dissolving method, the naphthol is pasted with a wetting agent (T. R. Oil) and a small quantity of hot water is added to make a smooth paste. The required quantity of sodium hydroxide (0.2 part for Naphthol AS-LB to 1.3 parts for Naphthol ASSW) is added and boiled until the milky dispersion is converted into a clear solution. The remaining quantity of cold water is added to achieve the desired solution concentration.

6.12.3.1 Naphtholation

The second stage of application of azoic colors is the treatment of textile substrates with naphthol solutions. Yarn may be dyed by open-beck method, tub-dip method or in package dyeing machine. In open-beck method, the bath is prepared with naphthol solution and common salt, maintaining suitable liquor ratio to ensure complete impregnation. Common salt, generally 20 g dm^{-3}, is added to the naphthol solution, except in the case of Naphthol AS-SW and Naphthol AS-G, where 10 g dm^{-3} and 15 g dm^{-3} NaCl, respectively, is used.

In package dyeing machine, the material may be treated with naphthol solution for 30–40 min, with flow reversals at intervals of 5–10 min. The material is hydro-extracted in the same machine or in an external hydro-extractor. The material is returned back to the same or another dye vessel for development.

6.12.3.2 Diazotization

Two general methods are used for diazotization of fast bases:

1. Direct method (all bases except Orange GR, Red B, and Bordeaux GP)
2. Indirect method (only Orange GR, Red B, and Bordeaux GP)

In direct method (suitable for bases whose hydrochlorides are water soluble), the base is pasted with concentrated hydrochloric acid and cold water. Hot or cold water is then added under constant stirring. The solution is then cooled to 10°C with ice and the sodium nitrate solution (1:1) is added with vigorous stirring, maintaining the temperature in the range of 10°C–12°C (5°C for Garnet GBC). Coupling will not occur if the pH is too low. Hence, after keeping the solution for 20 min, the excess hydrochloric acid is neutralized with sodium acetate. The alkali-binding agent, such as aluminium sulphate or acetic acid, is then added.

The solution is then filtered and made up to the required volume depending on the concentration of the fast base required. With Red RC and Red KB, aluminium sulphate is preferred to acetic acid.

6.12.3.3 Development or Coupling

The naphtholated material is treated with the diazonium solution in the development bath for 20–30 min when the final shade appears on the material. The development may be carried out in long liquor open beck or jigger or in short liquor as in tub or padding mangle. Common salt in concentration 25 g dm^{-3} is added in all development baths.

The actual quantity of diazotized base required in the development depends on the quantity of naphthol in the naphtholating bath and the coupling ratio. The coupling ratio is the quantity of fast base required for one part of naphthol. For example, when naphtholation is done with 5 g dm^{-3} Naphthol AS, the concentration of Yellow GC base in the development bath should be 5 × 0.95 = 4.75 g dm^{-3}, where 0.95 is the coupling ratio.

The developed material is washed repeatedly in cold water, then in hot water, and rinsed in acidified water (3 mL dm^{-3} hydrochloric acid, 32°Tw). The material is then washed again with cold water and carried forward for soaping. Soaping removes loosely fixed azoic pigments from the surface of the material, thereby increasing fastness to rubbing, light, and chlorine. Soaping is carried out at boil with 3 g dm^{-3} soap or detergent and 2 g dm^{-3} soda ash for a period of not less than 15 min, giving adequate agitation or squeezing to dislodge the adhering particles. It is then washed well and dried.

6.13 Dyeing of Wool Yarn

Acid dyes used for wool dyeing are arbitrarily classified into four groups (Lewis, 1992):

1. Level dyeing or equalizing acid dyes
2. Fast acid, half-milling, or perspiration-fast dyes
3. Acid milling dyes
4. Supermilling dyes

The above-mentioned list is in the order of increasing fastness and decreasing levelness, although there may be overlap between adjacent categories.

6.13.1 Equalizing Acid Dyes

They are usually applied from a dyebath containing 2%–4% (o.w.m.) sulphuric acid (168°Tw) or 1%–3% formic acid and 5%–10% (o.w.m.) sodium sulphate. The goods are entered at 50°C, pH is checked (2.5–3.0), and the dissolved dye is added in portions. The temperature is then raised to boil in 30 min and boiled for a further 45–60 min, rinsed, and dried. In case unlevel dyeing is encountered, the concentration of sodium sulphate can be increased to 15%–20% and boiling is continued for a further 30 min under such conditions.

6.13.2 Fast Acid Dyes

The goods are entered at 50°C into a dyebath containing 1%–3% acetic acid (80%) and 5%–10% Glauber's salt. The pH is adjusted to 4.5–5.5. The temperature is raised to boil in 30–45 min, kept at boil for a further 45–60 min, followed by rinsing and drying. If the exhaustion is insufficient, a little more acetic acid, or some formic or phosphoric acid, may be added. The leveling agent (1%) may be added, if necessary.

6.13.3 Milling and Supermilling Dyes

Dyeing starts at 50°C with 2%–5% ammonium sulphate or acetate or 2 g dm^{-3} sodium acetate. The pH is set at 5–7.5 with acetic acid. The dyeing pH depends on the depth of shade and nature of the substrate. Pale shades require a higher pH (6.0–7.5) to reduce the rate of dyeing; this is to achieve level dyeing. The temperature is raised to boil in 45 min, dyeing is continued at boil for a further 45 min, before rinsing and drying. If the exhaustion is incomplete, it may be necessary to add 1%–2% of acetic acid (30%) after boiling for 30 min.

6.13.4 (1:2) Metal–Complex Acid Dyes

The neutral dyeing 1:2 metal–complex dyes are chromium or cobalt complexes of azo dyes. Like 1:1 complexes, they are prepared from azo dyes, in particular from oo/-dihydroxyazo dyes.

Weakly polar 1:2 metal–complex dyes are sold under the names Irgalan (Ciba), Lanasyn (Clariant, now Archroma), Isolan K (DyStar), Ortolan (BASF), and so on. They display very good to excellent light fastness and very good fastness to wet treatments on wool in pale to medium depths. Weakly polar 1:2 metal–complex dyes, due to the absence of sulphonic and carboxylic groups, cover tippy wool very well.

The usual method of application of weakly polar 1:2 metal–complex is as follows:

The yarn or fabric is treated at 40°C for 10 min in a bath set with 2%–4% (o.w.m.) ammonium sulphate or acetate. Dissolved dye is then added and the temperature is raised to boil in 45 min. After 30–60 min at the boil, the bath should be exhausted to the extent of about 95%. The dyeing may also be carried out at high temperature (106°C–110°C). At lower temperature (80°C), there may be a tendency to skittery dyeing.

6.14 Dyeing of Silk Yarn

Owing to the high exhaustion rate of the dyes on silk, it is recommended that the temperature should be kept as low as possible at the beginning of the dyeing; this is to achieve level dyeing. For yarns, the dyeing temperature may be as high as 80°C–90°C for good penetration, but for more sensitive piece goods lower dyeing temperature of 60°C–70°C is recommended to prevent chafe mark. Weakly acid dyes, for example, Sandolan or Nylosan dyes (Clariant, now Archroma) are preferably applied in the neutral to weakly acidic conditions at pH 5–7 and for very dark shades at pH 4–4.5. It is best to use a buffer system like sodium acetate/acetic acid. It is also recommended to use dye-substantive leveling agent to reduce rapid exhaustion and to promote surface levelness. A guide recipe is as follows (Sandoz, 1985):

x% weak acid dye (Sandolan or Nylosan) or

x% metal–complex dye (Lanasyn or Lanasyn S)

5%–1.5% dye-substantive leveling agent

1–2 g dm^{-3} sodium acetate, acetic acid to pH 4.5–7

Dyeing at 80°C–90°C for 45–60 min

6.15 Dyeing of Nylon Yarn

Both staple and filament nylon yarns are generally dyed in package form with acid and 1:2 metal–complex dyes to obtain the necessary fastness for garments and carpets. Hand-knitting and carpet yarns are widely dyed in hank form.

Acid dyes exhibit significantly better wet fastness on nylon than on wool and their classification system, which was earlier based on wool, was found to be more convenient to modify. Considering the dyeing behavior and affinity of acid dyes for nylon, the dyes may be divided into three subgroups [10]—groups I, II, and III.

Group I: Dyes that have low affinity under neutral or weakly acidic conditions but exhaust well on nylon under strong acidic conditions.

Group II: The largest group of acid dyes that exhaust well on nylon within the pH range of 3.0–5.0.

Group III: Dyes that exhibit high affinity for nylon under neutral or weakly acidic conditions (pH 5.0–7.0).

Due to high crystallinity, nylon 6,6 shows slower rate of dyeing and slightly superior fastness in comparison with nylon 6 dyed with the same dye in equivalent depth. The dyes build up more readily on nylon 6 and show superior leveling properties.

The general procedure for exhaust dyeing is to adjust the pH of the dyebath between 3 and 5, according to the depth of shade and the nature of the dye. It is often advantageous to add a leveling agent. The addition of an anionic leveling agent like Lyogen PA 66 (anionic sulphonated oil derivative, Clariant, now Archroma) gives satisfactory level dyeing on barré material (having non-homogeneous chemical or physical structure) with leveling type acid dyes. The method may be used for other classes of dyes on non-barré material. Lyogen PA 66 has affinity for nylon, hence they saturate active sites in the fiber. It can, therefore, be termed as anionic blocking agent. The dyebath is set at 50°C, with 4%–2% Lyogen PA 66 liq. and depending on dye class (Groups I, II, and III respectively) mentioned in the following:

1. 1%–2% formic acid
2. 1%–3% acetic acid (80%)
3. 2%–4% ammonium acetate

(The amounts are o.w.m., where the lowest and highest amounts are for lighter and darker shades, respectively.)

1. For weakly acid-dyeing dyestuff, for example, Nylosan E and F (F = better wash-fast) (Clariant, now Archroma) in medium and dark shades.
2. For neutral-dyeing milling acid dyes (e.g., Nylosan N, Clariant, now Archroma), 1:2 metal–complex dyes (e.g., Lanasyn, Clariant), Solar (light-fast direct dyes, Clariant), or for light shades of Nylosan E and F.
3. For strong acid dyes like C.I. Acid Yellow 17, Red 1, and so on.

The goods are worked at this temperature for 15 min. If the barréness is pronounced, the temperature may be raised and kept at boil for 20 min followed by cooling to 50°C. The dye is dissolved in hot water, boiled for short time, if necessary, filtered, and added in the dye-bath. The temperature is raised to boil in 30 min and the dyeing is continued at boil for 60–90 min. The dyestuff for shading and further acid, if required, may be added without cooling, with prior addition of 0.5%–1.0% Lyogen PA 66.

Selected 1:1 metal–complex dyes can be applied on nylon. The dye liquor is set with:

$x\%$ dye

1 g dm^{-3} weakly cationic or pseudo-cationic retarding agent

3% ammonium acetate

5% acetic acid (30%)

(All percentages are o.w.m.)

The goods are added and the temperature is raised to the boil at the rate of $1°C$ min^{-1} and maintained at this temperature for 45 min. The goods are then cooled, rinsed, and softened. These dyes are normally used for bright shades and not normally after-treated with tannic acid or syntan.

It has been observed that the washing fastness of many existing anionic dyes on nylon can be improved by after-treating with tannic acid. The temperature is raised slowly to 90°C for nylon 6,6 or 70°C for nylon 6 and the goods are treated for 30 min. Two percent tartar emitic (potassium antimony tartrate) is added to the liquor and the treatment is continued for a further 20 min at the specified temperature. The fabric is thoroughly rinsed and dried.

Pure tannic acid is costly. The time-consuming and relatively expensive full back-tanning process can be substituted by a simpler after-treatment using synthetic tanning agents or syntans. These are essentially based on sulphonated phenol-formaldehyde condensation products, water-soluble, anionic in nature, and substantive to nylon under acidic situation. The products are less effective than full back-tanning treatment, but they improve the washing fastness of nylon dyed with most anionic dyes. Most syntans (e.g., Nylofixan PI liq. of Clariant, now Archroma) are applied to the dyed and rinsed material in the presence of formic or acetic acid (pH 3.0–5.0).

6.16 Dyeing of Polyester Yarn

The Society of Dyers and Colorists (UK) defines disperse dye as a substantially water-insoluble dye having substantivity for one or more hydrophobic fibers, for example, cellulose acetate and polyester, and usually are applied from fine aqueous dispersion.

Disperse dyes are sold under various brand names such as Dispersol (BASF, earlier ICI), Foron (Clariant), Palanil (BASF), Terasil (Ciba), Tulasteron (Atul), Navilene (IDI), Terenix (JDL), and so on.

While dyeing from aqueous media, the bulk of the disperse dye remains in suspension, but at least during the initial stage of dyeing, transfer of dye into the fiber takes place from a very dilute aqueous solution of the dye.

ICI disperse dyes (Dispersol) had been classified into A, B, C, and D classes (dyes without suffix are obsolete), whereas Clariant (now Archroma) disperse dyes have been classified into S (low-energy), SE (medium energy), and E (high-energy) classes.

Dyeing may be carried out in three ways:

1. Carrier method
2. High-temperature method
3. Dry-heat or thermosol method

Carrier method has several problems, and continuous thermosol method is not suitable for yarn.

The high-temperature method of dyeing polyester with disperse dyes is as follows:

The general method of dyeing is to prepare the dyebath with 0.25–2 g dm^{-3} of dispersing agent (Lyocol OI powder, Clariant, now Archroma). The pH of the bath is set at 5.0–6.0 with sodium acetate-acetic acid (approximately 1 and 0.25 g dm^{-3}, respectively) or ammonium sulphate (1–2 g dm^{-3}) formic acid buffer. A leveling agent, such as 0.5–1 mL dm^{-3} Lyogen DFT (Clariant, now Archroma), may also be added in case of light shades. Nonionic aliphatic polyglycol ether is substantive to disperse dyes. It acts as a leveling cum dispersing agent. In higher dosage, it acts as a stripping agent (5–10 mL dm^{-3} along with 1–4 mL dm^{-3} carrier, Dilatin EN [Clariant, now Archroma] at 130°C for 1 h). Any alkali used during scouring must be completely removed from the fabric, otherwise degradation of polyester may occur at high temperature. The dyeing should start at 70°C. The temperature is then raised slowly to 130°C and maintained for a period of 30–60 min. Shading dyes, if necessary, may be added at 120°C–130°C, allowing 20 min after each such addition.

On completion of dyeing, the bath should be dropped at a temperature not less than 85°C to prevent deposition of oligomers on the surface of the material. The oligomers present in the dye liquor will deposit crystals on any unpolished surface and in regions of high hydrodynamic shear (as in a pump) if crystals are already present. The oligomers deposited on the internal surfaces of dyeing machines are difficult to remove without severe cleaning. It is advisable to clean the machines at regular intervals of 100 h running time. The cleaning process must be more severe than that used for clearing dyed fabrics, and a stripping agent like Lyogen DFT (Clariant, now Archroma) (4 g dm^{-3}) may be added.

When heavy shades are dyed, it is advisable to give a reduction clearing treatment to avoid poor rubbing fastness. This is followed by rinsing and soaping with 1 g dm^{-3} Lissapol D paste at 70°C–80°C.

6.17 Dyeing of Acrylic Yarn

The basic or cationic dyes are historically interesting, as they are the first synthetic dye; Perkin's Mauve belongs to this class. Basic dyes like Magenta and Malachite Green are also among the earliest synthetic dyes.

The basic dyes are so-named because they are derived from organic bases. They are also called cationic dyes, as they ionize in water to produce colored cations.

The light fastness of conventional basic dyes is better on acrylics, although further improvements in dye structure have led to modified basic dyes of very good light fastness. These dyes contain a pendent cation away from chromophore with a nonresonating positive charge.

Cationic dyes show excellent build-up and fastness and many give brilliant shades on acrylics. They are usually less expensive than disperse dyes. However, their migration properties are, in general, poor. The dyeing should be carried out very carefully, so that the dye uptake is uniform from the beginning. The effects of dyeing conditions should be known and are to be carefully controlled. Material-to-liquor ratio does not exert much influence in case of basic dyeing processes.

The temperature range of dyeing should extend to about 30°C above T_g—between 75°C and 105°C. Migration and levelness are better at higher temperature. Different acrylic fibers, however, show varying degrees of mechanical instability at high temperature. There may be flattening of fibers, resulting in loss of bulk and handle. Maximum permissible temperature depends on fiber type, physical variables like tension, package density, and so on, and the characteristic of the dyeing machine.

The normal range of pH for dyeing of acrylic fibers is 3.5–6.0, at which both cationic and disperse dyes are very stable. At lower pH, leveling is better and the rate of dyeing and the equilibrium exhaustion are lower for cationic dyes. Most acrylic fibers show an increase of 10%–20% in fiber saturation value (A) and rate of dyeing for one unit increase in pH within the said range. This can be utilized to cope with saturation problems, for example, for very deep colors on fibers with low A value or while shading or redyeing. The evaporation of acetic acid may not be of much problem, until pH increases above 6, when dyes become unstable. It is better to use acetic acid–sodium acetate buffer system to maintain a pH of 4.5–5.5. For deep dyeing, pH may be higher and retarder is not used. Sulphuric acid (pH 2.0–3.5) is sometimes recommended for level dyeing of wool–acrylic blends, but many cationic dyes are unstable in this pH range.

The dye is dissolved by pasting with equal quantity of acetic acid (30%) followed by addition of 20–50 times boiling water. Some dyes require special methods and some warm water. A typical recipe for dyeing is as follows:

x% cationic dye

1–2 g dm^{-3} acetic acid (30%) to bring pH to 3.5–4.0

0–2.5 g dm^{-3} Glauber's salt or common salt

25%–0.5% nonionic detergent

y% cationic retarder

(All percentages are o.w.m.).

The value of y should be chosen according to depth of color, the type of dyes, and the fiber. Saturation should be avoided. There is very little adsorption below 75°C and the critical temperature is 80°C–100°C. It is, therefore, advisable to raise the temperature rapidly to 75°C and then allow 1 h for the liquor to reach the boil. The bath should be kept at the maximum temperature (specific to the dye) for 10–15 min after complete exhaustion (no visible change in the color of the dye liquor) for sufficient penetration and to avoid ring dyeing.

6.18 Conclusion

Yarn dyeing, particularly in package form, is a complicated and technically challenging process and has significant role to play in the coloration of textiles. Dyeing in yarn form is indispensable for certain end-uses. Dyeing in both package and hank forms may have their own advantages and disadvantages. The relative contribution of yarn dyeing in terms of quantity to the whole dyeing industry is not very significant. This field, hence, will not get much attention from the technological development point of view. Not much research nor machine development has been reported in the last few years.

References

Bayer A.G. (1976). *Bayer Farben Revue, No. 26* (Leverkusen, Germany: Bayer).
Clarke G. (1982). *A Practical Introduction to Fibre and Tow Coloration* (Bradford, UK: SDC).
Lange N.K. (2000). Biopreparation in action. *International Dyer*, 185, 18–21.
Lewis D.M. (Ed.) (1992). *Wool Dyeing* (Bradford, UK: SDC).
Niels P. (2000). TCC & ADR, 32 (5), 23–24.
Park J. (1981). *A Practical Introduction to Yarn Dyeing* (Bradford, UK: SDC).
Roy Choudhury A.K. (2006). *Textile Preparation and Dyeing*, Science Publishers (Hauppauge, NY).
Sandoz Ltd. (1985). *Silk and Colour* (Basle, Switzerland: Sandoz).

7

Woven Fabrics

Lindsey Waterton Taylor

CONTENTS

7.1 Introduction to Constructed Woven Textiles

The construction of a traditional woven textile is defined by warp and weft yarns that interlock at right angles to one another in the 0/90° direction to the fabric plane, as identified in Figure 7.1. In its simplest form a woven textile can be constructed on a tapestry frame and/or inkle loom. Tablet weaving and handlooms permit greater scope of pattern formations within the weaving process. A jacquard loom is the weaving technology to use when wishing to produce greater complexity in the interlocking of warp and weft directional yarns, in terms of: weave architecture combinations; single layer fabrics; multilayer multilevel wovens (shapes and solid woven fabrications). Each woven textile process has one thing in common, a shed. The shed is the operation by which the warp yarns are separated to then allow a weft yarn to be inserted—both the manual shedding operations (i.e., tablet weaving, handlooms) to the semi- and fully automated CAD power looms (jacquard and dobby) require a lifting plan. The sequential pattern within the lifting plan contains a minimum of two independent weft insertions to complete the desired weave architecture/pattern within the concluding woven textile. The sequential lifting of the weave architecture in manual weaving is aided by varying nonmechanical processes, such as:

- *Tablet weaving*: Requires the turning of card(s) by hand, with each card housing a number of warp yarns to create a shed.
- *Handlooms* (*table and floor*): Typically contains levers or treadles requiring lifting or pushing, respectively down to form a shed.

The sequential lifting of the weave architecture when using semi- and/or fully automated power looms is via mechanical principles allowing:

- Several simultaneous lifts of warp yarns that are located on the same shaft/heald frame, with a minimum requirement of two shafts resulting in repetitive elements within the final woven textile.
- Simultaneous lifting of independent warp yarns forming individual interlocking of the warp over and under the weft forming nonrepetitive and/or repetitive weave architectures within the final woven textile.

The shedding formation together with the method of weft insertion is fundamental to the classification of manual weaving, semi- and fully automated weaving. This

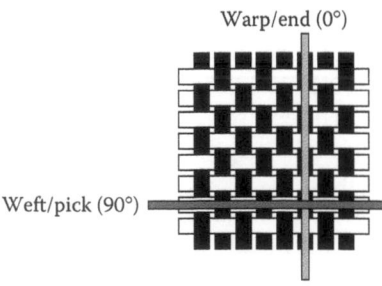

FIGURE 7.1
Constructed woven textile: plain weave architecture.

chapter introduces the manual and mechanical operational principles enabling an understanding of the categorization of traditionally constructed woven textiles by warp control and weft insertion. The fundamental operation of a loom includes shedding mechanism/motion; weft insertion mechanism; beat-up of inserted weft; let off of warp yarns; take up of woven fabric on loom. The weaving mechanism/motion can be categorized into primary, secondary, and auxiliaries. This provides a foundational overview of traditional weaving technologies and their operating principles; a prerequisite to understanding the innovative modifications and adaptations to weaving technologies for the construction of advanced technical weave architectures and concluding woven textile structures. Supporting the knowledge of traditional and advanced technical weave architectures/textiles, the array of weave architectures and their notation will be discussed in this chapter. An overview of weaving technologies process principles also requires an understanding of the following and will be presented in this chapter:

- Weaving technology capabilities and principles
- Warp and weft yarn preparation
- Loom preparation and dressing
- Warp and weft interlocking directions
- Woven textile handle-drapeability, sett, and quality
- Constructed woven textile end applications

7.2 Woven Construction, Apparatus, and Technology

To understand the progression of weaving technology and then in turn how this is utilized in woven textile production, it is wise to start from the beginning; the simplicity of the backstrap loom requires a post (or typically a tree), warp yarns, and sticks. The warp yarns are tied to sticks, known as back and front-warp bars, shown in Figure 7.2b. The back-warp bar is tied to a post via some means such as cord, as shown in Figure 7.2a. Weaving the plain weave architecture only, the warp yarns are divided alternatively one warp up and one warp down and repeated across all the warp yarns by warp shedding lease rods (Figure 7.2c and d). To enable the alternate lifting of the warp yarns in the plain weave architecture manner, a heddle rod (Figure 7.2e) would be inserted under and over the warp yarns in the alternative direction to the warp shedding lease rod (Figure 7.2d). The opening of the warp yarns is known as the shed. A tool, such as a stick would be used as a warp batten shedding rod (Figure 7.2f). The woven textile is produced by alternating between the heddle rod and the warp batten shedding rod to open the warp yarns, creating a shed for the weft yarn (Figure 7.2g). The warp batten shedding rod is also used to beat down the inserted weft yarn into the fell of the woven textile, which is the concluding woven fabric. Tension on the warp yarns is by the weaver leaning toward the post or backward away from the post as needed.

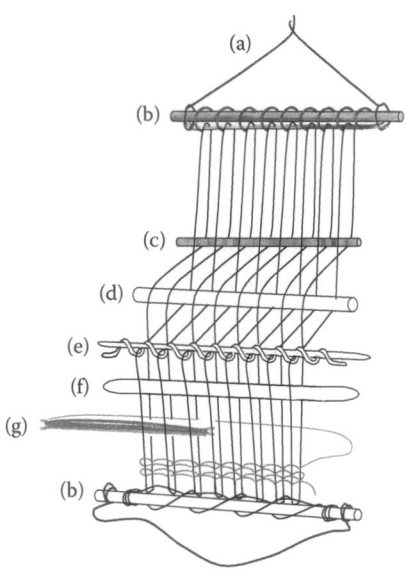

FIGURE 7.2
Backstrap weaving apparatus: (a) tie-up cord, (b) warp bar, (c and d) warp shedding lease rod, (e) heddle rod, (f) shed opening rod, and (g) shuttle.

Owing to the vast range of weave architectures and array of apparatus and technology utilized in their production, the classification of wovens incorporates the following:

- Craft (cottage industry) woven's, includes backstrap weaving; tapestry frame weaving; inkle loom; tablet weaving; hand looms (manual operation)
- Design (craft origins): CAD semi-automated handlooms (table and floor, levers, and treadle operative)
- Design and technical woven's for mass production (industrial, power weaving): CAD power looms (fully automated)
- Warp control mechanism and motion: dobby and/or jacquard
- Weft insertion mechanism: manual shuttle, power shuttle, power shuttleless (rapier, air, water)
- Weave architecture and woven surface (flat and pile structures)
- Single-layer woven's and compound woven's
- Two-dimensional-to-three-dimensional (2D-to-3D) woven's
- Warp and weft interlocking direction(s)

In the craft-cottage industry, apparatus in which a woven textile is constructed varies, with some of them listed earlier. These include the methodology of backstrap weaving, but also entails the use of a framework that houses more of the components required from the principles of weaving—they are still portable pieces of kit, like the backstrap loom (post not supplied). The apparatus framework on the whole permits only plain weave with the aesthetics being addressed through varying the color ways and proportions of the warp and weft yarns. The handle, the drapeability of such woven's, are then aided by the choice of fiber/yarns and ratios of in the warp and weft directions. When requiring diversity of weave patterning and color ways, tablet weaving allows the construction of

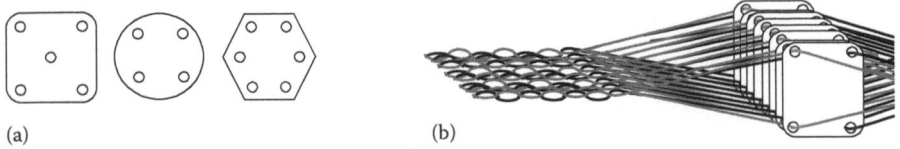

FIGURE 7.3

Tablet weaving apparatus: (a) varying shaped tablet cards with punched out holes for the warp yarn(s) and (b) square tablet cards with one warp end drawn through each hole of the numerous cards.

greater complex wovens. The components, however, are minimum in the principles of tablet weaving requiring two posts, or sources, of clamping at either end of the warp yarns and as many cards as the weaver desires. The cards come in a range of shapes and each with varying number of holes, which can be seen in Figure 7.3a and b. The warp yarns are each threaded through their own located hole on a card.

The weaving process is then enabled by rotating the cards to change the positions of the warp threads. A permanent, although rotational shed configuration is generated by the holes in the bottom and top sections of each card. After a weft insertion, and beat up of the weft into the fell of the woven textile, rotation of the cards allows the warp directional yarns to entrap and lock in place the previous weft yarn inserted. This then creates a new shed opening for the next weft insertion. If all the cards are rotated together in a particular direction each time the weft is passed through, then a characteristic diagonal pattern develops. A pattern that has the same number of rotations in both forward and backward directions overall will lie flatter.

More complex patterns can be made by rotating individual warp yarns separately: utilizing only one hole in the top and/or bottom of the card and rotating the card(s) back and forth, rather than around in a full clockwise direction. The principle of tablet weaving (the rotation of cards back and forth) has influenced adaptations to the mechanics of weaving technologies shedding mechanism; enabling complex warp and weft interlocked wovens for the technical textiles industry, discussed in Section 7.6.

The table top and/or floor handlooms advance on the weaving principles of the inkle and tablet construction processes. The handloom framework houses a number of shafts, also known as heald frames. The shafts, dependent on handloom type, are operated manually by either connecting levers or treadles. The levers and/or treadles are lifted according to the weave architecture. These require the weaver/operator to push down the levers and/or treadles according to the weave architecture/pattern required. Generally, the levers for controlling the movement of the shafts up and down are located on the side of the handloom; there is one lever per shaft. Alternatively, treadles are used to operate the lifting and dropping of the shafts. The treadles are located underneath the shafts, typically connected by pulley's/cords; there is one treadle per shaft. The weaver/operator is required to push the treadle down with their foot, which results in the lifting of the shafts. As the shafts lift, they form the shed within the warp yarns, shown in Figure 7.4a and b for the weft to be inserted. Figure 7.4b shows the formation of the plain weave architecture, requiring only two shafts— the greater the number of shafts a handloom contains, the greater the complexity of weave architecture(s) that can be generated in the construction of the concluding woven textile.

Other types of shaft/heald frame looms (hand and power looms) are equipped with mechanical pattern reading systems. This is where the information for the lifting sequence, the weave architecture, is stored on either wooden lags with holes representing all the shafts available on that loom and pegs inserted to inform which shaft is to be lifted or holes punched on paper/plastic/cards to inform which shafts are to be lifted. On the handloom

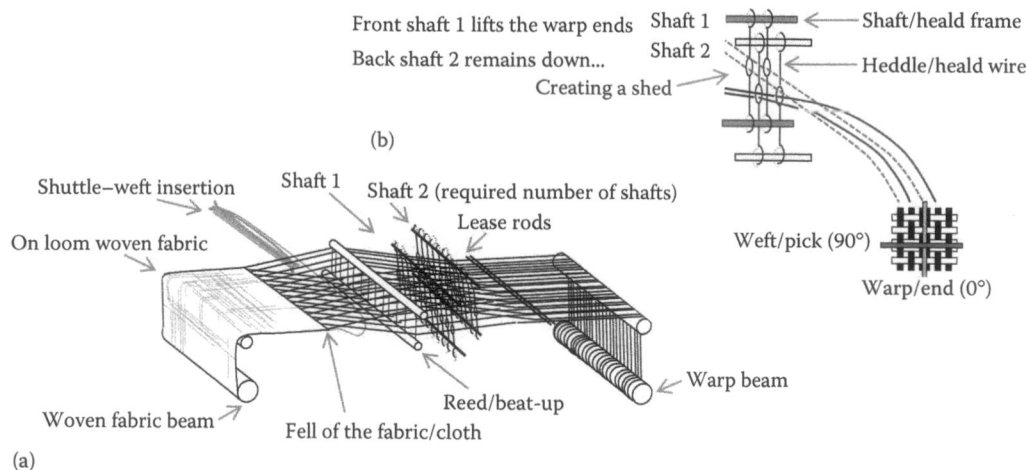

FIGURE 7.4
Basic weaving apparatus and primary principles: (a) shedding operation using two shafts and (b) shedding operation: two shafts for the production of plain weave.

with dobby mechanism (no power, and therefore a nonautomated weaving process), it is known as a dobby loom, such as the George Woods looms. The name dobby refers to the mechanical principles of the shaft lifting control device, which is a cog on the George Woods loom that is connected to both the shafts and one treadle underneath the shafts via pulley(s)/cord(s). On the cog are grooves, each of which provides a location for a lag to lock into upon the treadle downward movement. The lag has the same number of holes as the number of shafts on the loom. Each hole, therefore, represents a shaft. If a peg (similar to a small, slim piece of dowelling with a tapered end) is inserted into a hole, it informs the dobby, the cog, to lift that shaft. Only one treadle is required, as the dobby revolving cog mechanism that houses and operates the lags–pegs turns each time the weaver pushes with his or her foot the treadle down toward the floor (hence the terminology of such looms as floor looms). When the weaver/operator removes his/her foot from the treadle, it moves upward and the dobby-cog drops the previously lifted shafts; the warp yarns are then all lying down within the fabric plane. There is no maximum number to the number of lags that can be used, other than to consider the weight and support required if using many lags in the weaving of the woven textile fabric. The treadle is therefore the communicator to the operational pattern reading device, the dobby mechanism.

The dobby mechanism is an operational system device for the weaving of constructed textiles via a loom that houses shafts—this means that handlooms with levers are typically incorporated within the classification of looms as a dobby loom.

Advancing the handloom in terms of greater weave architecture complexities and production labor times, the dobby mechanism and CAD have been communicatively electronically combined together within semi-automated (hand/CAD) and fully automated (power/CAD) weaving technologies. These dobby looms, semi- and fully automated, are equipped with electronic pattern reading capability. The classification of semi-automated looms includes those which require a weaver/operator to lift the required shafts via lever(s) and/or treadle(s). These lifts are electronically completed by the reading of lifting sequence(s) generated by the required weave architecture of the woven textile to be produced. The woven textile design and its comprising weave architecture(s) are stored within, typically a specialized weave design program such as Ned Graphics, ScotWeave, and PointCarre. The completed weave design is then transferred via USB, wireless or, yes still the good old fashioned way, floppy disk. The weaver/operator, once the weave architecture has been transferred to the loom, can then operate the shedding and weft insertion accordingly. Semi-automated looms include:

- *Patronic (new generation)*: Shedding operation via the communication and transference of the desired weave pattern in CAD and connecting electronic treadle system.
- *AVL (little weavers)*: Shedding operation via the communication of the desired weave pattern in CAD and the connecting electronic forward and backward movement of a reed (tool to beat the weft into the fell of the cloth and to maintain the sett of the fabric), moved by the weaver/operator.

It is necessary for the woven textile designer and/or textile technologist to understand the fundamental working principles of weaving apparatus and technologies before *designing* the weave architectures and concluding woven textile fabric. Knowledge and understanding of the capabilities and limitations of weaving apparatus and technologies define the scope of woven textiles achievable. This also permits innovativeness, pushing the boundaries of the chosen weaving technology process, whether it is craft or industrial or an amalgamation of the two—craft weaving principles should not be dismissed, as they have and still can influence the engineering of warp and weft interlockings in new directions. The diversification of traditional wovens to technical wovens is achieved via thorough understanding of the weaving principles.

7.2.1 Weaving Principles: Primary, Secondary, and Other Elements

The fundamental weaving principles and operation of a loom includes shedding, picking, beat-up, let-off, and take-up. These are the weaving motions and are categorized into primary and secondary. Primary motions are the mechanisms that permit the weaving technology to generate a shed formation (shedding); the weft to be inserted (picking); the weft to be pushed forward and into place in the fell of the woven textile (beat-up). The *secondary motions* are the mechanisms that are required for continuous weaving of the desired fabric length, the let-off and take-up of the warp yarns. The schematic in Figure 7.5 is typical of

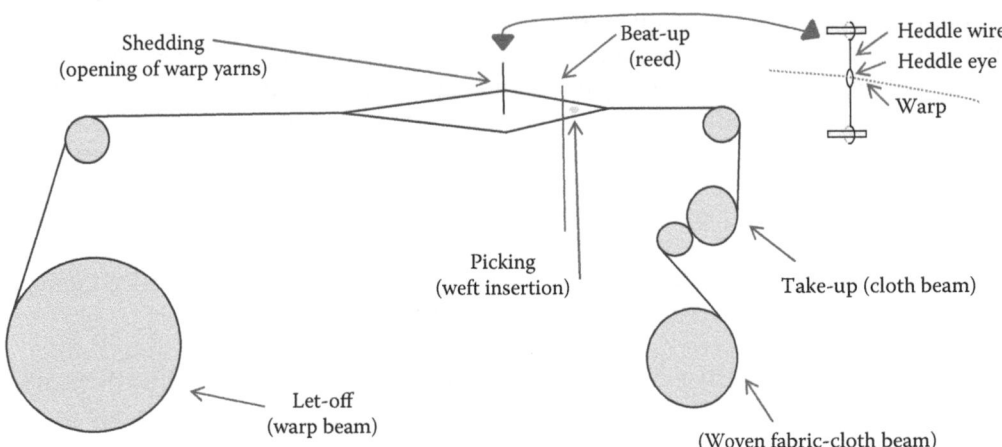

FIGURE 7.5
Weaving principles: primary and secondary.

all types of weaving technologies; the dobby (that has a dobby mechanism and also houses shafts) and the jacquard loom, which controls the warp yarns independently and contains no shafts (to be discussed further in this chapter). The principles of the primary and secondary motions also allow the operation of manual, semi-, and fully automated weaving technologies. The difference between the manual and the semi- and fully automated is as follows:

- *Manual*: The woven fabric beam/take-up beam and let-off warp beam would be turned manually once the area in front of the reed is too close to the fell of the cloth and is not providing a clear wide shed for the insertion of the weft yarn—the distance of the fell of the cloth to the reed is fundamental to the clear shedding sequence.
- *Semi- and fully automated*: Electronically driven woven fabric—take-up beam and warp let-off beam.

7.2.2 Warp Control and Shedding Mechanisms/Motions

The shedding mechanism/motion separates the warp yarns into an upper and lower layer, allowing a weft yarn to be inserted. The timing of the insertion is factored into the shedding mechanism/motion. As the warp yarns open, the weft is timed to be inserted, allowing for the closure of that shed upon its exit at the other side and the opening of the next shed. It also requires timing, on power looms, with the movement of the let-off and take-up beams. The shed formation is predetermined by the weaver/operator via the weave architecture; this is either manually performed or electronically with the weave design inputted within a CAD weaving software program.

7.2.2.1 Tappet and/or Cams (Shafts/Healds/Harness)

The tappet and/or cams shedding are the simplest mechanisms on power dobby weaving technologies and refer to the principle of operation by which a shed is formed by their use. It is typically used for plain weave only due to the minimum number of shafts housed on the dobby loom being 2, although on the majority of looms eight shafts can be controlled. Although limited in terms of weave architectures permitted to be woven, the ability to control up to eight shafts, eight warp ends, and eight weft picks in a repeat, the minimum number of warps and wefts required to complete the weave architecture pattern, derivatives of the elementary weaves twill and satin/sateen, can be produced (discussed in Section 7.5.1). To change a weave pattern, a different set of tappets/cams would be required and a change in the rotational speed of the cams may also be necessary according to the tension on the warp and the float lengths of the warp and weft within the weave architectures used; this therefore hinders its versatility as a commercial productive loom. The tappets/cams are termed as either positive or negative:

- *Positive cams*: Refers to a cam that drives the heald frame in both directions, top shed position and bottom shed position.
- *Negative cams*: Refers to a cam that drives the heald shaft in one direction only, usually down to the bottom shed position, some other means such as a spring mechanism will return the heald shaft to the opposite position.

7.2.2.2 Dobby (Shafts/Heald Frames/Harness)

The term dobby is a component/mechanism attached to a loom to support the process of shaft movement. The dobby requires a cam mechanism to drive the dobby and control the movement of the shafts to produce the required shedding configuration. The combined dobby cam system permits the weaving of more complex woven textiles than the tappet and/cam alone and therefore a greater number of shafts, allowing for more complex combinations of weave architectures. The dobby loom on the whole permits designs requiring up to 28 warp ends operating differently within a pattern repeat. Twenty-eight shafts are typically the maximum capacity of industrial power looms, although some loom manufacturers produce up to 40-shaft looms. The number of weft picks per repeat is only limited by the weave architecture, lifting plan design area within the specialized weave software. Specialized programs were introduced in the transferring of data to the loom as they permit greater numbers of weft insertions, allowing for longer nonrepetitive and repetitive weave architectures to the software that typically *comes with* the loom.

Further information on shuttle and shuttleless looms will be provided in Section 7.2.3, but to allow the differentiation between a positive cam and a negative cam, the weft insertion method is required to be briefly discussed here. Within the array of power looms available, the weft can be inserted through either a shuttle or by what is known as a shuttleless loom, which indicates that the weft is picked up at the side of the fabric and pulled through the shed via rapier projectile (single or double), water-jet or air-jet mechanisms. The manual shuttle looms require the weaver/operator to insert the weft yarn via a shuttle, which can be as crude as a stick with the yarn wrapped around or via *boat*-like devices that house a pirn, around which the weft is wrapped; upon the boat shuttles insertion, the weaver/operator pushes it through the shed and the weft yarn is released from spinning of the pirn through an eyelet in the shuttle. Power looms literally throw the shuttle so that it flies through the shed when traversing the warp yarns and exits the other side. The weft insertion approach dictates the dobby's cam mechanism as either positive or negative, as presented in the following:

- The dobby's shedding mechanism on a shuttleless loom is via a positive cam.
  - Whereby the dobby can send the shafts to both top and bottom shed positions from the true warp line (Figure 7.6). This is the position the warp lays in once drawn through the heddles that are positioned appropriately on the shafts, or attached to a jacquard head via a harness and appropriate number of cords.

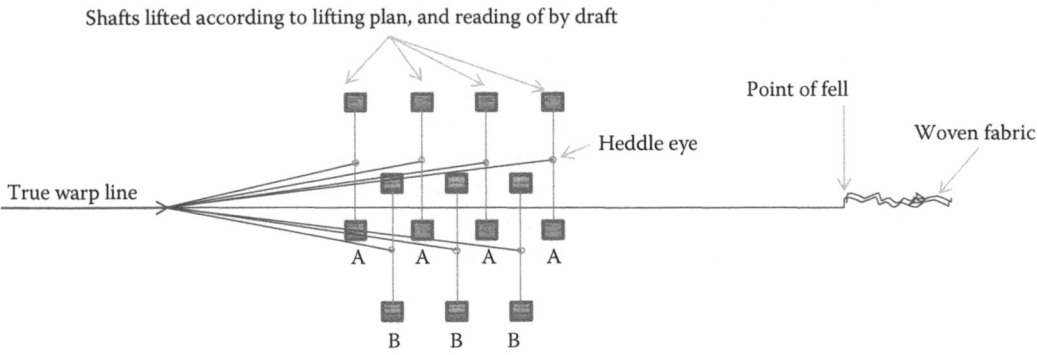

FIGURE 7.6
Warp position to the true warp line.

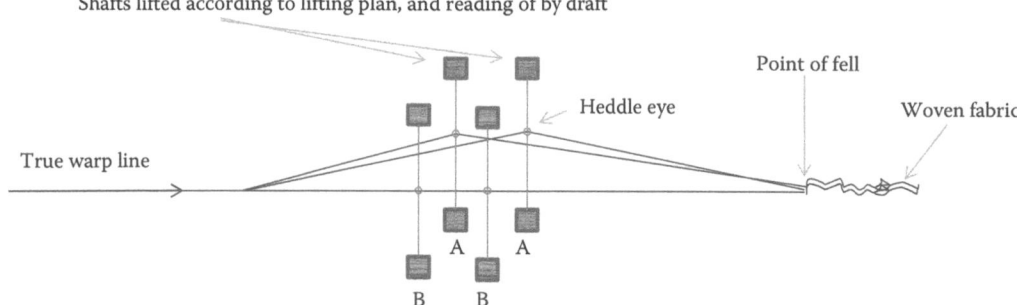

FIGURE 7.7
Warp position to the true warp line.

- The dobby's shedding mechanism on a shuttle loom is via a negative cam.
  - This is whereby the dobby moves the shafts in one direction, either above or below the true warp line, shown in Figure 7.7. The warp yarns are then drawn through the reed and tied to the front woven fabric/take-up beam. The negative cam dobby combination depends upon a spring return to allow all warp yarns to return to the true warp line position to lay flat within the 2D fabric plane; once the weft insertion has been completed and beaten up into the fell of the now completed woven fabric.
  - Both the negative cams mechanism and the dobby mechanism rely upon an alternative means of returning the shafts to allow the warp yarns to lay along the true warp line. They each require a different method and motion of shaft return. The negative cam system pulls the shaft(s) down to the bottom, underneath the true warp line, to create the required shed arrangement. The negative dobby generally lifts the shafts up above the true warp line to create the required shedding.

To change the weave design on a power dobby loom, a new weave architecture, or combinations, is to be completed. The weaver/operator creates a new design and concluding lifting plan in the specialized weave software. The new weave design still needs to take into account the maximum number of shafts in operation on the dobby and how the warp ends are drawn through the heddles on each shaft, known as the draft and discussed in Section 7.5. To support the understanding of how warp yarns are connected to the shaft(s), the schematic Figure 7.8 shows a shaft and a number of heddles, also known as heald wires, that are connected to the shaft. Each heddle is moveable along the shaft but only in terms of allowing the weaver/operator to push them to one side while preparing the loom for the drawing in of the warp yarns. Each warp yarn is then threaded-drawn through an eyelet of a heddle. Each warp yarn (or a number of warp yarns combined together acting as one warp yarn) has its own heddle on one of the shafts out of the total number of shafts to be used and available on the loom. The schematic in Figure 7.8 depicts eight shafts grouped together, with each housing a number of heddles.

The warp yarns are not drawn in randomly through the heddles and shafts, but follow a sequence of steps within a draft. The draft is the arrangement of and location of each individual warp yarn. Figure 7.9 identifies the location of 18 warp yarns in a draft configuration

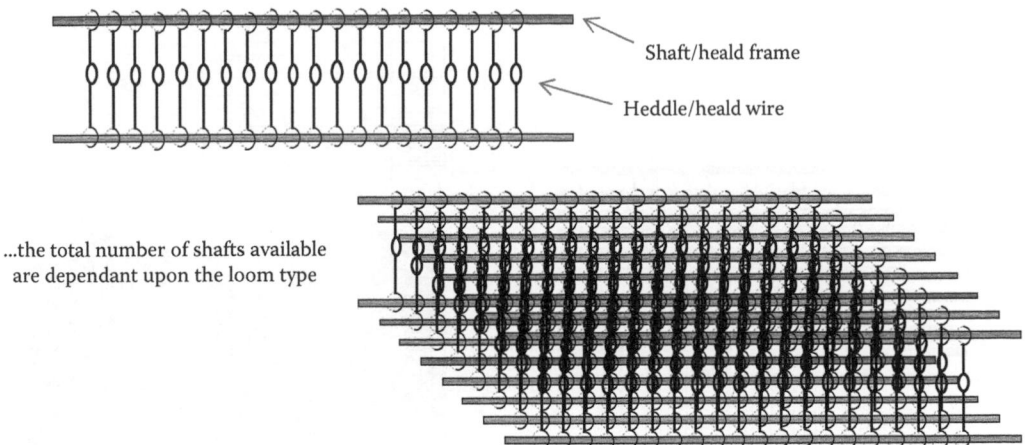

Shaft/heald frame

Heddle/heald wire

...the total number of shafts available
are dependant upon the loom type

FIGURE 7.8
Shaft frame and heddles.

known as a straight draft over eight shafts. As there are eight shafts, the *pattern* of locating
the warp yarns in a straight draft starts on the first available heddle on the left-hand side
on shaft number 1, when drawing in front-to-back. The second warp yarn from the left is
then placed on the next available heddle on the next shaft, shaft number 2. This process
of drawing in over a straight draft is repeated up to the eight shafts, using the first eight
warp yarns. In the straight draft, the straight sequence is repeated over a total number of
18 warp yarns. When stating the next available heddle on the next shaft, there are to be
no spare, empty, heddles in a position between warp yarns. Draft sequences will be discussed further in Section 7.4.

7.2.2.3 Jacquard (Harness but No Shafts/Heald Frames)

The jacquard shedding operation is dependent upon the type of compatible weaving
machine it is connected to; the maximum total number of hooks/warp ends in the weaving width; the shuttle type employed for the weft insertion; and the required shedding
geometry associated with the production of the woven textile: single layer; pile; leno; compound technical woven's. The various manual and automated mechanical and/or electrical operation of the jacquard shedding principles is far too complex to be covered in this
chapter. Therefore, this overview of the jacquard weaving technology allows for the differentiation in warp control to that of the dobby mechanism and/or motion in relation to
the scope and limitations upon the concluding constructed woven textile.

The jacquard is the preferred choice of weaving apparatus/technology for shaping
woven textiles within the 2D fabric plane or for the requirement of photographic imagery within a woven textile. The jacquard head houses a shedding mechanism attached
to a jacquard harness. The harness comprises a number of hooks, each of which connects to an individual cord that houses an attached heddle for the warp end to be
drawn through, identified in the schematic Figure 7.10. The bottom of the heddle is
generally then attached to a tension mechanism such as lingoes, elastic rubbers. The
jacquard harness houses a maximum number of warp yarns across the weaving width
and depth. The width and depth of the harness concludes in the maximum achievable
warp density that can be produced. This affects the warp sett of the cloth, which is

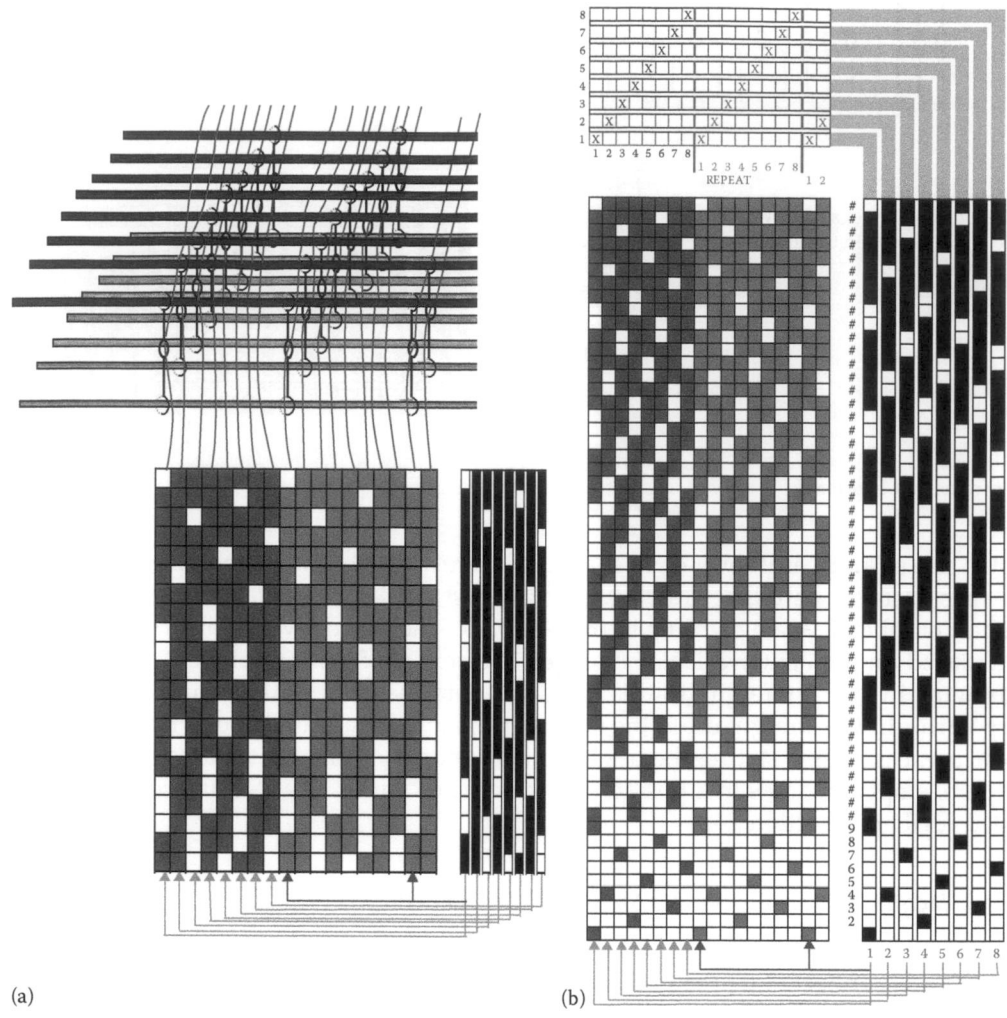

(a)   (b)

FIGURE 7.9
The relationship between the draft and the drawing in of warp yarns: (a) identification of each warp ends location through a heddle in accordance with a straight draft and (b) straight draft planned on graph/point paper–each X represents on warp end drawn through a heddle on that particular shaft.

the number of warp ends per inch (E.P.I) and/or ends per centimeter (E.P.CM). The warp sett may be altered to reduce the E.P.I/E.P.CM from the maximum concluding from the jacquard harness. The decrease in E.P.I/E.P.CM is achieved by casting out harness cords by disabling their controlling hook(s)—the simplest way to achieve this is within the specialized, commercial weave design software in the settings for the jacquard weaving technologies harness tie-up. Up to several thousand individual warp ends can be controlled independently for the creation of complex and intricate weave architectures/patterns advancing upon the constraints of the range of dobby weaving technologies. The scope of the pattern repeats, or nonrepetitive designs, however, is restricted to the division of the jacquard harness (Section 7.4.2). Once in place, the harness is very rarely changed due to the expense of changing the harnesses configuration,

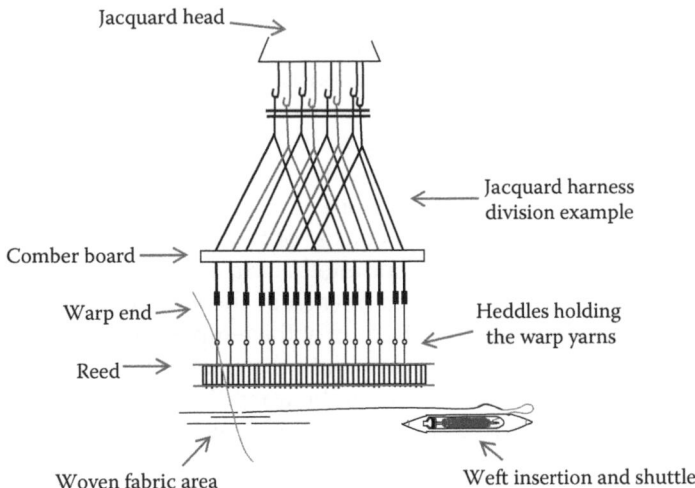

FIGURE 7.10
Jacquard harness.

which starts approximately from £5,000 upward and is dependent upon the maximum weaving width of the compatible weaving technology and the maximum sett of the warp yarns.

7.2.2.4 Tension (Beams, Lease Rods, and Temples)

In the successful production of a woven textile, the warp yarns are released/let-off and taken-up under a uniform tension from the back/let-off warp beam forward toward the take-up beam. Within the weaving cycle, whether employing manual or fully automated weaving technology, the woven fabric beam at the front takes up the released warp yarns from the back let-off warp beam. When the weaving apparatus/technology is operated manually, then the let-off and take-up processes are carried out separately to move the warp yarns forward through the heddles and reed toward the fell of the woven textile. This results in the warp yarns being released temporarily from the tensions and constraints of the weaving technology.

To support the warp alignment and individual warp positioning for warp re-tension before weaving can once again commence, lease rods are inserted through the warp at the back of the loom behind the shafts (identified in Figure 7.4). There are minimum of two lease rods; on fully automated weaving technologies, there can be numerous lease rods depending on weaving speed and yarn combinations within the warp. The lease rods are typically inserted in opposing directions to one another under and over the warp yarns to ensure their alignment and eliminate warp slippages and alleviate warp yarn friction. The alternating sequence of the lease rods under and over the warp yarns also aids the warp tension once weaving commences.

During the weaving cycle, tension is imparted in the weft direction. As the warp yarns close and interlock with the weft, the weft yarn contracts due to being locked under and over the warp yarns according to the weave architecture. This tightening within the fabric plane in the weft direction is due to the crimp factor imparted upon the yarn. It is also affected by the elongation characteristics of the yarn(s), which determine the stretch or greater contraction in the weft direction of both the bed of warp yarns before the fell of the cloth and the concluding on loom woven cloth. To support the retaining of straight woven edges in the warp direction and reduce tension and contraction upon both the weft and warp yarns, temples are positioned upon the woven cloth adjacent to the fell of the cloth. The temple is positioned across the weaving width in line with the first and last warp yarn drawn through the reed to reduce the narrowing, contraction of warp yarns inwards when the reed moves to and from the fell of the cloth.

The temples are typically wooden and consist of two interlocking rectangular lengths. The two wooden lengths interlock with one another via a pivoting action and secured into position via a locking system. At one end of each of the wooden lengths comprising the temple is a pin. The pins allow the maintaining of the temples position on top of the woven fabric, close to the last interlocked weft yarn. Both the ends of the wooden temple with the pins are positioned close to the outer edges of the warp. This enables all the warp yarns to remain in alignment to the 90° direction during the weaving cycle.

7.2.3 Weft Insertion Mechanism: Shuttle, Shuttleless, and Concluding Selvedge

As presented in the introduction, the shedding mechanism and control of the warp yarns together with the method of weft insertion are fundamental to the classification of manual, semi-, and fully automated weaving technologies, identified in Figure 7.11. Alongside the dobby and jacquard categories of warp control, weaving technologies/looms are commercially categorized on the basis of the weft insertion mechanism, shuttle and shuttleless. The shuttle and shuttleless weft insertion mechanisms shall be discussed and the concluding seams/selvedges in the weft direction on the final woven textile fabric.

The shuttle method provides a fabric with seamed edging known as the selvedge area. The role of the seam/selvedge is to secure the outside warp threads and prevent the to be woven fabric fraying or falling apart once removed from the tensions of the loom. As the weft traverses through the shed, left to right across the warp yarns and weaving width and beaten into the fell of the woven fabric, the next shedding operation is in motion. This then allows the shuttle to return to the other side, now right to left through a new and clear shed arrangement traversing through another shed configuration, the next weft pick

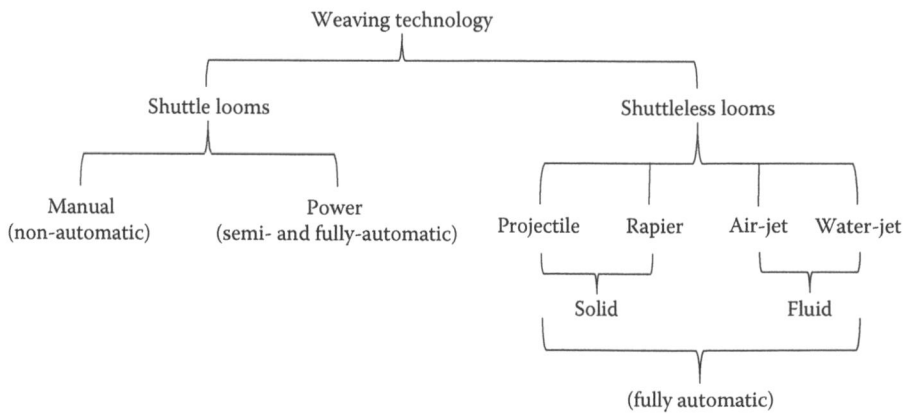

FIGURE 7.11
Classification of manual, semi- and fully automated weaving technologies.

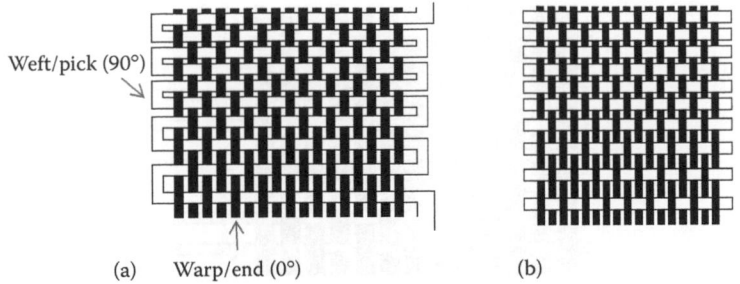

FIGURE 7.12
Woven selvedge's (continuous weft looped edges and fringed edge): (a) closed loop edges (selvedge) and (b) fringed edging (selvedge).

of the weave architecture/lifting plan providing a continuous weft insertion and closed loop edges, identified in Figure 7.12a. It is the only method of weft insertion to enable the weaving of tubular structures (such as fashion garments and accessories bags) without seams. When using a shuttle on a manual loom, the weft may be cut to produce a fringed edging, shown in schematic Figure 7.12b. This is also achievable on a semi- and fully automated loom via a set of cutters on both the entry and exit sides of the total number of warp yarns that make up the woven fabric. A new yarn, although this can be the same shuttle, is then inserted into the shed and again cut upon completion of traversing over the warp yarns.

In the shuttleless weaving technologies, there are a variety of discontinuous weft insertion mechanisms, including single rigid rapier, double rigid rapier, flexible rapier, air-jet, and water-jet. The discontinuous weft produces the following selvedges:

- *Fringed*: At one or both sides of the woven fabric and after the weft has left the yarn package and entered the shed, or after the weft has completed its route to the other side of the warp yarns, the weft is then cut. It can be cut on one or both sides of the completed woven fabric, shown in Figure 7.12b.
- *Tucked-in*: A predetermined length of weft yarn extends beyond the last warp end and pulled into the shed via a tucking device, shown in Figure 7.13.
- *Leno*: Involves doup heddles that allow pairings of warp yarns to twist around one another, entrapping the weft yarn, shown in Figure 7.14.

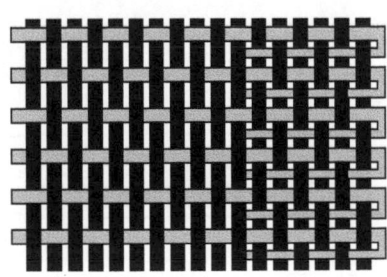

FIGURE 7.13
Woven selvedge's (tucked in).

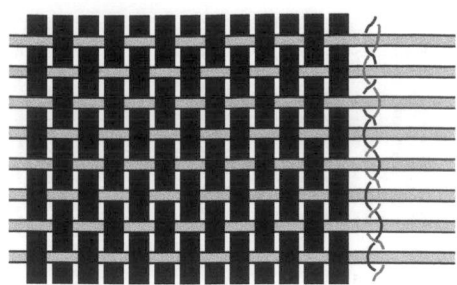

FIGURE 7.14
Woven selvedge's (leno edge).

7.2.3.1 Types of Shuttle

The weft yarn is carried by and/or within a shuttle. The term *by a shuttle* refers to the varied range of flat wooden sticks. These sticks tend to have a curved indentation at either
end to help support and house the weft yarn as it is wrapped around. These are only used
in hand weaving from craft weaving apparatus and technologies to manual and semi-
automated weaving technologies (Figure 7.15a). The term *within a shuttle* refers to the various boat-like shuttles that house a pirn/quill that holds a determined length of weft yarn
(Figure 7.15b).

7.2.3.2 Shuttleless: Single Rigid Rapier

In shuttleless weft supply within the weaving cycle, the method of weft insertion does
not have a compartment to house a supply of weft, such as a pirn. Instead, the weft is
supplied at the side of the weaving technology, to the left of the loom on a yarn package
holder known as a creel. There can be many weft packages supplied for the weaving
process, although typically eight weft yarn types and/or colors are used. The single
rigid rapier comprises of a long stiffened-rigid thin rod that connects to a gripping head
device at the end. The rapier head has a scissor-like action, whereby it grabs and pulls
the weft yarn from the package/creel supply to the left of the loom. It then pulls the weft
yarn through the shed, traversing the warp yarns until it reaches the outer edges of the
shed on the right-hand side of the sheet of warp yarns that will comprise the end woven
fabric. The single rigid rapier head then returns empty to the entry side of the shed.
Upon exit, the completion of weft insertion in the weaving cycle, the inserted weft yarn

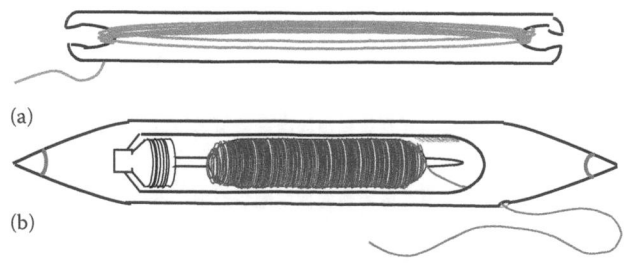

(a)

(b)

FIGURE 7.15
Shuttles: (a) stick shuttle and (b) boat shuttle.

is then beaten forward by the reed into the fell of the woven fabric. As the reed begins to travel backward away from the fell of the woven fabric, the next shedding movement and configuration is taking place and the next weft is grabbed and pulled to traverse the warp yarns. To be clear, the single rigid rapier head traverses the full weaving width of the warp.

7.2.3.3 Shuttleless: Double Rigid Rapier

The double rigid rapier comprises of two thin stiffened rigid rods, each of which is connected to a gripper/scissor style head. The heads are known as a giver and a receiver. Each of them enters the shed arrangement simultaneously and meets in the middle of the shed. The giver rapier takes the weft yarn from the left of the loom, from the weft supply on the creel, and upon insertion pulls the weft through the shed to the middle of the warp yarns. The receiver also enters this shed from the right-hand side to meet the giving rapier head in the middle of the shed, warp yarns. The giver places the weft yarn into the receiver, which then takes the weft yarn through the remaining half of the shed to the right-hand side of the weaving width. The weft inserted from left to right, typically, is cut once the taker and receiver rapier heads have completed that particular weft insertion; concluding in a fringed selvedge. This selvedge may be supported by the leno weave architecture for greater enclosure to the completed woven fabric, shown in Figure 7.14.

7.2.3.4 Shuttleless: Flexible Rapier

The flexible rapier is based on the principle of the double rigid rapier, but the stiffened rigid rods are now replaced by flexible steel or plastic tapes. These flexible steel/plastic tapes have the role of the giver and receiver. Unlike the double rigid rapier, however, the flexible steel/plastic tapes travel in a curved path each from underneath the *bed* of warps, one from the left-hand side and the other from the right-hand side. They each then follow a curve directional path once they travel out from underneath the warp sheet to then enter the shed configuration. Once again, the giver and receiver meet in the middle of the shed to complete the weft insertion in that shedding arrangement. Each flexible steel/plastic tape and attached head (giver and receiver) are driven-projected by crank gear drives in the weaving cycle.

7.2.3.5 Shuttleless, Other: Air-Jet, Water-Jet

Air-jet: Compressed air is used via a series of relay nozzles to project the weft yarn through the shed configuration in the weaving cycle. The air permits high-speed weaving of fine weft yarns—it is not suitable for weaving heavy, coarse yarns for the compressed air cannot hold up and project such yarns at the speed required to clear the shed in time for the next shedding operation to take place. The nozzles are known as relay nozzles, as they literally relay, *race* the inserted weft yarn in a timed sequence of air projection through each nozzle.

Water-jet: Nozzles are employed to disperse high pressurized jets of water to propel the weft yarn through the shed arrangement. This method is the fastest shuttleless weft insertion method. Water-jet weaving, however, is limited to weaving hydrophobic fibers/yarns, (water-resistant) such as nylon/polyester.

7.2.4 Beat-Up: Reed, Fabric Sett, and Shed Opening Support

Once the weft yarn has been inserted through the shed to enable the interlocking of the warp and weft yarns to form a stable woven fabric, the weft is pushed into the fell of the fabric. The fell of the fabric, also known as the fell of the cloth, is the area that starts after the previous weft insertion and its securing into a woven structure, shown in Figure 7.4. The use of a reed is the device that enables the pushing of, the beat-up, the weft into the fell area. The movement of the reed is controlled via a crank/cam on CAD-power looms and by hand on manual and semi-automated weaving apparatus and technologies.

The density, the sett of the woven fabric is also maintained by the reed. The reed consists of gaps, known as dents between typically metal thin bars, which are secured in place by an outer frame. The spacing of these dents is calculated in centimeter and/or inches. The number of dents per centimeter (D.P.CM) determines the ends per centimeters (E.P.CM). This is the number of warp ends within a centimeter, which provides the density, the sett of the warp. If using a reed with 28 D.P.CM and drawing in one warp yarn per dent, it would conclude in 28 E.P.CM. If the same reed was used (28 D.P.CM) and two warp yarns drawn in per dent, this would conclude in 56 E.P.CM. Generating a woven fabric of a balanced nature incorporates the same number of warp yarns drawn through each dent, and the same total number of the weft, the picks per cm (P.P.CM). Examples of balanced and unbalanced warp arrangements, in terms of the number of E.P.CM are shown in Figure 7.16a–c. Figure 7.16d is an example of an unbalanced arrangement of weft yarns (P.P.CM). In the schematic Figure 7.17 increasing and decreasing the E.P.CM

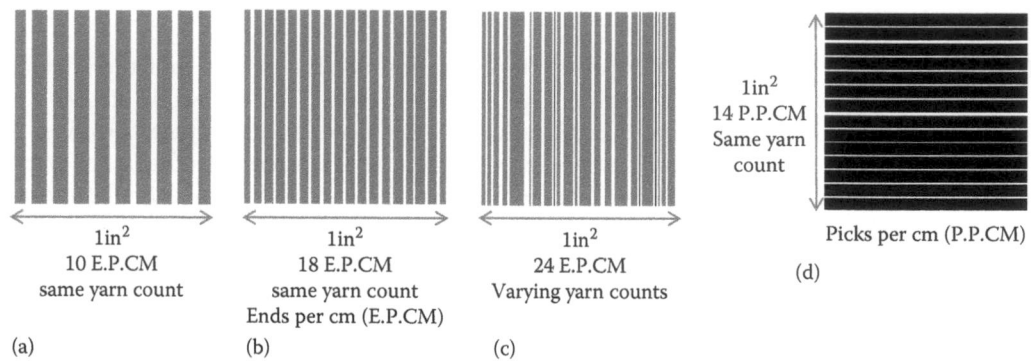

FIGURE 7.16
Ends and picks per centimeter (E.P.CM, P.P.CM): (a) balanced warp, (b) unbalanced warp, (c) unbalanced warp, and (d) balanced weft.

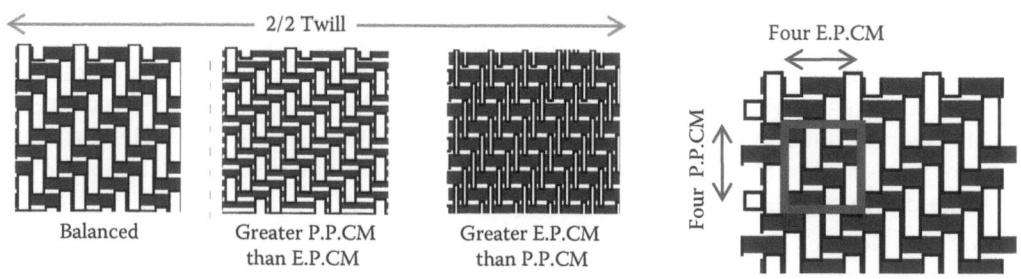

FIGURE 7.17
Warp and weft fabric sett.

and/or P.P.CM within a 2/2 Twill shows the visual effects that would be produced on the face of the woven fabric.

7.2.5 Let-Off: Back Warp Beam

The required numbers of warp yarns that make up the weaving width are located at the back of the loom either on a beam, known as the back-warp beam and/or weaver's beam, shown in Figures 7.4 and 7.5, or via a set of creels. During the weaving process, the warp yarns are evenly positioned, tensioned, and supplied by the let-off mechanism (beam or creels) into the weaving region (the front of the loom). Whether the let-off mechanism is a continuing movement of the back-warp beam in power weaving or as and when needed turned by hand in the manual weaving process, the tension must be constant in the releasing (let-off) of warp yarns during the weaving process.

The back-warp beam is classified as either negative or positive. A negative let-off system concludes in the back-warp beam and/or creels rotating in one direction only, forward, to supply the warp into the shed for the insertion of the weft. A positive let-off system concludes in the back-warp beam and/or creels rotating in two directions, forward and backward. This is ideal when weaving high extensible yarns such as elastomeric yarns. When weaving yarns with the same characteristics, such as extensibility, a back-warp beam is fine to use. If the intended woven fabric is to be constructed using a hybrid of yarns (varying characteristics and extensibility), either each yarn type has its own back-warp beam or is supplied by a creel let-off system. This prevents, or minimizes, warp breakages during the weaving process. The use of either back-warp beam and/or creels is also dependent upon the type of weaving apparatus/technology, the available floor space, and/or weaving of multiple warp and weft layers levels is desired.

7.2.6 Take-Up: Front Woven Fabric Beam

To enable the warp yarns to move forward in an even tension during the weaving process using fully automated weaving technologies, the front-warp beam, also known as the take-up beam, rotates forward to drive the newly woven fabric form at such a rate that the fabric will have the predetermined weft density. This mechanism also pulls the unwoven warp yarns situated on the back-warp beam forward to be interlocked with the next weft insertion in the weaving region toward the fell of the cloth.

7.3 Conventional Weaving Technologies: Preparation and Dressing

Before weaving can commence, the weaving apparatus/technology must be dressed, which includes the following stages: warp preparation: from chaining to beaming and creels; drawing in of warp yarns: from beam/creels through the required number of heddles and locations (on shafts, dobby or on cords, jacquard); drawing in of warp yarns from heddle to and through the reed; tie on of warp to front take-up beam.

There are a variety of approaches to forming yarns into a warp. This is very much dependent upon the type of weaving technology to be employed in the production of the woven textile. The following subsections, taking a handloom (manual weaving technology) and a fully automated loom into consideration, allow for an overview of their preparation and dressing for weaving production.

7.3.1 Manual Weaving Technology

The process of making a warp for handweaving is laborious but can also be therapeutic! A vertical floor standing rotating warping mill and/or wall-mounted warping board provides a framework to wind both the required warp length and total number of warp ends evenly around. The yarns used in the creation of the warp are supplied from yarn packages, typically housed on a creel. Tension is applied evenly to the warp yarns, as each is wrapped around the warping mill to aid the removal and transition of the finished warp to the back let-off beam. A technique to keep the prepared warp in order and in a manner that can be unraveled to support the subsequent setting up stages of the loom is known as chaining. Chaining the warp is carried out once the warp is complete—please see the chaining activity step sequence in Figure 7.18. Once the warp is chained, it is then carried over to the weaving apparatus/technology for beaming. Beaming requires all the warp yarns to be tensioned evenly and attention to be given to their placement as they are wound on to the warp let-off beam. To support an even distribution across the beam and alleviate any overlapping of warp yarns, a raddle is used to space the warp yarns to the desired warp weaving width. The raddle is a comb-like accessory. Each gap in the raddle is generally 0.5 cm wide. Therefore, the calculation of the warps E.P.CM translates to their location within the raddle. The raddle is typically secured to the back of the loom to allow the weaver-operator to wind the warp through the raddle while maintaining an even tension. It must be noted that the warp can be placed and wind on to the loom back-to-front and front-to-back. A good length of warp is left loose from the beam to be taken forward toward the front of the loom and temporarily suspended from the top framework of the loom. This allows the insertion of lease rods near the warp let-off beam. The intended weaving width is concluded via the E.P.CM. The next stage requires the weaver-operator to begin the drawing-in process of each warp yarn through a designated heddle on the allocated shaft with a reed hook. The drawing-in sequence of the warp yarns is informed by the designed draft that includes a location for each and every warp yarn. The final drawing-in stage sees the warp yarns drawn through the reed, known as denting using a denting hook. The reed, similar to the raddle, is an oblong style framework that houses wood/metal rods—each rod is spaced apart from its neighboring wire in accordance to the required E.P.CM known as the dents per centimeter (D.P.CM). The reed, unlike the raddle that is removed once the drawing-in stages of the warp through the heddles and reed is complete, is in situ during the weaving cycle. The reed maintains the E.P.CM and weaving width and beats each consecutive weft insertion into the fell of the cloth. The warp yarns are then tied in small groups to a rod/stick at the front of the loom. Tying either from the following directions: starting in the middle of the warp, alternating from side-to-side working, and tying outward to control the tensioning of the warp yarns; applying and maintaining an even tension upon the groups of warp yarns working from the outer edges of the warp and alternating side-to-side working and tying inward to the middle of the warp. The front rod is connected to an apron that is integral to the front take-up beam. Once the warp yarns are evenly tensioned and tied to the rod, the front beam weaving can then take place. To establish efficient weaving via a clear shed, the tension must be even and appropriate to the warp yarns elongation characteristics. For each variation in elongation when producing a hybrid woven textile, a separate let-off warp beam must be used. This ensures tension is appropriate for that particular type of warp yarn.

LEARNING ACTIVITY

Try the warp chaining process by following the step-by-step sequence of movements as shown in Figure 7.18.

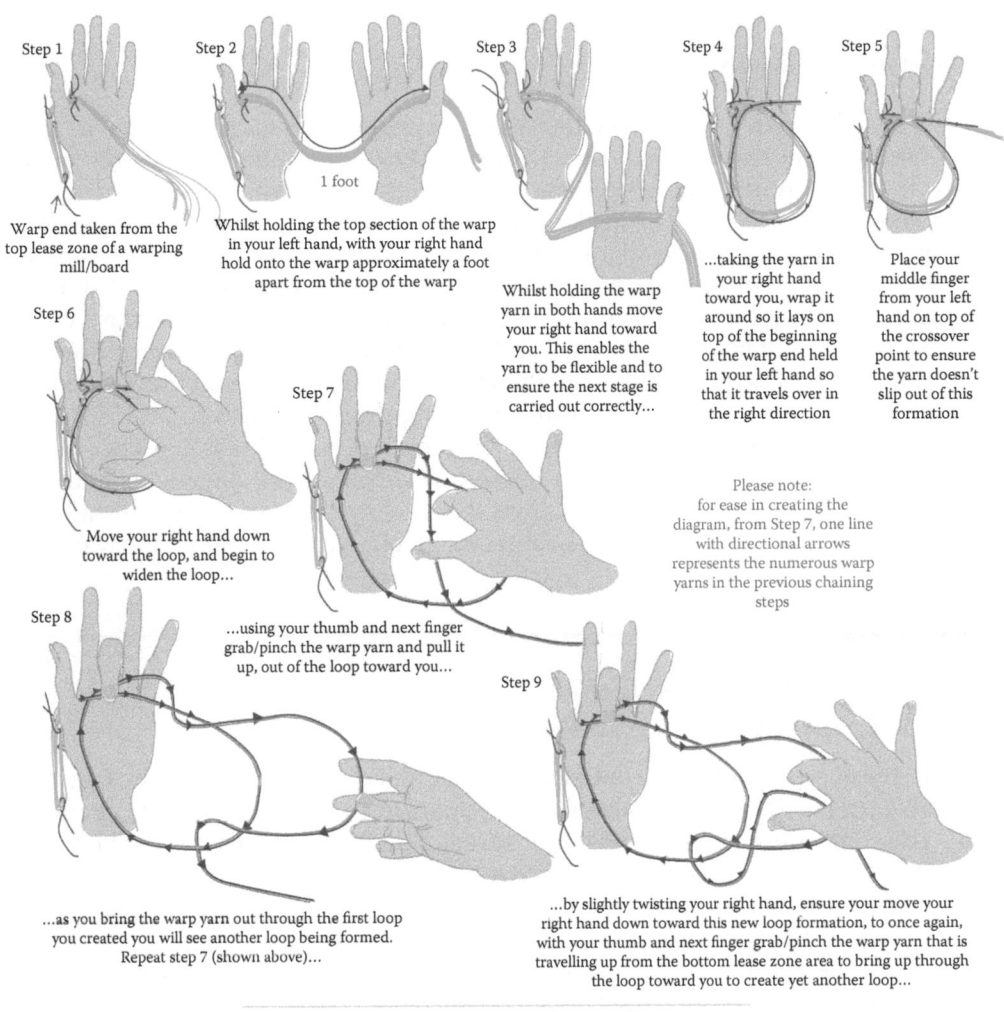

Step 1
Warp end taken from the top lease zone of a warping mill/board

Step 2
1 foot
Whilst holding the top section of the warp in your left hand, with your right hand hold onto the warp approximately a foot apart from the top of the warp

Step 3
Whilst holding the warp yarn in both hands move your right hand toward you. This enables the yarn to be flexible and to ensure the next stage is carried out correctly...

Step 4
...taking the yarn in your right hand toward you, wrap it around so it lays on top of the beginning of the warp end held in your left hand so that it travels over in the right direction

Step 5
Place your middle finger from your left hand on top of the crossover point to ensure the yarn doesn't slip out of this formation

Step 6
Move your right hand down toward the loop, and begin to widen the loop...

Step 7
...using your thumb and next finger grab/pinch the warp yarn and pull it up, out of the loop toward you...

Please note:
for ease in creating the diagram, from Step 7, one line with directional arrows represents the numerous warp yarns in the previous chaining steps

Step 8
...as you bring the warp yarn out through the first loop you created you will see another loop being formed. Repeat step 7 (shown above)...

Step 9
...by slightly twisting your right hand, ensure your move your right hand down toward this new loop formation, to once again, with your thumb and next finger grab/pinch the warp yarn that is travelling up from the bottom lease zone area to bring up through the loop toward you to create yet another loop...

...REPEAT this process until you reach the bottom of your warp. You have now completed the chaining process of your warp!

FIGURE 7.18
Warp chaining process.

7.3.2 Automated Weaving Technology

The setup of automated weaving technologies is a similar procedure to that of setting up the handlooms. Due to the mechanical operation and weight of various components ensuring the carrying out of the weaving principles, a range of auxiliary equipment is required, such as

- Creel system with individual yarn package tensioning for individual warp yarns
- Electrical lifting equipment, suitable for the maneuvering of warp and woven fabric beams
- Pirn winders for shuttle looms

- Weft package creels for shuttleless looms
- Warping technology/machine

The creation of a warp for automated weaving technologies requiring beams is via two approaches: indirect warping and direct warping. Direct warping permits the winding of yarns in small groups and/or sections around and onto a type of conical drum. This method allows the creation of striped warps, warp color variegation, and/or warps consisting of a hybrid of yarns. The alignment of warp sections around the drum is completed once the total number of warps is achieved, concluding in the required E.P.CM and warp/weaving width. Once the warp is completed around the drum, all the sections of yarns are transferred simultaneously, parallel, and under equal tension to a warp beam with flanges. The flanges allow the locking of the warp beam into a secured framework on the back of the weaving technology. The direct warping process, however, provides a faster method in creating a prepared warp beam. The required yarns (color and fiber type) are each housed onto a cone/package each housed on a framework known as a creel. The packages of yarns are then directly fed from the creel system to a warping head, which winds the yarns simultaneously, parallel and under equal tension around the warp beam.

7.4 Conventional Weaving Technologies: Woven Fabric Preproduction Setup Relationships

To enable the warp and weft to interlock and produce the desired woven fabric, each warp end/yarn needs to be operable via a mechanism that *reads* the weave architecture design. The controlling of warp ends differs from a dobby loom to a jacquard (discussed in Section 7.2 woven construction, apparatus, and technology). The dobby loom houses and operates a number of shafts. Each shaft is controlled by a cam mechanism/motion, enabling it to be lifted and dropped. Each shaft houses a number of heddles/heald wires (Figure 7.19b). The number of shafts is dependent upon the weaving technologies capabilities to both house and controls them. A group of shafts is depicted within the schematic shown in Figure 7.19a and b to provide an understanding of the relationship of the draft and shafts. The location of each warp end through a heddle and connected to its housing

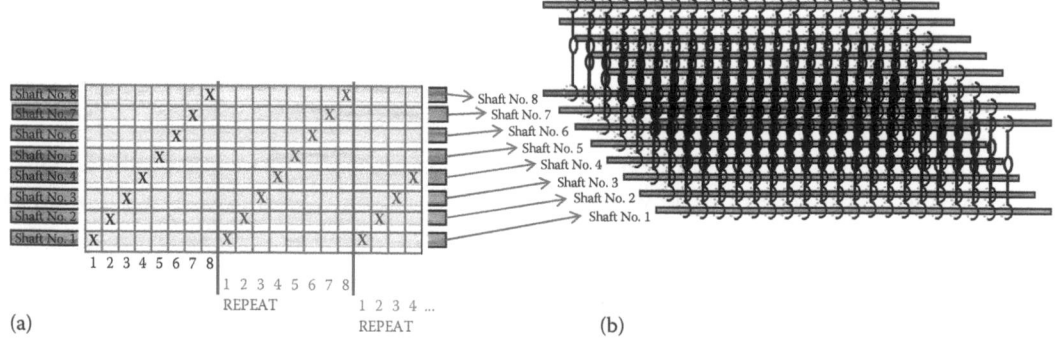

FIGURE 7.19
Draft and shaft relationship: (a) straight draft planned on graph-point paper and (b) schematic of eight shafts, each houses a number of heddles.

shaft is determined by and necessitates the desired woven fabric and comprising weave architecture(s). The weave designer, once knowledgeable of the weaving principles and process, can decipher the control of warp ends and therefore generate the draft. The draft is a representation of the shafts and heddles that correlates to the total number of warp yarns within the weaving width. If more than one warp ends are drawn through their own heddle on the same shaft, then they will be lifted together during the weaving process.

7.4.1 Dobby: Draft, Lift, and Face Relationship

Within the draft in Figure 7.20a and b, each row represents one shaft. Knowledge of the directional meaning of draft notation on point/graph paper will allow you to understand the drawing in of warp yarns on the many draft configurations permitted. The first row in the draft typically refers to the first shaft on the weaving technology nearest to the front of the loom (known as front-to-back drafting). Notating a warp yarn threaded through a heddle on the desired shaft is typically marked by a X as the indicator. Each column in the draft represents the total number of warp ends generating the weaving width and the location of each individual warp end on the number of shafts available/to be used; only one cross can be inserted per column, but as many crosses as necessary can be inserted in the same row. This is due to the principles of the shafts lifting. If you were to place two crosses in a column, as shown in Figure 7.20a, this would indicate the following:

- One warp end is drawn through a heddle on the sixth shaft from the back of the loom to the front and also drawn through a heddle positioned on the third shaft. During the weaving process, the warp end would lift any time shaft 3 and 6 were lifted, and not necessarily in accordance with the weave architecture, therefore creating a mistake in the completed woven fabric, the face of the fabric.

There are numerous ways to draw in the warp ends over the shafts within a draft, some of which are classified as straight, pointed, block, broken, intermittent, grouped, and combination.

To generate a straight draft, you would first outline around the required number of rows, each representing the number of shafts to be used in the weaving process. For clarity, numbering the shafts on the left-hand side aids the understanding of their lifting sequence when relating the draft to the lifting plan. The next step in the drafting process is to complete this outline around the total number of warp ends to complete the weaving width. If you were designing a draft requiring eight shafts, a total of 64 warp ends and in a straight

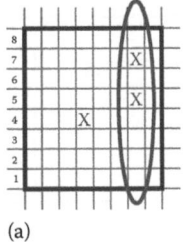
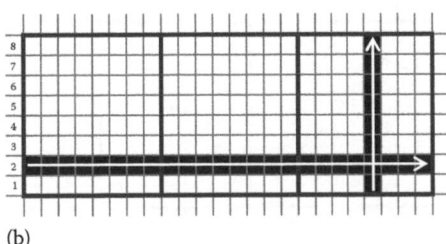

(a) (b)

FIGURE 7.20
Weave architecture notations: (a) one column represents a heddle and where a warp yarn is located and the allocated shaft the heddle sits on! and (b) one row represents a shaft. Notating a warp yarn threaded through a heddle on the desired shaft is typically by a X as the mark indicator.

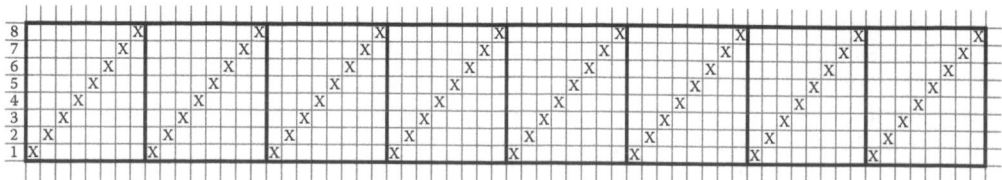

FIGURE 7.21
Draft (relationship of the draft to the warp yarns).

repeating configuration running from left-to-right/front-to-back, you would generate the draft shown in Figure 7.21.

The straight draft forms the basis for all others and proceeds in a repetitive diagonal sequence, left-to-right, shown in Figure 7.22a front shaft to back shaft, or right-to-left back shaft to front shaft shown in Figure 7.22b across the total number of warp ends. It can be used to include any number of shafts, requiring only a minimum of 2. The straight draft configuration is repeated until the last warp end has been allocated a heddle and shaft.

A pointed draft results from the running of a straight draft first in one direction, for instance left-to-right, from the first shaft to the back shaft and then from the back shaft to the front shaft to complete, as shown in Figure 7.23. The shaft at each point of reverse receives one warp end only on that particular shaft, the other shafts each carrying two warp ends for that particular concluding point.

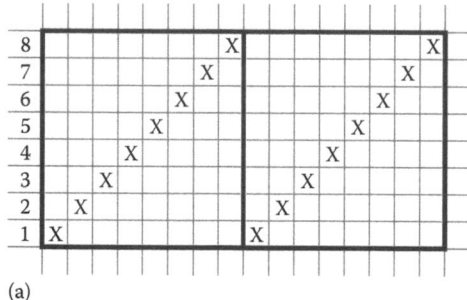

(a)

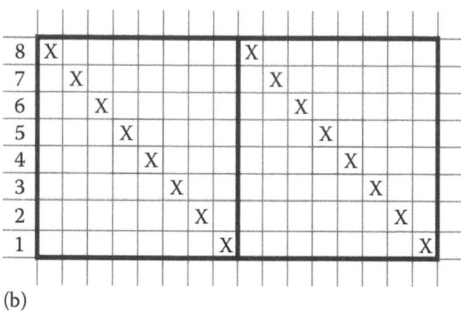

(b)

FIGURE 7.22
Straight draft: (a) left-to-right and front-to-back and (b) right-to-left and back-to-front.

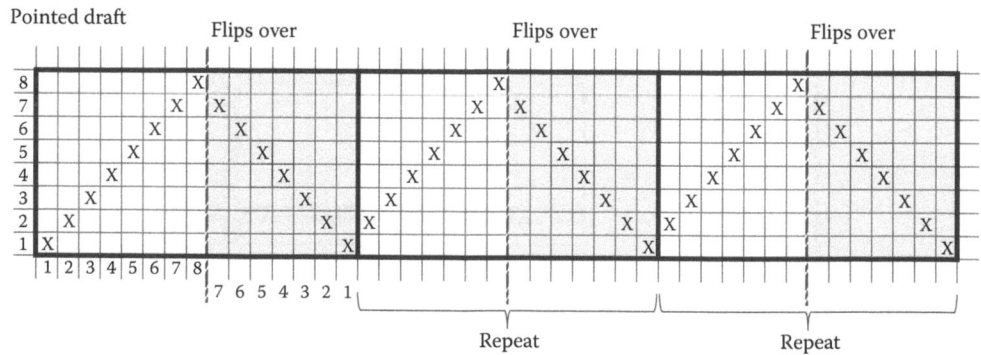

FIGURE 7.23
Pointed draft.

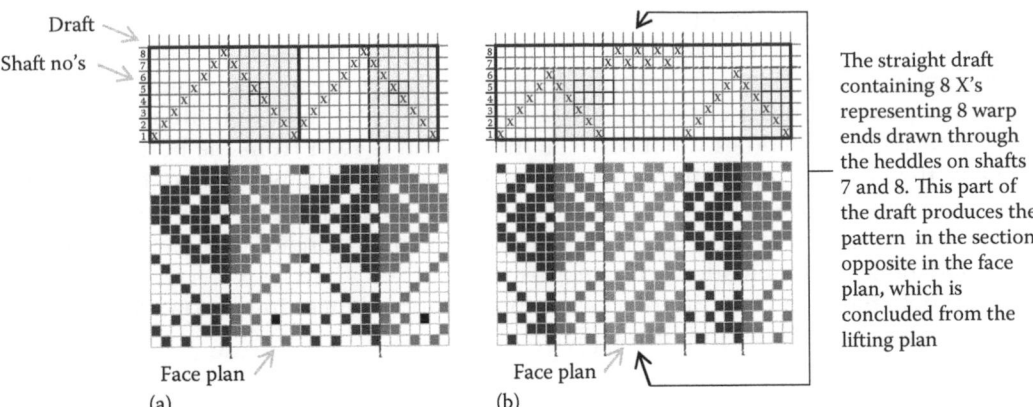

The straight draft containing 8 X's representing 8 warp ends drawn through the heddles on shafts 7 and 8. This part of the draft produces the pattern in the section opposite in the face plan, which is concluded from the lifting plan

Face plan

(a)   (b)

FIGURE 7.24
Pointed draft and combined point and straight draft: (a) pointed draft and (b) divided and combined straight and pointed draft.

When using a pointed draft across the total number of warp ends, as in Figure 7.24a, you can produce motifs/patterns that will lie side by side. If you wish the motif/pattern to have a *gap*, in terms of another weave architecture separating the motifs, then you would choose a draft similar to that in Figure 7.24b.

The schematic of a combined straight-point draft, lifting plan, and face plan in Figure 7.25 will aid your understanding of their relationship to one another and the weaving process, with the following overview of a draft, lifting plan, and face plan:

- *Draft*: Is how each individual warp end is drawn through heddles on a particular shaft. A warp end can only be drawn through one heddle on its housing shaft.
- *Lifting plan*: Is the complete weave architecture required to enable the weaver/operator and/or CAD software to produce the required lifting of shafts to produce the pattern.
- *Face plan*: Is the concluding face of the woven fabric that includes the total number of warp ends and weft picks.

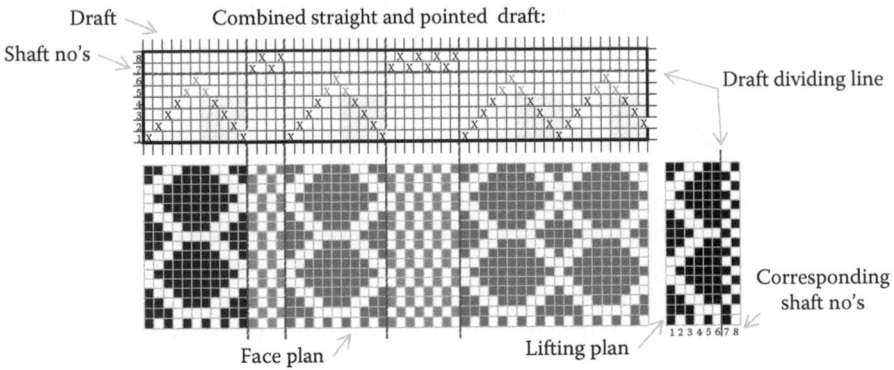

Draft dividing line

Corresponding shaft no's

Face plan   Lifting plan

FIGURE 7.25
Combined point and straight draft, lifting, and face plan.

As shown in Figure 7.25, the numbers on the bottom of the lifting plan represent each shaft within the draft. Within the lifting plan, a column represents a shaft and a row represents a weft pick (a weft insertion in the weaving process). Each mark indicator in the lifting plan (in this case, a colored in square) means that shaft will lift all warp yarns drawn through the heddles it houses. The lifting plan is read from the bottom-first row upward; the first row representing the first weft insertion in the weaving of the warp and weft interlocking yarns. The lifting plan also contains the minimum number of warp-weft interlockings (colored in squares), the lowest common denominator of the weave architecture used alone or combination with another.

Please note the following differences in representational meaning of rows and columns on graph paper for a draft and lifting plan:

- Draft:
  - A row represents a shaft. Each square on that row/shaft represents a heddle positioned on that shaft.
  - A column represents a warp yarn drawn through a heddle on a shaft (with each shaft numbered on the left-hand side for clarification of warp end location).
- Lifting plan:
  - A row represents a weft insertion. Each mark indicator in the lifting plan, typically a colored square, indicates that shaft will be lifted, and any warp yarns allocated to that shaft and its housed heddles.
  - A column represents a shaft. The first column, typically, is shaft number 1; the first shaft at the front of the loom. All shafts that are to be used in the weaving process are represented in the lifting plan.

Figure 7.25 shows how the shafts are represented in both the draft and lifting plan. Shafts 1-to-6 and 7 and 8 are separated by a red dividing line. This allows you to know that shafts 1-to-6 and 7 and 8 can be controlled independently from one another. This type of draft is known as a combination, and in this instance, combines straight and point draft configurations. Within the highlighted purple circles on the draft, you can see the warp end threaded through a heddle on shaft 2 stops here and the next warp end to be drawn through a heddle on a shaft is that heddle on shaft number 7. When looking at the face plan, if you look down along the column from shaft 2, you will see it is the same as the column number 2 in the lifting plan. In the second point draft configuration, separated by a straight draft over shafts 7 and 8, the point is completed by a final warp end drawn through a heddle on shaft 1. In the third pointed draft configuration, there is one warp end that is separating the two points—to ensure there are no *tram tracks*, whereby two or more adjacent warp yarns are lifted simultaneously, the division between the two points must only be one column containing one "X" on shaft 8, which can be seen in Figures 7.23 and 7.24.

By dividing and grouping the total number of available shafts on the weaving technology, the controlling of certain sections of interlocking warp and weft can be accomplished. This allows the combining of different weave architectures within the completed woven fabric. For example, Figure 7.26a shows the block draft configuration over eight shafts, and the shafts divided into equal parts; each part containing four shafts and employing the straight draft configuration. The weave architecture in Figure 7.26a can be woven using the block draft shown in Figure 7.26b. The concluding produced woven fabric would then consist of an alternative combination of plain weave and hopsack in the face of the fabric.

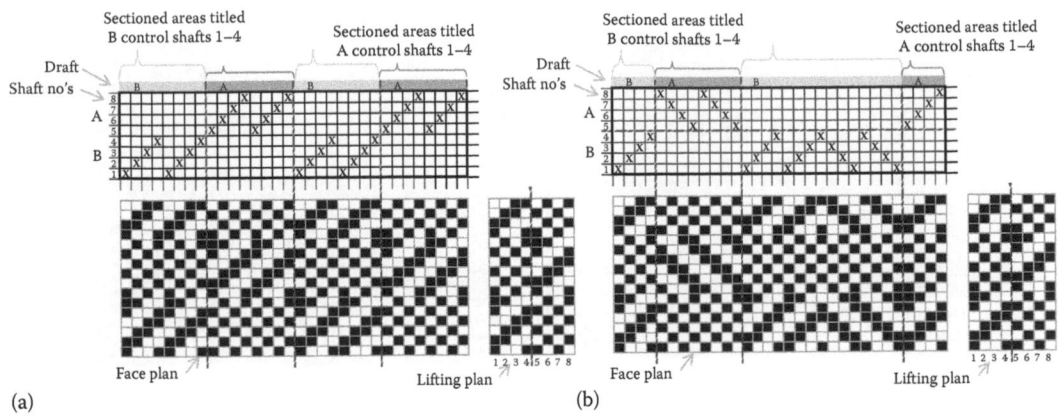

FIGURE 7.26
Block draft, lifting, and face plan: (a) straight draft with balanced division of 8 shafts creating a block draft and (b) straight block draft varying the direction: front-to-back and back-to-front.

The chequerboard effect within the lifting plan will see the common denominator of the plain weave architecture and hopsack weave being used in both the warp and weft direction for the desired number of lifts (weft insertions) before the weaves are swapped over in both directions.

Drafts come in long hand, so include the total number of warp ends, or shorthand, whereby the weaver will know the sequence of repetition of the desired draft configuration. Figure 7.27a shows the shorthand version of a block draft consisting of straight and pointed draft configurations. Through the analysis of the woven fabric the long hand draft is established, then shortened in to repetitive multiplications. When noting the areas of plain weave architectures in the face plan (Figure 7.27b) you will understand that these warp ends have been drawn in over shafts 1 and 2 (or require to be drawn in on the minimum number of shafts) in a straight draft sequence. When noting the drafting of areas

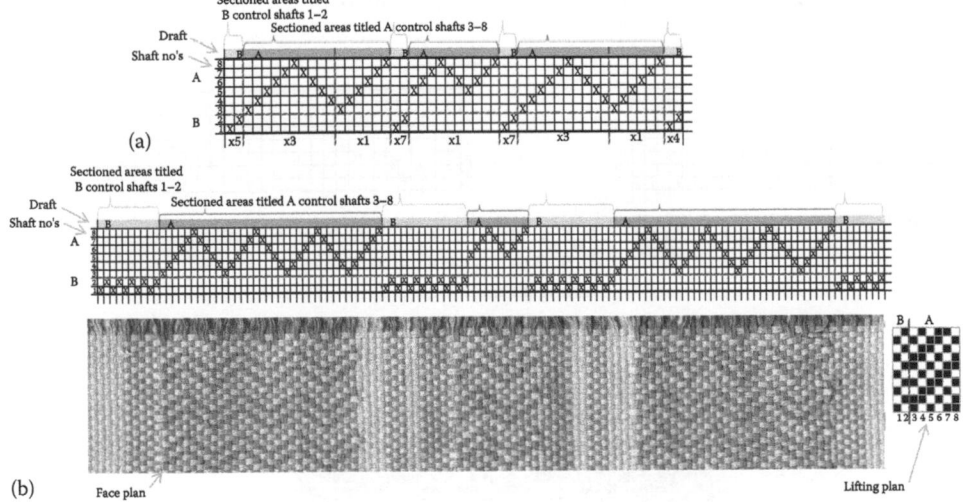

FIGURE 7.27
Block draft variations and associated woven fabric: (a) block draft: consisting of straight and pointed configurations and (b) resulting face plan (actual woven fabric).

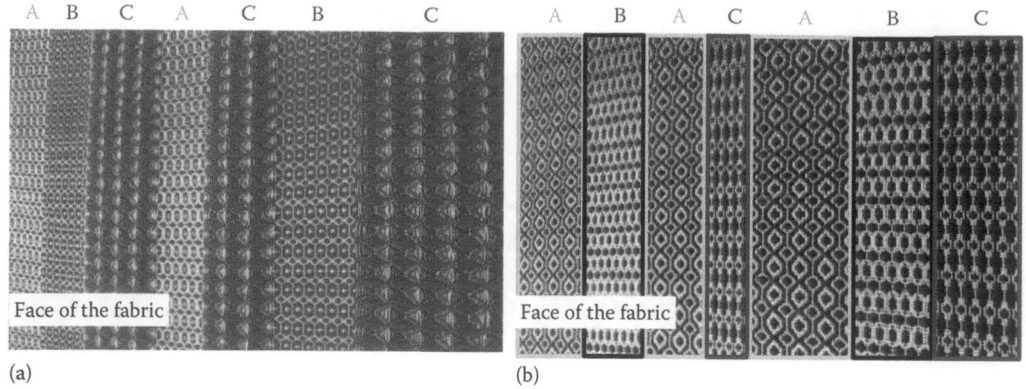

(a)   (b)

FIGURE 7.28
Woven fabrics produced on a block draft: (a) red weft, 3x weave architectures: resulting fabric face and (b) green weft, 3x weave architectures: resulting fabric face. Sectioned areas titled A, B, and C each control a number of allocated shafts. Therefore, three weave architectures can be woven side-by-side and also independently from on another. (Courtesy of Jessica McLaughlin, NTU Graduate.)

consisting of the pointed twill weave architecture in the face plan you will understand that the warp ends have been drawn in over shafts 3-to-8, Figure 7.27a and b. This also supports the understanding of the relationship between the draft, lift, and face plan.

Once the division of the draft/shafts and the lifting plan is understood, greater complex combinations of weave architectures can be achieved; only limited by the number of shafts the weaving technology controls and houses. Figure 7.28a and b would require at least 16 shafts to enable the three different weave architectures to be placed side by side across the weaving width and depth.

The greater the number of shafts housed by the weaving technology, the more complex the motif/pattern possibilities within the weaving cycle. One way to generate enlarged motifs/patterns, however, is to utilize the extended point draft technique. Widening the points within the weaving width *stretches* the motifs/patterns appearance and allows for varying motif geometries within the concluding woven fabric. In Figure 7.29, draft 1, a

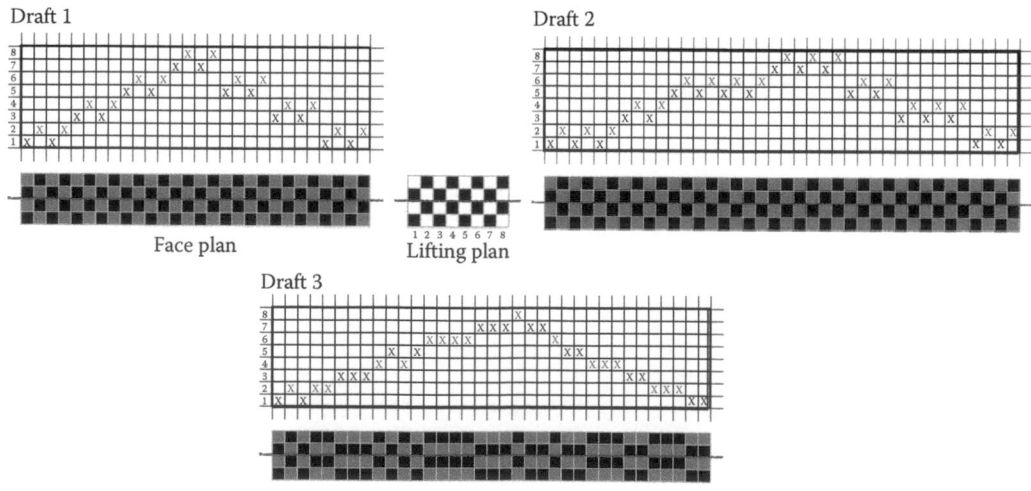

FIGURE 7.29
Extended point drafts.

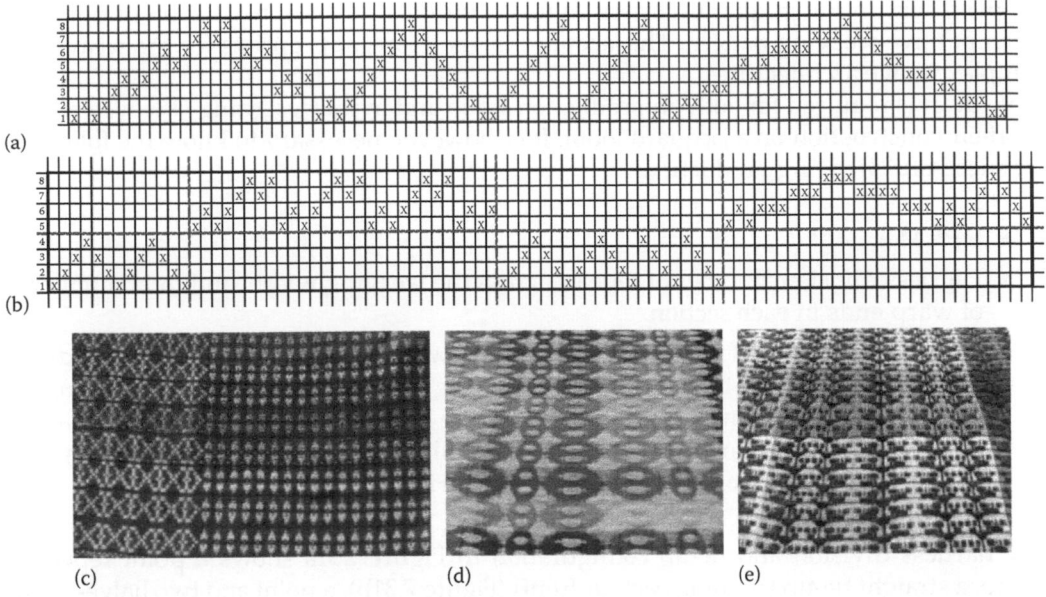

FIGURE 7.30
Combined and block drafts incorporating extended points with woven fabric exemplars: (a) combined extended point, pointed and straight draft over shafts 1–8, (b) block draft, sectioning off shafts 1–4 and 5–8 combined pointed and straight draft over shafts 1–4 ~ combined extended points over shafts 5–8, and (c–e) face of the fabric. (Courtesy of Anna Piper and Emily Brown, NTU Graduates.)

typical sequence of an extended pointed draft can be found, with two other variations, drafts 2 and 3. Using plain weave architecture in the lifting plans for all three drafts, you can see that draft 3 reads the lifting plan very differently from drafts 1 and 2 in the concluding face of the fabric. This is due to the sequence of odd and even warp threads being out of synchronization in draft 3. Greater diversity within a range of woven fabrics can be achieved within drafts 1 and 2, as they eliminate the tramline effect that occurs when utilizing draft 3 due to the arrangement of the warp threads.

The combination of point and straight draft configuration over all the shafts, shown in Figure 7.30a, is an example of how a motif can be halved, completed in full, enlarged, and broken up. The woven fabrics shown in Figure 7.30c–e have been produced utilizing the principles of extended points on both types of draft combinations shown in Figure 7.30a and b.

When creating a draft for the drawing in of warp ends and associated weave architecture combinations, it is advisable to generate a draft that has the flexibility to *read* many weave architectures within the lifting plan—this allows for a greater variation of concluding woven fabrics from the one set up and preparation of the dobby weaving technology.

7.4.2 Jacquard: Harness, Lift, and Face Relationship

When designing weave architectures and subsequent lifting plans for the jacquard loom, it is not essential to understand the relationship between the draft and weave generation. A comprehensive knowledge on the interrelationship of the warp and weft interlockings is important, however. Within the weave generation for the jacquard loom, the warp and weft lift arrangement is identical to the face plan (the face of the woven fabric) when designing single cloth constructions. The jacquard provides individual warp end control in the interlocking of the weft pick. The maximum number of warp ends connect to and supplied within the jacquard harness could be utilized in the construction of one large nonrepetitive weaving pattern. The jacquard is, therefore, appropriate for the weaving of complex weave architectures.

No draft is required for both preproduction setup and in the designing of woven textiles for their construction on a jacquard loom. It is, however, necessary to know the following for successful weaving production:

- E.P.CM.
- Harness division, number of repeats across the weaving width, and total number of warp ends in each section.
- Drawn in direction of warp ends, the tie-up of warp through the eyelet of a heddle, front-to-back or back-to-front, typically starting from the left-hand side. The direction, step sequence and repeat sequence of the harness cords within the jacquard harness tie-up-division dictates the individual warp ends location. Examples of harness tie-up-divisions can be found within Figure 7.31a–d.

The harness division and tie-up configuration in Figure 7.31a shows a point repeating tie-up, a straight tie-up (running back-to-front) (Figure 7.31b), a point and two halves tie-up (running front-to-back) is identified in Figure 7.31c, and a straight tie-up and two halves (borders) is shown in Figure 7.31d. The point tie-up allows for a design in one complete geometrical form. Within the straight tie-up, the same motif used in a point tie-up would be halved. Within the point and two halves, the full motifs will be in the two complete points and half of the motif will be in each of the two halves.

When drawing in the warp ends through the eyelets of the heddles, it is completed typically from left to right. The first warp end from the left-hand side of the harness at the front of the loom is drawn through the first available heddle. The heddles are connected to a chord above and through the first eyelet of the comber board, either at the back or front of the comber board. As the chord travels up from the comber board, it is located securely into the jacquard head. The second warp end (from the left-hand side of the jacquard loom) is then drawn through the second available heddle that is connected to a chord above and through the second eyelet on the comber board that is adjacent to the first eyelet in the column-wise direction. This is then repeated across the total number of warp ends.

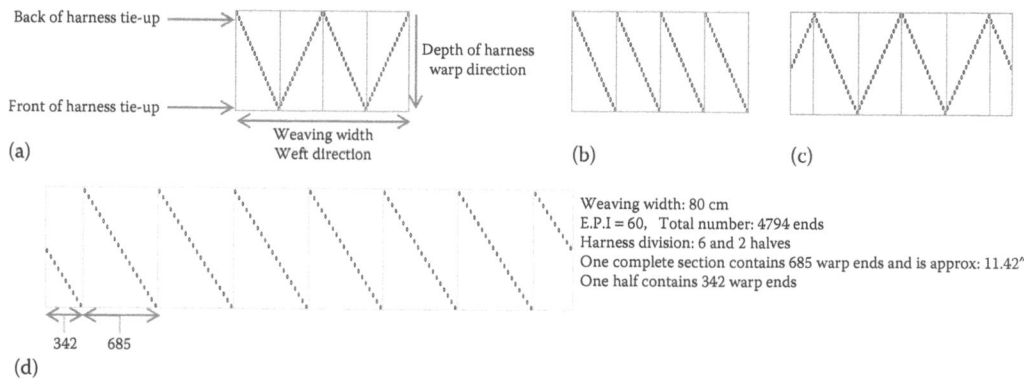

FIGURE 7.31
Jacquard harness tie-up-division configurations: (a) point, (b) straight, (c) point and two halves, and (d) straight with border.

Although the jacquard allows for intricate graphic images, care is required when amalgamating varying weave architectures in terms of the float lengths in the warp and weft direction. Consideration to float lengths in the weave architectures you wish to combine should be given, ensuring even tension as much as possible on the warp yarns as they come off the back-warp beam. Consideration to the warp yarn housing/let-off system is always required, whether creels and/or beams are used.

7.5 Weave Architectures and Notation

The notations of weave architectures are formatted on graph paper. Within Figure 7.32 each square represents the interlocking point of a warp and weft yarn. The column-wise direction represents the warp and row-wise direction represents the weft. When a mark is placed on the graph paper it represents, within the face of the fabric, a warp end that is raised and the weft yarn is inserted underneath. When a square is left blank on the graph paper, it indicates that the warp end is down and a weft yarn is travelling across the warp yarn on the face of the fabric.

To notate the plain weave architecture, first generate the weave in its nonrepetitive state. This entails the minimum number of warp and weft interlocking representations on graph paper and is known as the common denominator; for the notation of plain weave, two warp and weft interlockings are required, shown in Figure 7.33. The next stage, to create the lifting plan, the total number of shafts dictates the number of columns representing the warp direction. The rows represent the weft direction. The total number of rows required are those comprising the lowest common denominator, as shown in Figure 7.33. This is then used in the production of a plain woven fabric with the final weave architecture/lifting plan repeated as necessary.

Please note, when generating the weave architecture/lifting plan for a jacquard, the number of columns on the graph paper do not represent shafts due to single warp end control. Therefore, the total number of columns required in the lifting plan is established

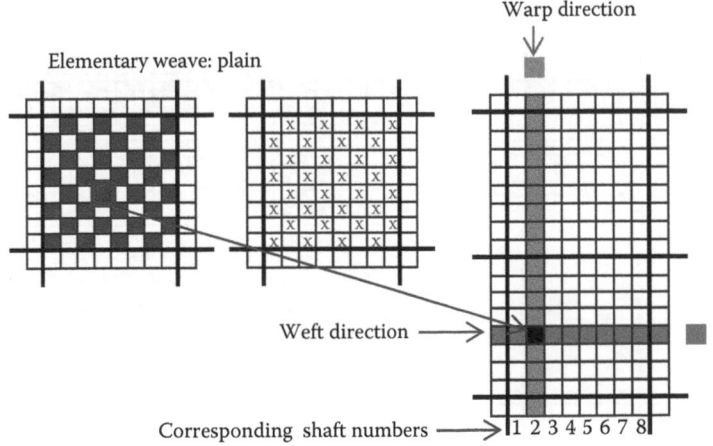

FIGURE 7.32
Plain weave architectures and notation on graph paper.

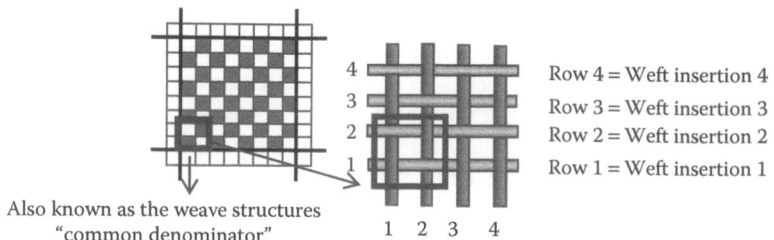

FIGURE 7.33
Weave unit cell of the plain weave architecture.

via the harness configuration: harness division; number of warp ends in each section; and repeat of sections. The number of rows will be derived from the lowest common denominator as in the creation of weave architectures for a dobby loom.

7.5.1 Single Layer: Elementary and Derivatives

The elementary weave architectures include plain, twill, and satin/sateen. No weave architecture is a *new weave*, but a derivative of the elementary weaves. Therefore, there is an exhaustive collection of derivative weave architectures, some of which are:

- Elementary weave plain:
  - The characteristics of elementary and derivative weaves are established from the first warp and weft interlacement of plain weave. It is the tightest weave available and the basis for the orthogonal structure, consisting of two ends and two picks within the unit cell. Derivative weaves of the plain weave architecture include basket, hopsack, and rib.
- Elementary weave twill: S and Z diagonal (which can be seen in Figure 7.34):
  - Twill weaves have many variations of float lengths, such as 2/2, 2/1, 1/2, 1/3, 1/4, and so on. The first number indicates the warp float length over the weft and the second number indicates the warp float length underneath the weft. The appearance of the fabric is that of diagonal twill lines formed by the float lengths. The minimum repeat within a 2/1 twill is over three ends and three

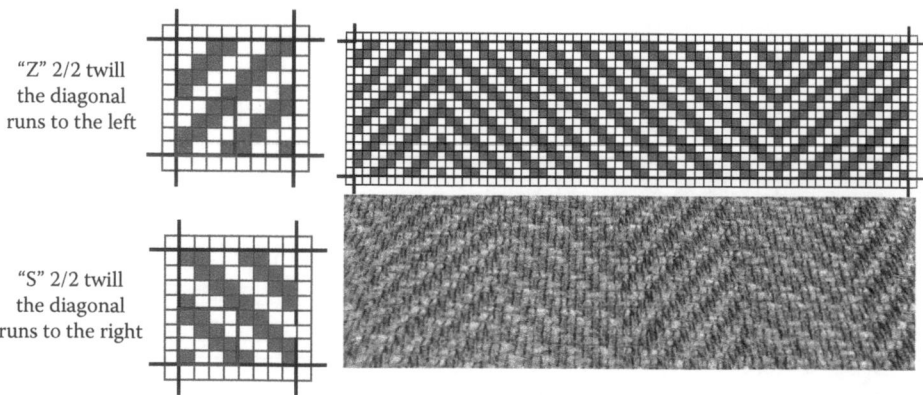

FIGURE 7.34
2/2 twill directions and combination.

picks with a float length over two warp and two weft yarns. Different appearances to the twill lines can be achieved by altering the float lengths within the weft and warp directions in the weave generation, creating derivatives of curvy, steep, and undulating twills. Derivative weaves of the twill include herringbone (Figure 7.34), pointed, diamond, and curved.

- Elementary weave satin/sateen:
  - Satin/sateen weaves are a reversible weave construction that produce a matt (sateen) weft faced appearance on one side and a shiny/silk (satin) warp faced appearance on the other. The weaves are described by their step number, the direction in which the next warp rises over a weft pick. A seven-end satin with a step number of 3 in the warp and weft direction can be seen in Figure 7.35, identifying the loose weave architecture within. The higher the step number over the number of ends and picks within the weave generation, results in a highly drapeable fabric. The satin/sateen weave architecture is created over an odd number of interlocking warp and weft yarns, although a balanced construction (the same number of yarns in both the warp and weft directions). Satin/sateen derivative weave architectures include scattered and satinette, which is generated over an even number of warp and weft directions.

The notation of a satin/sateen is achieved using the clock system, shown in Figure 7.36, which identifies the step sequence, when the warp yarn will be lifted. Therefore, if designing more than 7 warps and 7 wefts, the step number can be 1, 3, or 5, as it must always be an odd number. In Figure 7.36, the step number is 3. The clock therefore shows 7/3 to indicate the weave architectures number of warp and weft yarns; 7 is the total number of interlocking warp and weft yarns and 3 is the step sequence on each weft insertion (row direction) the warp yarn is to be lifted.

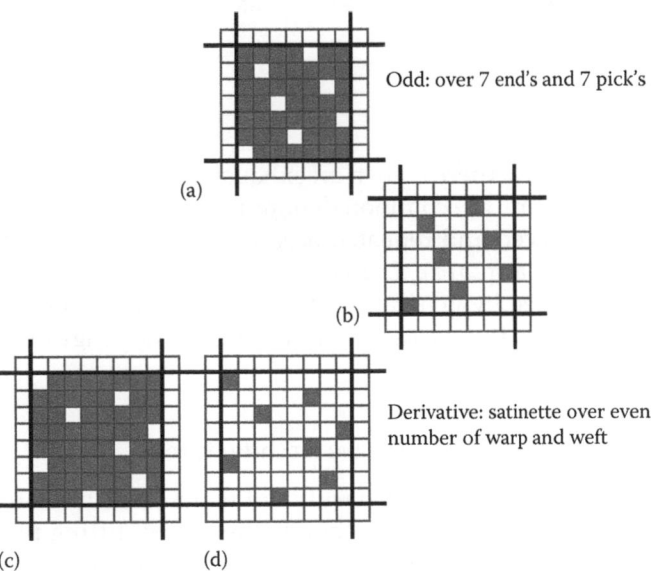

(a) Odd: over 7 end's and 7 pick's

(b)

Derivative: satinette over even number of warp and weft

(c) (d)

FIGURE 7.35
Satin/sateen weave architecture: (a) satin: contains an "a" = warp faced weave, (b) sateen: contains an "e" (or rather two) = weft faced weave, (c and d) satinetter over 8 end's and 8 pick's.

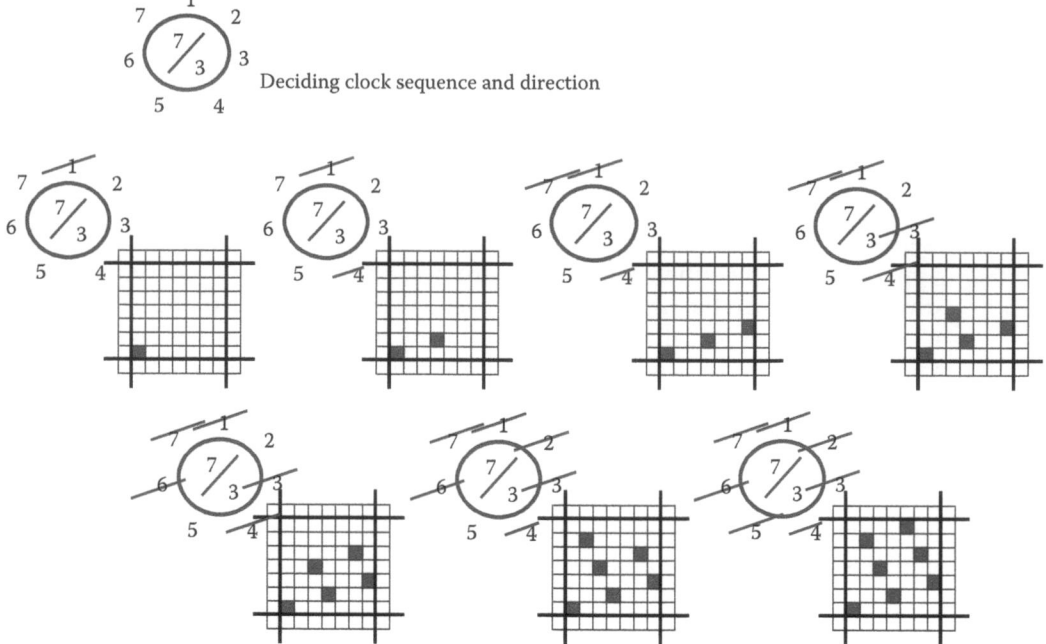

FIGURE 7.36
Satin and sateen clock step sequence.

LEARNING ACTIVITY

Try out the sateen clock sequence above to generate 7 ended sateen weave architecture as shown in Figure 7.36.

From the elementary weaves, there is then an exhaustive collection of derivative weaves. For every weave architecture, the common denominator and the pattern are repeated, as a weave unit cell is produced and repeated according to either the draft (dobby looms) or harness set-up/tie-up (jacquard). The complex weave architectures such as that in Figure 7.37 can be simplified by generating the weave unit cell and understanding how this will then repeat across the weaving width and up the length of the warp/woven fabric.

7.5.2 Figuring Yarns within Ground Weave Architectures

When motifs are required, the figurative weft and warp principles allow for geometric patterning within the woven textile. To generate the correct lifting plan, first design an appropriate draft, such as that in Figure 7.38a and then design the motif/figure over the number of shafts associated to the weaving technology and completed draft, as

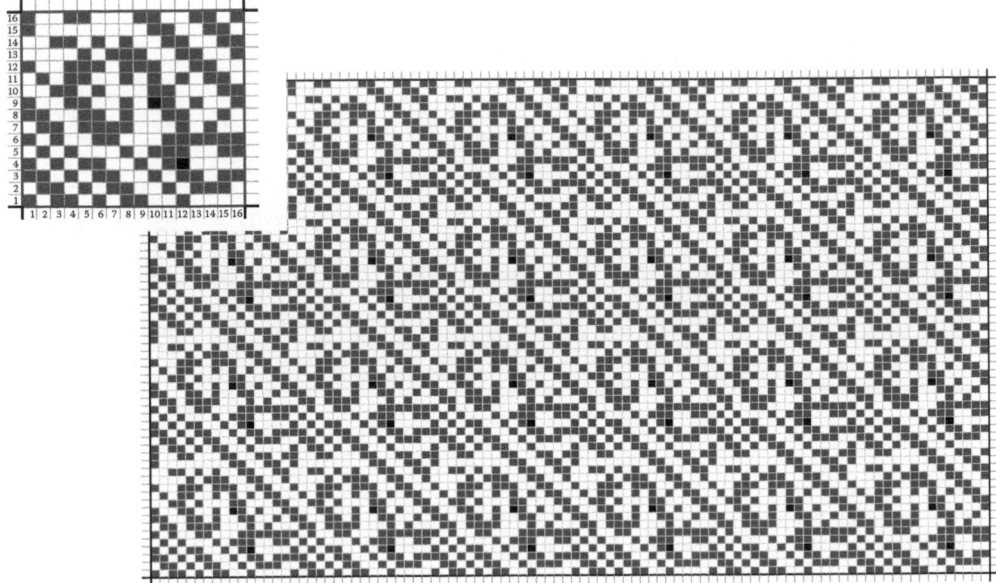

FIGURE 7.37
Weave unit cell of fancy complex weave architecture.

in Figure 7.38b. If producing figuring weft on a jacquard, the draft is not required. Design a weave architecture for the ground weave, as shown in Figure 7.38c, a plain weave architecture has been chosen. Then, over a weft ratio of one ground and one figuring weft amalgamate the two designs alternate within the lifting plan, identified in Figure 7.38d, enabling the fabric, as shown in Figure 7.38e, to be woven. Figuring warp is based upon the same principles; however, the figuring warp must be on its own warp let-off beam.

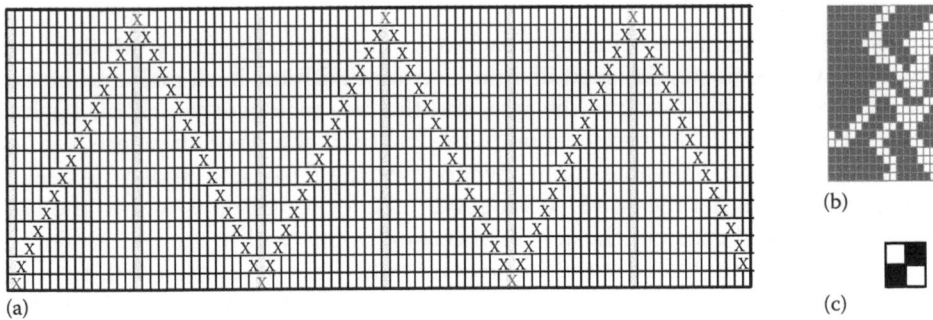

FIGURE 7.38
Figurative weft exemplar (draft, lift, and face): (a) pointed draft over 16 shafts, (b) figurative weft design is the generation of a motif over 16 shafts and 16 warp ends, (c) generate an appropriate ground weave architecture over the minimum number of warp and weft yarns. *(Continued)*

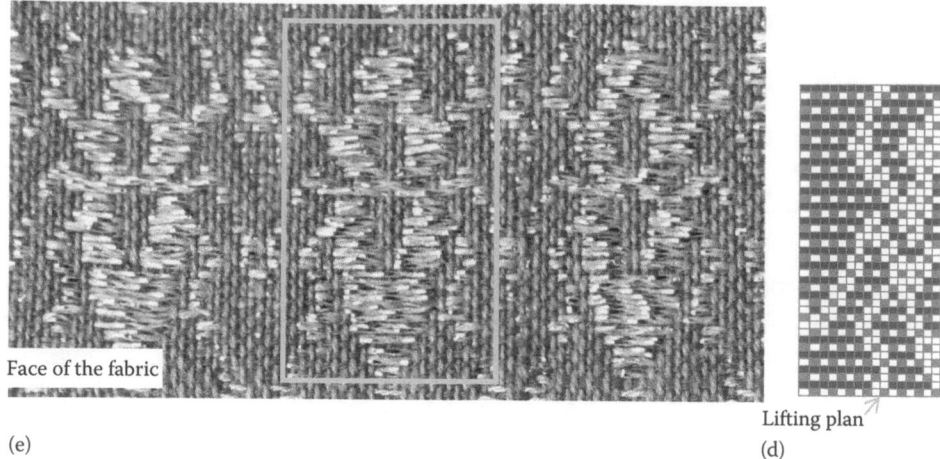

Face of the fabric

Lifting plan

(e)   (d)

FIGURE 7.38 (Continued)
Figurative weft exemplar (draft, lift, and face): (d) combine the figurative weft design with the required ground weave architecture into square paper, the lifting plan. Figurative weft insertion (red squares/rows). Ground weave weft insertion (grey squares/rows). Use a slightly thicker yarn than the ground weave architecture weft yarn. (Courtesy of Georgia Hughes, NTU Graduate.)

7.5.3 Compound: Structures and Amalgamation of Layers Levels

Designing a double cloth is the integration of two fabrics, each with its own warp and weft interlocking yarns, distinguished as the face cloth and back cloth to provide clarity when locating stitching possibilities. For aesthetic reasons, the stitches are covered by an appropriate yarn within the fabric level the stitching yarn is travelling to; therefore, compatible weaves are required. For instance, combining two 2/2 twill fabrics when lifting a back end up on the face pick will allow the back-end float to sit between two face ends, providing coverage for the back cloth. There are a variety of stitching possibilities, including the use of separate stitches in the weft or warp direction. The cloths can be integrated through interchanging the fabrics by bringing all the back ends up to weave with face picks for a desired length before returning to weave with the back picks. Interchanging a defined number of back ends to weave with the face picks will alter the appearance of the cloth.

Goerner (1989) states when bringing a back-warp up to join the face cloth that the back end should have previously finished its interlacement with the back pick and be floating above a back pick when lifted into the face, and the face pick over which the back end is raised must be within its defined location of the 2/2 twill at a position where it is floating underneath a face pick. The same applies when lowering a face end under the back picks. Therefore, when bringing the two levels together, the locating points within the two fabrics for stitching must consider the affects at certain points within the lifting plan on the appearance both on the surface of the fabric and on the reverse (Watson, 1947).

In Figure 7.39a–c, the weave generation details the typical approach to designing double cloths (the weave architecture and lifting plan only) and would have a cross section of either Figure 7.39d, as separate components or Figure 7.39e when looking in the weft direction on the side of the completed weave architecture known as the cross section.

Schematic in Figure 7.39a and b outlines the weave generation of a single cloth, with the individual cross section of the face and back cloth shown in Figure 7.39d, although this only shows the first warp end of each fabric layer. The complete cross section of

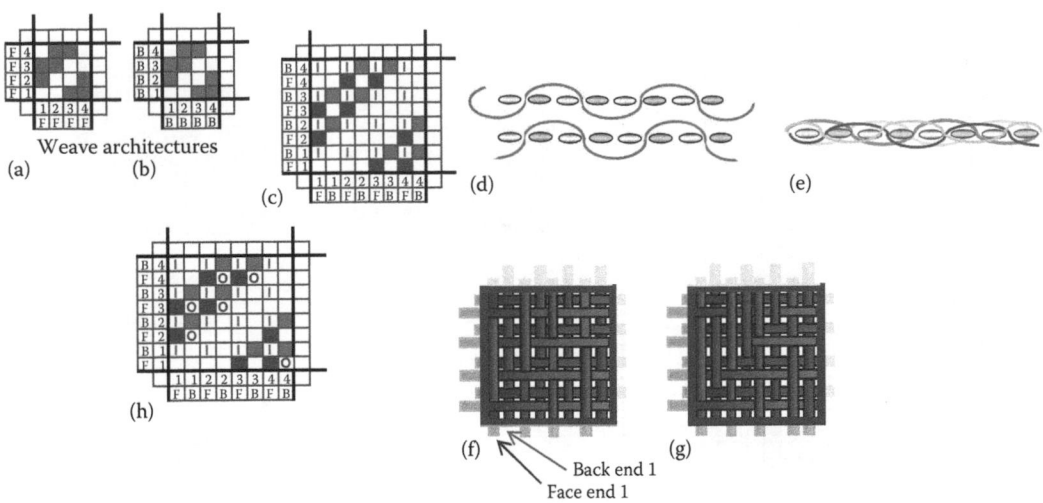

FIGURE 7.39
Detailed weave generation of a 2/2 twill double cloth construction: (a) face of the fabric/top cloth, (b) back of the fabric/bottom cloth, (c) integrating the top and bottom cloth together: The symbol: I represents the face warp ends to be lifted when a back weft pick is inserted to weave with a back warp end, (d) cross section showing the first warp end of the top cloth going under and over two weft insertions, and the first warp end of the bottom cloth, (e) cross section showing through the weft direction the first two warp ends of both the top and bottom cloth (totalling four warp ends), (f) schematic of the two fabrics, top and bottom not integrated, (g) schematic of the two fabrics, top and bottom fully integrated, and (h) integrating the top and bottom cloth together: The symbol: o represents the face warp ends to be lifted when a back weft pick is inserted to weave with a back warp end.

single cloth 2/2 twill, detailing all four ends within the single layer level construction, is given in Figure 7.39e. Figure 7.39f identifies the first end of both the face and back cloth. These cloths are still separate from one another as shown in the lifting plan in Figure 7.39c. In Figure 7.39g, the fourth back cloth warp end stitches up onto and over a face weft pick. The lifting plan for this is shown in Figure 7.39h. In the weaving process, this back-warp end up on the face stitch will be covered by the first warp end of the face cloth as the weave architecture repeats across the fabric width, and if both fabrics are of an equal balanced sett.

The schematic Figure 7.40 shows the development of the lifting plan and understanding of warp, weft interlocking, and stitches for a 2/2 twill triple layer woven fabric. As in the double layer woven, in Figure 7.39, each layer level of the triple cloth is allocated its own weave architecture (Figure 7.40a–c). These weave architectures are then combined to allow for a weaving ratio of 1:1:1, and the stitch mark indicators are then inserted (Figure 7.40d and g). Figure 7.40g and h shows the cross section (in the weft direction) of the triple cloths as separate components. Figure 7.40i shows the stitch indicators and how the integration of three woven fabrics will be completed. Figure 7.40f shows the face of the fabric and how the stitches are formed—once each weft is beaten into the fell of the woven fabric, the stitches will bunch slightly into one another and will not impact the aesthetics of the triple layer woven fabric.

The drafting of double cloths begins in the same way as the single woven fabric draft by outlining the number of rows/shafts and columns/warps required. The division of the available shafts can be achieved via many configurations, such as those shown in Figure 7.41.

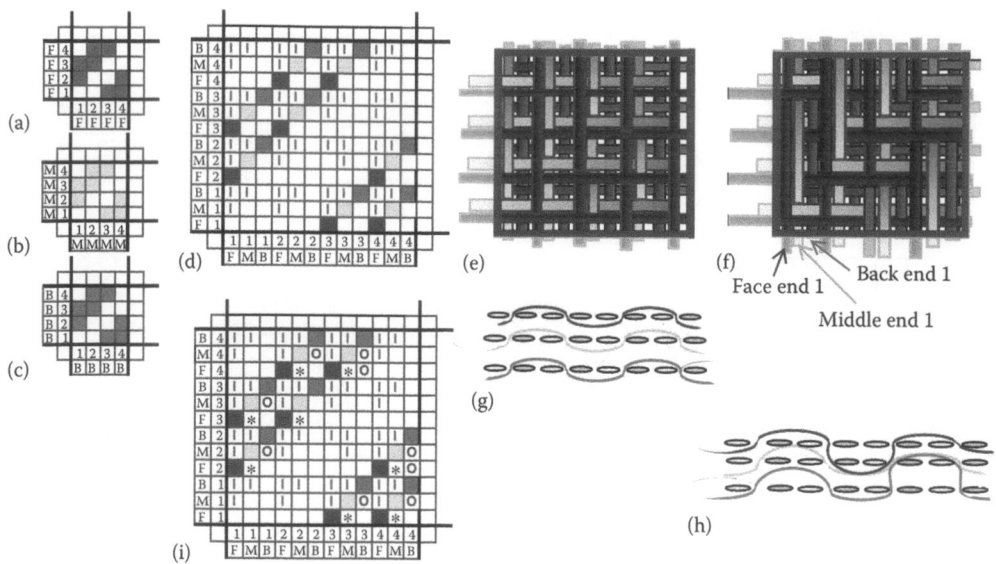

FIGURE 7.40
Weave generation of a 2/2 twill triple cloth construction: (a) 2/2 twill, (b) 2/2 twill, (c) 2/2 twill, (d) combining three layers-levels with warp interlocking (stitches), (e) three independent layers-levels, (f) three interlocked layers-levels, (g) cross section: combining three layers-levels with warp interlocking (stitches), and (h) cross section: combining three layers-levels with warp interlocking (stitches and next step).

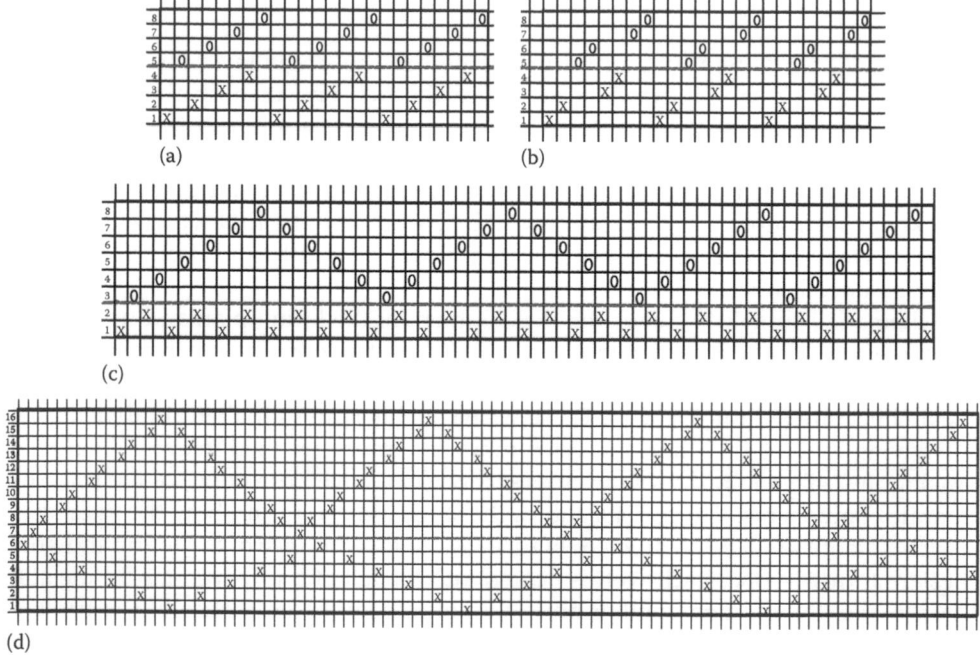

FIGURE 7.41
Exemplars of the division of shafts for compound structure drafts (double cloth): (a) division of shafts: straight draft, ratio 1:1, over shafts 1–4 and 5–8, (b) division of shafts:straight draft, ratio 2:2, over shafts 1–4 and 5–8, (c) division of shafts: ratio 1:1, over shafts 1–2 and 3–8, and (d) division of shafts: ratio 1:2, over shafts 1–6 and 7–16.

For ease of understanding the drafting process for compound structures (two or more integrated warp/weft-layer/levels) the examples provided in the following discussion focus upon straight and pointed drafts. Taking the draft in Figure 7.42a, the red dividing line is to show a draft containing four shafts (1-to-4) for the bottom warp yarns of the to be woven fabric and the draft over shafts 5-to-8 for the warps of the to be woven top cloth. This draft is also a ratio of 1:1. This means that when drawing in the warp yarns through the heddles on the shafts (setting up the loom) and to understand the lift plan, the first warp end of the bottom cloth is drawn through the first available heddle on shaft 1; the second warp yarn is then drawn in the first available heddle on shaft 5, which is the first yarn of the top cloths warp. This is then repeated on the draft configuration chosen, in this instance, a straight draft containing a ratio of 1:1. The draft shown in Figure 7.43a is still a straight draft, but it has a ratio of 2:2, two warp ends for the bottom cloth followed by two warp ends for the top cloth—remember that each column in a draft *houses* only one warp thread!

The drafts in Figures 7.41a–d and 7.42a contain a top warp and bottom warp ratio of 1:1 and are variations on combined point and straight draft configurations, and combined point draft configurations on differing number of shafts for the top cloth compared to the bottom cloth warp yarns.

Dividing the shafts further within a draft, but still only for the integration of two warp layers and two weft levels, the top warp could be drawn in over shafts 1-to-6, shown in Figure 7.44a and the bottom cloth warp yarns drawn in over shafts 7-to-16. This allows

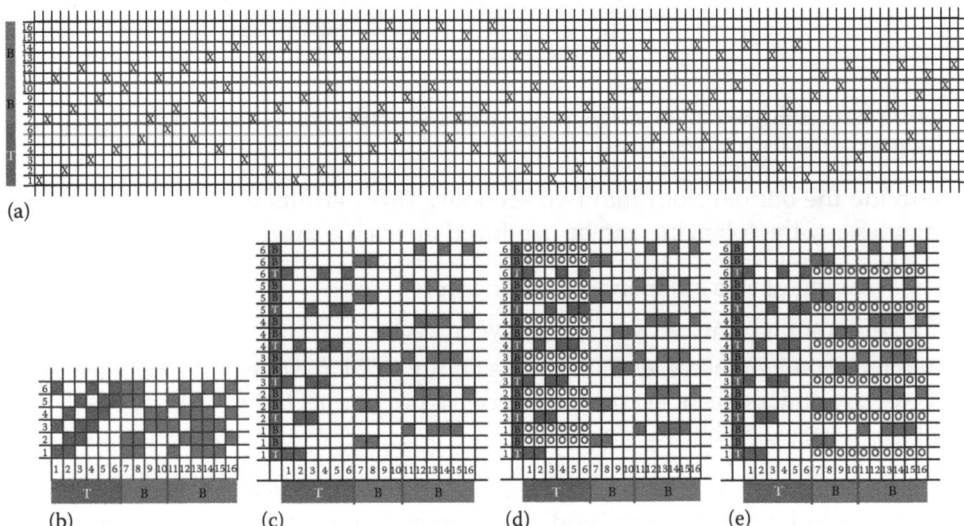

(a)

(b)　　　　(c)　　　　(d)　　　　(e)

FIGURE 7.42
The sectioning of shafts permitting the control of three independent weave architecture designs in the lifting plans within a compound structure (double cloth): (a) divided and combined draft, 1:1 warp ratio. Division of shafts for compound structure (for two integrated woven fabrics). This provides the opportunity to place two weave architectures side-by-side within the bottom fabric and/or weave them independently, (b) first Stage: plan out the weave architecture for each resulting division-section, (c) second Stage: beginning on the first row in the lifting plan transfer the weave architecture(s) from the first stage; weft ratio 1:1:1, (d) third Stage: Insertion of lifter marks, and (e) changing the fabrics positioning, TOP cloth now weaving as and on the bottom, BOTTOM cloth now weaving as and on the top.

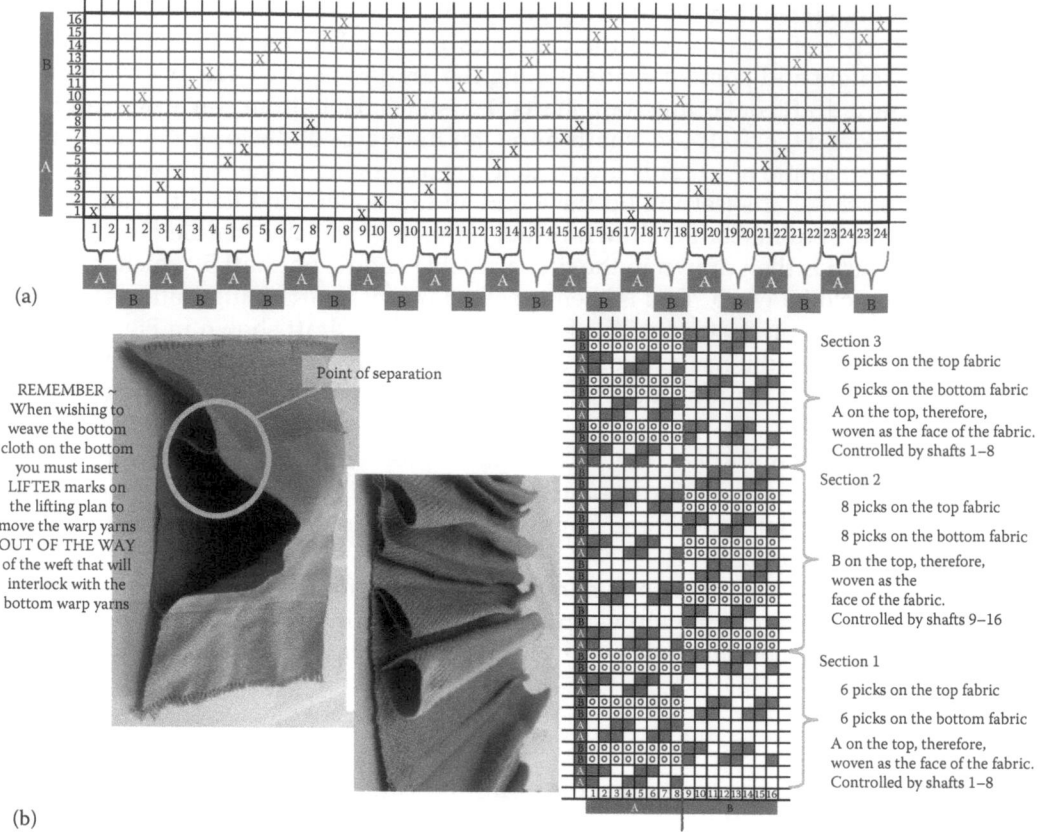

(a)

(b)

FIGURE 7.43
Exemplar of a 2:2 drafting ratio for compound structures (double cloth): (a) draft and (b) interchanging double cloth. (Courtesy of Elizabeth Murray-Jones.)

you to divide the bottom cloth into two sections. This permits flexibility in the weaving process, whereby the following can be produced on such a draft:

- The weaving of two independent fabric layers levels.
- The integration of two woven fabric layers levels.
- Two independent layers levels with the ability to weave one of the two layers levels as two independent sections; either apart or connected as one, but each with different weave architectures.

In Figure 7.42d and e, the process of generating a lifting plan for the weaving of the two warp layers as independent cloths and the bottom cloth containing two weave architectures is shown. First, design the weave architectures for the three sections as if it was a single cloth (Figure 7.42b). Second, using a ratio of 1:1:1: one weft insertion for the weave architecture on the top cloth; one weft insertion for the bottom cloth; a second weft insertion for the bottom cloth, insert the desired weave architectures into a lifting plan, as shown in Figure 7.42c. When weaving the two separate weft insertions on the to be woven bottom fabric, you must put lifter marks in all the shafts belonging/controlling the top warp yarns—this ensures the top warp yarns are not amalgamated into the design and weaving of the bottom cloth, as identified. The final lifting plan in Figure 7.42e allows you

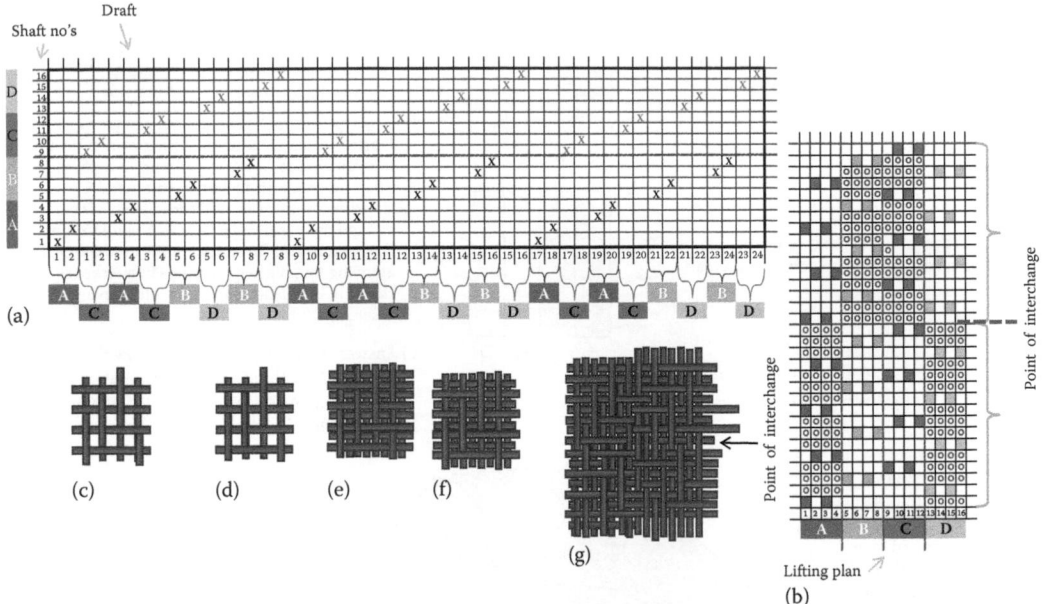

FIGURE 7.44

Compound structures: chequerboard (c) schematic of the completed woven layer level section A–B as a single cloth, (d) schematic of the completed woven layer level section C–D as a single cloth, (e and f) schematic (e) of the two completed woven's as single fabrics, whereby sections C–D are on the top; in (g) sections A–B is on the top, and (g) schematic (e) of the two completed woven's as single fabrics, whereby sections C–D are on the top; in (g) sections A–B is on the top.

to bring the to be woven bottom warp yarns from the bottom and weave as a top cloth; this then also allows you to take the to be woven top warp yarns from the top and weave as the bottom cloth. This allows you to interchange the two different warp/weft woven fabrics, shown in Figure 7.43b and c, whereby the two independent woven fabrics are connected by *switching* places and not with stitches (Figure 7.43d).

The integration of two independent woven fabrics can also be achieved within the weaving of a chequerboard effect fabric, shown in Figure 7.44. The draft to be used for weaving a chequerboard is the same as that used for the interchanging tubes. The only difference between the draft for the interchanging tubes and that for the chequerboard fabric is that it is now divided further into four sections. This is to allocate the lettering ABCD to the sections to enable simplification of understanding, and creation of an appropriate lifting plan. Shafts 1–4 are known as A, shafts 5–8 are known as B, shafts 9–12 are known as C, and shafts 13–16 are known as D. Within Figure 7.44, the gray area is known as the top cloth and the pink area is known as the bottom cloth. Enabling the understanding of the creation of an appropriate split draft and lifting plan you can maneuver various sections within the warp direction across the weaving width of the fabric.

The lifting plan in Figure 7.44b provides the principles of weaving a chequerboard fabric, which can then be used to advance knowledge and creation of further weave architectures. To begin the creation of the lifting plan, first decide upon a weave pattern; here plain weave is used in both the top and bottom fabrics. Then decided upon a weft ratio; in the lifting plans shown in Figure 7.44b, a ratio of 2:2 is used. For the first 16 weft insertions, rather than having a consistent gray weft and warp, or consistent pink weft and warp interlocking, as in the interchanging tubes, we now want to create a variety of areas of

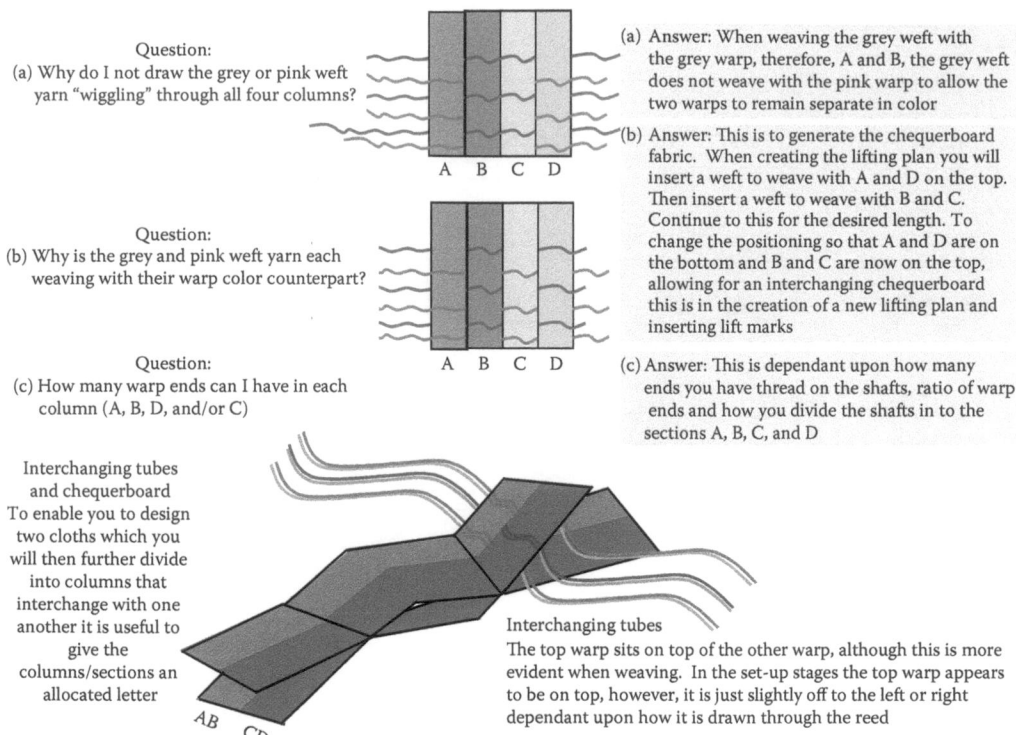

Question:
(a) Why do I not draw the grey or pink weft yarn "wiggling" through all four columns?

A   B   C   D

Question:
(b) Why is the grey and pink weft yarn each weaving with their warp color counterpart?

A   B   C   D

Question:
(c) How many warp ends can I have in each column (A, B, D, and/or C)

(a) Answer: When weaving the grey weft with the grey warp, therefore, A and B, the grey weft does not weave with the pink warp to allow the two warps to remain separate in color

(b) Answer: This is to generate the chequerboard fabric. When creating the lifting plan you will insert a weft to weave with A and D on the top. Then insert a weft to weave with B and C. Continue to this for the desired length. To change the positioning so that A and D are on the bottom and B and C are now on the top, allowing for an interchanging chequerboard this is in the creation of a new lifting plan and inserting lift marks

(c) Answer: This is dependant upon how many ends you have thread on the shafts, ratio of warp ends and how you divide the shafts in to the sections A, B, C, and D

Interchanging tubes and chequerboard
To enable you to design two cloths which you will then further divide into columns that interchange with one another it is useful to give the columns/sections an allocated letter

AB   CD

Interchanging tubes
The top warp sits on top of the other warp, although this is more evident when weaving. In the set-up stages the top warp appears to be on top, however, it is just slightly off to the left or right dependant upon how it is drawn through the reed

FIGURE 7.45
Compound structures: interchanging tubes and chequerboard.

gray and pink on the top and bottom cloth. One tip though, when weaving the gray warp, use a gray weft, when using a pink warp, use a pink weft; this enables a true chequerboard fabric to be woven and provides a clearer system when developing these technical skills, identified in Figure 7.45.

In the lifting plan in Figure 7.44b, the first weft insertion is for the weaving of warp sections labeled A and D as the top fabric and weaving warp sections C and B as the bottom fabric. Therefore, the insertion of lifter marks in the notation of the lifting plan are required, being the purple squares in the first weft insertion row in section D on shafts 7–8. This is so all warp yarns associated to area D do not interlock with a bottom warp/fabric weft insertion. The second weft insertion in Figure 7.44b shows the weaving of the top cloth still in the weaving process; however, section D is being woven separately from section A. Section D is originally a warp section from the bottom cloth, identified by the splitting of warp colors into stripes. The insertion of lifters in the second weft insertion row in area A are required to ensure the warp yarns do not interlock with a bottom warp weft insertion.

When weaving the third and fourth weft insertions, these are both a bottom warp interlocking yarn; interlocking with warp sections B and C. To ensure the warp yarns we wish to remain on the top, in this case warp sections A and D are out of the way and remain on top we insert lifters when planning the third and fourth weft insertion rows in warp sections A and D which are on shafts 1–8 and 9–16.

Designing triple cloth fabrics requires the understanding of the stitching possibilities and locating the top, middle, and bottom fabric layers levels and correctly weaving face ends on face picks, center ends on center picks, and back ends on back picks. The variations in interlocking the layers levels are the back fabric to center fabric, center fabric to face fabric, face fabric to center fabric, center fabric to back fabric, back fabric to center fabric, face fabric to center fabric (Watson 1947; Goerner, 1989). As in the double cloth, each layer level within the triple cloth can contain different weave architecture. The nature of designing compound structures, double and triple cloth constructions introduces the principles of weaving ratios. When weaving the face fabric, two picks could be inserted before inserting a center weft pick on a center warp end, followed by a back weft pick on a back-warp end, resulting in a weaving ratio of 2:1:1. Weaving a 1:1:1 ratio concludes in the weaving and interlocking of one face warp and one face weft; one center warp and one center weft; one back-warp; and one back weft. Then repeat. Increasing the density of the center fabric to that of the face and back cloths, locating the correct stitching points allows the weaving of spacer or hollow structures found within the technical textiles field.

Remember, when designing a lifting plan, *you* are in control of when a warp yarn is to be lifted!

7.5.4 Technical Weave Architectures

Knowledge of generating weave architectures and lifting plans for single layer woven's and compound structures provides the foundations to designing/producing multilayer multilevel weave architectures. The traditional way of integrating a minimum of two woven fabric layers levels together in the weaving process is to use stitches hidden underneath the face of the cloth, as discussed in Section 7.5.3. The hidden stitches are a requirement as to not distract from the desired aesthetics of the concluding woven fabric. Technical weave architectures were established due to their performance capabilities aided by suitable yarns and associated characteristics. Therefore, aesthetics were/are not a necessity in the generation of technical weave architectures. In replacement of warp/weft directional stitches, a through-the-thickness yarn now integrates the numerous warp layers and weft levels, providing a three-way yarn system: warp, X direction; weft, Y direction; and warp, Z through-the-thickness direction. The role of the through-the-thickness yarn is to integrate and produce a superior woven multilayer multilevel structure that meets all requirements of the intended end application.

7.5.4.1 Cross-Sectional Design and Notation

The cross-sectional weave provides a manual weave generation process for the visualization and tailoring of the warp, weft, and through-the-thickness yarn locations for 2D and 2D-to-3D woven's (Taylor, 2007). The cross-sectional process uses the same principles of unit cells, weave repeat (if required), woven fabric dimensions, and warp weft ratios to traditional generation of weave patterns.

To design the plain weave architecture using the principles of the cross-sectional process the first warp end would be identified within a schematic, Figure 7.46a. Details of the warp maneuvering over and under the weft and the lifts for the first column and defined number of weft picks and warp end 1 are given in Figure 7.46a–c. The face plan of the individual end within Figure 7.46c identifies the next design stage of locating warp end 2 travelling in the alternative direction to then complete the cross section and weave generation

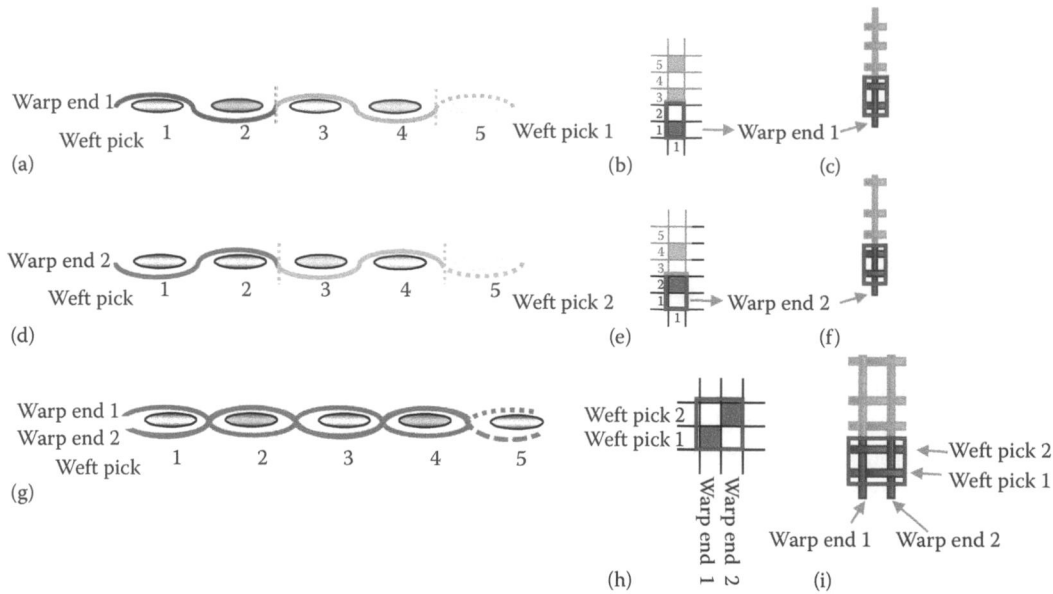

FIGURE 7.46
Cross-sectional slices, stepping arrangement, and concluding lifting plans: (a) first warp end, (b) generated lifting plan, (c) generated face plan, (d) second warp end, (e) generated lifting plan, (f) generated face plan, (g) cross-section of first and second warp end, (h) generated lifting plan, and (i) generated face plan.

shown in Figure 7.46d–f. Combining the two warp ends and picks produces the full weave unit cell repeat, identified in the cross section in Figure 7.46g and in the subsequent weave generation in Figure 7.46h–i.

When the idealized warp and weft yarn paths have been finalized within the cross section of the warp and weft, the formation of the interlocking linear ends and picks must employ a stepping arrangement within the resulting weave generation to produce an integral fabric. The employment of a step sequence of one allows subtle movement and integrity within a horizontal direction across the weaving width. In Figure 7.47a, the first cross section defines the roles of the first four warp ends and 48 weft picks. The following cross sectional arrangements derive from a step sequence of 1, which can be either up one and over to the right direction of the to be woven fabric; down one and over to the right direction of the to be woven fabric; and then generate the final weave generation, such as that shown in Figure 7.47b.

The ability to use the coordinate warp and weft system to pin point an individual yarn within the linear warp and weft of the cross-sectional weave enables the alteration, if required, of the located yarns positioning, both within the cross-sectional weave and within the large-scale 2D weave generation. This is achieved through numbering the weft/warp coordinates within each cross-sectional slice. These numbers are placed alongside one another to create the total number of ends and picks within the allocated segmentation. Once the 2D schematic has been defined within the total number of warp ends and picks, which concludes the weaving length and width of the nodal structure, then the coordinates of the yarns can be established from the concluding weave architecture. Figure 7.47c is the final weave generation detailing different marks that correspond to the interlocking of each end and pick within the complete cross section. It is repeated in Figure 7.47d in black and white to aid the understanding of which warp ends are lifted. It also indicates the complexity of designing within a 2D visualization format.

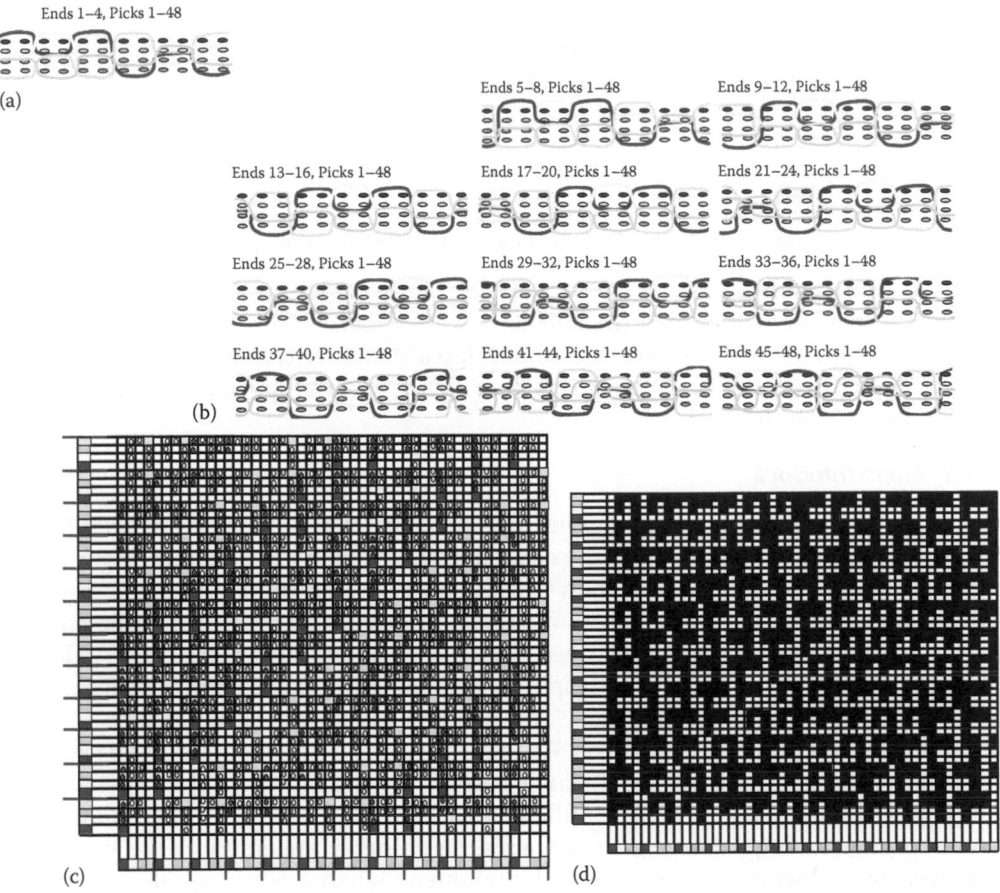

FIGURE 7.47
Cross-sectional slices, stepping arrangement, and concluding lifting plans: (a) completed cross-sectional weave slice, (b) using a step sequence of one the completed cross-section is completed once step (a) is required again, (c) completed lifting plan generated from the complete cross-sectional step sequence (ends 1–48), and (d) black and white lifting plan to aid warp ends lifting within the weaving process.

7.5.4.2 Orthogonal

The orthogonal woven architecture is categorized as a 3D structure containing warp, weft, and through-the-thickness yarns; a stronger weave architecture with minimal crimp within is produced, which can be seen in Figure 7.48a. The interlocking and resulting crimp is however on the outer top and bottom surface. The through-the-thickness yarn penetrates through the desired number of levels (weft direction) within the woven's wall thickness (the comprising number of integrated warp and weft layers and levels) for a desired length before reentering, binding the structure together. The orthogonal structure is a derivative of the plain weave, although instead of two warps locking around one weft, in the analysis of the cross section, two warps lock within the depth of the wall thickness (Figure 7.48a–c). The structure is compact with little room for movement and drapeability between the layers levels due to the high packing content of the weft. Modifying the

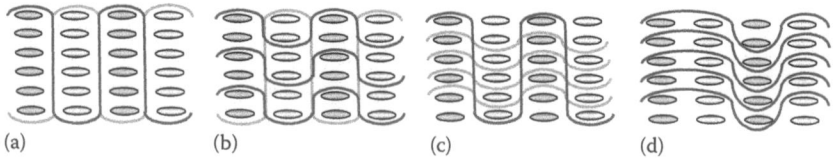

FIGURE 7.48
Orthogonal weave architectures: (a) binder warp through-thickness, (b) binder warp through to adjacent layer-level, (c) binder warp through-thickness plus stuffers, and (d) binder warp through-thickness.

orthogonal to contain a more even ratio of warp to weft (Figure 7.48d) will allow more flexibility within the woven wall thickness.

In Figure 7.48d, the orthogonal starts to develop a structure that is not as compact as its predecessor (Figure 7.48a) and provides a base weave to develop further long float orientated weave architectures.

7.5.4.3 Angle Interlock

Within the angle interlock weave architecture, the warp now interlocks the weft levels as the yarn travels at an angle through the desired woven wall thickness, allowing minimum bending and extensibility within the warp direction due to the length of elongation of the warp, which can be seen in Figure 7.49a. The properties are enhanced when a stuffer yarn, or stack of stuffer yarns, is placed between these binding warp yarns, identified in Figure 7.49b. When using the principles of the angle interlock within weave architectures classified as multilayer multilevel, the shaping of fabrics can take place—typically used in the production of 3D solid and 3D hollow woven structures. The angle orientation of the interlocking warp aids the forming of a 2D structure within the constraints of a loom within the weaving cycle to be pulled into a 3D woven structure once removed from the weaving technology. The angle orientation, stepping arrangement of the yarn among the woven wall thickness, elongates and straightens when force is applied to form the required 3D shape (Taylor and Chen, 2016). Figure 7.49c and d are derivatives of the angle interlock in Figure 7.49a.

The more weft layers within an angle interlock structure decreases the extensibility of the structure and increases the tensile stiffness (the degree of difficulty with which a structure can be extended by a tensile load); this is due to the straight weft picks taking the load. The breaking point when the warp yarns fully decrimp transfers the pressure to the already straight weft yarns. The breaking point relies on the correct placement of the yarns crimp, which will reduce the warp strength while easing the pressure on the weft, allowing the woven preform to take the load. The shear rigidity is related to the weave architecture, warp tension, and yarn type. Therefore, before the textile designer can begin to generate such complex weave structures, knowledge of all the characteristics of the weave architectures and loom capabilities are required (Taylor and Chen, 2015).

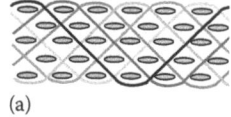 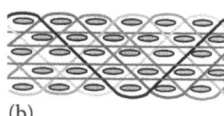 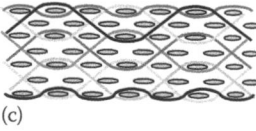 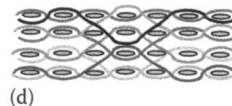

FIGURE 7.49
Angle-interlock weave architectures: (a) standard angle-interlock, (b) standard angle-interlock with stuffers, (c) angle-interlock variation, and (d) angle-interlock variation.

7.5.4.4 Multilayer Multilevel

Deriving from the principles of compound structures, the multilayer, multilevel weave architectures entail the through-the-thickness yarn to act as a stitching yarn, interlocking all or part of the structure together. Multilayer is the term associated to the numerous interlocking warp direction yarns. Multilevel is the term associated to the interlocking weft direction yarns. The stitching yarn traverses across the bottom, top, or through the desired layer level for a specified length before reentering the fabric layer level to return to its original layer level location. When designing multilayer multilevel fabrics, each layer level is dealt with, and designed individually, and then the designs are placed together with allocated stitching points. These binding/stitching warp yarns must be covered by the above weave architecture, so the appearance of the structure is not affected. In the manufacture of textile, however, preforms aesthetics is not the priority but the overall performance. Multilayer, multilevel weave architectures entail a combination of any yarn directional placement and interlocking of all or part of the warp and weft levels layers.

To establish the cross-sectional design of a multilayer multilevel configuration, each layer level is assigned the desired weave architecture. The location of the stitch is then introduced. Generation of the next warp interlocking of the linear weft picks commences by moving the linear warp ends in the desired stepping direction (up/down). The linear warp consists of all associated warp layers over the required number of linear weft levels. The weave generation in Figure 7.50a employing a step sequence one up over to the right

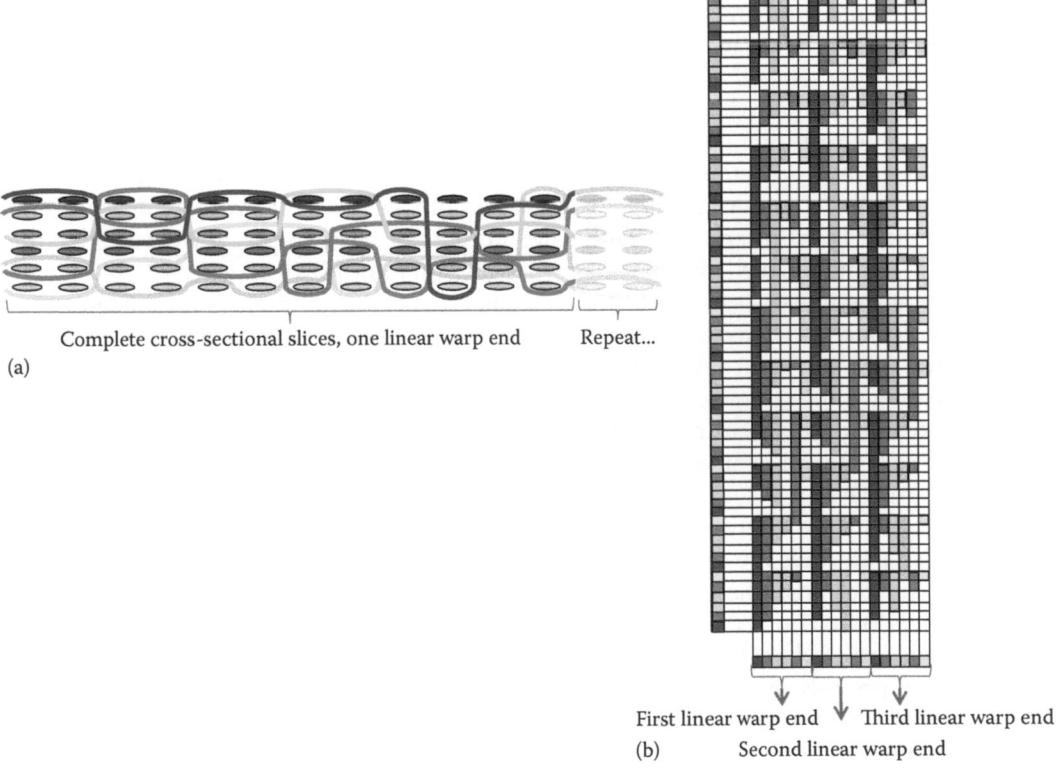

Complete cross-sectional slices, one linear warp end   Repeat...

(a)

First linear warp end   Third linear warp end

(b)   Second linear warp end

FIGURE 7.50
Linear warp and weft lifting plan and translation of in to a cross-sectional slice: (a) multilayer-multilevel cross section and (b) generated multilayer-multilevel cross section.

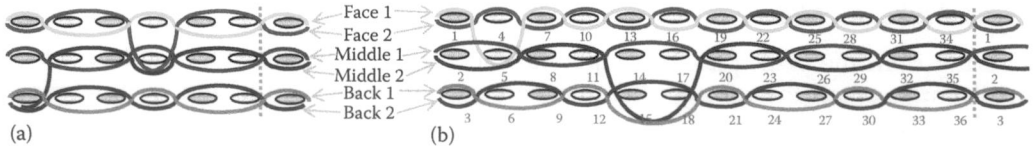

FIGURE 7.51
Cross-sectional design of multilayer multilevel weave architectures: (a) each layer level interlocks the same within the weave architectures of this cross-sectional slice and (b) each layer level interlocks differently within the weave architectures of this cross-sectional slice; the top layer level interlocks over and under each weft pick, known as plain weave.

would commence over 6 linear ends and 72 linear picks; the final weave generation would contain 72 linear ends and 72 linear picks. Figure 7.50b shows the first three linear ends of the weave generation.

Multilayer multilevel woven architectures allow the through-the-thickness yarns to travel all the way through the layers levels, from the top layer level to the bottom, to the adjacent layer level, or to any desired layer level within the fabric, all of which compound the structure together, as identified in Figure 7.51a and b. When combining multilevel, multilayer weave architectures, drapeability, conformability, and even a wrinkle-free appearance will derive from the length of the interlocking yarns within the depth of the woven wall thickness.

The schematic in Figure 7.51b details different weave architectures for each level. Therefore, the maximum number of ends and picks of each level within the cross section is required before the weave generation can be designed. The face fabric is plain weave, containing two warps and two wefts. The middle fabric is a 2/2 twill repeating over four warps and four weft yarns, and the back fabric is a 1/2 twill employing three warps and three wefts. The resulting number of ends required in each layer is two, four, and three. Therefore, the weave generation of minimum warp yarns within the cross section will be 12 ends, being the common multiple of the different weave architectures. The minimum number of picks is 36. When employing a step number of 1, the complete weave generation will require 72 ends, which are divisible by a common multiple of 12.

Visualizing the linear slices defined within the cross-sectional process can be applied to both the warp and weft direction. This, therefore, provides an understanding of the transition zone (Figure 7.52), the point at which the interlocking of layer-to-level changes direction. The transition zone within the weave generation is the point at which two weave architectures meet. Either side of this meeting point requires the modification of the warp and/or weft to ensure the yarns travel in an appropriate positioning within the woven wall thickness. It is therefore possible that the number of required ends and picks will be cut short of the concluding weave architecture and begin the interlocking of the adjacent weave. In this instance, suitable yarn pathways are required to be manually altered within the combination of the cross-sectional designs, whereby the float lengths and stitching paths can be restructured.

The utilization of both the warp and weft as interlockers originates from the tension variability associated with the combination of weave architectures within the weaving process. The process of generating a cross-sectional weave allows an in-depth view of the warp and weft yarns position. The cross-sectional process also enables the combination, visualization, and modification capabilities of varying layer-to-layer, layer-to-level, and level-to-level maneuvering, which is a fundamental requirement of weaving 2D-to-3D solid and hollow woven textile structures.

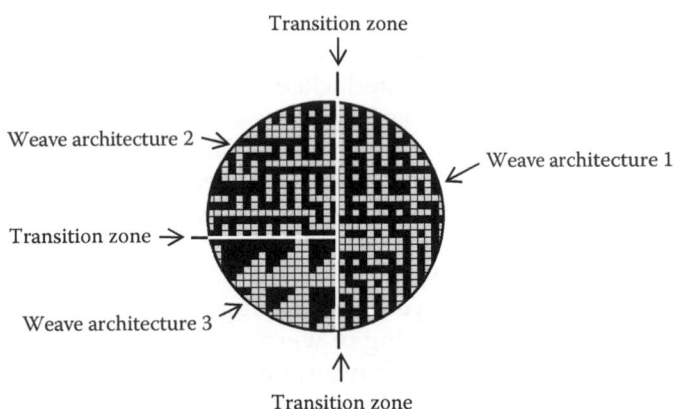

FIGURE 7.52
Weave architectures within a six layer level woven.

7.6 Classification Consideration: 2D Woven or 3D Woven

The shedding mechanisms, motions, and geometries are the primary principles in the differentiation of classifying weaving technologies. In association to the shedding principles, the let-off and take-up of warp yarns to woven textile concludes in the 2D and 3D configurations produced on conventional and specialized weaving technologies, respectively. The classifying of constructed woven textiles and associated weaving technology divides opinion and concludes in two groups.

The first group believe that producing a structure on a modified conventional loom flatly then pulled into shape can be categorized as a 3D woven structure. The addition of a third interlocking yarn elongates through the layers levels acting like a stitching yarn, but without penetrating the warp and weft yarns.

This enables the two yarns, the warp and weft, to lie in the horizontal plane of the fabric and the through-the-thickness warp yarn to enter and remain in the vertical direction to the integrated fabric plane. Pushing the boundaries of how these three directional yarns interlock with one another permits 2D-to-3D woven's. On loom shaping and folding permits the construction of a 3D woven component when pulled into shape once removed from the constraints and tensions of the weaving technology. During the weaving process on conventional weaving technology, only one weft insertion is permitted at any given time. This generates a building approach to the integration of layers levels. The concluding woven multilayer, multilevel interlocked weft yarns are positioned in an offset stepped configuration toward the fell of the cloth. The stacking up of weft yarns upon one another in a straight alignment upward toward the 2D fabric plane is possible. This weft arrangement is dependent on the take-up and let-off motions and associated speed of weft insertion.

The second group believes that only a dedicated loom, specially manufactured, can be employed to produce a *true* 3D woven structure, as it contains not only the 0/90° interlocking yarns and through-the-thickness yarns, but also ±45° yarns that are integrated within the woven structure (Khokar, 1999). The disadvantages to the *true*

3D weaving technologies are their limitations on the scope of woven configurations. Weaving technology that permits multiple simultaneous weft insertions conclude in the linear stacking arrangement of interlocked weft yarns upon one another. This linear stacking arrangement is believed by some to be the *true* 3D woven textile— however, others argue that the arrangement is not 3D, as it still consists of interlocking yarns within the 2D woven fabric plane. The benefit to the multiple simultaneous weft insertion is the speed at which multilayer multilevel woven configurations can be produced. The classification of what constitutes a true 3D woven structure therefore is not black and white.

Whatever the classifications employed, the weaving technologies still require a shedding mechanism/motion for the opening of warp yarns in a shed formation, whether for single or multiple simultaneous weft insertions. The shedding motion, concluding shed geometry and weft yarn carrier dictates the scope of shaping within the fabric plane, and ultimately the 2D-to-3D woven configuration once removed from the loom. The type of shedding motion and concluding geometry is therefore fundamental in the categorization of 3D weaving. The development of an alternative linear take-up mechanism such as tables and/or shaped surfaces rather than the take-up beam supports the woven configuration as it comes off the loom. This alleviates the compressing of the interlocked yarns and maintains their location within multilayer multilevel woven's, which is an issue inherent to the take-up beam system in the production of 3D woven textiles. Both are controlled by a let-off motion, which is controlled by the take-up mechanism that controls the pick spacing/picks per minute, identifying the timing connection for all mechanisms to allow the loom to weave efficiently, highlighting the relationship that the shedding mechanism has not only on the classification of woven structures but on the weaving operation of the loom. The shed timing works together in the positioning of the reed; as the reed is at its furthest point away from the fell of the cloth, the weft may be inserted; as the reed moves forward to beat the weft in position, the take-up draws the fabric forward, corresponding with the next opening of the shed. The height of the shed determines the maximum thickness of fabric that can be woven, when considering weaving solid 3D fabrics.

Remember the shedding of the tablet loom—although the tablet loom produces single-layer woven's, the principles of the card rotations and direction the warp yarns take generates an inspiring shedding motion. When Temple (1986) developed an apparatus for fabricating 3D woven's with through-the-thickness yarns to eliminate delamination, he utilized and modified the methodology of tablet weaving. Tablet weaving produces a 3D woven structure typically consisting of two levels, allowing the levels to pass from one level to the other. The methodology of the tablet weaving contains a solution to the problem of making a 3D woven on a 2D conventional weaving technology. The number of warp ends calculates the number of tablets used and the more holes within the tablet the thicker the fabric. The tablets are rotated forward to produce a shed and then backward for the next shed, and repeated. Moving the warp threads from the top then to the bottom of the fabric, then back again produces a thick fabric. Rotating the tablets in one direction only, however, produces a woven textile where the warp zigzags continuously backward and forward from top to bottom with no crimp in the weft and warp. This technique can be utilized on a conventional loom by thinking of a heddle as one hole within the tablet, the thicker the fabric requirements equates to more heddles. The shedding mechanism developed by Temple (1986) mimics the tablet weaving shedding motion with variable shedding heights and geometries.

7.7 Summary

Weaving apparatus associated to the woven cottage industry (inkle, tapestry, and tablet) and conventional manual and fully automated weaving technologies (dobby and jacquard) have many similarities in their weaving principles and motions. These can be employed and modified to produce innovative approaches in the yarn interlocking directions to and out of the 2D fabric plane, pushing the boundaries of constructed woven textiles.

LEARNING ACTIVITY: GLOSSARY OF TERMS

Within the weaving process and principles, there are an array of technical terms. Of those weave terms discussed in this chapter, consider what the following are:

- Warp and/or end
- Weft and/or pick
- Primary motions
- Secondary motions
- Dobby mechanism/motion
- Shafts/heald frame
- Heddles/heald wires
- Reed
- E.P.CM and/or E.P.I
- P.P.CM and/or P.P.I
- D.P.CM and/or D.P.I
- Lease sticks/rods
- Warp cross
- Shuttle weaving
- Shuttleless weaving
- Weaving ratios: 1:1 and 2:2
- Balanced woven fabric
- Through-the-thickness
- Apron

References

Goerner, D. (1989) *Woven Structure and Design: Part 2 Compound Structures*, British Textiles Technology Group, Leeds.

Khokar, N. (1999) A classification of shedding methods, *The Journal of the Textile Institute*, 90(4), 570–579.

Taylor, L. W. (2007) Design and manufacture of 3D nodal structures for advanced textile composites. PhD Thesis, School of Materials, The University of Manchester.

Taylor, L. W., and Chen, X. (2015) Nodal three-dimensional woven textiles. In: Chen, X., Ed., *Advances in 3D Textiles*. Woodhead Publishing, Cambridge, UK, pp. 99–122.

Taylor, L. W., and Chen, X. (2016) Generic production process for 3D woven nodal elementary and derivative structures, *The Journal of Composite Materials*, 50, 4103–4121.

Temple, S. (1986) *Large-scale Manufacture of Three-Dimensional Woven Preforms*, Cambridge Consultants, Mechanical Engineering Publishers (Institute of Mechanical Engineers), EC41/86, London, UK, pp. 133–140.

Watson, W. (1947) *Advanced Textile Design*, 3rd ed., Longmans, London, UK.

Recommended Reading

Hallett, C., and Johnston, A. (2014) *Fabric for Fashion*, Laurence King Publishing.

Holyoke, J. (2013) *Digital Jacquard Design*, Bloomsbury Academic.

Johnson, I., Cohen, A. C., and Sarkar, A. K. (2015) *J. J. Pizzuto's Fabric Science*, Bloomsbury Academic, 11th ed.

Marks, R., and Robinson, A. T. C. (1976) *Principles of Weaving*, The Textile Institute.

Ng, F., and Zhou, J. (2013) *Innovative Jacquard Textile Design Using Digital Technologies*, Woodhead Publishing Series in Textiles.

Posselt, E. A. (2017) *The Jacquard Machine Analyzed and Explained*, CreateSpace Independent Publishing Platform.

Sinclair, R. (2015) *Textiles and Fashion: Materials, Design and Technology*, Woodhead Publishing Series in Textiles.

Watson, W. (2014) *Textile Design and Colour: Elementary Weaves and Figured Fabrics*, Nabu Press.

8

Weft-Knitted Fabrics

Sandip Mukherjee

CONTENTS

8.1 Introduction

Knitting is the method of developing textile structures by forming a continuous length of yarn into vertically intermeshed loops. It is a method of interloping consists of forming yarns into loops, each of which is released only after a succeeding loop has been formed and intermeshed with it, so that a secured ground loop structure is achieved. It is one of the oldest and most popular techniques of fabric formation, which can be done either by hand or by machine. Knitting requires a relatively fine, smooth, and strong yarn with good elastic recovery properties. In the early days knitting was an art, because all knitted fabrics were fashioned by hand and the esthetic and performance of the end products were dependent entirely on the skill, dexterity, and artistic ability of the knitter. In the past few decades, the knitting industry has witnessed a number of rewarding technological advances that have transformed knitting from an art, requiring the skill and experience of the knitter, to a more efficient industrial technique. The scientific and technological approaches have not only resulted in better fabrics and machines but also made the development of knitting design technologies for producing a finished knitted fabric of desired characteristics and dimensions. Knitting is classified into warp and weft knitting based on the direction of yarn movement with respect to the fabric formation. In weft knitting the yarns run in the widthwise direction with reference to the direction of fabric formation and the knitted structures, thus formed, are known as weft-knitted or jersey fabrics and are characterized by the structural threads being perpendicular to the fabric selvedge. In the case of warp knitting the structural threads of the fabrics run along the fabric length approximately parallel with the selvedge. Most of the fabrics used in our daily lives as apparel are weft knitted. This chapter is solely concerned with the manufacturing and properties of weft-knitted fabrics.

8.2 Evolution of Knitting

The term knitting has evolved from the Saxon word *Cnyttan*, which was derived from the ancient Sanskrit word *Nahyat* [1]. Knitting by only fingers to produce open-loop structures may well have been practiced in ancient times as long as 1000 BC, before the use of knitting pins. It was done by using two sticks about 3000 years ago for making stockings, caps, and gloves by the women of royal families [2]. Although it is a slow process, these hand-knitting techniques are still existent and, of course, will continue in their own way. In 1589, Rev. William Lee of Calverton in Nottinghamshire invented the stocking hand frame, which was the first mechanized hand-driven flat-bed knitting machine. He developed the basic principles of mechanical loop formation, which means to form a row of loops to make the process at a faster rate using more needles. In 1730, the first pair of cotton hose was knitted on a stocking frame with fine yarns. Rib fabric was introduced by Jedediah Stroot of Derby in 1758. The Derby Hosiery frame was developed to produce a fabric having more widthwise extension. The rotary drive knitting machine was introduced in 1769. In 1847, the self-acting latch needle was patented by Matthew Townsend. Power-driven circular knitting frame with vertical bed and V-bed flat knitting were invented in 1850 and 1863 respectively. Jea Cock of Leicester patented a compound needle in 1856, which increased knitting speeds in the Tricot warp knitting machine built in 1915 in Germany. Intarsia knitting with around 42 color changes in single course was introduced in 2003 [2]. With the advancement in the area of technology and the introduction of modern computerized knitting machine manufacturers such as Shima Seiki, opened up a new era in knitting, creating a lot of new generation of knitted fabrics and garments with lots of design variations.

8.3 Properties of Weft-Knitted Fabrics as Compared to Woven Fabrics

The extensibility of knitted fabrics is higher than woven fabrics due to the loop structure. When the fabric is extended the loops are distorted and stretched out. For woven fabric, the extensibility would be to the extent of crimp in the yarn, and so they are more rigid. Knitted fabrics have poorer elastic recovery than woven fabric because of the distortion in shape, which arises due to lower tension. Knitted fabrics are highly crease resistant. Bending occurs mainly by loops, and fibers are not permanently deformed, so knitted fabrics are crease resistant or wrinkled and are preferred in case of casual wear. When a woven fabric is bent, the yarns and fibers constituting the fabric also bend and deform beyond the elastic limit giving creasing and wrinkling properties. Crease resistance is a disadvantage due to the difficulty in producing the desired creases required in the case of men's trousers. Weft-knitted fabrics are thicker than woven fabrics due to low twisted yarn. The loop-network structure means the number of air spaces is more in weft-knitted fabrics, and hence they are more air permeable than woven fabrics making them useful for undergarments, which are comfortable in summer. The tearing strength of weft-knitted fabrics is higher than woven fabrics. Knitted structures, being more extensible, distribute the stress throughout the entire fabric, thus the force is wasted to distort the loop. But in woven fabric, lesser force is required due to the more compact structure. Woven fabrics possess stronger tensile strength than knits due to the higher twist level of the yarns. They are stiffer than knitted fabrics and have higher initial modulus. The knitted fabrics have low bending length and soft drape. The feel of a fabric is related to flexural rigidity. As knits have lower flexural rigidity than wovens, they feel soft. Fullness depends on the bending modulus; weft-knitted fabrics are fuller than woven fabrics. The low twist in the yarn used to produce the knitted structures contributes to its softness and fullness.

8.4 Terms in Weft Knitting

8.4.1 Knitted-Loop Structure

The loop is the basic unit of a knitted structure produced by bending the yarn with the help of needles and sinkers, the main knitting elements. The structure is built up by the intermeshing of loops in consecutive rows. The newly fed yarn is converted into a loop in a needle hook, which draws the new loop through the old loop, which is formed in the previous knitting cycle. The old loop is released from the needle and is cast off, so that it hangs suspended by its head from the foot of the new loop whose head is securely held in the needle hook. A loop is called a face loop or a back loop in accordance to the direction of drawing the yarn from one loop through another during the knitting process. If the new loop passes from the back to the front of the previous loop made by the needle during interloping, or the yarn is pulled from back-to-front, the loop is called a face loop. When the new loop passes from front-to-back of the previous loop or the yarn is pulled from front-to-back, the loop is called a back loop as shown in Figure 8.1.

8.4.2 Course, Wale, Stitch Density, and Loop Length

A course is defined as a horizontal row of loops produced by adjacent needles during a particular knitting cycle. The number of courses per unit length of a weft-knitted fabric is expressed in terms of courses per inch (c.p.i.) or courses per centimeter (c.p.cm.). A course of loops is formed from a length of yarn known as course length and the number of loops formed in a particular course is equal to the number of needles employed for knitting. A wale is a vertical column of needle loops produced by the same needle in successive knitting cycles. The total wales along the full width of a weft-knitted fabric are equal to the number of needles incorporated for knitting. The number of wales per unit length of a weft-knitted fabric is expressed in terms of wales per inch (w.p.i.) or wales per centimeter (w.p.cm.). Stitch density denotes the total number of loops or stitches present in a unique area of a knitted fabric. It is expressed in terms of loops per square inch or loops per square centimeter and is equal to the product of courses per unit length and wales per unit length. Loop length or

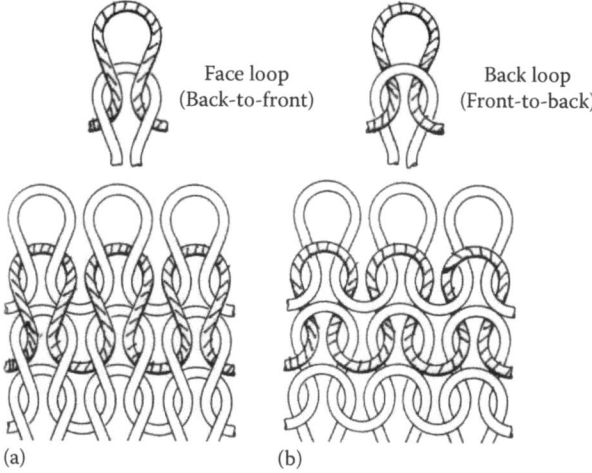

FIGURE 8.1
Knitted loop structure: (a) face loop and (b) back loop.

stitch length (*l*) in a weft-knitted structure is a measured length of a yarn present in a given loop. The loop length is measured by unroving a particular course, which is straightened by applying a suitable tension without stretching the yarn. The length of the yarn is then measured, which is then divided by the number of loops present in that particular course to obtain the loop length. Actually the operation is repeated for several courses to achieve an average value of loop length for that particular weft-knitted structure. The loop length is the fundamental unit of weft-knitted structure and its shape determines the dimensions of the knitted fabric. The configuration of a loop varies in different knitted structures and more than one loop is essential to define the complete loop structure of some knitted fabrics. Thus, the repeated unit in the knitted structure is termed as structural knitted cell (SKC).

8.4.3 Machine Gauge and Pitch in Weft Knitting

Machine gauge in the English system is defined as the number of needles present per inch in each bed of the weft-knitting machine. Machine pitch is defined as the distance between the centers of two adjacent needles in each bed of the machine. Both imperial and metric units can be used for denoting the pitch but the metric unit in *mm* is more conventionally used to indicate the pitch of a weft-knitting machine.

8.4.4 Tightness Factor and Loop Shape Factor of Weft Knitted Fabric

Tightness factor (TF) is defined as the ratio of the area covered by the yarn in one loop to the total area occupied by the loop. The use of TF to indicate the relative looseness or tightness of a weft-knitted fabric was first suggested by Munden [3]. It indicates the extent of the area of knitted fabric covered by the yarn and influences the compactness handle strength and dimensional properties of the knitted fabric. It is equivalent to the cover factor of a woven fabric.

Munden suggested the dimensions of a plain weft-knitted fabrics in a relaxed condition and defined knitted fabric constants as course constant Kc = c.p.i × l, wale constant Kw = w.p.i × l, and stitch density constant $Ks = Kc \times Kw = S \times l^2$.

The loop shape factor $R = Kc/Kw$ = c.p.i/w.p.i. The loop shape depends on the type of yarn used and on the fabric treatment, which determines the dimensions of the weft-knitted fabric.

If the loop length and diameter of the yarn is denoted by l and d (mm), respectively and stitch density by S then, the area covered by the yarn in the knitted fabric = $S \times l \times d$. Now, introducing the value of $S = Ks/l^2$, the area covering 1 cm² of fabric is: $Ks \times l \times d/l^2 \times 100 = Ks \times d/100l$. For comparison of the same knitted structures in a similar relaxation state a simplified formula is used to denote the value of TF in the metric system and is expressed as TF = $\sqrt{(tex)}/l$ where *tex* value denotes the measure of yarn count in the direct system. The value of TF in imperial units is expressed as TF = $1/l\sqrt{N}$ where N is the worsted count and l is the loop length in inches.

8.5 Weft-Knitting Elements

A knitting machine, either hand or power driven is an apparatus that applies mechanical movement and is used to convert a straight continuous length of yarn into loops by a process of interloping for fabric formation. The needle is the primary knitting element of the knitting machine. It uses the help of a sinker, which is the second primary knitting element during the process. The reciprocating movements of the needles are actuated by a device known as a cam, which is another important element along with a jack, a secondary element for weft knitting.

8.5.1 Needle

A hooked metallic needle is the main element of a knitting machine responsible for loop formation. They are arranged on a needle bed at regular intervals and are placed inside a groove known as trick, both in circular and flat bed weft-knitting machines for free movement along the axis. The needle is raised to clear the previous loop held in the needle hook from the latch prior to yarn feeding. After the yarn is being fed the new loop is held in the needle hook in the open state and it starts to descend, drawing the new loop down through the old loop, which is eventually cast off or falls over the needle with closed hook. Thus, all needles used for weft knitting must have a mechanism for opening and closing of the needle hook in order to catch the new yarn and to retain the new and release the old loop. The three types of needles used in weft knitting are (1) latch needle, (2) bearded needle, and (3) compound needle as shown in Figure 8.2.

1. The latch needle is the most versatile needle widely used for weft knitting. It is a self-supporting needle in which no extra element is required to open or close the needle hook during the loop-formation process. The latch, riveted on the needle stem can swing freely around its fulcrum and is gradually opened due to the thrust exerted by the loop held on the needle, on the inner surface of the latch during the upward journey of the needle toward the clearing position. The latch is closed due to the thrust exerted by the same loop on its outer surface during the downward movement of the needle. The swinging latch has a spoon or a cup at its end to fit the needle hook in the groove when the latch is closed, which enables better closing of the hook. The reciprocating movement of the needle is provided by means of a cam through the butt, which is an important component of the needle receiving the movement required for the formation of a loop. Latch needles are widely used for circular and flat bed/v-bed weft-knitting machines for single and double jersey fabrics.

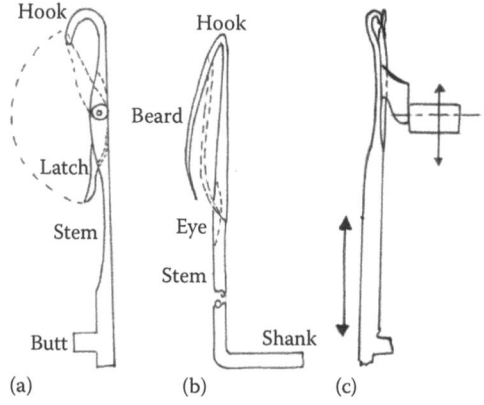

FIGURE 8.2
(a) Latch needle, (b) beard needle, and (c) compound needle.

2. The bearded or the spring needle is the oldest type of the needle and is a single unit made from a single piece of metal and of finer cross section suitable for very fine gauge weft knitting. It comprises a stem, head, beard, eye, and shank. The stem is converted into a hook at the needle head to draw the new loop through the old one, which is curved downward in continuation of the hook known as the beard, which is used to separate the new loop from the old loop as it falls off or cast off from the needle beard. The beard needle requires an external unit called a presser to close the beard to enable loop formation. A small eye or groove is cut in the stem of the needle to receive the point of the beard when it is closed. The shank is a bent bottom portion of the beard needle to connect with the knitting machine.

3. The compound needles are suitable for high-speed weft and warp knitting. The speed of a beard needle is restricted because of the presser (an external element), whereas the swinging action of the latch causes damage to yarns at a very high speed for knitting with latch needles. In a compound needle, the strain is not put on the yarn during the loop formation due to less movement of the versatile needle enabling operation at a very high speed, thus increasing the production rate to a high extent. A compound needle comprises two separate parts—the open hook and the sliding closing element known as the tongue. These two parts rise and fall together, but as the hook moves at a faster rate, it is open at the top of the rise and closed at the bottom of the fall. There are two types of compound needle used for knitting: tubular pipe and open-stem pusher. A tubular pipe needle consists of a hollow steel tube in which the tongue member is inserted and it slides inside the tube. The tongue slides down during the upward motion of the needle to keep the hook open for yarn catching and slides up during the downward motion of the needle for releasing or casting off of the old loop. There is a phase difference in the movement of the needle and the tongue actuated by twin cam races that are required to operate two moving parts and their movements are properly synchronized to enable the loop-formation process. In an open-stem pusher type of compound needle, the tongue slides externally along a groove on the edge of the flat hook assembly.

8.5.2 Sinker

The sinker is a second primary knitting element. It is a thin metal plate with an individual or collective action approximately at right angles from the hook side of the bed between two adjoining needles. The essential parts of the sinker are (1) throat, (2) belly, (3) neb, and (4) butt. The throat holds the yarn during formation of the loop; the belly provides the space for the old loops to reside after being cast off. The neb prevents the rising of the knitted fabric and the butt receives the motion from the cam system. The main function of the sinker is: (1) loop formation, (2) holding-down the formed loop, (3) knocking over, and (4) providing a surface over which the needle draws the loop. The butt of the sinker is actuated by the sinker cam, so that the sinker can move horizontally (backward and forward) at the appropriate time of the needle movement. The common function of the sinkers in modern machines is to hold down the old loops at a lower level on the needle stem than the new loops, which are being formed and prevent the old loops from being lifted as the needle rises to clear them from their hooks. On the bearded needle weft-knitting machine, the main purpose of the sinker is to sink the newly laid yarn into a loop as its forward edge advances between the two adjoining needles. The loop formation is not a function of

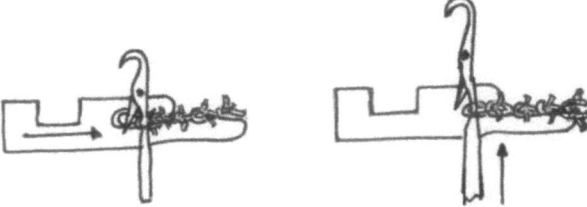

FIGURE 8.3
Needle positions along with sinker action.

sinkers in latch needle machines. The protruding neb of the sinker is positioned over the sinker loop, which is the piece of yarn that joins one weft-knitted needle loop to the next, of the old loops preventing them from rising with the needles. Figure 8.3 shows the needle positions along with the action of the sinker, which provides support and surface area for the newly formed knitted loops.

8.5.3 Knitting Cams

Knitting cams are devices that convert the rotary machine drive into a suitable reciprocating motion of the needle. Depending on the machine design, the cams are fixed, adjusted, or exchangeable. The axial movement of the needles to form a loop during knitting is produced in accordance with the arrangement of the cam system or the cam profile. The motions of the needles are actuated by the needle butt, which are placed inside the channel of the cam system and moved according to its profile. For knitting different structures the profile of the cams is made to produce precisely timed movement along with necessary dwell period and differs from machine-to-machine. The profile of the knitting cam system generally found in a circular single jersey weft-knitting machine along with different needle positions (1–6) is shown in Figure 8.6. The raising or clearing cam assists the needle to rise from tuck to clearing position and its rising motion is controlled by the guard cam. The downward movement of the needle is actuated by means of the stitch cam, which also determines the length of the loop produced during knitting. The stitch cam setting may be altered as per requirement of the loop length for a particular weft-knitted structure. The up-throw cam also guides the needle during the downward motion and takes the needle back to the rest position after the completion of loop formation. The guard cam maintains the level of the needle till the needle is in the position to rise for the next knitting cycle. Thus, the needle moves up and down in the cam system and forms the new loop in the fabric. The clearing cam is designed in a manner to provide clearing, tucking, and floating of the needles to form different varieties of knit designs. For double jersey structures, there are two cam tracks for cylinder and dial needles. For interlock structures having two sets of cylinder and dial needles, the butt for the short needle follows the upper and that of the long needle follows the lower cam track. In the case of rib structures in which only one set of needles is used for cylinder and dial needles, the needles pass through one track only in each bed. The profiles of the cam system incorporated in the front and back bed of the power driven, computerized flat-bed double jersey knitting machine of M/S Brothers, Japan of gauge 5.5 and width 40 inches were studied. The front-bed and back-bed cam systems along with their profiles and needle path are shown in Figure 8.4. The figure shows the underside of the cam carriage and the cams forming the tracks, which guide the needle butts for the knitting action [4]. A set of cams consisting of raising or clearing cam, guard cam, cardigan cam, and two stitch cams, as shown in Figure 8.4, force the needles to knit a course of loops in either direction of carriage traverse. The symmetrical camming

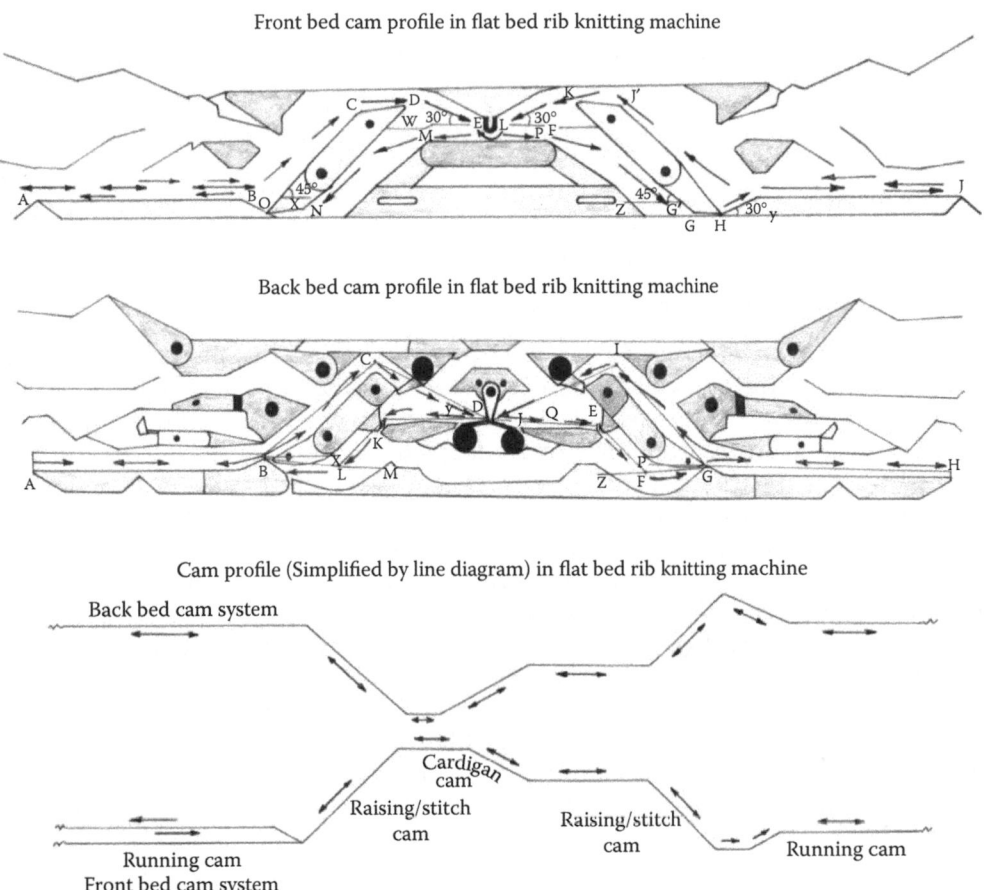

FIGURE 8.4
Cam profile and needle path in double-jersey knitting machine.

arrangement is typical of flat-bed machines as it enables a similar action to be achieved in both directions of carriage traverse. In flat knitting, the cam carriage moves from left to right and right to left irrespective of single and double jersey structures. In a V-bed machine, there are two carriages for the front and the back bed. The butt of the needle enters the traversing cam system from the right side during a left to right carriage traverse and from left during a right to left traverse. For each needle bed, there are two raising cams: two cardigan cams and two stitch cams. The raising and cardigan cams are generally spring loaded. The cardigan cam is the upper portion of the raising cam. The cam setting can change for each traverse to produce a knit, tuck, and float or miss stitch. A raising cam lifts the needle to the tuck position; the cardigan cam if in action lifts the same to the clearing position. For tuck stitches the cardigan cam and for miss stitches both the raising and the cardigan cams are taken out of operation.

8.5.4 The Jack

The jack is the secondary element for weft knitting which is used to provide versatility in the selection and movement of the latch needle particularly required for patterning.

The jack is positioned in the same trick below the respective needle. It has its own cam arrangement along with its operating butt following the necessary cam profile during the knitting operation.

8.6 Basic Weft-Knitted Structures

8.6.1 Plain-Knitted Structure

This is the simplest weft-knitted construction usually termed as plain fabric or single-jersey fabric. As all the knitted loops are interlooped in the same direction the fabric has a different appearance on the technical face and on the back side. The face side of the fabric is smooth, with the side limbs of the loops having the appearance of columns of inverted V's along the wale direction. On the back side of the fabric, the head of the needle loop along with the base of the sinker loop creates columns of interlocking semicircles [1] resembling a purl structure. Figure 8.5a and b shows the face and back of a two-color horizontal stripe single-jersey fabric produced on a flat-bed knitting machine along with the loop diagrams. The single-jersey structure can be produced both on a flat-bed and a circular-bed knitting machine using one set of needles along with sinkers. In flat knitting, the needles of either front bed or back bed along with sinkers and one set of cam assembly can be utilized to produce the structure. In circular knitting, the cylinder needles along with sinker and the corresponding cam assembly are used to produce the structure. Most of the single-jersey fabrics are produced in circular knitting machines using latch needles in which the cylinder with the needles placed in the tricks along with sinker revolves through a stationary knitting cam system. The yarn feeders in the form of cones are either placed on the top or side creel and the yarn is passed through the yarn tensioners, yarn guide eyes, stop motion, and the feeder, which is a small metallic plate with a hole in the center. The knitted fabric thus produced is tubular in shape and is withdrawn from the knitting zone in a downward direction from the cylinder by tension rollers and wound on the cloth roller. The process is known as the take-down mechanism.

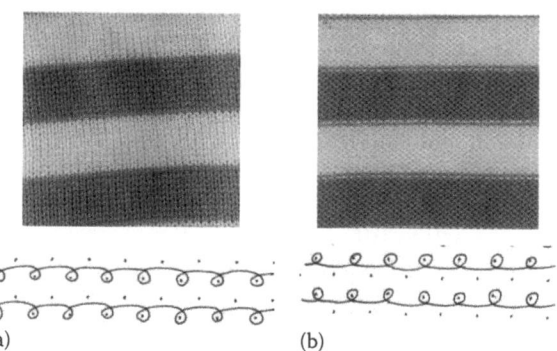

(a)   (b)

FIGURE 8.5
Fabric appearance and loop diagram: (a) face of single-jersey structure and (b) back of single-jersey structure.

8.6.1.1 The Knitting Cycle for Single-Jersey Fabrics

Figure 8.6 shows the knitting action for single-jersey fabrics using latch needles in a circular-bed knitting machine [1]. The needles are placed inside the tricks made on the cylinder needle bed and their butts follow the profile of the cam system to actuate the axial or reciprocating movement of the needles during loop formation. In the Figure 8.6, 1–6 shows the needle positions at different stages of a knitting cycle, that is, rest position, tuck position, clearing position, yarn feeding, or catching position, latch closing position, and knock over or cast-off position.

A complete knitting cycle for loop formation using latch needles as it passes through the cam system comprises six major stages and the needle positions in those stages are also shown in Figure 8.7.

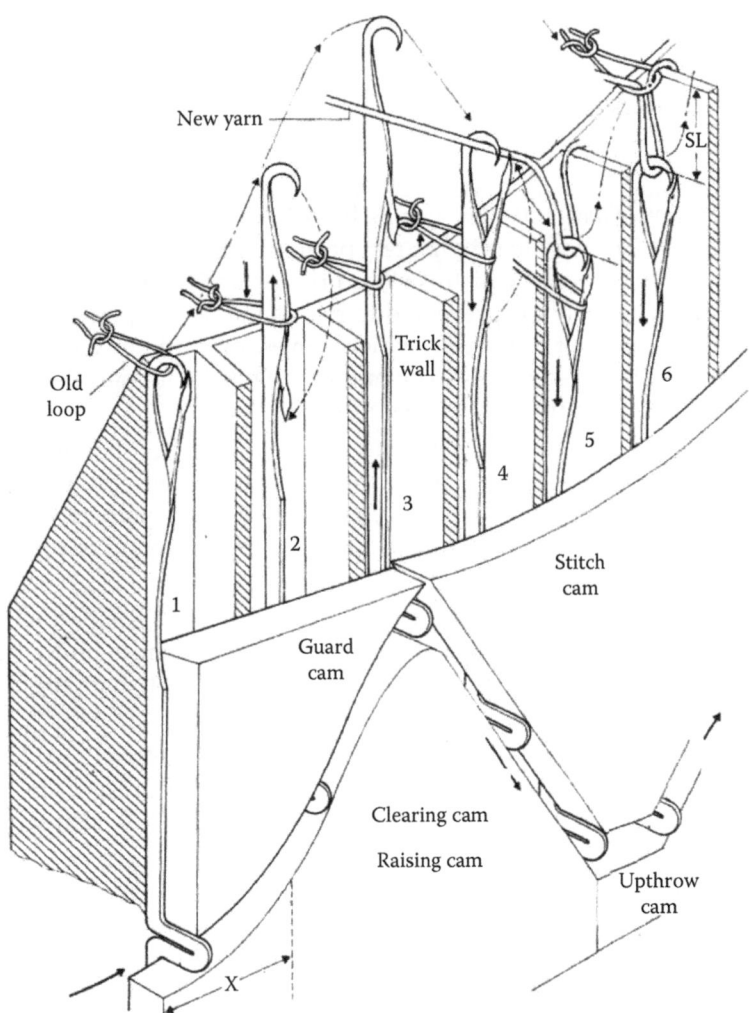

FIGURE 8.6
Knitting action for single-jersey fabric.

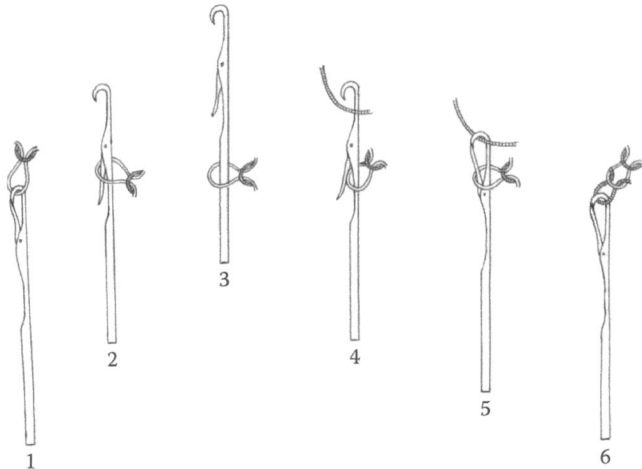

FIGURE 8.7
Needle positions in different stages of knitting cycle.

Stage 1: Rest position of the needle: The needle hook is almost in the same level with the top of the verge. The needle is in the rest position with the old loop formed in the previous cycle in its hook, which is closed by the latch. The old loop residing in the closed hook assembly is prevented to rise as the needle just starts its upward journey by the throat of the sinker, which moves forward between the adjoining needles holding the sinker loops. The butt of the needle follows the track formed by the guard and the running cam assembly.

Stage 2: Tuck position: The needle starts rising up with its butt following the clearing cam contour. The old loop held by the sinker, slides inside the hook exerting a thrust on the inner surface of the latch, gradually pushes and starts opening the same. As the needle rises up to the tuck position the old loop resides on the latch opened by it.

Stage 3: Clearing position: The needle is raised to the top most position following the clearing cam contour, the old loop clears the latch and resides on the stem of the needle. The feeder guide plate prevents the latch, which is hanging down from closing the hook.

Stage 4: Yarn feeding position: The needle starts descending following the stitch cam contour, the old loop residing on the stem moves up with respect to the needle. At this time, the needle hook catches the new yarn, which is fed through a hole in the feeder guide.

Stage 5: Latch closing position: The needle is lowered further following the stitch cam contour. The old loop comes in contact with the underside of the latch exerting a thrust and gradually closes the latch on to the needle hook. The new yarn is then trapped inside the closed latch–hook assembly.

Stage 6: Knock over position: The needle attains its lowermost position descending below the top of the trick and reaches the bottom of the stitch cam. The old loop slides over the needle hook, which is known as *casting off* and trapped the new loop inside the hook, which is drawn through the old loop completing the knitting cycle. The lowermost point is known as the knitting point. The loop length is determined by the extent of the downward needle movement and can be altered by varying the stitch cam setting. The cast-off loops reside on the belly of the sinker that completes its backward movement and pushes the knitted loops or fabric to assist the fabric take-down mechanism. After completing the knitting cycle the needle again reaches the rest position by rising up following the contour of the up-throw cam.

8.6.1.2 Characteristics of Single-Jersey Fabrics

This fabric has good extensibility in both course-wise and wale-wise direction. The course-wise extension is approximately twice that of the wale-wise extension, which is due to the constraint imposed on the loop by its interloping [5] in which the vertical extension of the loop is almost half as compared to the horizontal extension. The fabric feels smoother on the face as compared to the back due to the presence of side limbs of the loops on the face side. Single-jersey fabrics generally tend to curl at the edges when kept flat. The top and bottom curl in toward the face and the sides toward the back of the fabric, which is due to the unbalanced bending behavior of the yarns in the knitted structure. On the face side and the back side, bending easily occurs along the longitudinal and lateral directions, respectively. The fabric can be unroved from either end. If a yarn is broken, the wale disintegrates causing the loops, which are held by each other at the interloping points, in that line to unmesh. This structural breakdown is termed as laddering. The thickness of the fabric is almost double as compared to the diameters of the constituent yarns and is composed largely of air space, which is responsible for good air permeability and heat insulation properties, which is again dependent on the loop length. The loops distort easily under tension that provides comfort and fitting. Plain fabrics are produced in circular- and flat-bed knitting machines and are used in the manufacture of knitted outerwear, sweaters, and all types of fashion garments. During manufacture of garments by cutting pattern pieces followed by sewing, the difficulties faced due to the curling phenomenon can be minimized by heat setting during finishing. Circular-knitted fabrics used for garment manufacture do not need side seams if the correct machine diameter is available, producing body-size seamless garments with no uncomfortable side seams. This type of machine is known as a circular garment length machine.

8.6.2 Rib-Knitted Structure

Rib fabric has a vertical cord-like appearance composed of loops formed in opposite directions. The face loop wales tend to move over and in front of the reversed loop wales [1], thus when viewed from one side, both face and back loops are apparent. The simplest rib structure is the 1 × 1 rib, which is formed by alternating wales of face and back loops. As the face loops shows a reverse-loop intermeshing on the opposite side, it has the appearance of the technical face of single jersey on both sides and hence is a reversible fabric. The reversed loop wales in between, which are the back loops, are visible only after stretching the fabric along the widthwise direction. Here, one vertical column of wales is meshed in the opposite direction to the other column and if one column of loops knits to the front, the other will knit to the back. Thus, the loops of a rib structure lie in two separate planes and make it a double-jersey fabric. Figure 8.8a–c shows the structure, loop diagram, and fabric appearance of 1 × 1 rib fabric, respectively.

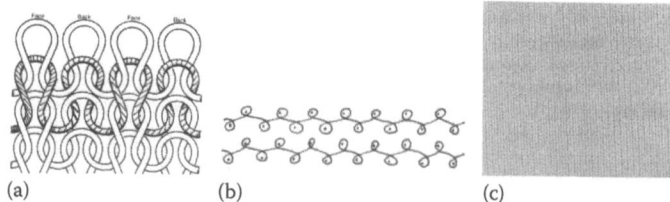

FIGURE 8.8
(a) 1 × 1 rib structure, (b) 1 × 1 rib loop diagram, and (c) 1 × 1 rib fabric appearance.

Rib fabrics are used for sweaters owing to their exceptional elasticity and greater weight. They are used in waist bands, collars, tops of socks, and the cuffs of sleeves. They are more opaque, applicable for swimwear and stretch to fit garments, and are used where portions of garments are required to cling to the shape of the human form and also be capable of stretching. It has good lengthwise extensibility and excellent widthwise elasticity and is almost double that of a single jersey. As there are face and back loops side by side, the fabric does not curl at the edges and can be spread evenly like a woven fabric. A relaxed one-by-one rib is almost twice as thick and half the width of an equivalent single-jersey fabric and has twice as much widthwise recoverable stretch. The fabric normally relaxes around 30% as compared with its knitting width. Other constructions of rib structures commonly used are 2 × 1 rib, 2 × 2 rib, 3 × 3 rib, and 6 × 3 rib (Derby rib). Figure 8.9a and b shows the fabric appearance of 2 × 2 and 6 × 3 rib structures produced on a V-bed knitting machine.

8.6.2.1 The Knitting Cycle for 1 × 1 Rib Fabrics

Figure 8.10 shows the knitting cycle and loop formation of a 1 × 1 rib structure produced on a circular double-jersey knitting machine having a cylinder and dial. Here, two sets of needles are used known as cylinder needles and the dial needles arranged on cylinder bed and dial bed, respectively, mutually perpendicular to each other, and acting along two mutually perpendicular planes to form a rib loop unit comprising face and back loop formed by the cylinder and dial needles, respectively. In the course of knitting, the neighboring cylinder and dial needles may form the loop simultaneously by drawing the yarn from the supply package and reaching their respective knitting points at the same time or with a phase lag, which is in accordance to the synchronized and delayed timing, respectively. The cylinder needles follow the cam track mounted on the cylinder and the dial needles follow a separate cam track on the dial. The position (a) shows the needles at the beginning of the loop formation at rest positions holding the old loops in the closed needle hooks.

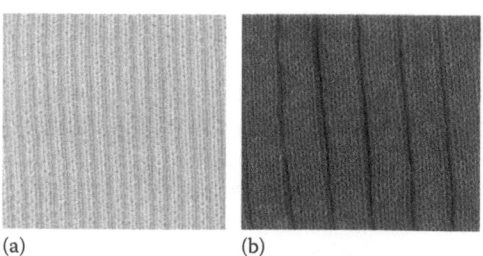

FIGURE 8.9
Fabric appearance: (a) 2 × 2 rib fabric and (b) 6 × 3 rib fabric or *Derby Rib*.

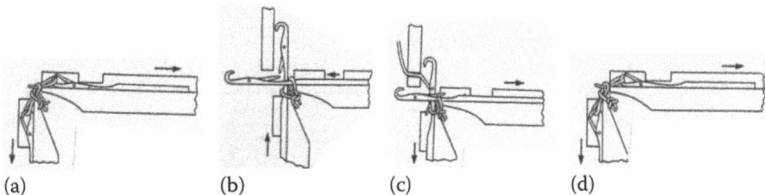

(a) (b) (c) (d)

FIGURE 8.10
(a) Rest position, (b) clearing position, (c) yarn catching position, and (d) knock-over position.

While at (b) the adjacent needles move to the clearing height, following the raising or clearing cam contour and the old loops residing on the needles open and clear the latch. At (c) the needle starts the return movement by following the stitch cam contour and at the same time catches the new yarn into the needle hook simultaneously. Position (d) shows the complete knock over of both cylinder and dial needles operating together simultaneously to produce the loops. In synchronized timing for rib knitting, the tension in the yarn is less resulting in bigger loop lengths, less chance of yarn breakage, less compactness, and lower tightness factors as compared to delayed timing. In delayed timing the cylinder loop is knocked over or cast off prior when the new loop is drawn through the old loop, with the cylinder loop being knocked over and that the dial needles have not completed their knock over. This timing is known as delayed timing, which produces tighter fabric due to the robbing back phenomenon in which some yarn is taken or robbed from the previously knitted loop to make the current loop.

In flat knitting, rib fabrics are produced in a V-bed rib-knitting machine. The knitting action showing the positions of yarn catching and loop formation is shown in Figure 8.11. The yarn fed from the yarn package passes through the yarn guide and tensioners and then threaded through the yarn carrier the tip of which is set between the two needle beds. The yarn carrier moves with the cam carriage on each bed. The fabric is made in a small gap between the two beds and is taken downward by means of a take-down load. Rib fabric is produced by the needles of both beds, which alternately ascends to a clearing position and then descends to form their loops [5]. To produce a rib fabric the cam boxes in both the beds are active and are symmetrically designed for knitting a course of loop on

FIGURE 8.11
Knitting action of V-bed knitting machine.

both the front- and back-bed needles during a traverse from right to left and for the second course during the return, a traverse from left to right.

8.6.3 Interlock Structure

An interlock structure is obtained by interlocking two 1 × 1 rib structures in such a way that the face loops of one fabric are directly in front of the back loops of the other fabric in one wale line and in the next wale line the order is reversed, which produces two layers in the fabric, face, and back. Simple interlock is a reversible structure used in undergarments, sports jackets, ladies sweaters, and so on. It has the technical face of plain fabric on both sides, but its smooth surface cannot be stretched out to reveal the reversed meshed loop wales because the wales on each side are exactly opposite to each other and are locked together. Figure 8.12a shows the needle set out of the machine, with long and short needles alternating on the cylinder to produce an interlock fabric. The dial needles are set out exactly opposite to those on the cylinder. This sort of gaiting is called interlock gaiting in which the long cylinder needles face the short dial needles and the short cylinder needles face the long dial needles, but only one set of needles will knit at each feeder. There are two separate cam tracks with one controlling the short needles and the other the long ones both for cylinder and dial. Minimum two feeders are required to produce an interlock fabric. At the first feeder only, the long needles will knit and at the second feeder only the short needles will knit. Figure 8.12b and c shows the loop diagram and structure of interlock fabric.

Interlock fabric is a balanced, smooth, and stable structure having symmetry in its construction, which lies flat without curling. Hence, it is suitable for cutting and sewing and is an excellent fabric for making knitted garments. Due to the crossings of sinker loops when passing from wale-to-wale, the fabric is rigid in both directions, which gives it a sturdy and luxurious look being reversible; it has an appearance of plain jersey structure on both sides and relaxes by about 30%–40% compared with its knitted width. It has elasticity in the widthwise direction and the pull between the wales causes the fabric to have a strong tendency to come back to its original width if stretched in a widthwise direction. The fabric ravels only from the last knitted end and that only if two feeds are used at the same time. It is more run resistant, thicker, heavier, and narrower than plain or rib fabric having a soft handle, and requires a finer, better, and more expensive yarn.

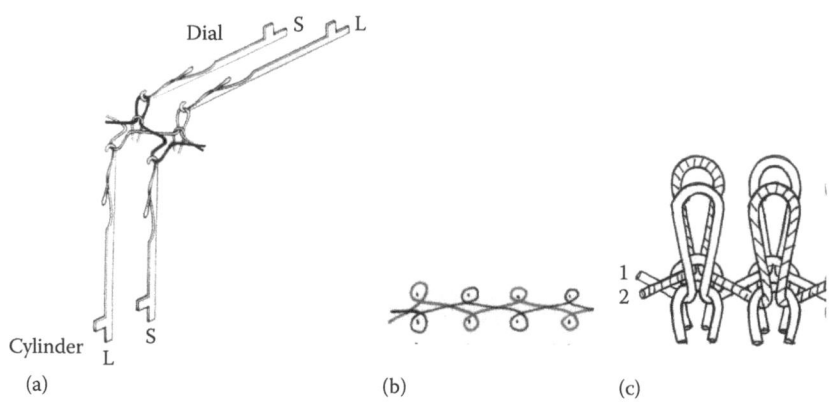

FIGURE 8.12
(a) Interlock knitting, (b) interlock loop diagram, and (c) interlock structure.

8.6.4 Purl Structure

Purl fabric has got an appearance that is similar to pearl droplets. The structure has both face and reverse loops along the same wale line, which is achieved by means of knitting with a double-ended latch needle. The loops of one course in a purl structure are intermeshed in one direction and the loops of the succeeding course are intermeshed in the opposite direction. Thus, alternate courses are knitted in opposite directions resulting in the formation of 1 × 1 purl structure. Figure 8.13 shows the knitting action of a flat-bed purl machine, which has tricks in each of the needle beds. The needle beds are set in a straight line with a gap in between acted upon by a double-ended latch needle. There is a cam carriage, which moves from right to left and left to right alternately above the needle beds. The needles are operated by auxiliary devices called sliders. The needle beds are in line with one another to enable the needles to transfer from one bed to the other. Sliders positioned in each trick control the movement of the double-ended latch needles.

Position (1) shows the needle knitting in the front bed under the control of the slider in that bed. In position (2) the needle has been moved to the centre, with both the sliders trying to engage the needle hook. The slider in the back bed is pressed down by a cam at the point *x*, thus engaging with the corresponding needle hook. The one for the front bed is not in position with the corresponding needle hook. In position (3) the sliders start to move back, and the slider in the back bed controls the needle movement and the yarn is fed in and then held by the hook of the other needle. Position (4) shows that the slider in the back bed has moved the needle to the knock-over position to complete the formation of a purl stitch. For the next course, the stitch is formed by the other hook of the same needle. Thus, if back loops are produced in the first course, then face loops are produced in the second one making a 1 × 1 purl structure. The lengthwise elasticity of a purl fabric is double than widthwise and it has a full, lofty, and soft hand. The back loops are visible on both sides of the fabric and only on stretching it in the lengthwise direction are the face loops visible. As it has a balanced structure, the fabric has no tendency to curl. It has got a high bulk, good cover with the thickness being double that of plain single-jersey structure, and provides an excellent thermal insulation. Purl fabrics are used in children's wear, sports shirts, socks, and so on.

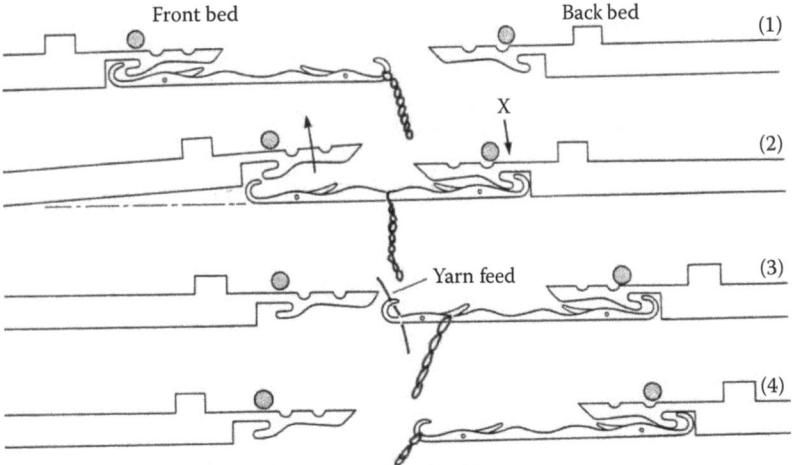

FIGURE 8.13
Formation of purl structure.

8.7 Tuck and Float Stitches

The weft-knitted structure is generally composed of loops, which are formed when the needle clears the old one and after receiving the new yarn, the old loop gets cast off over the new yarn being held by the needle hook, or in other words the new yarn in the form of a loop is drawn through the previous loop, completing the knitting cycle. But for design variations, stitches are deliberately produced by varying the sequence of the needle loop intermeshing. The two most commonly produced by controlling the height of the needles along with individual selection are tuck stitch and float or miss stitch. Figure 8.14 shows the three different needle positions to produce a normal knit, tuck, and a float stitch. The latch needle producing a normal knit, moves up to the clearing height, which clears the loop from its latch and eventually catches the yarn while descending. The needle, to produce a tuck stitch is raised, but not enough to clear the loop from its latch, but its position is sufficient to catch the new yarn. The needles forming the float or miss stitch do not rise from its rest position, thus, neither the old loop can clear the latch nor can the new yarn be fed or taken up by the needle hook.

8.7.1 Tuck Stitch

A tuck stitch is produced when a needle catches a new loop without clearing the previously formed loop below the latch. Thus, the needle has two loops in the hook resulting in the formation of a tuck stitch when it knits in the subsequent course. A tuck stitch is made of the previously formed knitted loop called the held loop, which the needle had retained, and the loop or loops joining it as tuck loop, which always lies behind the held loop. The tuck loop is not intermeshed with the previous loop but is tucked behind it on the reversed side of the stitch and when the needle is cleared on some consequent course the new loop is drawn through the held and the tuck loop from the back side of the fabric. Figure 8.15 shows the formation of a tuck stitch by a latch needle in which both the held and the tuck loop are clearly visible on the front of a fabric, the loop diagram, and a graphical

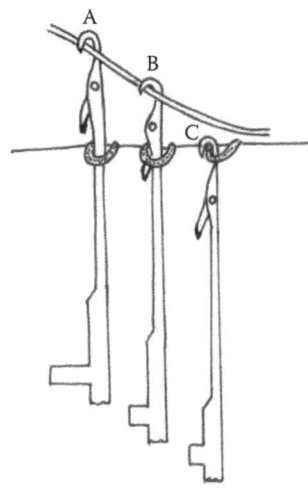

FIGURE 8.14
Needle positions for producing knit, tuck, and float stitch.

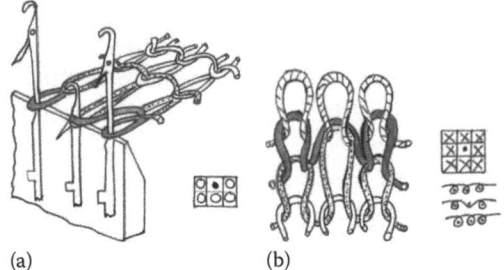

FIGURE 8.15
(a) Formation of tuck stitch and (b) structure, loop diagram, and graphical representation of tuck stitch.

representation of the knitted structure along with a tuck stitch. The tuck stitch creates a textural effect that is dependent on the needles selected to be tucked [6].

The number of consecutive tucks on any needle is restricted by the quantity of yarn that the needle hook can accommodate, up to four successive tucks, without yarn breakage or needle damage. The tuck loop assumes an inverted V- or U-shaped configuration and the side limbs are visible on the face between two adjacent wales that makes the structure wider due to the distortion of the loop shape at the tucking point. Fabric with tuck stitches is thicker and less extensible due to the accumulation of yarn at the tucking place and a smaller loop length, respectively. The fabric is heavier and the structure is open and more porous. They are used for producing different designs in weft-knitted structures and different colored yarns are used to achieve fancy effects.

8.7.2 Float Stitch

A float or miss stitch is produced when a needle holding the previously formed loop is unable to rise and catch the new yarn. The yarn then passes as a float or miss stitch and lies freely on the reverse side, joining together the two nearest knitted loops produced by the adjacent needles. Under normal condition, the maximum number of floats on the same needle is four and six; adjacent needles can be used for a continuous float without disturbing the knitted structure. Figure 8.16 shows the formation of a float stitch by a latch needle in which both the held and the float stitch are clearly visible on the front of a fabric, the loop diagram, and a graphical representation of the knitted structure.

A float stitch has a U-shaped appearance on the reverse of the fabric exhibiting faint horizontal lines. It makes the basic fabric thinner due to less yarn accumulation and narrower

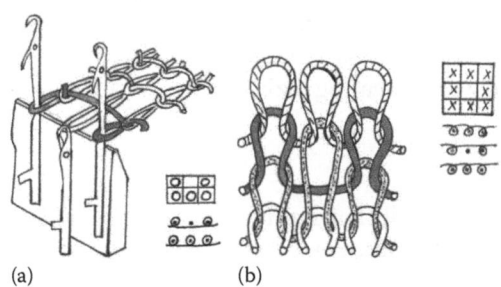

FIGURE 8.16
(a) Formation of float stitch and (b) structure, loop diagram, and graphical representation of float stitch.

as the wales are pulled closer together, reducing the widthwise elasticity and improving the fabric stability. It also makes the fabric less extensible, less rigid, and lighter in weight due to minimum yarn consumption in making the float structure. By incorporating a float or miss stitch, a more uniform texture is achieved and a lot of design variations in weft-knitted structures can be obtained.

8.8 Design Variations in Weft-Knitted Fabrics

A lot of new structures with a special appearance and characteristics creating design variations can be obtained for weft-knitted fabrics by incorporating tuck and float stitches with the help of a needle selection device.

8.8.1 Ornamentation of Weft-Knitted Fabrics

The ornamentation of knitted fabrics to create new designs may be done by using yarns of different colors, materials, and twists along with different types of stitches using suitable needle selection devices. Alternate courses with S or Z twist can give a zigzag path along the wale direction on the surface of the fabric. The loop length of the knitted fabric can be altered by varying the stitch cam setting. On decreasing the stitch length the fabric becomes more dense or compact and on the other hand on increasing the stitch lengths means the fabric becomes more open. This variation due to alteration of the loop length can create an effect. Fancy, soft twisted yarns, lustrous yarns, neppy yarns, metal/wire threads, crepe yarns, and so on can create interesting effects in the fabric. By varying the yarn count, ornamentation can be achieved. The combined effect of color and needle selection enriches the design and esthetic value of weft-knitted structures [2]. The dyeing process provides the possibility of differential and cross-dyeing at any stage of manufacturing from fiber to the finished product. Printing techniques introduce color designs on plain color surfaces and finishing processes, including heat treatment can transform the appearance of a structure. The simplest designing technique in weft knitting is horizontal striping, which is achieved by feeding different colors of yarns from different feeders. This can be achieved both for circular-bed and flat-bed knitting machines. Individual needle selection is the most versatile technique for knit designs based on the relative position of the needle hooks during clearing. Latch needle weft-knitting machines are mostly suitable because their individually tricked and butted elements offer the possibility of independent movement and depending on machine cam arrangement, needles may be selected to produce one or more of the following stitches: knit, tuck, float, plated, plush, loop transfer, and purl needle transfer. In V-bed flat knitting, designs can be created with a needle bed racking by distorting the wales creating a zigzag appearance on the fabric.

8.8.2 Intarsia

Intarsia is a decorative colored design knitted into a solid color fabric. The design areas are created with colored yarns, producing large motifs which are a development of part, of course, horizontal striping, composed of areas of pure color each consisting wholly of loops of one color without floating threads. Here, two or more colors are used to make a motif along the course and the knitting width is divided into adjoining blocks of needles.

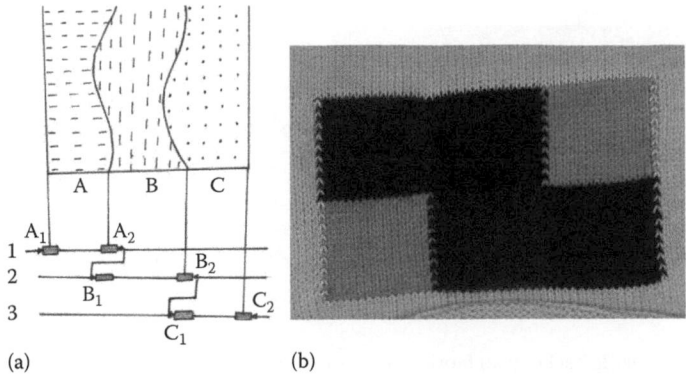

FIGURE 8.17
(a) Yarn carrier positioning for intarsia with three colors and (b) fabric appearance for intarsia.

Each block, knitting a separate colored area is exclusively supplied with its own particular yarn supplied by its own carrier, which traverses a limited and measured distance in accordance with the design requirement. The structure is cohesive, which is obtained by the minimum overlapping of adjoining areas and interloping in each wale applicable for plain, 1 × 1 rib, purl, or cable structures. An intarsia effect with three colors and the fabric appearance is shown in Figure 8.17a and b respectively. Intarsia are mostly obtained in V-bed knitting, including modern computerized machines such as Shima Seiki, Stoll, and so on producing different designs, including Argyle sweaters with their recognizable diamonds [6] of color and socks with different geometrical patterns, in addition to fully fashioned sweaters using expensive yarns to ornate the surface can be knitted by using the traditional method of intarsia.

8.8.3 Plating

Plating is widely used for single-jersey, plush, openwork float, and interlock fleecy structures. Colored yarns are used for embroidery plating [1]. Plating requires great precision and offers limited color choice. And ………………………………..?????

8.8.4 Single-Jersey Jacquard

Single-jersey jacquard produces weft-knitted fabrics by incorporating knit and miss stitches. The widthwise extensibility of these fabrics is comparatively less due to the floats utilized in the structure. Odd numbered needles are selected to form knit and miss stitch, whereas the even ones knit at every feed. Colored yarns are used to ornate the structure producing different types of motifs and designs mostly used for making woolen cardigans, pullovers, socks, and so on. The fabric appearance of such fabric with all needle backing along with the loop diagram of a complete repeat of the pattern is shown in Figure 8.18. One of the most popular design is *Fair Isle*, which is an ornamentation effect obtained on the face side of the fabric by using two or more differently colored yarn. A needle knits the yarn of one color along two wale lines in the face but the same yarn floats at the back along the next two wale lines. In the next course, a different yarn may be taken and floats at the back along the first two wale lines and knit along wale line. This goes on for four courses (say) and then the order is reversed. This produces a plaid effect resembling the pattern of a chess board. The loop diagram of Fair Isle structure is shown in Figure 8.19.

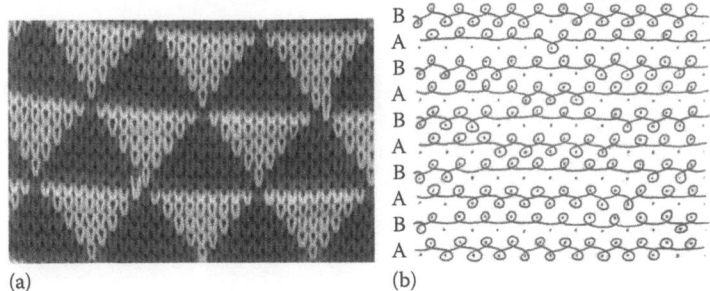

FIGURE 8.18
Knit jacquard with all needle backing: (a) fabric appearance and (b) loop diagram.

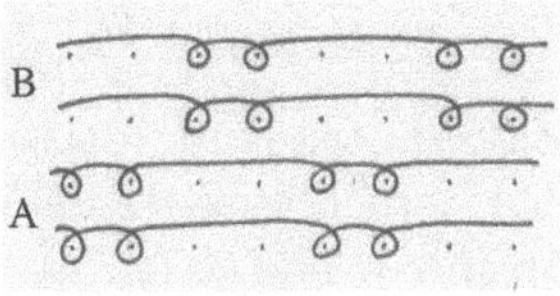

FIGURE 8.19
Loop diagram of a Fair Isle structure.

8.8.5 Accordion Fabrics

In accordion, fabrics with long floats are held in place on the technical back by tuck stitches. They were originally developed using knit and miss pattern wheel selection. It is a single-jersey structure produced mostly on woolen fabrics containing knit, tuck, and float stitches in the same course. According to the procedure of needle selection, there are three types of accordion: straight, alternate, and selective accordion. Figure 8.20 shows the loop diagram of straight and selected accordion fabric.

8.8.6 Circular-Knitted Single-Jersey Fabrics Using Tuck and Float Stitches

Tuck and miss stitches along with knit are used to produce different varieties of single-jersey weft-knitted structures, which are further ornamented by using different types and colors of yarn. The incorporation of these types of stitches along with the help of a needle selection device produces new structures with special and versatile characteristics and appearance suitable for various diversified applications. They are extensively used to

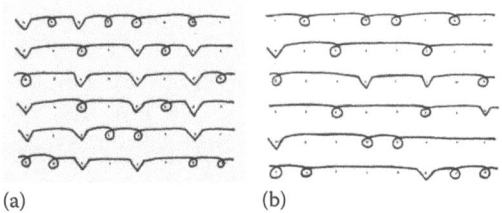

FIGURE 8.20
Loop diagram of accordion fabric: (a) straight accordion and (b) selected accordion.

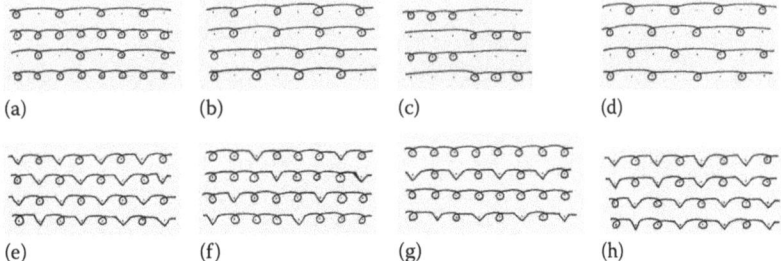

FIGURE 8.21
Single jersey fabrics: (a) weft lock knit, (b) birds eye, (c) mock rib, (d) cross miss, (e) single cross tuck, (f) simple crepe, (g) lacoste, and (h) double lacoste.

produce knitted garments, including T-shirts and slacks for men and women. The graphical representations of such structures are shown in Figure 8.21a–h.

8.8.7 Derivatives of Rib Fabrics

8.8.7.1 Cardigan Structures

The two major derivatives of 1 × 1 rib structure are half-cardigan or royal rib and full-cardigan or polka rib. They are produced by the incorporation of tuck stitches causing the rib wales to set apart, so that the width of the body disperses outward to a greater extent than the rib border, increasing the fabric thickness and making the structure heavy and bulky that are used for making sweaters in flat-bed knitting. They are repeated in two courses. Half cardigan is produced on a 1 × 1 rib base having tuck stitches along with normal loops on the back bed on alternate courses. The next immediate course is produced by a normal 1 × 1 rib structure. It is an unbalanced structure with a different appearance on each side. They are wide and thick and the tuck stitches provide elasticity to the fabric. Full cardigan has the same appearance on both sides as it is a balanced structure. It incorporates tuck stitches on both front and back bed along with normal loops in two consecutive courses, completing the repeat. Figure 8.22a and b shows the loop diagram, and the fabric image of half- and full-cardigan fabric produced on a V-bed knitting machine. If different color yarns are knitted with alternate courses a *short rib* will be produced, which in the relaxed state will show one color on one side and the other color on the opposite side.

The 2 × 2 rib version of half cardigan is termed *Fishermans Rib* and the full cardigan as *Sweater stitch*.

8.8.7.2 Milano Structures

Milano rib is a derivative and diversified rib fabric, which is produced in both flat-bed and circular-bed knitting. On a V-bed flat knitting, it is produced by selecting the needles of both front and back bed in the first course when the cam carriage moves from left to right followed by selecting the back-bed needles only in the second course when the cam carriage moves from right to left and front-bed needles in the third course in which the direction of the motion of the carriage is similar to the first course, thus forming a three-course repeat. Thus, the first course comprises a 1 × 1 rib structure, the second and third one having only back and face loops, respectively. In the case of half milano rib, the structure

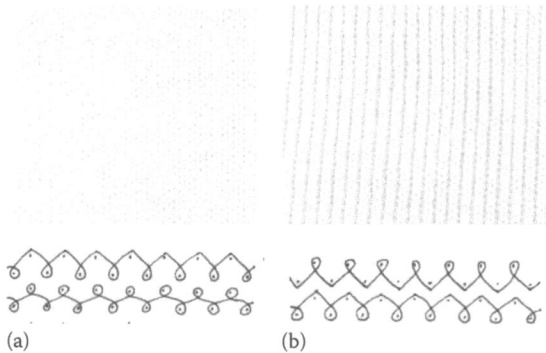

FIGURE 8.22
Fabric appearances and loop diagrams of (a) half cardigan and (b) full cardigan.

is produced in the same way as that of milano except the repeat completes in two courses, where in the second course only the front-bed needles are selected to form face loops. In alternating half milano the repeat ends in four courses in which the first and the third courses comprise 1 × 1 rib structure, followed by back loops and face loops in the second and fourth course, respectively. Another diversified rib structure is the rib ripple in which the repeat completes in four courses in which the first course comprises 1 × 1 rib structure followed by face loops in the next three consecutive courses, forming horizontal ridges resembling the configuration of wave or ripples. These structures can be produced with one color or two or more colors to get a desired multicolored effect using more feeders. They have a greater widthwise extensibility and are used for making sweaters in flat knits and fashioned collars in circular knits. Figure 8.23a–c shows the loop diagrams and the fabric images of different types of milano fabrics and Figure 8.24 shows these for a rib ripple fabric produced in a V-bed knitting machine.

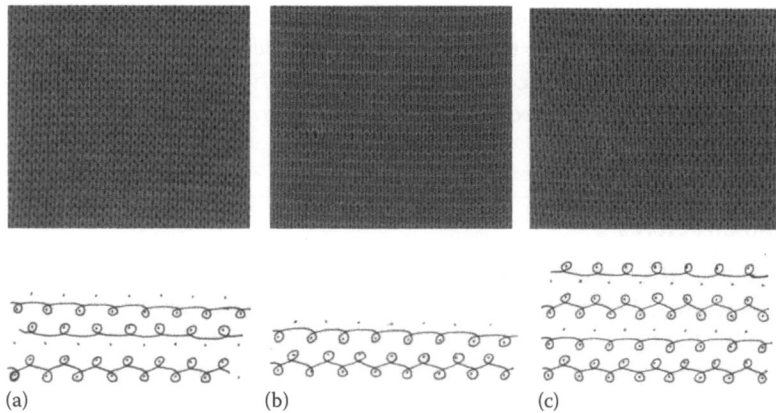

FIGURE 8.23
Milano structures showing fabric appearance and loop diagram: (a) milano, (b) half milano, and (c) alternating half milano.

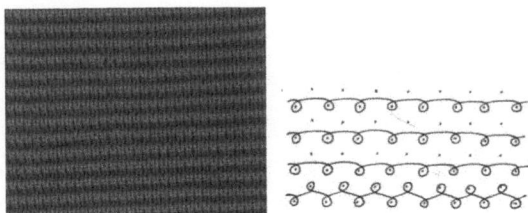

FIGURE 8.24
Fabric appearance and loop diagram rib ripple structure.

8.8.7.3 Racked Rib Structures

Racked rib structures are generally produced in a V-bed flat knitting machine by distortion of wales creating a zigzag effect. This is achieved when the needle bed is racked by one or more needle tricks either toward the right or left, past the other needle bed, after each traverse of the carriage as and when required. The structure is rib based along with tuck stitches as in cardigans and a loop on a needle in one bed is shifted or racked past a loop of the same course on a needle in the other bed creating a zigzag effect. Figure 8.25 shows such a structure with the front needle bed being shifted or racked by two pitches toward the right. The distortion is increased with the increase in the racking or shift of one needle bed with respect to the other by more needles. Cardigan structures consisting of courses of knitted loops on one bed and tucked loops on the other provide a more appropriate base for the racked structure.

8.8.8 Cable Structure

Cable structures are formed by the technique of transferring the loops from one needle to the other in the same needle bed. The three-dimensional cable design is achieved when two groups of needles interchange their places, so the wales formed by them cross each other [7]. A two-needle cable is formed by interchanging or switching of a loop with another. When a pair of loops switches places with another pair, a four-needle cable structure is formed. The cables are generally knitted on one needle bed and the base on the other. The cable stitch fabric has an appearance of a plaited rope or cable running in the direction along the wale line and a basic fabric structure is a rib knit widely used in sweaters [8]. Figure 8.26a shows a six-needle cable design produced on a flat knitting machine in which three loops switch or change their places with another set of three. Figure 8.24b shows a cable structure simulated on a Shima Seiki computerized knitting machine.

FIGURE 8.25
Fabric appearance of racked rib structure.

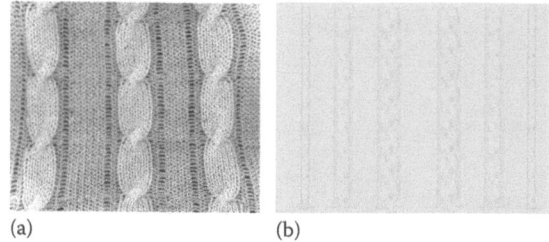

(a)   (b)

FIGURE 8.26
Cable design in (a) flat knitting and (b) simulated in Shima Seiki computerized knitting.

8.8.9 Lace and Leaf Patterns Produced by Loop Transfer

Weft-knitted fabric with open work designs having lace or leaf patterns is produced by the loop-transfer technique. The loop can be transferred from one needle to another by a special transferring instrument having a flat handle with an eyelet on a fine metal point, from one needle to another in the same bed or in different beds. When the transfer takes place in the same bed, the corresponding stitch is known as plain-loop transfer stitch and when the transfer takes place from one bed to the other in the corresponding stitch is called rib-loop transfer stitch [1]. Lace or leaf designs are produced by means of plain-loop transfer stitches. Lace can be used to form motifs and the fabrics are very attractive and provide infinite possibilities of design when used in a garment, the edges are required to be neat and stable by means of a stockinette stitch [6]. As it provides holes, this type of fabric is ideal for threading ribbons and other materials. Figure 8.27a and b shows the fabric image of lace and leaf pattern produced by plain needle loop-transfer stitches in which the transfer takes place in the front bed and Figure 8.27c shows an image produced by rib-loop transfer stitch.

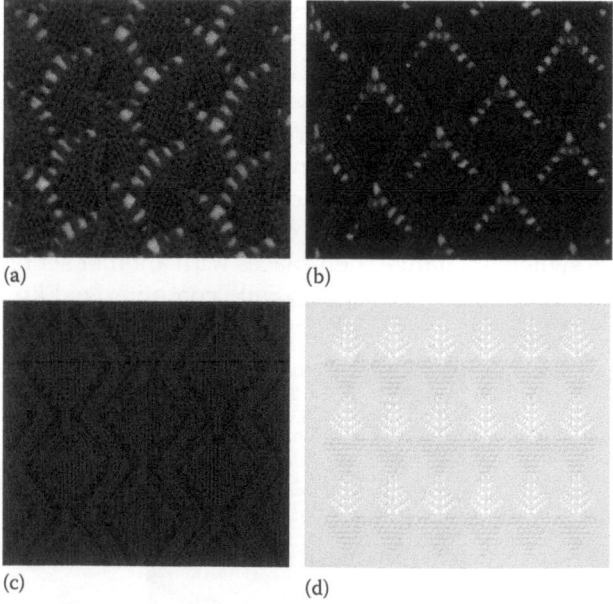

(a)   (b)

(c)   (d)

FIGURE 8.27
(a and b) Pattern by plain needle-loop transfer, (c) pattern by rib-loop transfer, and (d) lace pattern simulated in Shima Seiki computerized knitting.

8.8.10 Double-Knit Fabrics

Figure 8.28 illustrates the loop diagrams of different variations of nonjacquard double-jersey structures produced in a circular knitting machine having an interlock gaiting. They have a six or eight feeder sequence in which only alternate needles in one bed function in a particular course. Here, tuck and miss stitches are incorporated along with the normal knit stitch. A brief description of each is given below:

- *Bourrelet* is a typical circular-knitted double-knit fabric produced on a machine with interlock gaiting, having a ripple stitch or corded effect on the fabric surface. They have pronounced horizontal cords at regular intervals produced by knitting excess courses on the cylinder needles. The cord courses may be in a different color to the ground course. The repeat of the design completes in six courses.

- *Single pique or cross-tuck* interlock was one of the first interlock variation having tuck-knit interlock structures with a repeat completing on six feeders. Only alternate needles in one bed are in action in a course. The tuck stitches are given on the first and fourth feed, which make the structure fairly run resistant, approximately 15% wider than a normal interlock and increases the fabric weight.

- *Texi pique* is wider and bulkier and shows the same pique effect on both sides. The repeat completes on six feeders providing tuck stitches on both cylinder and dial needles in the first and fourth feed.

- *Piquette* is a reversible structure produced by knit and miss stitches in two pairs of consecutive courses with a 1 × 1 rib structure in between them. The repeat completes in six feeders that shows a light cord effect.

- *Pin tuck* incorporates tuck stitches in long- and short-dial needles on fifth and sixth feeders of a six-feeder repeat. The fabric is almost 15% wider due to the tuck stitches and the surface uniformity is broken and helps to mark feeder stripiness, which increases the fabric weight.

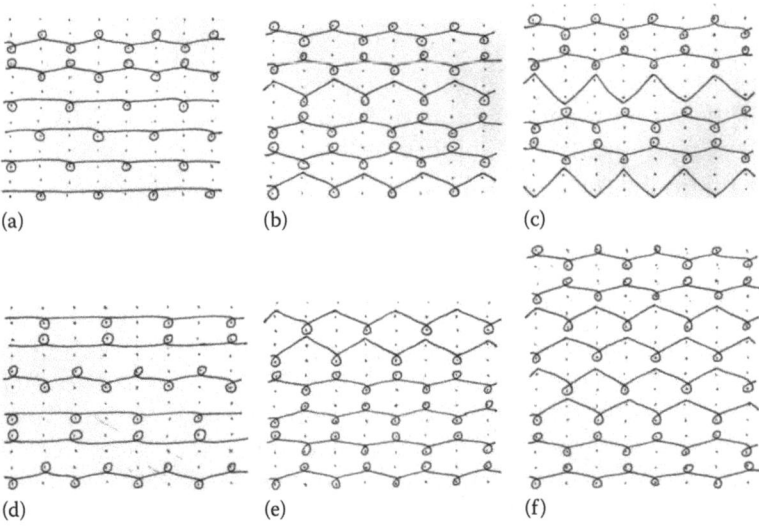

(a)   (b)   (c)

(d)   (e)   (f)

FIGURE 8.28
Double-jersey fabrics produced in a machine having interlock gaiting: (a) bourrelet, (b) single pique, (c) texi pique, (d) piquette, (e) pin tuck, and (f) ottoman rib.

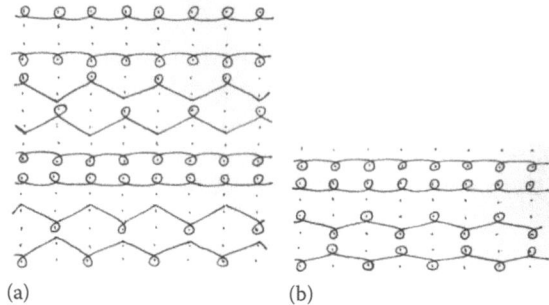

FIGURE 8.29
Double-jersey fabrics: (a) Evermonte and (b) Punto di Roma.

- *Ottoman rib* also known as horizontal ripple fabric is a double-knit fabric having pronounced ribs across the width of a fabric, which is achieved by knitting more courses per unit length than the other [8] and tends to be heavier with a less pronounced cord effect. The repeat completes in eight feeders with tuck stitches incorporated in four courses on alternate short- and long-dial needles. These four courses are embedded between a pair of 1×1 rib structure on either side.

- *Double pique* is a stable knit miss rib gaited fabric, which is narrower and has a less pronounced pique appearance than single pique and tends to be heavy.

- *Evermonte* has a row of tuck stitches on one side after each tubular course, which produces a slight ripple effect. The repeat completes in eight feeders with tuck stitches on four courses on alternate dial needles. In between two such courses a course of face and another of back loops are incorporated.

- *Punto di roma* has replaced double pique as the most popular nonjacquard double-jersey fabric. It belongs to a group of structures, which are reversible and have a tubular sequence of only dial and cylinder knit with a repeat completing in four feeders. It is usually produced in one color and is very elastic having a slight horizontal stripe effect [8]. Figure 8.29a and b shows the loop diagram of Evermonte and Punto di Roma.

- *Milano* is the rib equivalent of Punto di Roma with greater elasticity and width. It is used in the production of fashioned collars.

- *Birds eye* is a double-knit structure that creates a salt and pepper color effect on the back of a double-knit fabric.

8.8.11 Speciality Weft-Knitted Fabrics

Speciality weft-knitted fabrics include fleecy, plush, and high pile knits and knitted velour produced in some unique type of knitting machine. The surfaces or textures of the fabrics are made during the process of finishing on the technical back side of the fabric [1]. In *fleecy fabrics*, the fleece yarn, which is usually in the form of inlaid yarn is entangled and invisible and is tucked into the back at every fourth wale to be meshed with the binding or base yarn. The *plush* is distinguishable from the base and lies at an angle from the base surface. This fabric is popular for leisure wear and sportswear. The elongated sinker loops of plush are formed over a higher knock over the surface as compared to the normal ones, which show a raised or pile effect between the wales on the back of the fabric. The fabrics are softer, more flexible and are used in robes, coats, and so on. *High-pile knit* is a type of fabric produced by feeding staple fiber in the form of a sliver into the knit material while the yarns pass through the knitting needles forming a normal weft knitted jersey structure. The fiber is caught in the knit structure and is thus held within and between the plain loops. Special finishing treatments are given to produce a fur-like effect. High-pile knits are used in fur coats, coat lining, children's snow suits, footwear linings, hat linings, and so on [8]. *Knitted velour fabrics* are knitted with two yarns, a loop yarn and a ground yarn, feeding simultaneously into the same knitting needles. When the fabric is knitted the loop yarns are pulled out by special devices forming the loop pile, which is cut by a process called shearing. The velour-knitted fabrics are soft, flexible, and have a suede-like texture used in dresses, jogging suits [8], and other sports and leisure wear.

8.8.12 Weft-Knitted Spacer Fabrics

The knitted spacer fabrics are structures comprising two separately produced fabric layers [9], joined back-to-back during manufacturing. The two layers are produced from different materials. The yarn joining the two-faced fabrics can either fix the layer directly or space them apart forming two complementary slabs of fabric with a third layer tucked in between, giving a three-dimensional structure. Currently, spacer fabrics, also referred to as 3D fabrics, are being utilized in a myriad of applications varying from mattresses to seat cushions, automotive applications, active wear, extreme sporting apparel, and intimate wear. Spacer fabrics have several attributes that make them ideal for a variety of uses, including strength, insulation, breathability, and durability. Spacer fabrics or two-faced fabrics represent a class of knits, comprising two main layers connected by yarns transformed in tucks or loops. Weft-knitted spacer fabrics can be produced on circular double-jersey machines as well as electronically controlled flat-bed machines. Proper selection of spacer yarn and order of tuck-loop formation is essential to achieve proper features and properties of spacer fabrics in weft knitting. Electronic weft-knitting machines with two needle beds have the ability to create two individual layers of fabric, held together by loops or tucks, or by separate fabrics as connecting layers. The need for comfortable garments has been constantly a strong reason for developments, and double-layer knits made of natural and synthetic yarns, have been always recommended for producing clothes with a large range of thermal adjustment [10]. The knitted products are preferred to be worn in direct contact with the skin, because of their high extensibility, soft touch, and superior thermal comfort properties compared to woven fabrics. Recently, a lot of attention has been given to weft-knitted spacer fabrics due to their good transversal compressibility and excellent air permeability. Their good comfort properties [11] make them suitable for apparels and medical care. Weft-knitted spacer fabrics found fast development into a large range of products with applications in all areas of industry [12]. The compression rigidity of weft-knitted spacer fabrics made from shrinkable and nonshrinkable acrylic fibers [13] with knit–tuck structures are higher than that made from knit–miss structures. The esthetics and handle of textile products are mainly dependant on the bending behavior [14]. Compression and bending stiffness of the weft-knitted spacer fabrics make them relevant and suitable, especially for functional clothing and cushioning.

8.8.13 Fully Fashioned Weft Knits

Fully fashioning is a method of shaping a weft-knitted fabric during knitting, which involves the movement of a small number of loops at the selvedge ends of the

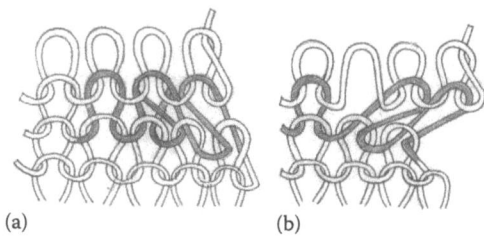

(a)   (b)

FIGURE 8.30
Wale fashioning: (a) single-needle narrowing and (b) single-needle widening.

knitted fabrics, which reduces or increases the total number of loops being knitted. Such types of movements are known as narrowing and widening, which are collectively termed as fully fashioning [5]. Fully fashioning is done on a flat-bed full fashioned garment-length knitting machine, which is capable of producing shaped weft-knitted garments by increasing or decreasing the total number of needles or wales in the shaping zone in which the individual pieces are more engineered, so that each garment piece is made with no extra fabric and the pieces are basically knitted together at the seams. In this case, each piece is shaped using techniques specific to knitwear by using special stitches, making lines of loops shrink and grow depending on where more length is needed and where not. The knitted structure is widened by increasing the number of needles in action or narrowed by decreasing the number of needles [8]. In narrowing, the innermost loop of the group being moved combines with the loop adjacent to it. Figure 8.30a shows two loops being moved by one loop space, thereby [5] losing one loop at the edge. In fully fashioning this is known as a single-needle narrowing. In two-needle narrowing instead of one the outer group can be moved by two needles. In widening, the outward movement creates a space adjacent to the innermost needle of the group in which a new wale may start. The empty space, followed by a tuck stitch formed in the subsequent course leaves a hole in the knitted fabric. Figure 8.30b shows a single-needle widening.

The shaping angle can vary by changing the number of plain courses between each fashioning course, aided by the possibility of single-needle, two-needle, or four-needle narrowing. Where narrowing occurs, a fashion mark appearing as distorted stitches results from the loops being transferred to the adjacent needles. The fashion mark is the transferred loop effect due to the transfer of a block of loops, clearly visible in the garment away from the selvedge, which is a hallmark of fully fashioned garments.

Fully fashioned shaped knitwear has got the following important characteristics:

- Shaped knitwear is engineered to size and shaped at the point of knitting.
- It is very distinctive and easily identifiable by the *fashioning marks*, which normally run parallel to the garment seams.
- The garment panels are assembled using *cup seaming* and *linking* in which usually the garment sides, sleeves, and underarms are cup seamed and the shoulders and collars are linked.
- The difference between linking and seaming is that with linking a stitch per stitch joint results, whereas cup seaming stitches the edges of the fabrics together.
- Fully fashioned knitwear includes sweaters, cardigans, ladies suits, jackets, coats, tights, stockings, lingerie, and so on.

References

1. Spencer DJ (2001), *Knitting Technology*, Woodhead Publishing, Cambridge, UK.
2. Ray SC (2012), *Fundamentals and Advances in Knitting Technology*, Woodhead Publishing, New Delhi, India.
3. Munden DL (1959), The geometry and dimensional properties of plain knit fabrics, *Journal of Textile Institute*, 50, T448–T471.
4. Mukherjee S (2010), Some studies on dimensional properties of double Jersey (rib structure) fabrics produced in both circular and flat bed machines, PhD Thesis, Maharshi Dayanand University, Rohtak, India.
5. Brackenbury T (2005), *Knitted Clothing Technology*, Blackwell Publishing, Oxford.
6. Elliott SA (1989), *Creative Machine Knitting*, Barron's Educational Series, New York.
7. Raz S (1993), *Flat Knitting Technology*, Meisenbach GmbH, Bamberg, Germany.
8. Cohen AC, Johnson I (2010), *J.J. Pizzuto's Fabric Science*, 9th ed., Fairchild Books, New York.
9. Anand S (2003), Spacers-at the technical frontier, *Knitting International*, 110 (1305), 38–41.
10. Farima D (2007), The ventilation capacity of the stratified knitted fabrics, *Tekstil ve Konfeksiyon*, 17 (3), 215–216.
11. Liu Y, Hu H (2011), Compression property and air permeability of weft knitted spacer fabrics, *Journal of the Textile Institute*, 102 (4), 366–372.
12. Araujo M, Fangueiro R (2011), Weft-knitted structures for industrial applications, in *Advanced in Knitting Technology*, K. F. Au (Ed.), Woodhead Publishing, Cambridge, UK, pp. 136–170.
13. Bakhtiari M, Shaikhzadeh Najar S, Etrati SM, Khorram Toosi Z (2006), Compression properties of weft knitted fabrics consisting of shrinkable and nonshrinkable acrylic fibers, *Fibers and Polymers*, 7 (3), 295–304.
14. Qun Du Z, Zhou T, Yan N, Hua S, Dong Yu W (2011), Measurement and characterization of bending stiffness for fabrics, *Fibers and Polymers*, 12 (1), 104.

9

Warp-Knitted Fabrics

Sadhan Chandra Ray and Mirela Blaga

CONTENTS

9.1 Introduction

The history of warp knitting is closely associated with two names—William Lee, who mechanized the process of flat-bed weft knitting in 1589 and Karl Mayer, the farsighted businessman and mechanic who exhibited his first warp knitting machine in an international trade fair in 1947. Karl Mayer's simple machine with only two guide bars used to run at 200 rpm using bearded needles and marked the beginning of an era full of innovative knitted fabrics. Karl Mayer GmbH introduced the company's own first Raschel machine paving the way for the *Super-Rapid* era in 1953 using latch needles.

In warp knitting, fabric is made by forming loops from yarns coming in parallel sheet form, which run in the direction of fabric formation (such as the warp in weaving). Large numbers of yarns in parallel sheet form are supplied from warp beams. Hence, warping is essential in warp knitting. Warp-knitting machines are flat and the needles fitted on the needle bar make the loops to simultaneously use the yarns coming from a number of beams through a large number of guides fitted on a number of guide bars [1–4]. The more is the number of guide bars, the more will be the scope of producing diversified warp-knitted structures, as relative order of movements (swinging and shogging) of the guide bars in successive knitting cycles is the key parameter in deciding the structure of the fabric. The movements of guide bars are traditionally controlled mechanically but electronic control devices are very common in modern machines. A few of the popular warp-knitted structures are locknit, sharkskin, queenscord, double atlas, velour, and so on. Locknit is the largest product of warp knitting and is popular in the United States as a jersey knit (Figure 9.1).

In the recent past both Tricot- and Raschel-type warp-knitting machines have been modified with double-needle bars keeping in view the production of speciality fabrics, which have opened up the new horizon in warp knitting. The two needle bars work back-to-back and are arranged in such a way, so that some of the guide bars can move through the needles of both the beds for swinging purpose.

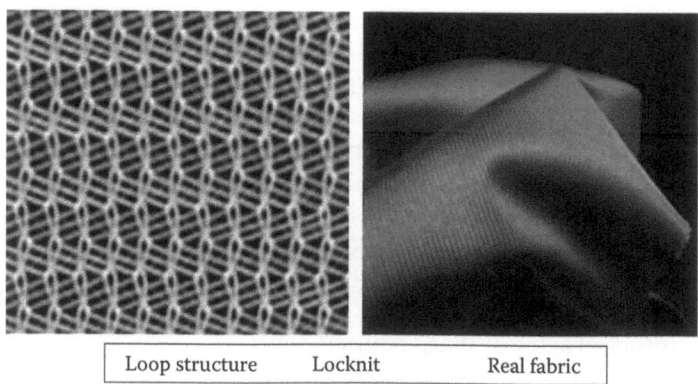

Loop structure   Locknit   Real fabric

FIGURE 9.1
Locknit warp-knitted structure.

The dream of knitting-shaped garment in warp knitting was materialized only when a double-needle bar warp-knitting machine (both Raschel and Simplex) came into existence [1,4]. The Karl Mayer HDR 16 EEW machine was introduced in 1970 for producing a range of simple garments such as seamless panties, brassieres, and pocketing. In 1999 the company delivered the first RDPJ 6/2 machines with 138 inches working width in gauge E24 with an electronic (Piezo) Jacquard system. In 2008 the company introduced the so-called *Seamless Smart* DJ 4/2 [1], a machine with a 44-inch working width in gauge E24. Further in 2009, Karl Mayer introduced its sister model the DJ 6/2, 44-inch machine in gauge E28 with an electronic pattern drive (EL) system A DJ 6/2 EL. Garments such as ladies' dresses with a drawstring waistband and stretch top are being produced in these two machines at high speed and in one-piece construction.

9.2 Types of Warp-Knitting Machines

The warp-knitting machines are classified into two categories: Tricot and Raschel. Both Tricot and Raschel may be made with either single-needle bar or double-needle bar. In the past, it was usual to distinguish between Tricot and Raschel by the type of needle used in each machine type. Tricot machines were equipped with bearded needles, whereas Raschel machines only used to knit with latch needles. With the production of modern warp-knitting machines, the compound needle has replaced the bearded needle in Tricot and penetrated into the Raschel as well. However, using either a special attachment fitted on the machine or due to the speciality of the product of any machine, the Tricot or Raschel machine can further be classified as pile warp-knitting machine, garment length warp-knitting machine, and so on. In addition, two more types of warp-knitting machine, that is, Simplex and Milanese are also available for producing speciality fabrics. The major differences in the features of Tricot and Raschel machines and their knitting principles are given in Table 9.1.

TABLE 9.1

Comparison between Tricot and Raschel Warp Knitting

	Tricot	Raschel
1	Bearded or compound needles are used	Latch needles are commonly used
2	Sinkers control the fabric throughout the knitting cycle	Sinkers only ensure that the fabric stays down when the needles rise
3	Less number of warp beams and guide bars (2–8)	More number of warp beams and guide bars (2–78)
4	Warp beams are positioned at the back	Warp beams are positioned at the top
5	Gauge is defined as needles per one inch	Gauge is defined as needles per 2 inches
6	Machines are made in finer gauges (commonly 28–40 needles per inch, may be lower also)	Machines are made in coarser gauges (commonly 24–64 needles per 2 inches, may be much lower also)
7	Mechanical design allows less accessibility on the machine	Mechanical design allows more accessibility on the machine
8	The angle between needle and fabric take-down is about 90°	The angle between needle and fabric take-down is about 160°
9	Comparatively simple structures are produced	Simple to complex structures can be produced
10	Mainly suitable for filament yarns	Mainly suitable for spun yarns
11	For guide bar movement the chain link numbering are 0, 1, 2, 3, 4, and so on.	For guide bar movement the chain link numbering are 0, 2, 4, 6, and so on.

9.3 Knitting Elements

In both Tricot and Raschel machines, yarns coming from the beam as a parallel sheet are converted in to fabric by loop formation before being wound in an open-width form on the cloth roller. Although the two types of machines differ to certain features, their loop formation technique is almost similar. The important functional elements required for the purpose of knitting are discussed in brief as follows:

- *Needles and needle bar*: The typical construction of all the three types of needles (latch, bearded, and compound) as shown in Figure 9.2 are used in warp knitting; however, the actual shape, particularly the stem part varies from machine-to-machine. Whatever may be the type of needle, all the needles move up and down together for loop formation at the same time, that is, all the loops in a course are made simultaneously. So, instead of giving motion to the individual needles, all the needles are connected/fixed to a bar called the needle bar (Figure 9.3) and the needle bar is lifted up and lowered down by means of a cam fitted outside the machine, generally at the driving side.

- *Presser bar*: In order to close the hook for casting off the old loop in a Tricot machine, some closing element, that is, presser bar is required. The pressure bar needed in a Tricot machine is set across the full width of the machine and gets motion from a cam or crank fitted on the main shaft. The presser bar closes the hook of the bearded needle when the same moves downward after catching off the new yarn for loop formation.

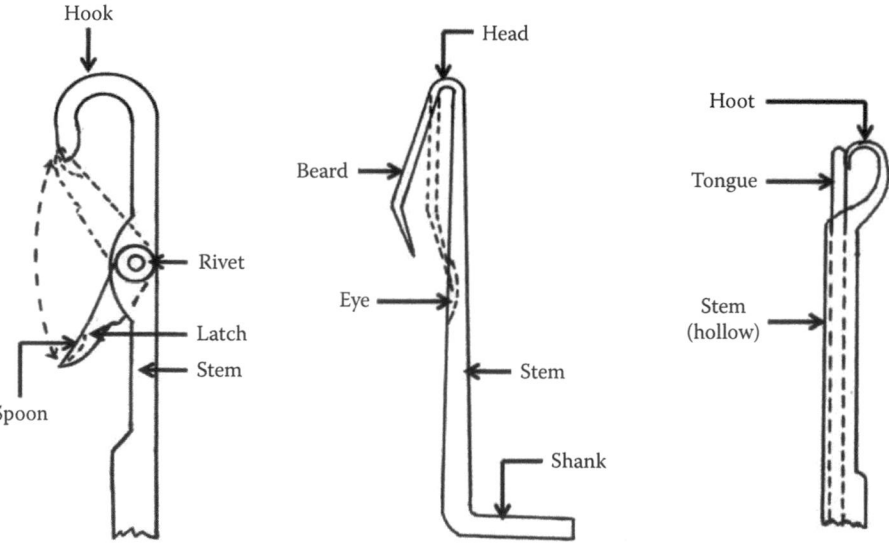

FIGURE 9.2
Types of needles used in warp knitting: (a) latch needle, (b) bearded needle, and (c) compound needle.

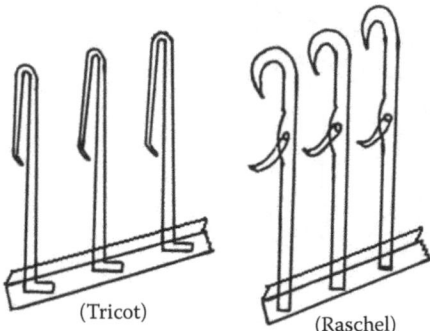 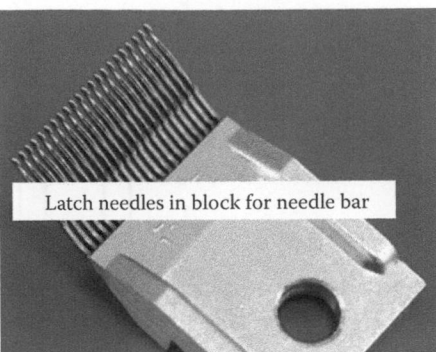

(Tricot)

(Raschel)

Latch needles in block for needle bar

FIGURE 9.3
Needle bars.

- *Sinkers and sinker bar*: The sinker is a thin plate of metal, which is placed between every two needles. The sinkers are usually cast in units of 1-inch long, which in turn are screwed into a bar called the sinker bar (Figure 9.4). The sinkers are given almost linear horizontal (forward and backward) motion through the sinker bar. The drive, generally, comes from a crank or eccentric arrangement. The neb and the throat of the sinker are used to hold down the fabric, whereas the belly of the sinker is used as a knocking-over platform.

- *Guides and guide bars*: Guides are thin metal plates drilled with a hole in their lower end for drawing a warp end through it. The guides are held together at their upper end in a metal lead of 1-inch width (Figure 9.5) and are spaced in it to the same gauge as the machine. The leads, in turn, are attached to a horizontal bar to form a complete guide bar assembly, so that the guides hang from it with each one occupying a position at rest midway between two adjacent needles. As the needle hooks face either the knitter or the back, the needle hooks do not catch the warp yarns during downward motion from the top position. The needles can only catch the warp yarns in their hooks if the guides wrap or lap the yarns across the needles. For this purpose, the guide bars are given a compound lapping movement. The number of guide bars in a machine is equal to the number of warp beams and each guide bar contains guides equal to the number of yarns in each warp beam.

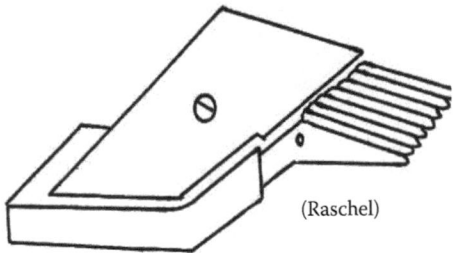

(Raschel)

FIGURE 9.4
Sinker bed.

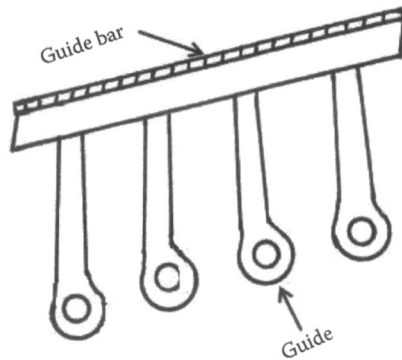

FIGURE 9.5
Guide bar.

- *Warp beams*: The required numbers of yarns are wound as a parallel sheet of warp (2) on a flanged beam (1) under uniform tension for supplying yarn in the knitting zone at a constant rate and tension (Figure 9.6). The warp beams in knitting are similar to the beams used in weaving but the technique of preparation may differ. There is no need of sizing but application of certain amount of oil/wax on the warp may improve the knitting performance. Both sectional warping and direct warping are applicable depending on the nature of the warp to be produced. Utmost care should be taken particularly for staple yarns, so that variation of yarn diameter and the presence of defects such as slubs and knots should be minimum in the final beam. The beam width (flange to flange) is generally equal to the width of the needle bar. The number of full-width beams in the machine is equal to the number of guide bars. The beams are situated at the top of the back side of the machine.

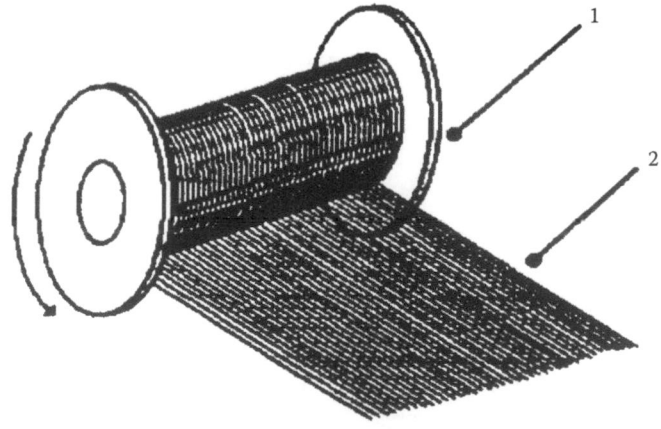

FIGURE 9.6
Warp beam.

9.3.1 Knitting Machine and Knitting Zones

With the help of a line diagram the different elements and zones of a typical warp knitting machine are shown in Figure 9.7 and only the knitting zones of Tricot and Raschel machines are shown in Figure 9.8.

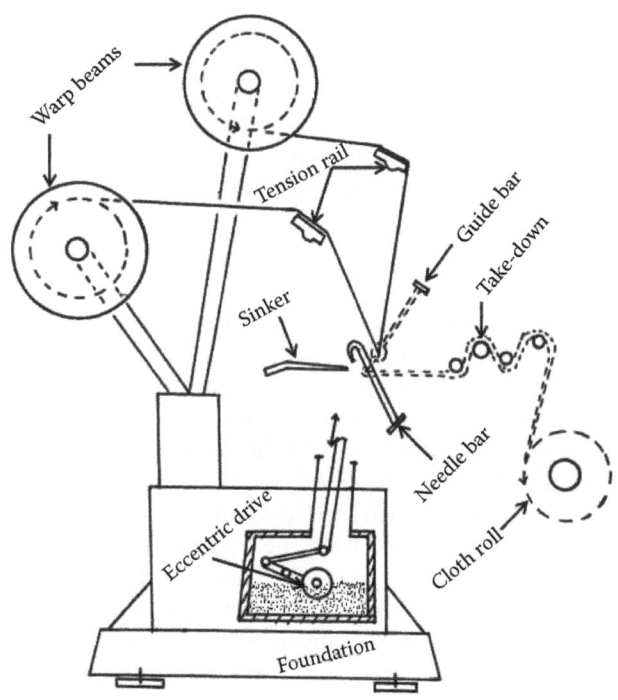

FIGURE 9.7
Line diagram of different elements and zones of warp-knitting machine.

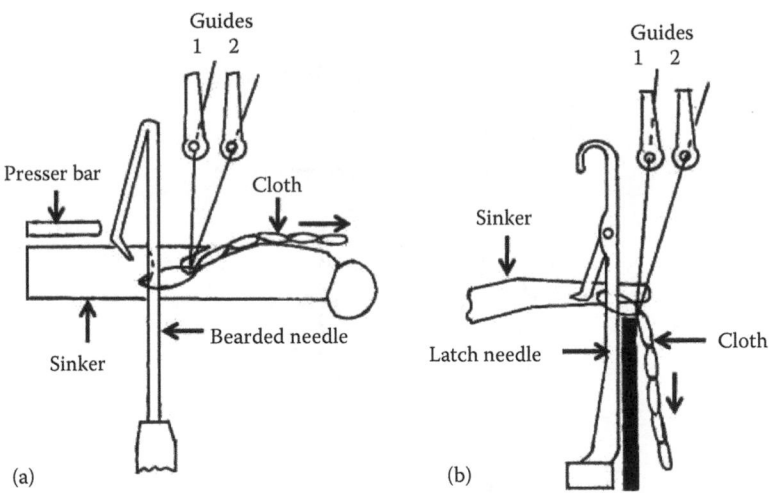

FIGURE 9.8
Knitting zone: (a) Tricot and (b) Raschel machines.

9.4 Technologies of Producing Warp-Knitted Fabrics

9.4.1 Needle-Bar Movement

The needle bar is lifted up and lowered down for the purpose of loop formation. During the upward movement, the old loop is cleared and a needle catches the yarn wrapped around it by the guide and forms the new loop during the downward movement. Such movement is imparted on the needle bar by means of a cam or eccentric fitted on a shaft called eccentric shaft. The shaft extends to the full width of the machine and the cam is located outside the machine, generally, at the driving side. The cam is kept in an enclosed oil bath in order to have less vibration, noise, heat generation, and higher life.

9.4.2 Guide Bar Movements

The guide bars are given compound lapping movement (Figure 9.9) for wrapping the feed yarns around the needle as well as for connecting the adjacent wales [1–4]. This compound lapping movement is composed of two separately derived motions—swinging and shogging. The swinging motion of the guides takes place either from the front of the needles to the back or from the back of the needles to the front, which occurs between adjacent needles. The swinging motion of the guides takes place either from the front of the needles to the back or from the back of the needles to the front. The two swinging movements produce the two side limbs when combined with the overlap shog.

The shogging movement of the guide bar is the lateral motion of the guides, which occur parallel to the needle bar. The shogging movement of guides may be from left to right or from right to left. The extent of shog, moreover, may vary from cycle-to-cycle as well as from bar-to-bar. The shogging of the guide bar may occur either in the front of the needles or at the back of the needles and accordingly produces the overlaps or underlaps. A shogging movement can occur when the guides have swung clear of the needle heads on the

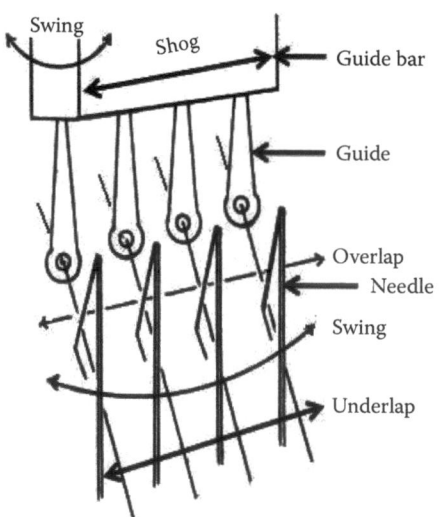

FIGURE 9.9
Guide bar movements.

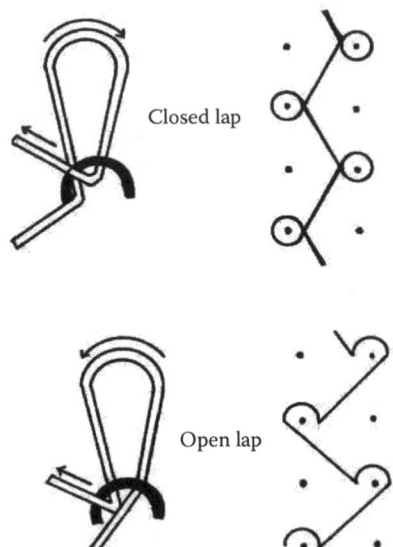

FIGURE 9.10
Types of loops.

back or front of the machine. The occurrence, timing, direction, and extent of each shog are separately controlled for each guide bar by its pattern chain links or pattern wheel attached to a horizontal pattern shaft driven from the main cam shaft.

A swinging motion and a shogging motion act at right angles to each other in order to form an overlap and underlap. The combined effect of underlap and overlap is the lapping of yarn around the needle. Depending on the relative direction of underlap and overlap, there are two types of laps—closed lap and open lap. The loops made of closed and open laps are shown in Figure 9.10. A closed lap is produced when an underlap follows in the opposite direction to the overlap and thus laps the thread around both sides of the needle.

9.4.3 Lapping Diagram

In order to produce various warp-knitted structures for various end uses, various types of lapping have been developed [1–4]. The lapping movements of the guide bars are composed of one or more of the following lapping variations (Figure 9.11) at any time during the knitting of fabrics:

1. An overlap followed by an underlap in the opposite direction (closed lap).
2. An overlap followed by an underlap in the same direction (open lap).
3. Only overlaps and no underlaps (open lap).
4. Only underlaps and no overlaps (laying-in).
5. Neither overlaps nor underlaps (miss-lapping).

In the last two cases, overlapping of another guide bar is required to hold them into the structure.

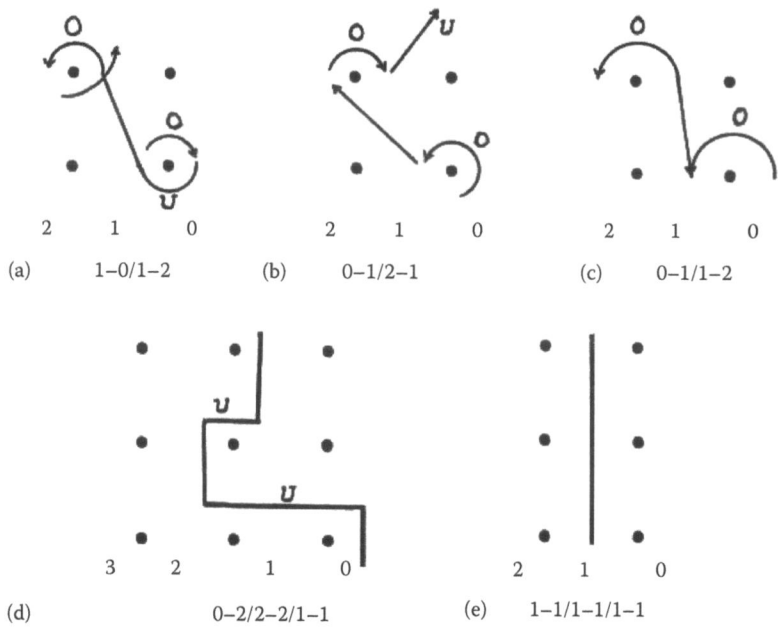

FIGURE 9.11

Types of lapping: (a) closed lap, (b & c) open lap, (d) laying-in, and (e) missed lap.

9.4.4 Fabric Take-Down and Warp Let-Off

In every knitting cycle, a small length of cloth is formed, which is pulled down from the knitting zone for rolling the same onto the cloth roller. This motion is known as fabric take-down. This taking down of the fabric also influences the quality and appearance of the fabric. The take-down is always positive as well as continuous. In order to allow fabric take-down, the required amount of warp is released from the warp beams. The releasing of warp from the beam is known as warp let-off. The rate of warp let-off from the warp beams is influenced not only by the rate of fabric take-up but also by the extent of traverse of the individual guide bars [1]. The parameters (courses and wales per inch, loop length, tightness factor, GSM, thickness, etc.) of warp-knitted fabric and ultimately the properties (air permeability, rigidity, drape, feel, etc.) of the same can be controlled by varying the rate of fabric take-down and warp let-off. It is obvious that unequal lengths of yarn are consumed from different beams for loop formation. Although both negative and positive type let-off motions are found in warp-knitting machines for feeding of the warp to the knitting zone, it is preferred to use only positive warp let-off, particularly for maintaining uniform yarn tension throughout the knitting.

9.4.5 Patterning Mechanisms

As mentioned in the earlier, the guide bars are given two types of motions—shogging and swinging. As the ultimate pattern or structure of the fabric depends on the nature (direction, relative position, and extent) of movements of the guides, the control of the nature of movements of the guides is very important. The following pattern-controlling mechanisms are generally used in warp knitting machines for imparting the necessary motions to the guides [1,3]:

1. Pattern wheel
2. Pattern chain links
3. Electronic jacquard

9.5 Warp-Knitted Stitches and Structures

9.5.1 Types of Stitches and Structures

The popular warp-knitted structures are mainly produced with two full guide bars. The structures are based on a two-course repeat cycle and the direction of lapping changes in every course. The two guide bars should invariably make different lapping movement; otherwise, the resultant structure would be equivalent to the structure produced with a single guide bar. The proportion of yarns in the fabric is influenced by the extent of underlap and overlap of the guide bars. The presence of yarns in the face or back side of the fabric depends on the controlling guide bar. Under normal conditions, the threads of the front guide bar dominate on both the face and back sides of the fabric. The nature of the guide bar lapping movement, considering two guide bars (front guide bar and back guide bar), is shown in Figure 9.12 for producing some popular warp-knitted structures.

9.5.2 Laying-In

During knitting of certain structures, for technological, designing, or commercial reasons, yarns of some guide bars are not knitted into fabric. Instead, the warp yarns passing through those guide bars are only inserted into the fabric, which is known as laid-in or inlay. To inlay warp yarn into the fabric a special lapping movement is required, but no additional special equipment is needed.

The laying-in technique is adopted keeping in view technical, designing, and commercial aspects. It is mainly done to modify one or more of the following properties [1] of the knitted structures:

1. Stability
2. Handle
3. Surface interest
4. Weight
5. Visual appearance
6. Elastic stretch and recovery

9.5.3 Full-Width Weft Insertion

Weft insertion during warp knitting is carried out to achieve both esthetic and technical advantages [1–4]. The weft-insertion technique was developed by Karl Mayer in the 1960s

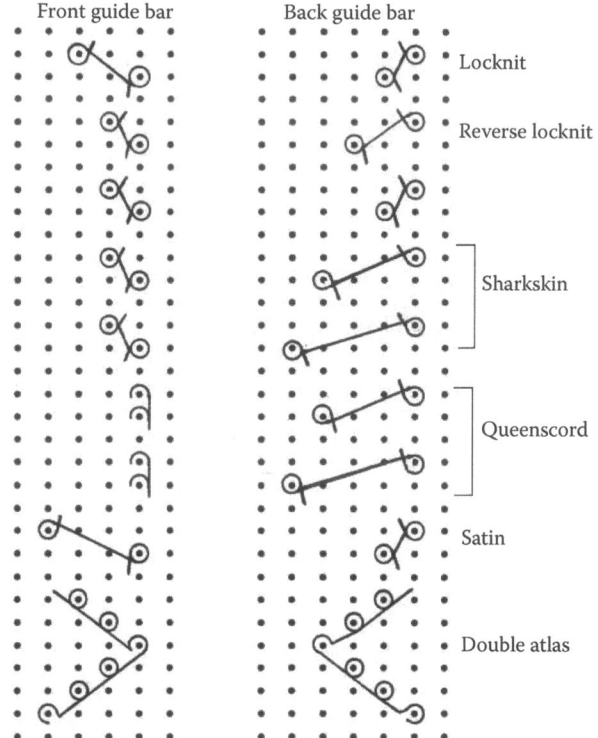

FIGURE 9.12
Popular warp-knitted structures.

for knitting fabrics with dimensional stability and some other special properties required in apparel. In a similar technique to laying-in, the weft yarns introduced into the fabric do not enter the hooks of the needles and so do not form loops. It is, therefore, possible to insert different yarn types, which are otherwise not suitable for knitted fabric production. Figure 9.13 shows the locked-in position of the weft yarns inside the warp-knitted structure. The inlaid weft yarns run horizontally from selvedge-to-selvedge as weft in woven fabric, as a result, the widthwise shrinkage and other properties of the warp-knitted fabric are similar to those of woven fabric.

The main objects [1] of weft insertion may be summed up as follows:

1. To stabilize the fabric in the width direction
2. To improve fabric cover
3. To introduce fancy yarns and yarns of multiple colors in the width direction
4. To introduce materials or lower quality yarns, which could not normally be used in warp knitting
5. To introduce functional property in the fabric required for special end uses

Weft insertion during knitting requires a special attachment/device termed as a weft-inserting element to be fitted in the knitting machine along with other knitting elements. In addition to the weft-inserting element, the other requirement is the supply of weft yarn

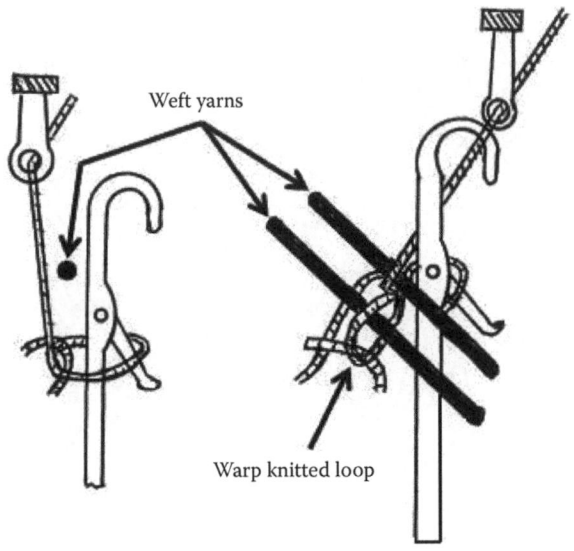

Weft yarns

Warp knitted loop

FIGURE 9.13
Weft yarns insertion.

from a creel positioned at the back. Nowadays, both Tricot and Raschel machines are available with a weft-insertion facility.

9.5.4 Milanese Fabrics

Milanese fabrics are generally acclaimed as the superior quality warp-knitted product, in a class of their own but expensive [3]. However, due to their peculiar construction they cannot be made on ordinary warp-knitting machines. The term *Milanese* has originated from Milan, the city in Italy, the center of production of warp-knitted silk fabric. Originally, it was a fine gauge warp knitted silk fabric characterized by superb hand and smooth texture. The appearance of Milanese fabric gives a resemblance to the plain weft-knit structure. The Milanese knitting machines (both continental and English) are specially constructed with limited design and stitch variation possibilities. A Milanese fabric is equivalent to a two-bar Tricot fabric, effectively constructed from two sets of warp threads, but the lapping movements for Milanese fabrics are so arranged that each warp thread traverses across the full width of the fabric from one selvedge of the fabric to the other, and not over a limited number of needles as on other types of machines. The warp threads are divided in to two sets, and one set of threads is traversed across the fabric from left to right, whereas the other set traverses at the same speed in the opposite direction (Figure 9.14). At each side of a straight bar Milanese machine, the threads are transferred from one set to the other to preserve the alignment of the threads with the needle bar. The machine is, therefore, provided with two sets of warp threads, which are moving continuously in opposite directions across the needles, and, with the usual-set threading, each needle will receive two warp threads at each course. As the warp threads traverse across the whole width of the fabric, warp cannot be supplied from a stationary beam. Movable sectional beams or flanged bobbins are used in the warp creel for supplying the warp. Fine bearded needles are used for a straight bar Milanese machine and latch needles can be used in circular Milanese machine.

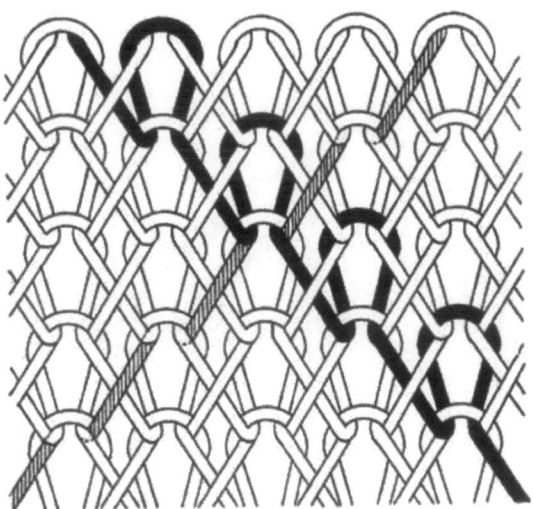

FIGURE 9.14
Milanese structure.

There are basically two different types of Milanese fabrics and they differ from each other in the manner in which the threads are traversed across the fabric. When the cotton lap is used the warp threads of both sets traverse across the machine by passing from one needle to the next as each successive course is knitted. The lapping movement for the silk lap, however, is such that each warp thread makes overlaps on alternate needles only, and thus moves two needle spaces at each course, passing under one needle and over the next. Another peculiarity of these overlaps is that they form open laps. The cotton lap is used when knitting spun yarns, whereas the silk lap is used for continuous filament yarns.

9.5.5 Tulle Fabric

Tulle is lightweight, extremely fine, and surprisingly strong and durable, machine-made netting, usually with a hexagon-shaped mesh effect. It can be made of various fibers, including silk, nylon, and rayon, although polyester yarn is the first choice. The production of tulle started with bobbinet technology, was invented in Britain in the early nineteenth century, but presently such fabrics are mostly made in warp knitting using 50 D 100% polyester filament yarn [5]. End uses include dance costumes, veils, ballet tutus, and gowns, particularly wedding gowns, but also widely used for undergarments, sportswear, and for lining purposes. So, the most common uses for tulle nettings are in garments (Figure 9.15). Tulle is often used as an accent, to create a lacy, floating look.

9.5.6 Production of Stretch Warp-Knitted Fabrics Using Tencel Fiber

The production of stretch warp-knitted fabric (Figure 9.16) from eco-cotton and the elastomer started only in the new century, and the interest of the people was attracted only when the first recyclable bra made of such fabric was shown at the Nutec trade fair in November 2008 in Frankfurt [6]. The next phase of development took place in 2014, when Lenzing's

FIGURE 9.15
Tulle fabric and garment.

FIGURE 9.16
Stretch warp-knitted fabrics from Tencel.

tencel fibers were used instead of eco-cotton. The following types of fabrics are nowadays produced from the combination of tencel and elastomeric yarns:

- Smooth, two-way-stretch fabrics having a dense surface and soft handle on an RSE Raschel machine
- Lightweight, all-over-patterned lace produced on a lace Raschel machine
- Tulle having two different performance profiles, depending on the percentage of elastane used, and exhibiting a distinct, natural look, also produced on an RSE Raschel machine
- Raschel-knitted fabrics with geometric patterns
- Fabrics with a ribbed construction on the surface and two-way stretch

Very recently, such stretch warp-knitted fabrics are getting popular in the manufacture of lingerie, swimwear, and sportswear due to their soft and supple handle as well as better comfort and form-fitting properties.

9.5.7 Delaware Stitch and Modified Delaware Stitch Tricot Fabrics

Delaware stitch is a very popular warp-knit structure. It has a woven-like appearance, low stretch, and is used in men's and women's outer garments, and is extensively used in women's lingerie. This stitch was developed by M/S DuPont DeNemours and Co [7]. There are wide variety of Delaware stitches (DSs), the simplest DS offers a warp-knit fabric produced with the front bar knitting a three-needle float stitch (2–3, 1–0) and the back bar knitting a chain stitch (1–0, 0–1), which appears similar to a woven fabric. By varying the number of needle float, warp-knit structures can widely be changed in terms of appearance, weight, drape, and other characteristics.

The DS-type fabrics, compared to the Jersey stitch construction, offer woven-like low stretch, greater liveliness, and new esthetics, including satin-type luster. They also offer greater fabric stability, higher air permeability, and greater rigidity. Moreover, special color effect can be produced on the two sides of the fabric by using different colors in the front bar and back bar. In spite of such advantages, DS has some demerits such as shifting of wale and stitch distortion.

9.5.8 Warp Knitted Nets

Different types of nets are the most interesting products of warp knitting, particularly using a Raschel machine. Different techniques such as inlay, weft insertion, fall plate, and tulle are used for producing nets with attractive designs and mesh openings. The selection of technique fully depends on the end application of the nets and the design and mesh size are produced accordingly. The inlay technique is mainly utilized for producing marquisette curtain net and the tulle net is used as a ground structure for general curtain fabrics. The traditional nets used for foundation wear are the power nets. Power net is a four-bar structure in which both ordinary and elastomeric yarns are used. The elastomeric yarns become straight and lie vertically in the fabric when knitted under high tension and forces the loops of other yarns to distort. Production of net is basically the creation of surface interest on the fabric. This surface interest can be produced either by variation in threading in one or more guide bar(s) or by variation in the extent of underlapping. Net formation in warp knitting is much easier than weft knitting. No special requirement is needed for this purpose on the machine, only two bar frames are sufficient. Generation of suitable lapping movement in conjunction with two partially threaded guide bars are the minimum requirement for knitting nets. The manipulation of the abovementioned parameters may result in a mesh or opening size to be very fine to too coarse and at the same time meshes of different shapes such as diamond, hexagonal, or nearly circular can be produced (Figure 9.17).

FIGURE 9.17
Warp-knitted nets of various designs.

9.6 Double Needle Bar Warp Knitting

9.6.1 Double Needle Bar Tricot Knitting

Tricot machines with double needle bar are called Simplex machines. Double needle bar Tricot machines was, traditionally, used in the glove industry; however, these machines can produce fabrics for outerwear and innerwear as well. Although the gauge (needles per inch) range is 28–34, the most popular gauge is 32 and the machines are used for producing finer fabrics. The two needle beds are not parallel but are approximately at an angle of 45° to each other. The speed of Simplex machine may be up to 300 courses per minute while knitting simple designs with quality yarns.

9.6.2 Double Needle Bar Raschel Knitting

Different types of double needle bar Raschel machines are built today for the production of a wide variety of products ranging from simple sacks to complicated artificial blood vessels. Seamless tubular fabrics, pile fabrics, and many other speciality fabrics can easily be produced in double needle bar machines by using 4–6 guide bars. But by using more guide bars (16 and above), production of branching tubular fabrics such as artificial blood vessels, patterned panty hose, and so on, and shaped innerwear is now easily achieved.

In the case of Raschel, the two needle bars are vertical as well as parallel to each other and the needles in the two beds face back-to-back. Moreover, the gap between the two beds is adjustable. The arrangement of the knitting elements as well as the knitting action is not complex like Simplex machines. At the same time attachments such as a fall plate, creeping motion, Weft inlay, and so on can be fitted for producing speciality products [1,2]. The creeping motion is used to disengage any one of the needle beds for a specific number of knitting cycles. These machines are generally made with coarser gauges compared to Simplex machines. The first (clearing) phase of the knitting action of the front needle bar is shown in Figure 9.18. A similar action is observed also for back needle bar.

- The front needle bar is lifted up for clearing the previous overlaps from the hook and latches. The back needle bar holds down the last row of loops of the fabric at the idle position.
- The guide bar swings from the back of the needles to the front of the machine and then shogs for producing the overlap and again swings back.
- The needle bar moves down and the new loop passes through the old loop and the old loops are cast off.
- The guide bar swings over the front needle bar, so that the back needle bar can be lifted for commencing the knitting action.

9.6.3 Pile and Spacer Fabrics on Double Needle Bar Machine

Pile fabrics (both cut pile and point pile) are produced on double needle bed Raschel machines. Cut pile is achieved by knitting a separate base fabric on each needle bed but joining the two together by a lapping movement of the pile, which is later slit to produce the two cut pile fabrics. Cut pile fabrics are employed for a wide range of end uses such as simulated fur and skin fabrics, upholstery, and coat/jacket lining. Each bed knits alternatively and has a cam shaft, needle bar, trick plate, sinker bar, and two guide bars with no

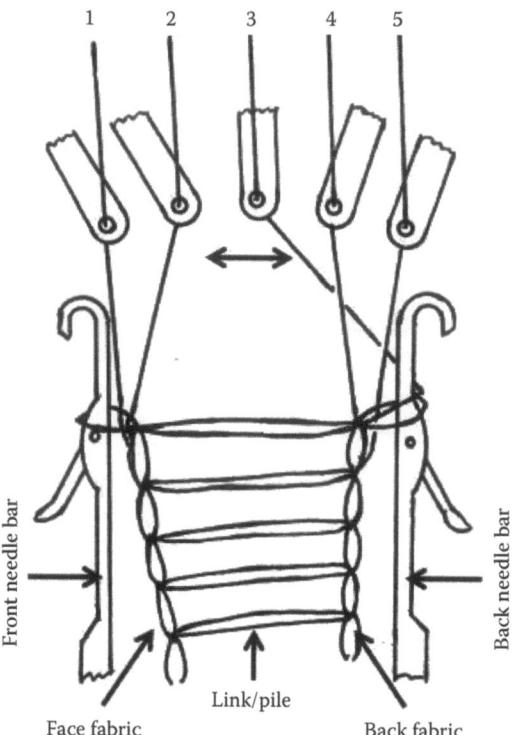

Front needle bar

Back needle bar

Face fabric

Link/pile

Back fabric

FIGURE 9.18
Double-needle bar Raschel machine.

swinging action. The needle bar and trick plate swing through these two guide bars to produce the base structure on that particular needle bed. The middle (pile) guide bar has normal swinging facilities for lapping the pile alternatively on each needle bed. As the pile is severed in the centre, its height is half the distance between the two trick plates. This distance may be altered to produce a range of pile height. The effect produced is determined by a combination of type of fiber, denier, lapping movement, and finishing treatment applied. In point pile, the loops are produced at a right angle to the common base fabric knitted together by both the needle beds. The pointed or projected piles are sometimes cut or sharpened to achieve a softer feel. Such fabrics are mainly suitable for floor covering and carpeting. The technique of producing spacer fabric in double needle bar warp-knitting machine, as shown in Figure 9.18, is also applicable for producing cut pile fabric. The spacer fabrics are mainly suitable for winter garments and medical textiles [1].

9.7 Warp-Knitted Cloth and Clothing

Textile fabric or *cloth* is defined as—a two-dimensional plane-like structure formed out of textile materials, having reasonable strength, elongation, flexibility, and so on for different applications, particularly for apparels. *Clothing* or apparel (also called *garment*) is manufactured textile material worn on the body. The amount and type of clothing one wears depend on physical requirements and local culture. The most obvious function of clothing is to improve the comfort of the wearer, by protecting the wearer from the elements, including rain, snow, wind, and other weather, as well as from the sun and to make the wearer modest by hiding the personal zones of the body. Clothing protects against many things that might injure the uncovered human body. Clothes also reduce risk during activities such as work or sport. Some clothing protects from specific environmental hazards, such as insects, noxious chemicals, weather, weapons, and contact with abrasive substances. Similarly, clothing has seasonal and regional aspects; clothing performs a range of social and cultural functions, such as individual, occupational, and sexual differentiation, and social status. In many societies, norms about clothing reflect standards of modesty, religion, gender, and social status. Clothing may also function as a form of adornment and an expression of personal taste or style.

Moreover, from users' point of view the following characteristics of the fabric passed on to the garment are very much important:

- Drapeability
- Wrinkle resistance
- Dimensional stability
- Durability
- Porous structure
- Soft handle
- Lightweight
- Ease of care

In today's context, the environmental aspects should also be incorporated in modern warp-knitted fabric and garment technology.

In order to fulfill the abovementioned requirements of the clothing in the end application, it is very much essential to engineer the basic textile material, that is, the textile fabric to be used in the manufacture of the clothing. Out of the various techniques of fabric manufacture, warp knitting is very versatile and provides sufficient scope for engineering for the knitted structure by suitably manipulating the various input parameters and desired raw material as mentioned as follows [1]:

- Fineness, quality, fiber composition, color, finish, and so on of the yarn
- Number of warp beams and guide bars
- Thread density and relative gaps of yarns in various guide bars
- Extent and direction of swinging and shogging motion of guide bars in successive knitting cycles controlled by the patterning mechanism
- Nature of lapping of yarn around the needle
- Relative rate of consumption of yarn from different beams during knitting
- Type of knitting machine as well as number of needle bar in the machine

In fact, a very wide range of clothing—intimate to outer—for both women and men whether young or old can be either manufactured with warp-knitted fabric or can be directly produced in warp knitting.

In warp-knitted fabrics, generally, the face is the loop side and the other side is the float or back side. It is important to identify the face and back of fabrics for garment purposes. In a warp knit, either the loop side or float side could be used as the face of the garment. For warp knits made with 100% nylon or polyester, the loop side is more attractive and used as the face side in garments. When warp stitch is made from nylon or polyester in the front bar and spandex on the back bar, due to contraction of spandex, the float or back side becomes more attractive and can be used as a face in garments.

Most warp knit fabrics tend to curl, and hence appropriate heat treatment is essential for thermoplastic yarn warp knit fabrics in order to result in minimum trouble in cutting and sewing of garments.

The warp-knitted fabrics (particularly Tricot) were traditionally considered as stiff or rigid fabric and unsuitable for apparel other than dancing frocks. Moreover, warp-knitted fabrics made out of ordinary filament yarns were not skin friendly. Due to the various developments in warp-knitting machinery and developments of newer and speciality yarns have resulted in the production of fabrics with desired properties for apparel manufacture. At the same, new types of measuring instruments can nowadays objectively quantify the handle, drape, comfort, and other surface properties (roughness, softness, etc.) required for garmenting as well as for the end application. In cases in which some of the properties of the warp-knitted fabric differ with the desired ones, these can be modified by means of the latest types of finishing treatments.

The continuous exploration of the potentials has established the suitability of warp-knitted fabrics not only in apparel and furnishings but also in most of the technical textile applications. In fact, a very wide range of clothing—intimate to outer—for both women and men whether child or young or old can be either manufactured with warp-knitted fabric or can be directly produced in warp knitting. The types of apparel commonly made nowadays out of warp-knitted fabrics are as follows:

- *Ladies apparels/dresses*: Top (long, short, fancy, sleeveless, casual, and off shoulder), Skirt (pleated, gypsy, beach, and frill), Kurti (full sleeve, sleeveless, designer, fashion, printed, and short), T-shirt, shirt, trousers, capri, leggings/jeggings, pajama, gown, jackets and hoodies, blouses, tunic, shorts, night dress, jumpsuit, party wear suit, coat, sportswear, Islamic abaya and hijab, maternity clothing, thermal wear, dungarees, ethnic dress, bra, panty, lingerie, swimsuit, and so on.

- *Gents apparels/dresses*: Formal and casual shirts, T-shirt, trousers, blazer, jacket, house coat, nightgown, sportswear, pajama, briefs, vest, swimwear, and even ethnic dresses.

Figure 9.19 represents the display of widely used ladies tops in shopping malls.

FIGURE 9.19
Display of ladies top made of warp-knitted fabric in shopping mall.

9.8 Garment Making on Double Needle Bar Warp Knitting

The dream of knitting a shaped garment in warp knitting was materialized only when double needle bar warp-knitting machine (both Raschel and Simplex) came into existence with the capacity of producing tubular fabrics with branches [1,4,8]. Tubular fabrics are generally produced on double needle bar machines using 4, 6, 8, or even more guide bars in such a way that the side connections are identical to the body structure for forming a seamless tube. The technique of producing branched tubular fabric has opened a new horizon for producing seamless-shaped panty hose and many other types of garments and artificial blood vessels (cardiovascular tube). In 1970, Karl Mayer introduced HDR 16 EEW machine for producing a range of simple garments such as seamless panties, brassieres, and pocketing. The first double needle bar Raschel machine with jacquard system was built in 1980 by Nippon Mayer, Japan, a Karl Mayer subsidiary, mainly for the production of panty hose. The existence of two needle bars in the machine makes it possible to produce tubular fabric along with branching. An electronically programmable shaping device helps to vary as well control the diameter of the branches at any point. Knitting starts with one tube and as and when required branching takes place. The length of the base tube as well as of the

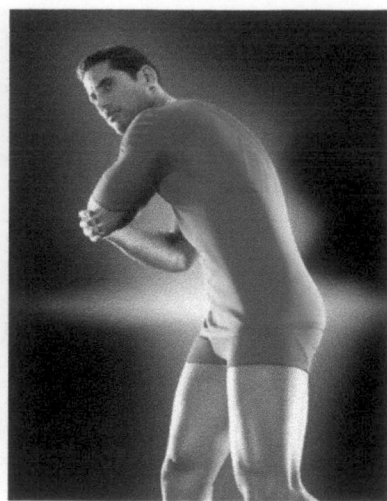

FIGURE 9.20
Seamless sportswear.

branches can be controlled precisely. Moreover, electronic control of the selection and patterning devices facilitates the shaping of the tubular fabric, which is required for garment making. Double needle bar warp-knitting machines are, in general, slower than the single-needle bar machine, but the recent double-needle bar machine with an electronic patterning device can produce a lady's dress with a drawstring waistband and stretch top at a comparatively higher speed and in one-piece construction. In 2008 and 2009, the company successively introduced the so-called *Seamless Smart* DJ 4/2 and DJ 6/2. These machines were developed to provide versatile knitting structures in order to produce a wide variety of products ranging from innerwear such as seamless foundation wear and panty hose up to outerwear. Two such seamless sportswear garments are shown in Figure 9.20.

9.9 Application of Computer in Warp Knitting

The recent warp-knitting machines are of high speed and smoothly operating, technologically advanced, and computer controlled. The developments with regard to electronics and microprocessors for warp knitting were focused mainly on two objectives: (1) to improve the patterning possibilities and (2) the production efficiency. In the case of Tricot machines, the developments have been concentrated on the beam control, the patterning mechanism, the fabric take-down, and the pattern design software solutions. For the multiguide bar and Jacquard Raschel machines, the challenging objective is the time-consuming patterning process. Thus, the machine builders channel their efforts into electronic patterning aids [4]. The main computer-controlled developments have, generally, envisaged improvements in the following aspects:

1. Electronically controlled yarn feeding unit
2. Electronically controlled patterning mechanism
3. Electronically controlled fabric take-up mechanism
4. Pattern design software

9.9.1 Electronically Controlled Yarn Feeding Unit

The electronically controlled warp beam drive (EBC), produced by the Karl Mayer company, for the positive feeding of the yarn sheet, has been implemented for the high-speed Tricot machines, and it is recognized by certain advantages, such as: the constant and accurate yarn feeding, precisely controlled warp tension, low time for pattern changing, and reproducible patterns. But the most significant advantage remains the multispeed feeding sequences, up to 85, which can be accurately programmed. The system has found its utility especially in the production of elasticated swimwear in which a combination of different feeding sequences and lapping movements is required to run-in together [4]. The EBC system (Figure 9.21) presented by the Liba company [9] consists of a microprocessor (3), used to control the beam drive (1). The programming of the system is carried out directly on the monitor (2), the value necessary for the operation (the inside and outside beam circumference, number of revolutions, number of courses in each sequence, and amount of run-in per rack per sequence) are fed into the system. The constant control of the beam circumference is made by the rollers (4), and all information is sent back to the computer to calculate the decrease in warp circumference during each sequence of the knitting process.

9.9.2 Electronically Controlled Patterning Mechanism

In order to meet the requirements of a large pattern repertoire and flexibility, in both stretch and nonstretch fabrics, these machines are equipped with the electronic guide bars (EL). Electronically controlled patterning mechanisms [4,10] provide a variety of possibilities in regard to the production of warp-knitted fabrics, with the same threading arrangement of the guide bars, by simple commands the guide sideways displacement is produced.

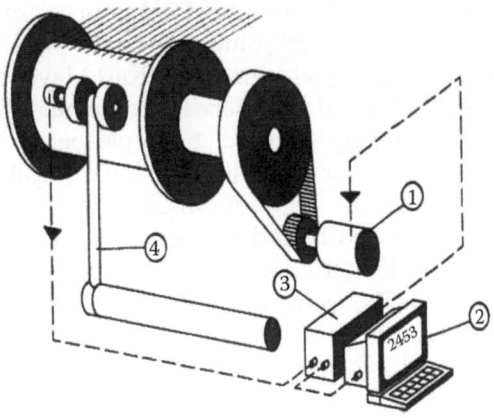

FIGURE 9.21
EBC electronic beam control. (Courtesy of Liba.)

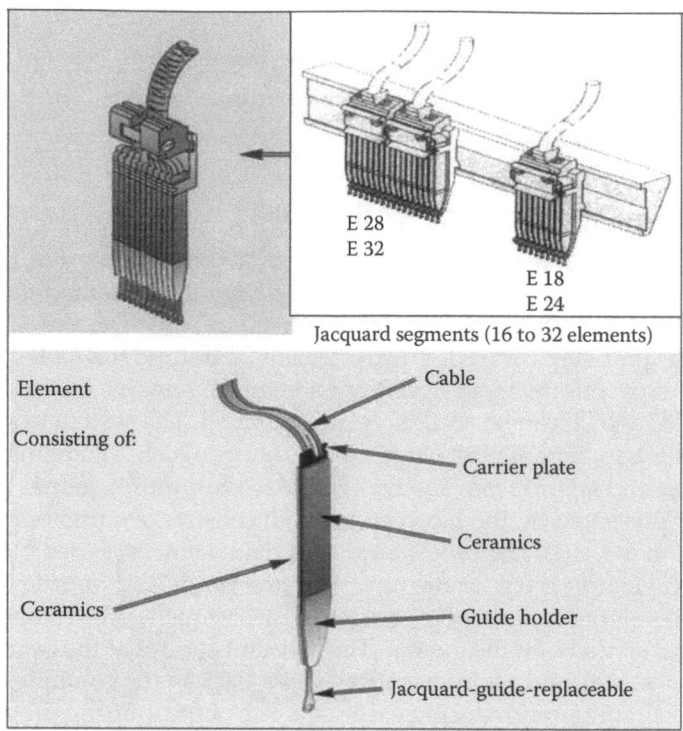

FIGURE 9.22
Jacquard element. (Courtesy of Karl Mayer.)

This way, very quick pattern changing is achieved, much more flexibility in production planning, and an efficient and safe pattern data storage is achieved. Improvements in patterning techniques such as jacquard have provided sophisticated design potential for widening the range of end uses beyond the conventional guide-bar lapping facilities.

On the electronic machines, the jacquard head has been replaced by a computer control that is simply linked by a cable to the combined selection element and jacquard guide (Figure 9.22), which are one unit. The latest solution to obtain the jacquard guide deflection movement, is the piezo technology, developed by the Karl Mayer company.

When an electrical voltage is applied to a piezoelectric material, it expands or contracts as a function of the polarity. Each jacquard guide has a piezoelectric ceramic strip on either side that positively moves the guide one needle space left or right when required.

The following lapping possibilities are exemplified for one jacquard unit, which comprises two courses, noted with I and II, and one space between two adjacent wales. The set of signals given for the guide deflection consists of the T information, which means left-side shifting (←) and H, for the right-side shifting of the guide (→), controlling both overlapping and underlapping movements. Thus, for one jacquard unit of two rows, four sets of information are required, such as those exemplified in Table 9.2.

A large variety of designs can be derived from using the piezo jacquard selection, mostly used to create sophisticated structures for laces and curtains, such examples are shown in Figure 9.23.

TABLE 9.2

Jacquard Lapping Options

Lapping	Row	Electrical Signal	Description
	II	H	Normal Tricot lapping in both rows, no deflection
		H	of the guides from the normal position.
	I	H	
		H	
	II	H	In the second row, the normal lapping is kept by the H
		H	set of information.
	I	T	In the first row, the T information deflects the guide to
		T	the left and the overlapping moves to the left with
			one needle.
	II	H	In the second row, the T information deflects the guide
		T	to the left; the overlapping is performed, then due to
			the H signal, the guide moves further to the right.
	I	H	Normal lapping in the first row.
		H	
	II	T	Similar situation in the second row, all the movements
		T	are shifted to the left.
	I	T	The set of information T–T, generates the guide
		T	deflection to the left, and after overlapping its
			maintenance in the same position.

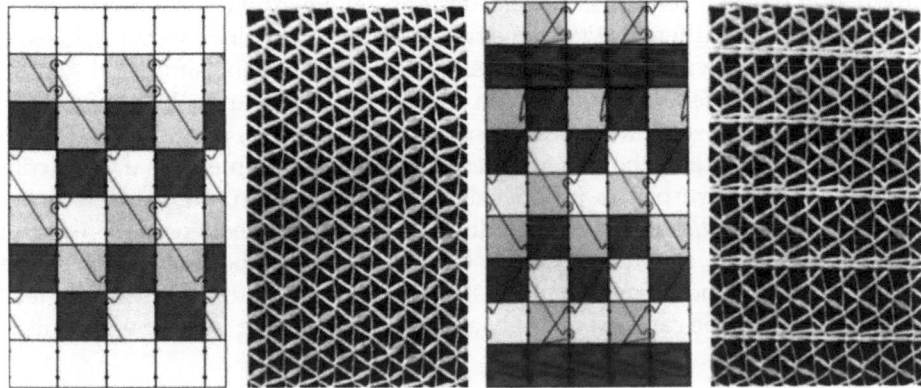

FIGURE 9.23
Piezo jacquard patterns. (Courtesy of Karl Mayer.)

9.9.3 Electronically Controlled Fabric Take-Up Mechanism

In order to produce more elaborate Tricot designs, an electronically controlled fabric take-up mechanism, capable of multispeed operation, known on the market as electronic access control (EAC), has been integrated into the modern equipment. The required fabric take-up speed is programmed directly onto the input panel of the computer, together with the beam and yarn information [4].

 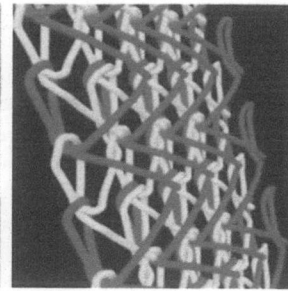 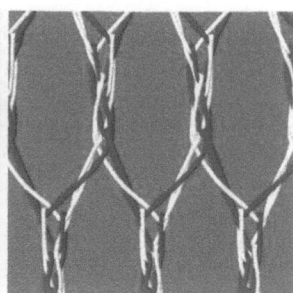

FIGURE 9.24
ProCad warpknit 3D system. (From Texion, www.texion.de, ProCAD system, September, 2016. With permission.)

9.9.4 Pattern Design Software

Apart from these solutions developed by the machine builders, the CAD system producers are struggling to offer companies, competitive CAD solutions and/or top-quality 3D simulation, especially for the high-complex structures, such as laces, curtains, and double-needle bar fabrics. Warp-knitting machine builder Karl Mayer has worked closely with its system partner Texion Software Solutions to develop Texion's ProCad [11] warpknit 3D (Figure 9.24), the latest addition to the ProCad family of software solutions for manufacturers of warp-knitted fabrics.

The display and evaluation of three-dimensional textures at the development stage is made possible by ProCad warpknit 3D. Technically, sophisticated algorithms enable realistic reproduction of elastic effects in pattern designs. Single- and double-needle bar fabrics can be calculated with different stitch densities, thread let-off values, yarn thicknesses, colors, and elasticity. The various lapping arrangements can also be simulated and used to calculate and simulate textiles produced on single- and double-needle bar warp-knitting machines with varying stitch densities, yarn feeds, yarn counts, colors, and stretch levels. ProCad warpknit 3D obtains the data for calculating the three-dimensional fabric construction from the ProCad warpknit's product database. In addition, the yarn characteristics obtained from the *yarn atlas*, the fabric take-down and yarn feed sequences are also included in the calculation model, so that a physically correct picture can be obtained on the basis of the yarn parameters. Under normal circumstances, these are extremely complex technical procedures, which can be carried out easily and intuitively using ProCad warpknit 3D.

9.10 Design Technology Aspects of Warp Knitting

9.10.1 Run-In per Rack and Run-In Ratio

The rate of yarn consumption of each guide bar is called run-in and a working cycle of 480 knitted courses is called a rack, hence the amount of warp yarn consumed for 480 knitting cycles is known as the run-in per rack. The yarn quantity can be measured with special measuring equipment or simply by marking one of the yarns close to the warp beam and then measuring its position after 480 cycles (1 rack). From the run-in per rack, it is possible to calculate the average loop length. In fact, the loop length in warp-knitted fabric cannot be determined by unraveling the knitted yarn from the fabric as done in weft-knitted fabric.

The size of the loops is changed by feeding different amount of yarn into the knitting zone; a longer run-in per rack produces a slacker fabric with big loops, whereas a shorter run-in per rack produces small and tight loops. Run-in per rack should always be recorded when a new fabric is produced. The run-in per rack may vary from the guide bar-to-guide bar, the relative amount of yarn fed from each beam is also very important. The relation is called run-in ratio and the same is different for different fabric constructions and qualities. As reported by Raz [4], in a typical two-bar machine the run-in ratio (front bar: back bar) is in the range of 1.19 : 1.0 to 1.47 : 1.0, from the values it can be stated that the loop produced by the front guide bar is bigger than the loop produced by the back guide bar. So, the length of the warp to be wound in the beams shall vary according to the run-in ratio.

9.10.2 Yarn to Fabric Ratio

One of the important parameters of the warp-knitted fabric is the yarn-to-fabric ratio. The ratio between the length of yarn consumed by the knitting machine and the length of fabric that is produced can be calculated as

$$\text{Yarn-to-fabric ratio} = \frac{(\text{run-in per rack} \times \text{courses per unit length})}{480}$$

This ratio may vary from guide bar-to-guide bar as well as with variation in courses per unit length. The length of warp required to produce a definite length of fabric can easily be obtained if the ratio of yarn-to-fabric is known.

For example, the yarn-to-fabric ratio for a particular guide bar when knitting a fabric of 20 courses per cm and run-in per rack of 240 cm is

$$\text{Yarn-to-fabric ratio} = \frac{(240 \times 20)}{480} = 10$$

So, 10 cm of warp is needed for knitting 1 cm of fabric.

9.10.3 Tightness Factor of Warp Knitted Fabrics

The compactness of knitted fabric is expressed by the tightness factor (TF), which is equivalent to the cover factor of the woven fabric. It was established by Munden [12] that the TF (K) of weft-knitted fabric can be determined from the expression

$$K = \frac{\sqrt{(tex)}}{l}$$

where:
 l is the loop length in millimeter
 tex is the yarn linear density

This relationship can simply be applied to warp-knitted fabric produced on a machine with a single guide bar. But for warp-knitted fabrics produced with two or more guide bars, the TF should be obtained by adding the TFs contributed by the individual guide bars. So, the TF of a two guide bar fabric is given by the following relationship:

$$K = \frac{\sqrt{(tex_f)}}{l_f} + \frac{\sqrt{(tex_b)}}{l_b}$$

where:
  Suffixes f and b refer to front and back guide bars
  l is the stitch length equal to (run-in per rack)/480 and is measured in millimeters

If the same *tex* is used in both the bars, then

$$K = \frac{\sqrt{tex}}{\left\{ \left[\left(\frac{1}{l_f} \right) + \left(\frac{1}{l_b} \right) \right] \right\}}$$

Further, if the loop length is same for both the guide bars, then

$$K = 2 \times \frac{\sqrt{tex}}{l}$$

For most commercial two guide bar full set fabrics, the value of K is in the range of 1–2 with a mean TF value of 1.5.

9.10.4 Yarn Count and Machine Gauge

The yarn thickness that can be used in warp knitting is limited by the size of the needle's hook and the space between the needle and the knock-over trick or sinker. A too fine yarn for the machine gauge only forms a mesh-like structure, whereas a too thick yarn will be chopped up by the descending needles into the tricks, or the yarn may cause damage to the needle itself. When more than one guide bar is used, the resultant count of all the yarns wrapping any needle should be considered.

The maximum yarn count for each machine gauge is given in Table 9.3. The Raschel knitting elements allow a somewhat coarser count to be knitted.

TABLE 9.3

Typical Yarn Count for Tricot and Raschel Machines of Different Gauges

Machine Gauge (NPI)	Tricot Machine		Raschel Machine	
	Tex	Ne	Tex	Ne
16	–	–	88	7
18	–	–	72	8
20	52	11	60	10
22	44	13	50	12
24	36	16	42	14
26	29	20	35	17
28	25	24	29	20
30	21	28	24	25
32	16	37	19	31
36	10	59	–	–
40	6	100	–	–

9.10.5 Fabric Specifications

In order to find out the specifications of any fabric, it is necessary to carry out analysis of the same. However, analysis can also be performed for other purposes mentioned as the following:

1. To know the various fabric parameters/particulars
2. To verify whether the desired design has been made with materials of desired quality
3. To estimate the various requirements for reproducing the similar fabric
4. To settle commercial disputes
5. To investigate crime

During recent years, the purchasers/consumers have been becoming conscious regarding precisely specifying the standards of construction and performance of warp-knitted fabrics. Hence, in order to ensure the specifications, a lot of improvements have taken place in warping, knitting, and subsequent dyeing and finishing operations. As mentioned by Palling [3], a typical composition of specifications for a widely used warp-knitted fabric (say locknit or jersey knit) may be as follows:

- Courses and wales per inch or centimeter (Figure 9.25)
- Yarn used in both guide bars (type, denier, luster, twist, etc.)
- Number of guide bars and their lapping movements
- Number of warp beams and color pattern of warp
- Technical face and technical back of the fabric as well as upright direction
- Finished fabric construction (e.g., courses and wales per inch or centimeter)
- Loop structure
- Run-in per rack and run-in ratio as well as loop length
- Tightness factor

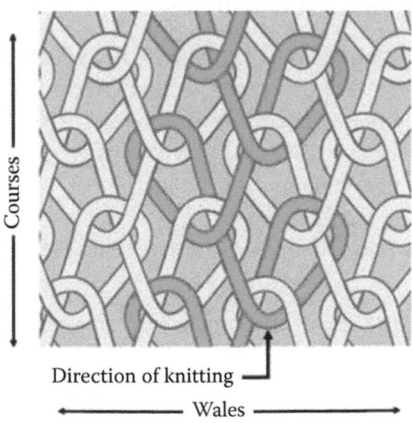

FIGURE 9.25
Courses and wales in a warp-knitted fabric.

- Finished yield (length per unit weight and tolerance, GSM, etc.)
- Requirement of yarn quantity for manufacturing certain area
- Dimensional stability (based on 30 min boiling or some specified washing)
- Color fastness to washing, light, and rubbing
- Possible machine type along with gauge
- Possible fabric type/commercial name along with end uses
- Identification of the fiber along with blend proportion, if any
- Bursting strength
- Type of finish
- Fault rate
- Additional for special fabrics (extensibility and recovery, air permeability, etc.)

9.10.6 Determination of Technical Face of Warp Knitted Fabrics

- In warp-knitted fabrics, the needle loops are visible on the face side, whereas the floating on the surface underlaps are dominant on the backside. The top or head of the needle loop is toward the direction of the last knitted fabric (upright direction). Needle loops can be identified easily from the appearance of the loop legs.
- Under normal conditions, in fabrics made with two guide bars, the threads of the front guide bar tend to dominate the face as well as the back of the fabric, as a result the position of the front guide bar thread is observed at the top of the technical back of the fabric.
- Design with two or more colors will be prominent on the face side only.
- As unraveling of the yarn is not possible, position of loops inside the knitted structure is to be observed minutely with the help of a magnifying glass.

9.11 Evaluation of Warp-Knitted Fabrics

In order to assess the quality, the knitted fabrics are also subjected to physical testing similar to the woven fabrics. However, the types of test carried out for knitted fabrics may to some extent differ from the tests conventionally done for woven fabrics as the structures and end uses of knitted fabrics are different to those of woven fabrics. The tests are mainly done keeping the following in view:

- To evaluate the quality aspects, including the comfort properties, feel, and handle of the fabrics
- To assess whether the quality of the knitted fabric matches the required quality of the end product
- To assess the performance of the fabric during the end use

9.11.1 Traditional Testing of Warp-Knitted Fabrics

The following tests are carried out, traditionally, for knitted fabrics before converting in to garments:

1. Fabric appearance
2. Thickness
3. Fabric pilling
4. Fabric extension
5. Air permeability
6. Abrasion resistance
7. Bursting strength

In addition, some other tests mentioned as follows in the undergoing may be carried out for speciality fabrics according to their end applications:

- Thermal insulation value (TIV)
- Compressibility (compression behavior)
- Tensile, tearing, and impact strength
- Water permeability
- Water repellency/proof
- Fire resistance

9.12 Summary

This chapter deals with the presentation of the warp-knitted fabrics, considering the huge potential of this knitting technology for the markets of textiles (apparel and technical garments) and nontextile products. This chapter has been organized in a logical manner, offering the summarized information for a large group of readers, from the beginners to the most experienced specialists in the field. Thus, surveying the content, one can discover the short history of this technology, the principles of producing various loops and structures, and the main categories of the available machines. This way, the large area of products, from light fabrics out of filament yarns, to the very wide range of women's and men's apparel applications has been demonstrated. Moreover, with the understanding of knitting science it is very much possible to engineer the specific warp-knitted fabric suitable for fulfilling the desired physical and functional properties in the end application of the garment. In addition, the latest *seamless smart* (double-needle bar Tricot and Raschel) machines with electronic pattern drive system can directly convert yarn into shaped garments eliminating the cutting and sewing operations. Particular attention is given to the developments with regard to electronics and microprocessors, considering that the recent warp-knitting machines are of high speed and smoothly operating, technologically advanced, and computer controlled.

References

1. Ray S. C., *Fundamentals and Advances in Knitting Technology*, Woodhead Publishing, Cambridge, UK, 2012.
2. Spencer D. J., *Knitting Technology*, Woodhead Publishing, Cambridge, UK, 2001.
3. Palling D. F., *Warp Knitting Technology*, Columbine Press, Buxton, 1970.
4. Raz S., *Warp Knitting Production*, Melliand Textilberichte, Heidelberg, Germany, 1987.
5. www.en.wikipedia.org/wiki/Tulle-(Netting), September, 2016.
6. www.knittingindustry.com/swiss/tencel, September, 2016.
7. Gajjar J. B., Advances in warp knitted fabric production, in *Advances in Knitting Technology*, Au K. F. (Ed.), Woodhead Publishing, Cambridge, UK, 2011.
8. E-Leaflets on Warp Knitting of M/s. Karl-Mayer.
9. www.liba.de, EBC, *Software Manual*, Liba, Naila, Germany, 1998.
10. www.karlmayer.de, Pattern possibilities for tricot machines with guide bar controlling via EL.
11. www.texion.de, ProCAD system, September, 2016.
12. Munden D. L., The geometry and dimensional properties of plain knit fabrics, *Journal of Textile Institute*, 50, T448–T471, 1959.
13. Kothari V. K., *Progress in Textiles: Science and Technology, Vol – 1: Testing and Quality Management*, IAFL Publication, New-Delhi, India, 2000.

10

Nonwoven Fabrics

Muhammad Tausif and Parikshit Goswami

CONTENTS

10.1 Introduction

The internationally accepted, BS EN ISO 9092:2011, definition of nonwovens is *Nonwovens are structures of textile materials, such as fibers, continuous filaments, or chopped yarns of any nature or origin, that have been formed into webs by any means, and bonded together by any means, excluding the interlacing of yarns as in woven fabric, knitted fabric, laces, braided fabric or tufted fabric*

NOTE: *Film and paper structures are not considered as nonwovens.*

Nonwovens fabrics are directly made from fibers/filaments. The term *nonwovens* itself explains that the versatility of such fabrics is defined by what they are not. The use of one word *nonwovens* is preferred over *non-wovens*, so the wrong message of these fabrics *only*

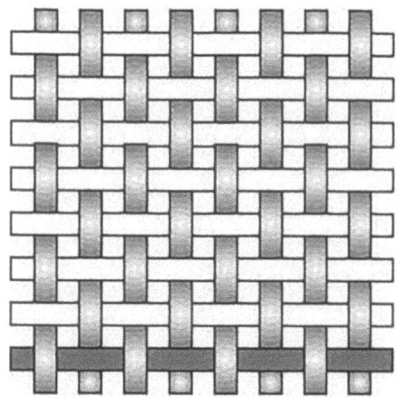

FIGURE 10.1
Plain weave. (Courtesy of Colan Products Pty Limited, http://www.colan.com.au/compositereinforcement/ resources/manufacturing/weave-information/plain-weave/. With permission.)

not being wovens could be avoided as this could cause further confusion when compared to knitted fabrics. It will rather be more intuitive to define these fabrics on the basis of structure than that of technology. The structure of woven and knitted fabrics can be geometrically depicted by a unit cell. For example, a plain weave (Figure 10.1) and a weft knit (Figure 10.2) can be shown by a unit cell. On the contrary, the structure of exemplar nonwoven fabrics cannot be geometrically depicted (Figure 10.3).

In summary (Figure 10.4), nonwoven fabrics are directly made from fibers (limited length, generally 15–250 mm but typically less than 100 mm) or filaments (endless length). The manufacturing of such fabrics involves a range of web formation and web-bonding methods. The web formation methods involve conversion of fibers/filaments in to a fibrous sheet and this sheet is consolidated by the selection of appropriate web-bonding method(s). The postprocessing methods of fabrics (consolidates webs) include dry finishing, wet finishing, and combination with other materials (conventional textiles, nonwovens, films, etc.) to produce multilayer assemblies.

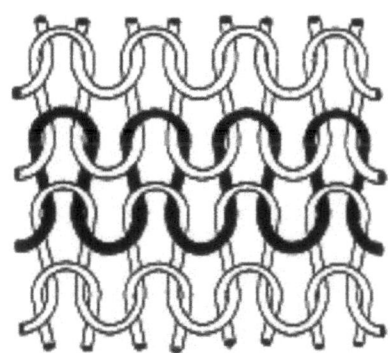

FIGURE 10.2
Weft knit. (From Duhoic, M. and Bhattacharyya, D., Composites Part A: Applied Science and manufacturing, 37 (11), 1897–1915, 2006.)

FIGURE 10.3
Hydroentangled nonwoven.

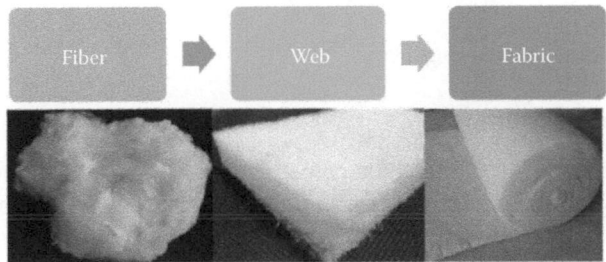

FIGURE 10.4
Summary of nonwoven process.

10.2 Nonwovens Technology

The combination of a range of fiber/filament types, web formation, web bonding, and postprocessing methods allows to engineer a myriad of nonwoven structures (Figure 10.5). The manmade/manufactured fibers dominate the nonwoven market with polypropylene (63%), polyester (23%), and viscose (8%) being the mostly consumed fibers. The choice of raw materials is dependent on processing technology, target application, properties, sustainability, and economic factors.

The two key components of nonwoven technology are web formation and web-bonding methods (Figure 10.5). These two methods in conjunction with selected raw materials and postprocessing form the fabric. These fabrics are usually produced in rolled good form and higher width for further use. In some cases, there could be additional conversion steps, online or off-line, which converts a rolled good product in to a final form. For example, rolled viscose nonwoven fabric is converted and applied with lotion before packaged as wet wipes. Hence, the choice of raw material, web formation, web bonding, and postprocessing allows producing a range of nonwoven fabric structures. The choice and control of these variables can help to engineer the application-specific nonwoven fabric structure. For example, polypropylene (PP) is a low-cost polymer and widely employed in polymer and textile industries. PP could be processed to form a shopping bag, coverstock layer of a diaper; on the other hand, the same polymer type with appropriate materials functionalization and technology could be used to produce blood filter media.

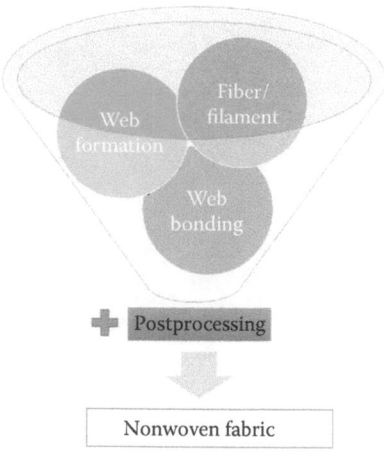

FIGURE 10.5
Key variables of nonwoven manufacturing.

10.2.1 Web-Formation Methods

The web formation methods include the conversion of fibers to a fibrous sheet. The web-formation methods can be categorized into dry-, wet-, and polymer-laying (Figure 10.6). The key methods in each category are explained in brief detail.

10.2.1.1 Dry-Laid

Dry-laying involves the mechanical principles for separation and uniform distribution of fibers across the width of the web, continuously produced and usually fed to in line web-bonding process. The preparation of fibers for web formation includes fiber-opening lines, which convert a fiber bale to an opened form. The opening of fibers is similar to fiber opening for yarn manufacturing and is covered in Chapter 3. The two key types of dry-laid webs are *carded* and *air-laid*.

10.2.1.1.1 Carded Webs

The input of the carding process is opened fiber, which is fed in the form of an uniform sheet and is further opened and mixed by the interaction of toothed rollers situated

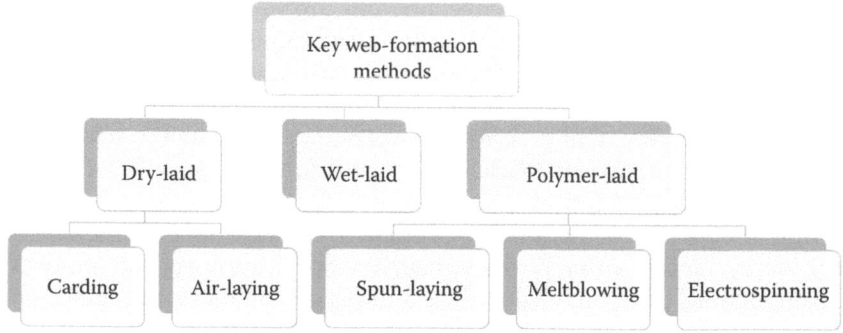

FIGURE 10.6
Key web-formation methods.

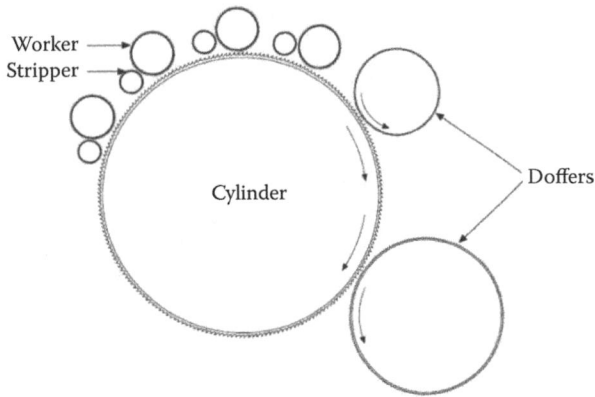

FIGURE 10.7
Schematic of carding process. (From Russell, S.J., *Handbook of Nonwovens*, Woodhead Publishing, Cambridge, UK, 2006. With permission.)

throughout the carding machine (Figure 10.7). The output of carding machine is a web/batt (a loose sheet of fibers), which can be collected directly on to a continuous layer to form a parallel-laid web, and a series of parallel-laid webs can be superimposed to achieve desired mass area density (grams per square meter) of the webs. More frequently, the card output is directly layered side-to-side onto a continuous layer to form a cross-laid web (Figure 10.8). In case of cross-laid webs, the speed of continuous layer, and traverse lapping speed can be adjusted to control the mass area density and width of webs. The cross-laying helps to achieve widths many times greater than the width of the card itself. In parallel-laid webs, fibers are predominantly oriented parallel to machine-direction (MD). In case of cross-laid webs, there are more fibers in cross-direction (CD) compared to MD. Hence, both parallel- and cross-laid webs exhibit in-plane (MD–CD) anisotropic behavior. The modern nonwovens carding equipment may employ different arrangements to control the MD-to-CD ratio of the produced webs.

There is no standard carding machine configuration and the key parts of a carding machine are feed roller, cylinder, worker, and stripper rollers (work in pair) and doffer (Figure 10.7). The cylinder is the largest roller and is known as the heart of a carding machine. The two key interactions between wired carding rollers are point-to-point and point-to-back (Figure 10.9). In point-to-point (working/carding) action, the points of the

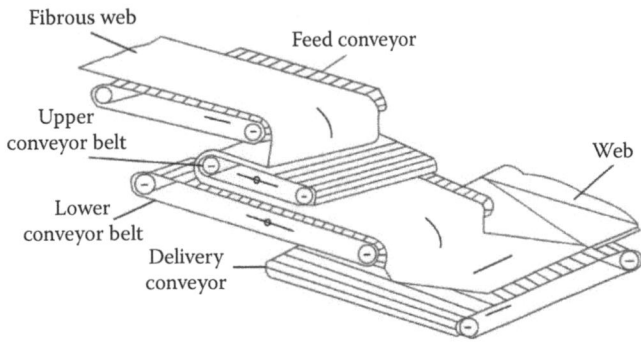

FIGURE 10.8
Principle of cross-laid web formation.

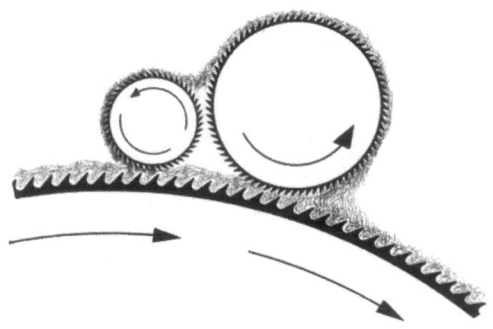

FIGURE 10.9
Working and stripping actions on a carding machine. (From Russell, S.J., *Handbook of Nonwovens*, Woodhead Publishing, Cambridge, UK, 2006. With permission.)

teeth of two rollers directly oppose each other and one surface move faster than the other surface. The interaction between worker roller (slow-surface speed) and the cylinder (high surface speed) separates the fibers trapped between the teeth of two rollers and working/ carding action takes place. In point-to-back (stripping) action, the points of the teeth of fast moving roller interact with the back of teeth of the slow moving roller. The stripper– worker and cylinder–stripper interactions are the examples of point-to-back action.

10.2.1.1.2 Air-Laid Webs

The uniformly dispersed fibers (both short and long staple) in air are collected by a moving permeable conveyor with suction (Figure 10.10). The fiber transport to permeable conveyor in the stirred air and the suction under the screen is important for web formation. The transport of fibers to web-forming zone can be achieved by free-fall, air suction and/or compressed air, and closed air circuit. A large range of machine configuration exists with different design but the technology is named after the transport of fibers in air flow from the last opening roll/feed. The degree of fiber opening is crucial in this process to achieve uniform webs

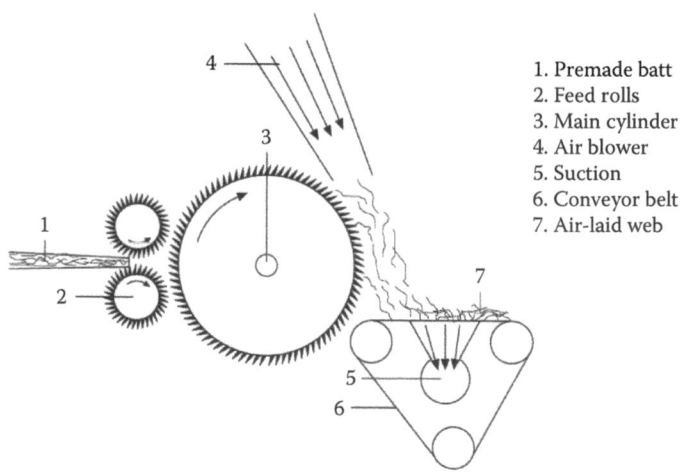

1. Premade batt
2. Feed rolls
3. Main cylinder
4. Air blower
5. Suction
6. Conveyor belt
7. Air-laid web

FIGURE 10.10
Principle of web formation in a simple air-laying process. (From Russell, S.J., *Handbook of Nonwovens*, Woodhead Publishing, Cambridge, UK, 2006. With permission.)

and especially is more crucial in low-mass area density webs. It is pertinent to mention that fiber orientations in the collected webs are by chance and hence generally results in webs with randomly oriented fibers (compared to preferentially oriented fibers in carding). Hence, the orientation and resultant properties (such as tensile strength) exhibit improved in-plane (MD–CD) isotropic behavior. The raw material, its processing, machinery layout, and the process variables are the important influencing factors. The type and characteristics of raw materials are vital such as the low bending rigidity of finer fibers makes their processing difficult. Similarly, the blending of fibers with different densities could result in separation and the deflector shields can be used to mitigate the separation. The production of lightweight webs (<100 g m^{-2}) with uniform quality can be challenging and is addressed by using a combination of carding and air-laying principles. A range of machine configurations exist and the description of these machines is beyond the scope of this book.

10.2.1.2 Wet-Laid

The wet-laid nonwovens are prepared from fibers dispersible in a liquid (generally water). These dispersed fibers are filtered on a wire mesh and the web is consolidated and dried in a continuous process, Figure 10.11. The binder, can be in the form of fibers as well, can be added for the bonding of the webs. The bonding of webs can be achieved by hydrogen bonding, thermal, latex, and mechanical (hydroentanglement) bonding means. The webs achieved possess random arrangement of fiber segments, which lead to isotropic characteristics. The technology of wet-laid nonwovens is adapted from the papermaking technology and employs longer fibers (generally about 10 mm long) compared to that in the paper (cellulosic fibers of about 2.5 mm length). The homogenous dispersion of single fibers in the fluid is vital for the processing of wet-laid webs. The key factors affecting fiber dispersion include length, length-to-diameter ratio, crimp, and wettability of the fibers. The wettability of fibers, especially in the case of synthetic fibers, can be improved by the use of wetting agents (surfactants) and other auxiliary agents can be added for specific purposes.

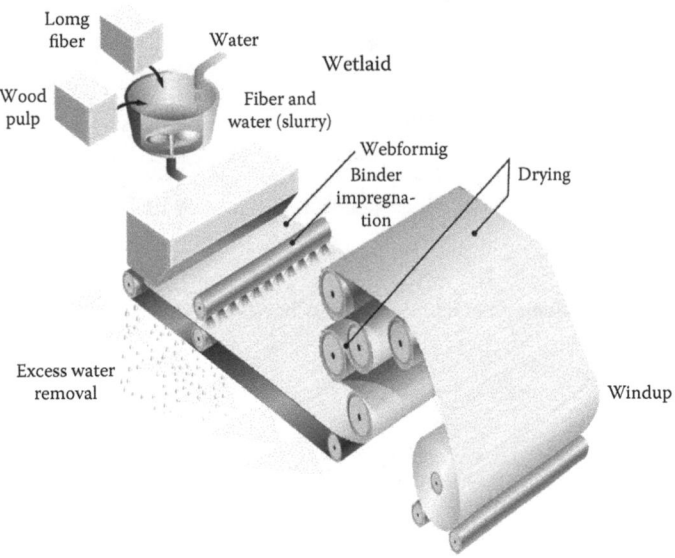

FIGURE 10.11
Description of wet-laid web formation. (Courtesy of EDANA.)

10.2.1.3 Polymer-Laid

Polymer-laid technology involves the extrusion and direct collection of filaments, followed by a consolidation step in some cases. As this method eliminates the intermediary steps, it results in very high production rates. The three main types of polymer-laid technology are spun-lagging, meltblowing, and electrospinning. The first two are widely employed at commercial scale with spun-laid (spun-bonding) being one of the most common nonwoven technology. The electrospinning technology is one of the most active research areas in textiles/nonwoven technology but did not have considerable commercial success yet.

10.2.1.3.1 Spun-Laid Webs

Spun-bonded fabrics are prepared by filament extrusion, drawing, laid down (Figure 10.12), and subsequent bonding of the laid webs in a continuous process (Figure 10.13). Hence, the first three steps can be grouped under spun-laid web formation, whereas the bonding operation is covered in Section 10.2.2. The input to

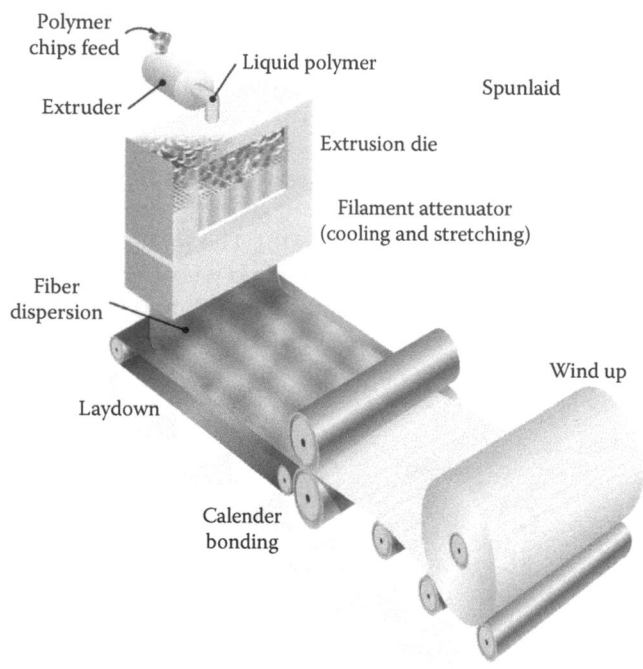

FIGURE 10.12
Schematic description spun-laying and bonding process. (Courtesy of EDANA.)

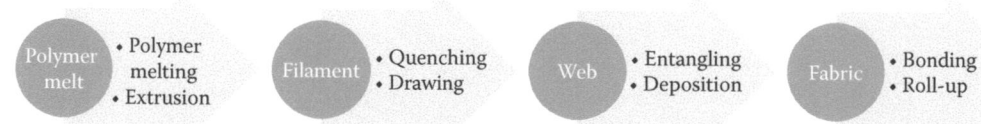

FIGURE 10.13
Process sequence in the production of spunbond (spun-laid + bonding) fabrics.

spun-bond process is polymer resin (in pellet form) and the output is consolidated fabric. Polypropylene is the most commonly employed polymer to produce spun-laid webs. The low cost, low density, good chemical resistance, good strength, low melting point, and hydrophobicity make it a polymer of choice for a myriad of applications. The filament extrusion techniques include melt, wet, dry, gel, and centrifugal spinning, as covered in Chapter 2. To produce widewidth webs, many spinnerets are grouped together across the width of the machine.

The melt extrusion is most commonly employed in the spun-laid process. The extrusion of filaments is carried into air or alternative gas, which allows cooling, solidification, and the drawing of the filaments. The drawing of filaments is commonly carried out aerodynamically compared to roller drawing for melt spun filaments. The aerodynamic entangling of filament is also achieved, which improves mechanical properties in the cross direction and the suction under mesh belt helps in filament deposition. A range of polymers can be processed and the spinneret geometry can be altered to prepare multicomponent filaments. For example, a PP/PET bicomponent sheath–core filament can be extruded with a high-melting point PET core and low-melting point PP sheath. On application of thermal energy, the sheath could melt and bond, whereas the core provides the mechanical strength. The bonding of spun-laid webs can be achieved by thermal, chemical, and mechanical means and will be discussed in Section 10.2.2.

10.2.1.3.2 Meltblown Webs

The meltblown fabric technology involves the extrusion of filaments through a die with row of spinnerets and convergent hot air streams at high velocity from both sides to draw the filaments to fine diameters (usually in the range of 1–5 µm). The surrounding air cools and solidifies the filaments, which are then collected onto a vacuum-assisted perforated roller. This results in webs that are self-bonded by entanglement induced during turbulent flow and cohesive sticking, and do not require any subsequent bonding step. The schematic description of the process is shown in Figure 10.14. The viscosity of the melt needs

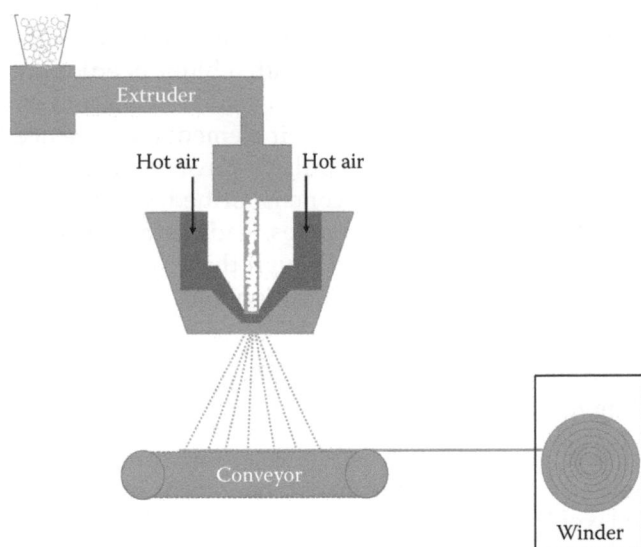

FIGURE 10.14
Schematic description of meltblowing process.

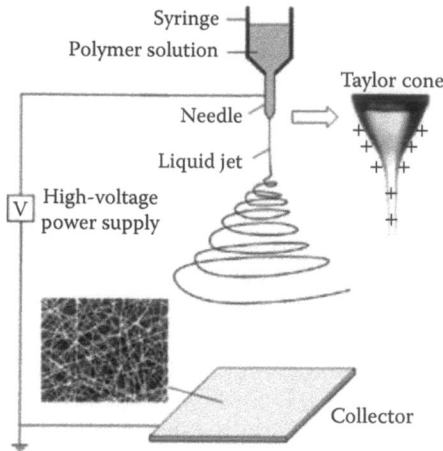

FIGURE 10.15
Schematic description of electrospinning process. (From Li, D., and Xia, Y., *Adv. Mater.*, 16, 1151–1170, 2004. With permission.)

to be very low for achieving fine diameters. Unlike spun-bond process, the filaments are not endless as are broken down by the turbulence generated below the nozzle. The fibers are randomly collected due to turbulent flow and some directional bias is imparted by the moving collector in the machine direction. The meltblown fabrics exhibit low-to-moderate strength due to limited drawing of filaments.

10.2.1.3.3 *Electrospinning*

Compared to melt-spinning and melt-blowing processes, electrospinning can produce nanofibers (generally the produced fibers are <500 nm, whereas as per definition nanofibers are with an average diameter of less than 100 nm), which offer advantages of extremely high surface areas as are about 100 times thinner than human hair. The key elements of the electrospinning process are a high-voltage power supply, reservoir with conductive element (most commonly syringe with a blunt needle), and a grounded collector (Figure 10.15). The polymer is dissolved in a suitable solvent or melted and high-voltage electric field is applied between reservoir element (syringe needle) and collector. The jet is formed when the electrostatic repulsive forces overcome the surface tension of the polymer solution/melt and a Taylor cone is formed. On its travel to the grounded electrode, the jet stretches and solvent evaporates, leading to fine diameter filaments randomly collected on the grounded surface. Though these webs have high per unit mass strength but still are weak for practical application and are usually used with suitable substrates.

10.2.2 Web-Bonding Methods

As discussed previously, the web is formed in sheet form and lacks structural integrity. Hence, the bonding step is crucial for the use of these webs in final applications. In some cases, such as carded wool (with ability to felt), meltblown webs, and electrospun web may/not require any additional bonding steps. The webs can be bonded by thermal, chemical,

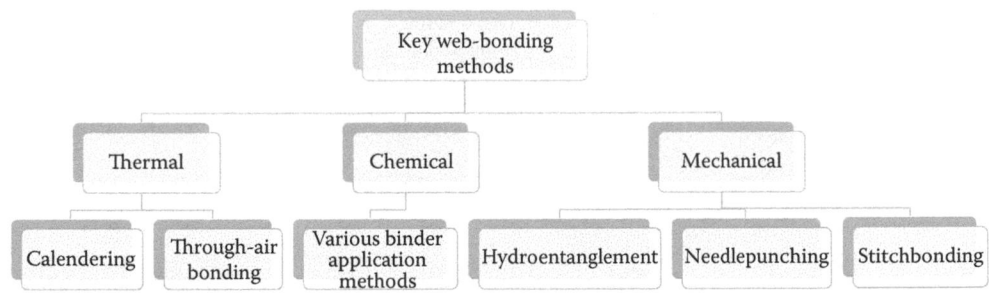

FIGURE 10.16
Key web-bonding methods.

and mechanical means (Figure 10.16). It is pertinent to mention that more than one bonding method can be employed to achieve desired properties.

10.2.2.1 Thermal Bonding

Thermal bonding uses heat energy to soften or melt at least one thermoplastic component in a web to impart stability or strength to form a bonded fabric. The polymer melts flows by capillary action and surface tension to fiber-to-fiber crossover points to form a bond point on cooling. Usually, a low-melt (binder) fiber is blended with high-melting point thermoplastic or a non-thermoplastic (base) fiber. In case of bicomponent fibers, such as core–sheath morphology, each crossover point could result in a bond point. The mode of heat transfer could be any or combination of conduction, convention, and heat radiations.

10.2.2.1.1 Calendering

The formed web, with at least one thermoplastic component, is passed through the nip between two heated rollers and the trio of temperature, pressure, and time influences the bonding (Figure 10.17). The choice of smooth rollers results in area bonding in which the thermoplastic binder component melts at all crossover points. In point bonding, at least one of the rollers is engraved with the desired pattern and the design of this roller can be varied to achieve a variety of patterns on the final fabric surface. The size, frequency, and pattern design allows achieving a different level of bonded areas and consequently different fabric designs as well as mechanical properties can be tailored by the choice of point-bonding variables. Owing to its lesser bonded area, point-bonded fabrics are soft and permeable as compared to area-bonded fabrics. Calendering is more suitable for light-to-medium weight fabrics as heavier webs could insulate heat.

10.2.2.1.2 Through-Air Bonding

The hot air is drawn through a pre-formed web, made of thermoplastic component or bicomponent fibers and is supported by a highly permeable drum/belt, by means of a suction fan. The schematic of a through-air rotary (with drum) oven is shown in Figure 10.18. The temperature and airflow are two key variables of the process. The high air flow results in high fabric strength by improved number of bond points but is also liable to cause structural disturbances in the web.

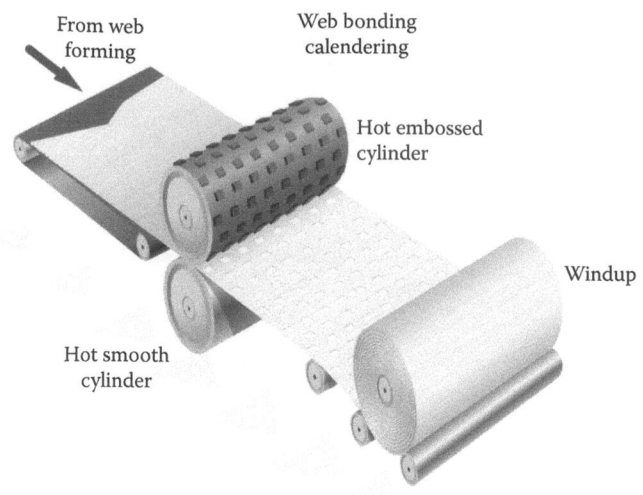

FIGURE 10.17
Process description of calendering. (Courtesy of EDANA.)

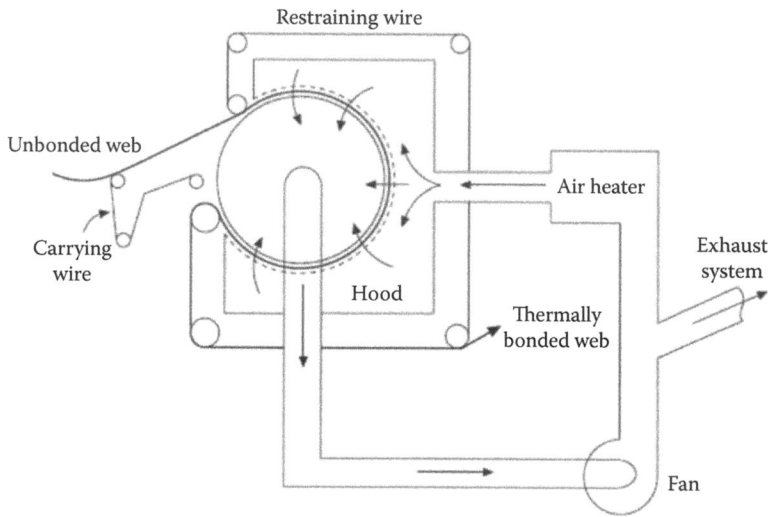

FIGURE 10.18
Schematic of through-air bonding process. (From Russell, S.J., *Handbook of Nonwovens*, Woodhead Publishing, Cambridge, 2006. With permission.)

10.2.2.1.3 Infrared and Ultrasonic Bonding

Infrared bonding employs electromagnetic waves that do not require a medium to transport heat energy, and the web structure is undisturbed. In ultrasonic bonding, high frequency vibration of the horn generates heat and subsequent melting and localized bonding of thermoplastic component in the web, rested on an embossed roller.

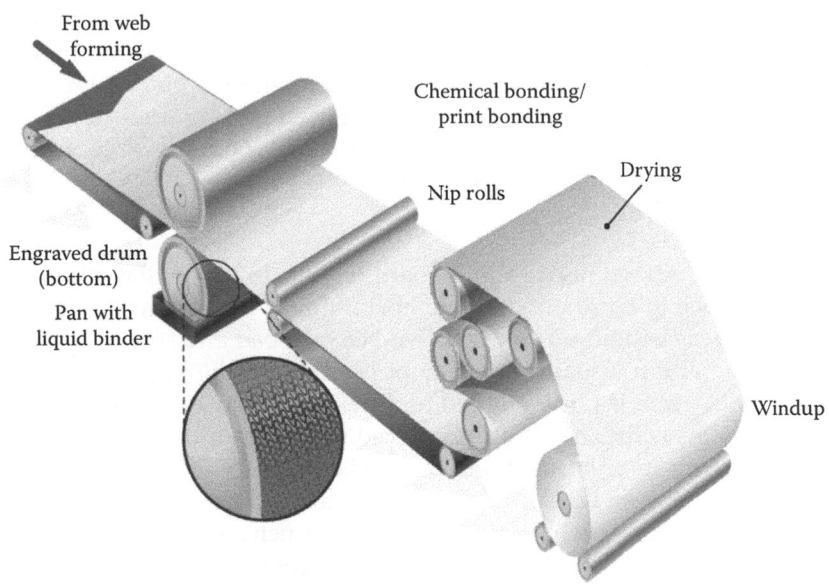

FIGURE 10.19
Description of chemical (print) bonding. (Courtesy of EDANA.)

10.2.2.2 Chemical Bonding

The preformed webs, in some cases pre-bonded as well, are subject to binder impregnation, drying, and curing. The binder can coat fibers; it can form film on the surface, but for good bonding the binder is required to migrate to fiber crossover points and form an adhesive bond. The strength of this bond depends on the intermolecular forces between the fiber and the binder layer. The binder polymers can be dissolved in a solvent (water being the most common) or alternatively can be applied as dispersions or emulsions. Latex-based binder systems are most commonly applied and the commonly used polymers are acrylate polymers and copolymers, styrene–butadiene copolymers, and vinyl acetate ethylene copolymers.

The various methods to apply binders include saturation, foam bonding, spray bonding, and print bonding. The saturation method involves total immersion of the web in a binder bath, by employing horizontal or vertical padder, and the excess binder is removed by passing the web through a roller nip or vacuum slots. Foam bonding involves the dilution of a binder in a mixture of air and water, which helps in efficient drying. As evident by the name, spray bonding involves the spraying of fine particles of the binder and helps to produce voluminous nonwovens. Print bonding involves the application of binders in predetermined areas and remaining areas are left unbonded (Figure 10.19). The choice of pattern affects the strength, drape, and fluid-handling properties of the fabric. The drying and curing can be achieved by convection-, conduction-, and infrared-based dryers. During drying, the solvent evaporates and binder particles form the film and cross-linking may take place.

10.2.2.3 Mechanical Bonding

The use of mechanical means to integrate fibrous assemblies includes twisting and wrapping in yarn manufacturing, interlacing in woven and knitted fabrics, and fiber-to-fiber bonding in nonwoven fabrics. The rearrangement of fibers in preformed webs by the use of barbed needles (needle punching) or high-velocity water jets (hydroentanglement) can produce mechanically integrated fabric. In mechanically bonded nonwovens, fiber entanglement leads to resistance to fiber movement. The principal governing mechanisms in fiber entanglement are capstan effect and contact pressures. The capstan effect is introduced by the mutual crossings of fibers, whereas contact pressures are developed by encircling of fibers around other fibers. In the case of stitch-bonded fabrics, additional yarns may be used to stitch or knit the nonwoven web.

10.2.2.3.1 Needle Punching

In needle punching, barbed needles in a board (Figure 10.20) oscillate through the thickness of a moving web to cause out-of-plane fiber segment deflections, leading to entanglement and in effect the bonding. The web is fed between bed and stripper plates with holes corresponding to pattern (usually an array of needles) on the board. The whole process is continuous as the web is positively fed and the fabric is collected on to a take-up roll. Commonly, multiple needle boards are employed in machines that punch on both sides of the fabric. Multiboard machines include double boards (up/down), twin boards (up and down, same plane), tandem boards (up and down in alteration), and quad punch (double twin board). The key process control parameters are punch density (punches/cm^{-2}) and the needle penetration depth. Pre-needling is the initial consolidation of the web and is also employed in conjunction with other bonding methods as it provides enough strength to web, so that it can bear the stresses of the subsequent process. Flat finish needling is used to produce high strength fabrics and elliptical needling offer higher advance per stroke.

The important part of the whole process is the barbed needle itself, which carries the fibers with every stroke (Figure 10.21). The key parameters in the selection of needle are the type of reduction, barb spacing, and needle blade cross section. Single-reduction needles are stiffer than that of double-reduction needles. Regular barb spacing is commonly used and the closed bard spacing could be used for more aggressive action in smaller penetration depth. The needles with triangular cross section, three apices, and three barbs per apex are standard. Other configurations include star-blade needle with

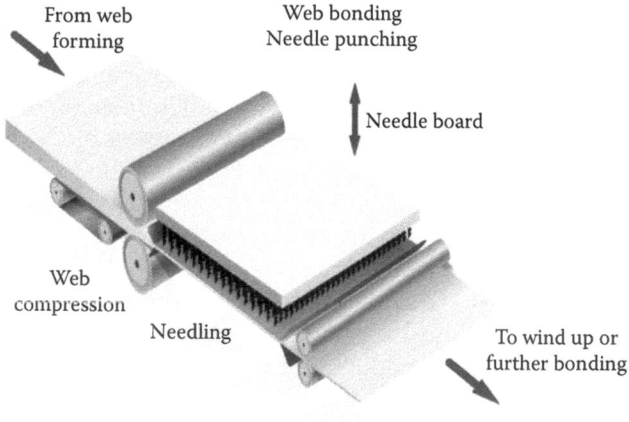

FIGURE 10.20
Description of needle punching process. (Courtesy of EDANA.)

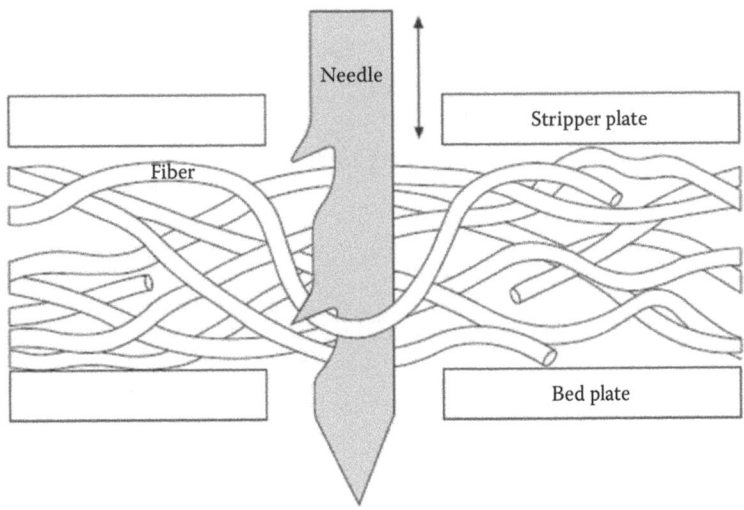

FIGURE 10.21
Action of a barbed needle. (From Russell, S.J., *Handbook of Nonwovens*, Woodhead Publishing, Cambridge, UK, 2006. With permission.)

four apices, which help to carry an increased number of fiber segments per stroke. Fork needles, as evident by the name, have a fork-like opening that helps to carry a large number of fibers in patterning and structuring applications. Needle wear and tear is bound to happen, rate of which depends on the use, and needles are frequently changed to maintain the quality of the products. At a time, one-third of the needles are changed to achieve consistent quality.

10.2.2.3.2 Hydroentanglement

Hydroentanglement is also commonly known as spunlace. The process involves entangling of fibers by means of high-velocity water jets. The energized water interacts with the web and porous support surface, which can be a flat conveyor or a cylindrical surface, to induce displacement, twisting, rearrangement, and entanglement of fibers/filament segments in the web to produce an integrated fabric, held together by fiber-to-fiber friction. Figure 10.22 illustrates a typical hydroentanglement process. The water is pumped through cone-shaped capillary nozzles to create collimated water jets and the fabric structure is formed by the effect of water jets and turbulent water in the web, which intertwines neighboring fibers. The water passing through the conveyor belt or drum is recycled. Normally, multiple injectors are used for hydroentanglement. Fiber properties such as modulus, wettability, dimensions, and fiber type influence the degree of bonding. Water jet pressures up to 60 MPa and nozzle diameters in range of 60–120 µm are employed. The process does not essentially require any additional chemicals to affect bonding, which contributes to sustainability and cost-effectiveness. The hydroentanglement technology offers improved physical properties compared to other bonding methods, such as a esthetics, softness, strength, flexibility, hand, drape, conformability, and absorbency.

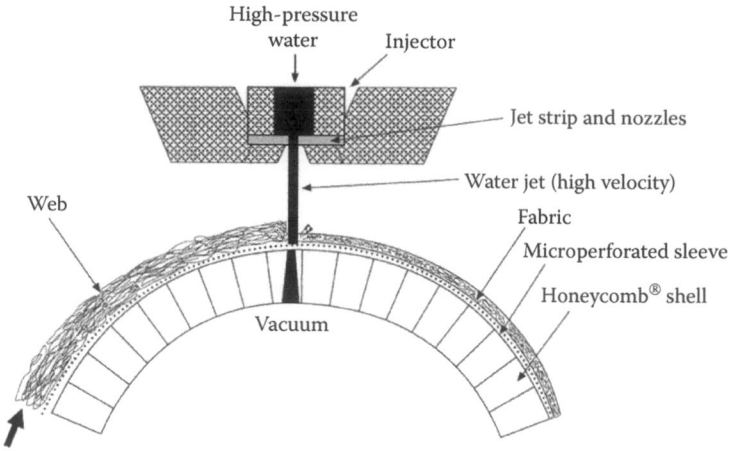

FIGURE 10.22
Key elements of Hydroentanglement process. (From Russell, S.J., *Handbook of Nonwovens*, Woodhead Publishing, Cambridge, UK, 2006. With permission.)

The transfer of kinetic energy from water jets to the web is expressed as the specific energy consumed by a unit mass of fibers in the web (J kg^{-1}). The increase in specific energy results in improvements in strength until a critical point is reached. Further increases past this point may lead to deterioration of the structure and consequently the strength. The ratio of energy applied to two sides of the fabric i.e. pressure profile is also important, and different pressure profiles result in different fabric structures. The bonded web is dewatered by suction and subsequently dried. It is common to form patterns or designs by employing an embossed cylindrical surface. Hydroentanglement is also used for the splitting of bicomponent filaments to achieve microfiber nonwovens.

10.2.2.3.3 Stitch Bonding

Stitch bonding involves the use of stitching or knitting elements to consolidate a web. In case of Maliwatt stitch bonding, the web is fed in a vertical direction and needles on a needle bar penetrate the web. After penetration through the web, yarn is laid in to the hook and pulled back through the web in to previous loop and so on. The other type of stitch bonding includes Malivlies, Malimo, Malipol, Voltex, Kunit, and Multiknit systems.

10.2.3 Postprocessing Methods

Nonwovens are versatile technical fibrous assemblies and their structures are fully dependent on the end use of the product. In addition to a range of raw materials, web forming and bonding; a range of postprocessing methods can be employed to meet the end use requirements. Such methods can include an additional bonding method, joining of nonwovens to other nonwovens/other materials, and dry/wet finishing of the nonwoven fabrics.

10.2.3.1 Finishing Methods

Finishing methods are aimed to improve the functionality or appearance of a fabric. For nonwovens, these methods mostly stem from the conventional textiles finishing and coloration methods. The finishing methods can be categorized into dry and wet finishing.

Dry finishing methods include splitting, winding, perforating, ultrasonic welding, softening, calendering, singeing, polishing, flocking, raising, and shearing. These methods are applied to either achieve a certain effect or convert the nonwoven fabric for further end use. For example, ultrasonic welding is commonly used to make bags from spun-bond polypropylene nonwoven.

Wet finishing methods for nonwovens include washing, coating, finish application, and coloration of nonwovens. The application of finishes on textile substrates and their method of application are covered in Chapter 13. The coloration of textile includes dyeing and printing routes, which are covered in Chapter 11. Nonwovens are usually anisotropic materials and are easily stretched and exhibit poor elastic recovery. In such cases, the application of wet methods could further increase the risk of structural deformations. For example, needle-punched nonwovens exhibit major structural deformations when dyed by jigger dyeing as compared to that of package dyeing methods. Hence, necessary changes are necessary for fabric processing through conventional routes.

10.2.3.2 Composite Nonwovens

Composites are heterogeneous materials that are a combination of two or more distinct materials that can be recognized by a discrete interface separating them. The two or more distinct materials can also be of significantly different physical forms of the same material. The use of nonwovens can be grouped as either (1) composite nonwovens or (2) nonwoven-reinforced composites. The so-called composite nonwovens can be multilayer nonwovens, component nonwoven layer(s) with other suitable material(s), and a homogenous blend of two or more fibers. Note that the first two types of composite nonwovens are mostly laminated structures. A common example of a nonwoven composite is a baby diaper, which is composed of three layers: (1) the inner nonwoven layer, (2) the middle super absorbent polymer, and (3) the outer film layer to prevent leakage.

Nonwoven-reinforced composites can further be classified into flexible and structural composites, and nonwoven reinforcements are used which either can be single or multilayer. The flexible and structural composites are generally impregnated or coated with an elastomeric and thermosetting/thermoplastic polymer resin, respectively. The resin is termed as the polymer matrix of the composite and acts as a binder for the fibers as well as a stress transfer medium. An example is wet-laid nonwovens of glass fibers impregnated with thermosetting resin for printed circuit boards

10.2.4 Characterization and Testing

There is a need to better understand and represent the structure of nonwoven fabrics as the individual fibers cannot be controlled during manufacturing and the fabric cannot be represented by a geometrically repeating pattern. The spatial arrangement of fibers in the structure and their properties affects the physical properties of the nonwovens. Fabric mechanical properties, that is, in-plane and out-of-plane, are affected by the fiber segment orientation distribution. In addition, pore size distribution, geometrical and hydraulic properties are also influenced by fiber segment orientation distribution.

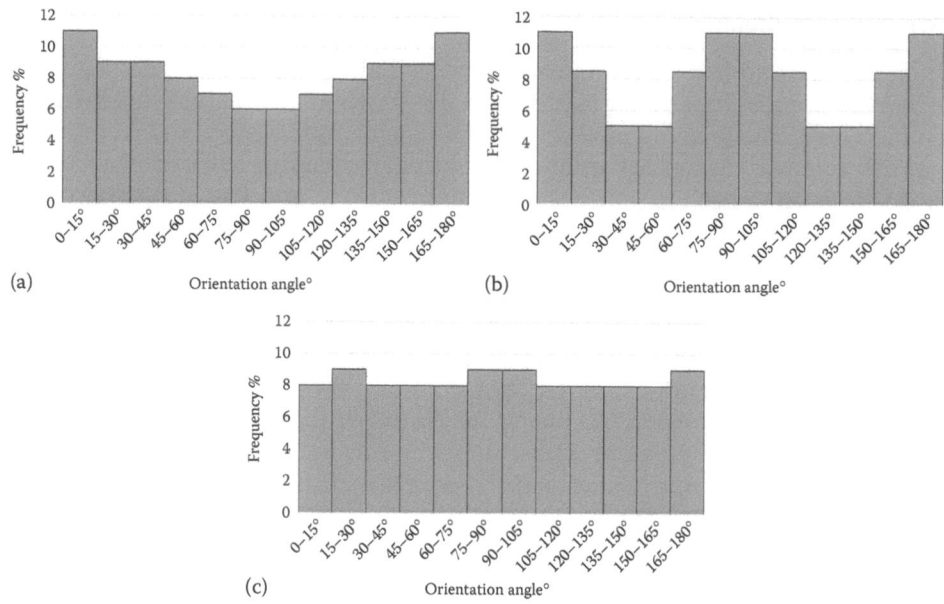

FIGURE 10.23
Characteristic orientation frequency distributions for key nonwoven structures: (a) parallel-, (b) cross-, and (c) air-laid (0° = Machine Direction).

10.2.4.1 Fiber Orientation

Fiber orientation can be described by an angle formed of a vector on fiber axis with respect to fabric axis (usually machine direction). The orientation of the whole fabric can be shown by a frequency distribution of orientation angles of all the fiber segments. Two-dimensional image processing approaches are commonly employed to characterize nonwovens for orientation. Typical frequency distribution for key nonwoven structures is shown in Figure 10.23.

Nonwoven fabrics have a substantially planar fiber arrangement but depending on the method of web formation and bonding used to produce the fabric, a small fraction of fiber segments may be aligned out-of-plane. For example, out-of-plane orientations are critical in the case of mechanically bonded nonwovens. Recently, digital volumetric imaging (DVI) and X-ray microtomography (XMT) have been employed to characterize three-dimensional orientation distribution of fiber segments in a nonwoven structure.

10.2.4.2 Pore Size Distribution and Porosity

Pore size distribution in a nonwoven fabric can be measured by optical methods, image analysis, porometry, and porosimetry methods. In porometry, a fully wetted (employing a low surface tension liquid) sample is placed in the sealed sample chamber and gas is allowed to flow on one side of the sample. When the pressure overcomes the capillary pressure of the fluid within the largest pore, the bubble point (largest pore) has been found. The pressure is further increased and the flow is measured until all pores are empty. The flow rates through dry sample are determined as well and pore size distribution is estimated in reference to applied pressures and wet/dry flow rates. Sieving techniques, for apparent pore size measurement, determine the amount of spherical glass beads/sand particles passed through a nonwoven fabric.

Porosity of a nonwoven fabric can be described as the overall total pore volume of the fabric. It can be computed by the ratio of fabric density to fiber density (Equation 10.1).

$$\text{Porosity \%} = \left(1 - \frac{\rho_{fabric}}{\rho_{fiber}}\right) \times 100 \tag{10.1}$$

where ρ is the density (kg m^{-3}).

10.2.4.3 Testing

Testing of nonwovens is end use specific and the most commonly tested properties include thickness, fabric mass areal density, tensile strength, bursting strength, air permeability, absorbency, and wicking. Major international testing bodies (International Organization for Standardization—ISO, European Committee for Standardization—CEN, British Standards Institution—BSI, German Institute for Standardization—DIN, American Society for Testing and Materials—ASTM) include standard methods for nonwoven fabrics. However, nonwoven standards are most comprehensively covered by the Nonwovens Standard Procedures (NSP), jointly prepared by International association serving the nonwovens and related industries (EDANA) and Association of the nonwoven fabrics industry (INDA). The complete list of NSP is available at http://www.edana.org/docs/default-source/default-document-library/toc-cover-preamble.pdf?sfvrsn=1 (accessed January 26, 2017).

Nonwovens Standard Procedures

10.3 Nonwovens Applications

Nonwoven industry manufactures and converts versatile and engineered fibrous assemblies with numerous existing applications and continuously innovating products for new applications. The classification on the basis of employed technology can be challenging as numerous combinations of raw materials, web formation, web bonding, and postprocessing methods lead to nonwoven products for a wide range of application areas. This degree of freedom is the key to offer nonwovens with unique properties to suit a myriad of application areas. The application areas of nonwoven fabrics can be segmented into, but not limited to, hygiene, wipes, apparel, medical, health care, personal care, automotive, electronics, filtration, agriculture, horticulture, furnishings, construction, and packaging. It is important to mention that a single product in an application area can be manufactured by employing different technologies. The examples of use for key application areas are tabulated below:

Applications	Examples of Use	Technologies Employed
Absorbent hygiene products (AHP)	Baby diapers, feminine hygiene products, and adult incontinence products	Air-laid, carded nonwovens, spunmelt (SMS, spunbond)
Agriculture and horticulture	Crop covers, plant protection Seed blankets, weed control fabrics, greenhouse shading, root control bags, biodegradable plant pots, capillary matting, landscape fabric	Needle punched, spunbond
Automotive	Headliner, dashboard insulation, carpets and flooring, seats, interior trim, cabin air filters, airbags, wheel housing, dashboard insulation, hood insulation, filtration, molded bonnet liners, heat shields, parcel shelf, boot liners, boot floor covering, boot carpets	Needle punched, spunlace, thermally bonded, spunbond
Building	Insulation (thermal and noise), house wrap, roofing, covers for acoustic ceilings, air infiltration barrier, vapor barrier, flooring substrates, facings for plaster board, pipe wrap, concrete molding layers, foundations and ground stabilization, vertical drainage	Dry-laid, needle punched, spunlace, thermobonded, spunmelt
Geotextiles	Road and rail building, dam, canal and pond lining, hydraulic works, sewer lines, soil stabilization and reinforcement, soil separation, drainage landfill, filtration, sedimentation and erosion control, weed control, root barriers, sport surfaces, asphalt overlay, impregnation base, drainage channel liners	Dry-laid, needle punched, spun-laid
Filtration	Engine air, oil, fuel, cabin air, HVAC—industrial heating, ventilation, and air-conditioning, industrial consumer products, clean rooms, food and beverage, pharmaceutical/medical, water, blood, hydraulic, antimicrobial, biopharmaceutical, dust, odor	Air-laid, electrospun, meltblown, needle punched, spunbond, spunlace, thermobonded, wet-laid
Household	Abrasives, bed linen, blinds/curtains, carpet/carpet backings, covering and separation material, detergent pouches/fabric softener sheets, flooring, furniture/upholstery, mops, table linen, tea and coffee bags, vacuum cleaning bags, wall covering, wipes	Needle punched, spunbond, wet-laid
Medical	Disposable caps, gowns, masks, scrub suits and shoe covers, drapes, wraps and packs, sponges, dressings and wipes, bed linen, contamination control gowns, examination gowns, lab coats, isolation gowns, transdermal drug delivery, shrouds, underpads, procedure packs, heat packs, ostomy bag liners, fixation tapes, incubator mattress, sterilization wraps, wound care, cold/heat packs, drug delivery	Dry-laid, spunlace, meltblown, spunbond
Personal care wipes	Baby wipes, facial wipes, cleansing wipes, hand and body wipes, moist towelettes, personal hygiene wipes, feminine hygiene wipes, antibacterial wipes, medicated wipes	Air-laid, spunlace, wet-laid
Further information	http://www.edana.org/discover-nonwovens/products-applications	

10.3.1 Nonwovens for Apparel and Fashion Applications

Woven and knitted fabrics are widely employed in apparel and fashion applications. Nonwoven fabrics find some application in disposable sector and are mostly employed as components of durable applications. The key areas of applications include interlinings, protective clothing, shoe linings, and synthetic leather. Going into the history of disposable dresses, Scott Paper Company introduced paper dresses in 1966 and sold more than half a million in less than a year. The dress was actually part of an advertising campaign and included yarn reinforcement. Following the trend, other companies also launched paper products, including evening dress, wedding gown, water proof rain coats, bikinis, and paper suits; some of the examples with commercial success are mentioned here. Miratec® fabrics by Polymer Group International are textured hydroentangled fabrics and offer applications in durable apparel. DuPont introduced the spunbonded olefin, Tyvek®, which is a lightweight protective apparel and has achieved a lot of commercial success. Tyvek®-based jackets are also offered but not in the mainstream. Evolon® by Freudenberg is another notable example. Evolon® is made from splittable bicomponent fibers, which are split during hydroentanglement resulting in microfilament fabric and the fabric is claimed to be processed like a traditional fabric. Another interesting example is Spray-on© fabric by Fabrican, which uses liquid suspension sprayed by the use of either a spray gun or an aerosol can to create an instant nonwoven fabric. In 2004, School of Design at the University of Leeds experimented with a range of nonwoven fabrics to manufacture nonwovens garments (an overview of work is given at http://www.nonwovensnetwork.com/fashion/ [Figure 10.24]).

FIGURE 10.24
Nonwovens for fashion application—School of Design, University of Leeds. (Courtesy of The Nonwovens Network, http://www.nonwovensnetwork.com/fashion. With permission.)

Though nonwovens have some degree of success in commercialization of limited-use apparel but, in general, struggle to meet the properties required for the durable applications. Fabric drape, handle, elastic recovery, and abrasion resistance are important for applications of nonwovens in main stream apparel. Nonwoven technology has advanced and fabrics with desirable properties for apparel applications are being developed. For example, hydroentangled fabrics provide better drape characteristics than other available nonwoven fabrics as the entanglement and twisting mechanisms during hydroentangling are superficially similar to the twist in yarns. Furthermore, the use of postprocessing methods, especially coloration approaches, functional finishes, and joining techniques, are the key to develop nonwovens for durable apparel applications.

Bibliography

1. Russell, S.J., 2006. *Handbook of Nonwovens*. Cambridge, UK: Woodhead Publishing.
2. Turbak, A.F., 1993. *Nonwovens: Theory, Process, Performance, and Testing*. Atlanta, GA: Tappi Press.
3. Albrecht, W., H. Fuchs, and W. Kittelmann, 2006. *Nonwoven Fabrics: Raw Materials, Manufacture, Applications, Characteristics, Testing Processes*. Weinheim, Germany: John Wiley & Sons.
4. Sinclair, R., 2014. *Textiles and Fashion: Materials, Design and Technology*. Cambridge, UK: Elsevier.
5. Tausif, M., 2013. *Characterisation of Bi-layer Hydroentangled Nonwovens*. Leeds, UK: University of Leeds.

11

Fabric Dyeing and Printing

Asim Kumar Roy Choudhury

CONTENTS

11.1 Introduction

Fabric dyeing is most popular because this is the final stage of manufacturing in the textile mill after which the material is sold in the market. Fabric is the most stable form of textile materials. It is easier to handle and is available in long length, so that high production continuous processing is possible. Fabric dyeing has maximum flexibility and is capable to respond to the market rapidly.

However, during dyeing in fabric form, maximum levelness of dyeing is to be ensured. Dyes having high wash fastness have generally poor migration property. As a result leveling properties of such dyes are poor. Unlevelness of the dyed fabric cannot be rectified easily—leveling type is to be selected even sacrificing wash fastness to some extent.

Fabrics have to be properly prepared before dyeing by removing unwanted substances to make them uniformly absorbent, so that they can pickup and retain dyes in sufficient quantities and in a uniform manner. The advantage of fabric form is that it can be converted into a very long strand of material in rope form or in open width form. Hence, fabrics can be pretreated in large batches or in continuous form.

11.2 Preparation of Cotton Materials

Cotton is mostly processed in fabric form and rarely in fiber form. Some quantities of cotton material are prepared and dyed in yarn form. The different stages for processing of cotton fabrics are shown in Figure 11.1.

11.2.1 Inspection and Mending

Gray fabrics are first inspected to check whether they are in conformity with the standard. They are sorted into lots. Fabrics in each lot should be of the same width and similar weight per unit length, so that they can be processed in the same manner. All the pieces of the lot are sewed end-to-end with loops, which can be easily removed after finishing. In some textile units, gluing or bonding of fabric is used instead of sewing in order to reduce the amount of rags. The fabric after sewing end-to-end is passed slowly over an inspection table with inclined glass top. The table is illuminated with a fluorescent lamp fitted below. The illuminated fabric is visually inspected for various weaving faults, damages, and so on, which are suitably marked with colored pencils. The marked fabric is then repaired for missing end (warp) or picked (weft) by mending, or the damaged portion is cut off, which may otherwise be entangled with machine parts causing excessive damage.

In addition to the naturally occurring impurities, a number of impurities are embedded accidentally on the textile materials. The most common accidental impurities are stains caused by various agents, which are easily visible in the embedded portions. Most of the stains should be removed by local spotting with an efficient stain remover followed by thorough rinsing and soaping.

11.2.2 Shearing and Cropping

After inspection and mending, small projected yarns may remain on the surface of the fabric, which obstructs penetration of dyes during dyeing and printing processes. They may also be entangled with machine parts causing damage of the materials. These projected yarns are removed by a process called shearing and cropping, during which the projected fibers or yarns are cut from the fabric surface. In shearing operation, dry or occasionally slightly dampened fabric is drawn between a shearing bed or table and the shearing device, which consists of a shearing cylinder and a ledger blade. The action is similar to that of a lawn mower.

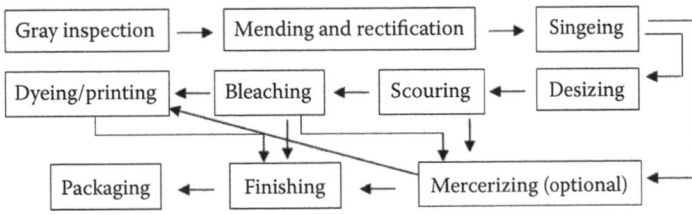

FIGURE 11.1
Sequence of operations in cotton fabric processing.

11.2.3 Singeing Process

When yarn is made from short staple fibers, the two ends of the fiber may not be embedded inside the yarn and are projected due to the twist imparted during spinning. For multifilament yarns, some filaments may break or cut in places and are projected from the surface. Singeing is a very effective method for removal of projected fibers.

- Many cotton materials are valued for their smooth appearance. When yarn or fabric is made, the surface appears fuzzy or hairy due to the presence of projected fibers or yarns. The protruding fibers obstruct subsequent dyeing and printing process in an uncertain manner. It is almost essential with goods intended for printing.
- In case of polyester/cellulosic blended fabrics, singeing is the best method for controlling pilling. Pills or beads are formed on the surface of these fabrics during use. Sometimes double singeing is done to minimize pilling.
- The process is not essential for all types of fabrics.
- The object of singeing is to burn off projecting hairs without scorching or otherwise damaging the body of the fabric. Improper singeing may cause irreparable faults.
- Before singeing, the fabric should be perfectly dry, as wet fabric is scorched much more readily than dry fabric.
- Uneven singeing can cause streaks when the fabric is dyed, or bubbles when the fabric is finished.
- Improper singeing may lead to as much as 75% tensile strength loss in warp direction.
- The fabric to be singed should not contain any acid-releasing salt
- Singeing may cause hardening of size and can lead to difficulties in desizing.
- During singeing of synthetic fabrics, the protruding fibers melt and recede from the flame and form beads. In order to avoid this problem, these fabrics are singed after dyeing.

The types of singeing machines commonly employed are

- Plate singeing machine
- Rotary cylinder or roller singeing machine
- Gas singeing machine

Plate and roller singeing are obsolete nowadays, because of insufficient singeing and a typical hardness on the fabric produced by them. Although gas plants are known since long, the type and size did not suit the processing houses. With the advent of suitable gas units, the old contact singeing methods soon became outdated.

In a gas singeing machine, the fiber ends should be loosened from the body of the sized yarn by prebrushing and subsequently passed over an open flame or heat radiated from ceramic (heated by gas). When the flame with a temperature of about 1300°C touches the cold and moist material, an air or steam buffer zone is created between the flame and the fabric. This may hinder singeing operation. The flame should have sufficient and controlled thermal and mechanical energy to reach to the base of the material. The situation is further complicated by the fact that the pyrolysis of cotton is an exothermic process, whereas that of polyester is endothermic. Polyester ignites at 480°C–500°C, whereas it melts at a much lower temperature (250°C–270°C). In order to avoid melting of polyester before burning, the energy is to be supplied in a shock form and pyrolysis should be similar to an explosion igniting the whole length of the projected fiber instantly (Karmakar, 1999).

11.2.4 Desizing Process

The yarns, particularly the warps running lengthways throughout the fabric, are subjected to a high degree of abrasion during weaving. To prevent breakage or damage of warp yarns due to abrasion, size is imparted to the warp yarn. The presence of size on the fabric makes it stiff and renders its treatment difficult with different liquors used in dyeing and finishing. Therefore, one of the initial steps in wet processing is the elimination of size and water-soluble admixtures, the operation being called desizing.

The typical sizing materials are

1. Natural starches from potatoes, maize, rice, or topioca.
2. Chemically modified starches (ethers or esters).
3. Organic polymers, for example, polyacrylates, carboxymethylcellulose, methylcellulose, polyesters, or polyvinyl alcohol.
4. Solvent-soluble materials, for example, copolymers of methyl methacrylate.

About 75% of the sizing agents used throughout the world consist of starch and its derivatives because of low cost and high effectiveness as a sizing material on cellulosic textiles. Starch is difficult to remove, as it is not soluble in water or in normal scouring liquor. Removal of starch before scouring minimizes the work required in subsequent cleaning processes; hence, reduced concentrations of the chemicals are required in subsequent scouring and bleaching processes.

Under favorable conditions, starch can be progressively hydrolyzed to the following products with varying solubility in water:

Starch (insoluble) → dextrin (insoluble) → dextrin (soluble) → maltose (soluble) → α-glucose (soluble).

In desizing, the hydrolysis reaction is carried out up to the stage of only soluble dextrin and not up to the stage of α-glucose to avoid degradation of cellulose.

Starch desizing methods can be classified as follows:

1. Hydrolytic methods
   a. Rot steeping—The fabric is padded with warm water (40°C) and is squeezed to about 100% expression. The fabric is then allowed to stand for about 24 h at 35°C–40°C or overnight at 60°C in an open space or in an open box.
   b. Acid steeping—The fabric is impregnated with 2.5 g/L acid solution at room temperature (30°C) followed by storage for about 6–8 h.
   c. Enzyme steeping—The fabric is padded with 2–4 g/L α-amylases enzyme (Bactosol TK liq.), 1–2 g/L trisodium phosphate (pH 8.5–9) and wetting agents. It is then batched for 4–12 h at room temperature followed by hot and cold wash.

2. Oxidative methods

   a. Chlorine desizing—The fabric may be impregnated with a dilute solution of sodium hypochlorite or bleaching powder containing 1.5–2 g/L of available chlorine, squeezed and stored for 1 h at 30°C. It is then washed and antichlored.

   b. Chlorite desizing—A concentration of 10 g/L of sodium chlorite is sufficed to complete desizing within an hour. Sodium chlorite may be activated by a mixture of acetic acid and sodium acetate (set at pH 4) at 80°C or with ammonium sulfate (quantity equal to that of chlorite) at boil.

   c. Bromite desizing—The fabric is padded with 0.3% sodium bromide, a non-ionic wetting agent and 0.1% sodium carbonate at room temperature. After padding, the material is stored for 6–20 min or more at room temperature or slightly higher. The fabric is washed and treated with caustic soda solution and rewashed. Best pH for bromite desizing is 10.

   d. Peroxy compounds—The possibility of using sodium persulfate or hydrogen peroxide has been suggested but their commercial use to date is small. The material is to be saturated with hydrogen peroxide at 60°C or above with 80%–100% pickup, preferably with surfactants for better diffusion. It is then steamed for 8–10 min. 2% o.w.m. caustic soda is preferred, but can be increased to 6% for combined desizing and scouring.

11.2.5 Scouring and Bleaching

Scouring and bleaching preparatory processes have been discussed in Chapter 6. These processes should be carried out before dyeing. Scouring is must before dyeing, but bleaching may not be necessary for dyeing in dark shades. The processes are same for both yarn and fabric forms, but the machines differ. For small batches in winches and jiggers, scouring and bleaching processes are done in the same machine. But for large batches, these processes are done separately before dyeing. Scouring and bleaching do not require much precision as dyeing. In the past century, these processes are carried out in large lots (tons) in special machines such as Kier, j-box, vapor-lok, or pressure-lok machines.

Kier is a big closed cylindrical vessel with a perforated bottom in which fabric is uniformly piled and hot scouring liquor is circulated for many hours at high pressure or at atmospheric pressure.

Other two machines are continuous in nature. J-box is having mirror shape of the letter J. The fabric is impregnated with scouring liquor and passed through the preheated chamber of this machine at a certain speed, so that the fabric in rope form is held for a specific time in the chamber (dwell period) under atmospheric pressure. Due to the presence of air (oxygen) during treatment, scouring is much faster than that in kier. In vapor-lok or pressure-lok machine, the fabric is passed through a pressurized chamber in open width for a very short duration. All these machines have now lost importance due to their inherent problem—the most important being the nonuniformity of scouring throughout the whole length of fabric.

The saturator or impregnator device (Figure 11.2) is an extremely important part of a bleach range. It must evenly and rapidly impregnate the fabric with the treatment chemicals. The most common open-width application is a single box with four or six dips and nips before final squeezing to the desired wet pickup. Two-stage impregnation units force initial impregnation into the fabric using a roller nip with 10-tons pressure, and a second series of dips before expression with a nip under a pressure of 5 tons.

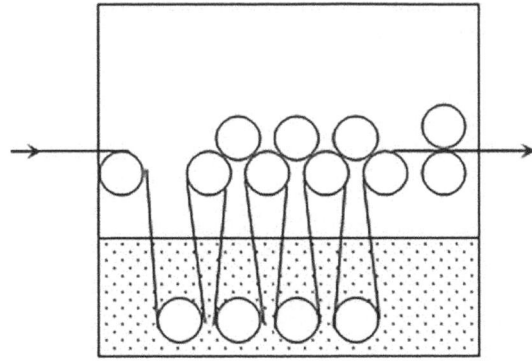

FIGURE 11.2
Saturator or impregnator.

The open-width bleaching processes have been developed in recent years at a rapid pace. These processes are of particular importance for the polyester–cotton blends, which are very sensitive to creasing. The quality of processed goods may suffer in these machines due to uneven heating and the energy consumption is high. But these machines have revolutionized bleaching processes as treatment time has been brought down very significantly. Roller steamers with 50–200 meters capacity allow very short reaction time because of rapid and uniform heating. In roller steamer systems, the impregnated cloth is passed through a closed steamer over a number of rollers. The machine is ideally suited for highly crease-prone fabrics. They suffer, however, from high chemical cost and the high cost of frequent replacements of various machine parts. The conveyor steamers with 200–500 meters capacity permit the processing of more relaxed fabrics by plaiting it without tension on the conveyor; longer dwell time results in more efficient use of chemicals. The scouring and bleaching recipes for cotton fabrics in three continuous steamers having varying dwell periods are presented in Table 11.1 (Mahapatro et al., 1991).

Most of the recent installations are a combination of roller–conveyor steamer (Figure 11.3) (Mock, 1985). The enlarged roller preheater and conveyor sections, needed to prevent packing creases, can be used separately or in tandem. The fabric path may be changed without

TABLE 11.1

Scouring and Bleaching Recipe for Three Continuous Steamers

Chemicals (g/L)	Dwell Time in the Steamer (minutes)		
	20–30	10–15	2–3
Scouring			
Caustic soda	20–30	30–40	80–100
Scouring aid	3–8	5–10	10–15
Wetting agent	2–3	2–3	2–3
Bleaching			
H_2O_2 (35%)	20–30	30–40	40–60
Sodium silicate	5–10	12–15	15–20
Caustic soda	4–6	5–7	6–8
Wetting agent	0.5	0.5	0.5

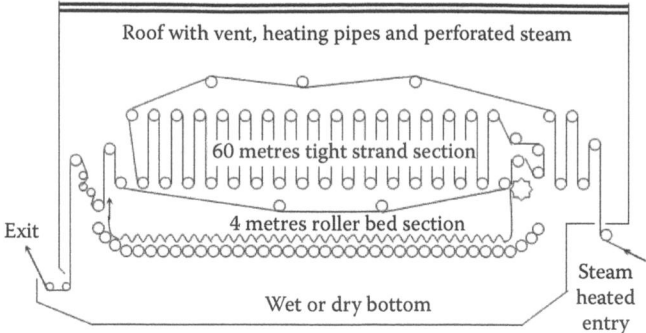

FIGURE 11.3
Combo roller–conveyor steamer.

rethreading. Thus, there will be maximum flexibility with regard to retention time. The fabric passes over a number of rollers on entry to the heated chamber to ensure level heating and to start the reaction in a flat, creaseless condition. After prestabilizing in this way, the fabric is rippled on to a conveyor consisting of a large number of driven rollers, all of which are close together. Such a conveyor, also called roller bed, keeps the fabric pile continuously moving, so that the folds in the fabric are never long enough in one position for a permanent crease to form. The contact of the bottom of the fabric pile with each roller is also not long enough for drying to occur. Such types of steamers are also known as roller bed steamers.

11.3 Preparation of Woolen Fabric

Three steps of preparation of woolen fabric are

1. Crabbing
2. Scouring
3. Carbonization

11.3.1 Crabbing

Before scouring, woolen fabric is to be set with a hot aqueous solution to prevent shrinkage during scouring and milling resulting in cockling and crows footing, that is, width distortion of the design and weave. For greasy material, alkaline solution is preferred. The process is known as crabbing. In the simple crabbing process (Figure 11.4), the fabric previously wound under tension in a roller (A), is wound on to a second roller (B) by passing under tension through a trough of hot water (H). The second roller moves in touch with another roller fitted on its top (C). As the roll of fabric builds up on the lower roller, pressure is applied from the upper roller, which operates in slots to allow for the increasing diameter of the lower roller. Pressure is applied on the upper roller with the help of levers, wheels, weights, and so on. The process may be repeated in a second trough. The application of heat and pressure in the moist condition is responsible for setting the fabric in a smooth and even manner. The duration of immersion in hot water may be 5–15 min, followed by natural cooling or by running through cooled water. The crabbing process

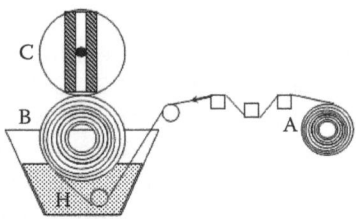

FIGURE 11.4
Crabbing machine.

is often replaced or followed by a faster blowing process in which the fabric is wound on a perforated roller through which steam is blown for 1–3 min under a pressure of about 0.1–0.5 MPa (1–5 atm). The process is repeated after rewinding the fabric, so that the external layers form the interior of the roll. Machines are also available in which crabbing and blowing can be carried out consecutively.

11.3.2 Scouring

Although the oil content of the wool is reduced to 0.4%–0.6% before spinning of woolen yarn, it is customary to add olive oil or specially prepared mineral oil during spinning. Manufactured products may therefore contain about 5% oil in worsted yarns, to as much high as 20% or more in carpet yarns, together with dirt picked up during manufacture.

The scouring in rope form is mostly done in Dolly scouring machine (Figure 11.5) resembling a winch dyeing machine (Trotman, 1968). The machine consists of a pair of heavy squeeze rollers (R) and guide rollers (G). A trough (T) is situated under the squeezing rollers to collect the liquor expressed by them. Two valves (W) are fitted into the trough, so that the collected liquor can either be returned to the vessel (V), or be discharged. The discharge is necessary when the scouring process is nearing completion and the liquor is considerably soiled.

The pieces of fabrics (about 60–70 meters for worsted fabric) are made into an endless chain, which is threaded, in rope form, through the squeeze and guide rollers. Twelve or more ropes can be processed simultaneously keeping the ropes side-by-side and separated by dividing rods (D). The liquor-to-material ratio may be as low as 2:1, which is much smaller as compared to that employed for scouring of cotton. For certain varieties, however, a higher ratio is maintained, so that the fabric floats in the liquor, which opens it somewhat, and prevents watermarks.

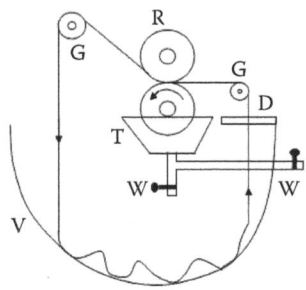

FIGURE 11.5
Dolly scouring machine.

Scouring liquor composition depends on the nature of the oils to be removed. Lubricants, prepared from hydrolysis of natural oils and fats, require sodium carbonate (about 3%–4% on the weight of material) in the scouring bath at 35°C–40°C. Low quality woolen goods contain as much as 15%–20% oil and require treatment with 4%–5% of sodium carbonate. It may be necessary to repeat the operation with lesser alkali before final rinsing. Worsted goods generally contain very low amount of oil. About 0.5%–1.0% o.w.m. of soap or synthetic detergent is usually adequate. To ensure alkalinity, sometimes 1% sodium carbonate or ammonia is added.

11.3.3 Carbonization

Carbonization of woolen fabric may be carried out before or after dyeing. The advantage of carbonizing before dyeing is that the acid can be carried forward to the dyebath as dyeing is done in acidic pH. However, carbonized goods do not always dye evenly. If carbonization is done after dyeing, considerable care is necessary to ensure that the shade does not change after acid treatment.

Carbonization may be done for unscoured wool, but penetration will be poorer and higher amount of acid will be required. Generally, woolen fabrics are carbonized between scouring and milling processes. Carbonized wool takes longer time to full or mill. However, for acid milling, the acid from the carbonization process may be allowed to remain in the goods. After milling, the burrs and other water-insoluble impurities are fastened in the fabric and are difficult to remove.

The cellulosic matters can be converted into hydrocellulose in two ways:

1. Dry carbonization with hydrochloric acid gas.
2. Wet carbonizing either by
   a. Immersion in 5%–7% sulfuric acid solution (5–6°Tw) in the presence of non-ionic wetting agent at room temperature, followed by squeezing to 75%–80% liquor retention, giving an acid uptake of 5% (o.w.m.). Typical running speeds in continuous impregnation tank are 5 m/min for heavyweight and 25 m/min for lightweight fabrics. The fabric should be dried to 10% moisture content at temperatures up to 90°C and baked for 3 min at 130°C. Crushing is done discontinuously for 60 min.
   b. Immersion in aluminum chloride (12°Tw) or magnesium chloride solution (10%–15%), drying followed by heating at 120°C.

11.4 Degumming of Silk

Raw silk does not possess the luster and softness for which this fiber is known. The gummy substance called sericin (about 20%–25% in mulberry silk and about 15% in wild silk) covering the fibrous material, fibroin, imparts a harsh handle and must be removed in order to bring out the supple and lustrous qualities. Although sericin and fibroin, the two components of raw silk are both proteins, they differ considerably in their relative compositions of various amino acids and accessibility. Degumming also removes accompanying substances such as fats, oils, natural pigments, and mineral components. The thorough and uniform removal of these impurities are important not only for uniform dyeing, but also particularly for good printing results. The process for removal of sericin is known as degumming, boiling off, and less commonly scouring.

The conventional working method for degumming or boiling off mulberry silk is as follows:

1. Soaking overnight or for 6 h in a bath with 3–5 g/L Marseilles soap and 1–2 g/L wetting agent.
2. Boiling off for 2–6 h at 90°C–95°C in a bath containing 8–10 g/L Marseilles soap at pH 9.5 with soda ash.
3. Repeat the treatment in a second bath similar to the first one for 2 h, if necessary.
4. Rinsing first at 50°C with ammonia and then followed by two rinsing, one at 40°C and another at room temperature.

A rapid method of degumming with soap is to treat for 1–2 h at 90°C–95°C at pH 9.5–10 in the following bath:

8–10 g/L Marseilles soap

1–2 g/L wetting agent

1 g/L soda ash

1 g/L sodium tripolyphosphate.

The wild silk (e.g., Tussah or Tussar) requires an initial scouring in a liquor: material ratio of 30:1 (minimum) and 10% (by weight of silk) soda ash.

11.5 Preparation of Synthetic Materials

The fabrics manufactured from synthetic fibers require a special preparatory operation called heat-setting. During this operation, the material is subjected to thermal treatment in tensioned condition in hot air, steam, or hot water medium. Unless this treatment is done, these materials will shrink when treated with aqueous solution in loose condition, and as a result the dimensions and the shape of ready articles change. The optimum conditions for setting various textile fibers in water, steam, and dry air are presented in Table 11.2 (Peters, 1967).

TABLE 11.2

Optimum Temperature and Time for Setting of Textile Fibers

Fiber	In Water		In Steam		In Hot Air	
	°C	Minutes	°C	Minutes	°C	Minutes
Triacetate	–	–	–	–	190–220	0.5–2
Nylon 6	100	120–180	108–121	10–30	175–190	0.25–0.5
Nylon 6,6	100	120–180	115–130	10–30	190–215	0.25–0.5
Polyester	100	120–180	140	10–30	180–220	0.33–0.66

TABLE 11.3

Heat-Setting Conditions for Different Polyester and Blended Fabrics

Type of Fabric	Temperature (°C)	Time (Seconds)
100% Polyester	180–220	20–40
Texturized	160–180	30
Polyester–cotton	190–210	20–40
Polyester–wool	170–190	30
Polyester–acrylic	190–200	30–60
Cationic dyeable	170–180	20–40

Different varieties of polyester fabrics are normally set by dry heat. The sequence, scour–heatset–dye is the safest and the most satisfactory for all types of goods. However, setting being in the central position of processing sequence, the expense of drying would be higher as the fabric is to be dried twice.

For materials that are perfectly clean in loom state, it is possible to use the sequence, heatset–scouring–dyeing. The sequence is very useful for warp-knitting industries. When heatsetting is a final process, as in the routine scour–dye–heatset, the conditions of the setting may alter the handle of the fabric.

Clip stenters and pin stenters are used for heatsetting, but the latter is preferred for the purpose, because of its versatile dimensional control. The conditions chosen for stentering vary considerably between different types of polyester materials as shown in Table 11.3 (Rao and Gandhi, 1991).

11.6 Fabric Dyeing Machines

Of all forms of cotton materials, the fabric is the one, which is dyed to a much larger extent than in any other form. The fabrics are woven or knitted in various machines in pieces of short length and these pieces are sewn together to make a continuous length of fabric. The dyeing may be carried out in rope form or in open width. The machines in which dyeing is carried out in an open width are jigger and padding mangles, whereas in winch and jet dyeing machines, dyeing is done in rope form. The continuous padding machines are used for continuous dyeing, whereas most of the other machines are batch machines, that is, they handle a finite length of fabric at a time.

11.6.1 Jiggers

The jig or jigger is one of the oldest types of machine for dyeing woven fabrics, which must not be creased during dyeing, for example, most taffetas, satins, poplins, ducks, and suiting. This is most suitable for materials in which the dyes do not exhaust well as the machine operates with low volume of liquor (i.e., low material to liquor ratio of 1:5 to 1:6).

The jigger (Figure 11.6) consists of a V-shaped trough (100 to 150 gallon capacity) with two rollers called draw rollers fitted above it. There is an additional roller, which may be placed on the fabric-wound roller. The purpose of this roller is to squeeze out liquor from the fabric at the end of the treatment. The rollers may be made of stainless steel or ebonite. Fully immersed guide rollers are provided in the trough. The fabric in open width

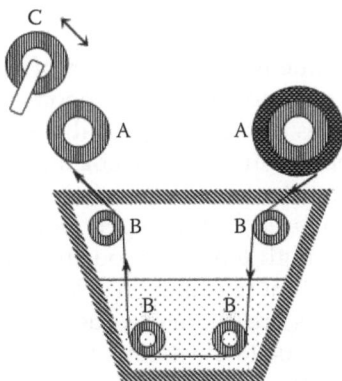

FIGURE 11.6
Jigger machine.

is unwound from one of the draw rollers, passed into the liquor contained in the trough with the help of guide rollers, and is wound on the other draw roller. When the second roller is full, the direction of fabric movement is reversed and the fabric is rewound on the first roller. This reversal of motion may be manual or automatic. In modern jigs, the length of run is set after the first passage through the machine and thereafter the direction of travel is reversed automatically. Both take-up and let off rollers are driven, thereby reducing lengthways tension on the fabric. The manufacturers have developed various devices for driving the rollers to achieve fairly constant fabric speed (about 40–100 m/min) at a moderate tension. Although many fabrics are adversely affected by substantial warp-ways tension under hot wet condition, some fabrics, particularly heavy cottons, can withstand tension as high as 100 kg (Duckworth, 1983).

Most modern jiggers have rollers 1.8 or 2.0 meters long. On old jiggers, roll lengths of 500 or 1000 meters were customary, but modern ones can accommodate rolls of up to 1 meter diameter, representing about 5000 meters lightweight taffeta fabric. Old jigs with small rolls had relatively small surface and liquor capacity of 200 liters. These were advantageous for vat dyeing, as the chances of oxidation were less.

When the whole fabric passes from one roller to the other, the passage is called one end or turn. The number of ends given depends on several factors such as size of fabric batch or the time required for one end (usually 10–15 min) and the depth of shade required.

The fabric should be wound on a roller with coincident selvedges, that is, no part of selvedge should come out of the roll, as that portion will be dyed differently. With vat dyes, the portion will be quickly oxidized and pick up more color in the next turn. On most jiggers, one of the rollers can be traversed to a limited extent to assist in straightening a roll. Some jiggers have the facility to traverse over a small range automatically to minimize selvedge built-up on certain varieties of fabric. For loading, the leading end of the fabric is threaded under the two free-running rollers located at the bottom of the trough, pulled upward and then wound around one draw roller. When winding is complete, the trailing end is then secured around the second draw roller.

Heating is usually by low-pressure steam, through a pipe having perforations at the lower side of the pipe. The pipe runs over the whole width of the machine at the bottom of the trough. Close-coil heating may be used—it is slower than direct steam injection, but offers better control of dyeing temperature and liquor ratio. The temperature may be controlled by a thermostat through a pneumatic valve on the steam line.

11.6.2 Winches

The winch or beck dyeing machine is the oldest form of piece dyeing machine. The construction is comparatively simple and therefore economical to purchase and operate. It is suitable for practically all types of fabric, especially lightweights, which can normally withstand creasing when in rope form such as woolen and silk fabric, loosely woven cotton and synthetic fabrics, and circular and warp-knitted fabrics. The winch imposes much less lengthways tension as compared to the jigger; hence, it is suitable for delicate fabrics, which are damaged under high tension. Scouring efficiency is high due to greater mechanical action caused by constant reformation of lengthways folds. Many fabric varieties, such as tubular knitted fabrics are, therefore, successively scoured and dyed in these machines. Crimps are developed due to the greater mechanical action combined with low tension. Winch dyeing results in thicker fabric with fuller handle, more fabric cover, and better crease recovery.

Winches are made of high-quality stainless steel (AISI type 316 or equivalent) with welded joints, ground and polished. The shape and size of the vessel and reel vary considerably depending on the type of fabric to be processed. Most of the winch machines are about 2.5 meters in length and 0.5 meter (single rope) to 4.5 meters (40 ropes) in width. Most winches are fitted with an overflow duct at the back, so that rinsing can be carried out in liquor flow from front-to-back.

In winch machines, a number (1 to 40) of endless ropes or loops of fabrics of equal length (about 50 to 100 m) are loaded with much of their length immersed in folded form inside the dyebath. Pegs separate individual ropes from each other. The upper portion of each rope runs over two reels mounted over the dyebath. One 15 cm diameter reel called fly roller or jockey (J) is at the front and a much larger driven reel called winch reel (W) is at the back for pulling the ropes round. The winch reel not only controls the rate of movement of the fabric rope, but also the configuration of the rope in the dyebath. The reel does not grip the fabric positively, but by the weight of the wet fabric and the friction between the reel and fabric. The reels are made of stainless reels with corrugated and broken surface for increased frictional forces. The reels may be wound with polyester or polypropylene tape at a narrow angle across the reel. Tape-covered reels retain a much greater degree of friction even at high temperature. However, dye contamination may occur from lot-to-lot through the tapes unless sufficient precautions are taken.

Winches are of two types:

1. Deep-draught winches.
2. Shallow-draught winches.

Deep-draught winches (Figure 11.7) usually have circular or slightly elliptical reels with diameter ranging between 20 and 50 cm. It pulls the fabric and lifts it out of the dyebath and over the jockey reel. On leaving the winch reel, the fabric falls straight into the dyebath with very little plaiting action.

For woolen and heavy cotton fabrics a deep-draught vessel with a sloping back is preferred. The depth enables long length of the bulk fabric to be accommodated and the sloping back enables piled-up fabric to be pushed easily toward the front of the machine. The dye liquor is usually 1-meter deep and the comparatively small surface area minimizes steam loss when dyeing at high temperature and oxidation of vat dyes during dyeing.

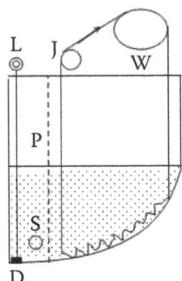

FIGURE 11.7
Deep-draught winch.

As the fabric falls into the liquor from the winch at the back of the machine, it tends to bunch up and lie in a heap for a short time. This creates creasing problems for man-made thermoplastic fibers. To overcome this, a long shallow machine with an elliptical winch has been designed. Shallow-draught winches (Figure 11.8) handle lighter fabric better and have elliptical winches of about 2 meters circumference and 2:1 axis ratio. Elliptical winches provide more mechanical action and a plaiting action as the fabric falls into the dyebath—the material is folded in wide layers, which are free from random creases as it falls into the dye liquor. The size of plait increases with reel size. The larger the reel, lesser is the mechanical action and is advantageous for delicate fabrics. Reels with adjustable cross section are also available.

The dyeing of filament viscose, acetate, and nylon fabrics require shallow-draught winches to minimize creasing. The depth of liquor is reduced (to about 75 cm) to decrease pilling pressure and the bottom of the machine is flattened. There is no great weight of water pressing on the fabric to make creases permanent.

The winch reel is driven from one side of the machine through a V-belt system permitting a variation in reel speed between 40 to 80 m/min, depending on the nature of material to be dyed. The greater the speed or movement of the fabric, higher will be the uniformity of dyeing. However, woolen fabrics are felted and delicate fabrics are damaged when run under high speed and a slow speed is preferred.

The jockey or fly roller (J) acts as a guide roller. It is mounted at the front of the machine in free-running bearings, and is rotated by the pull of the fabric. Mostly they are circular in cross section, but for providing higher mechanical action on the fabric, sometimes the cross section is made polygonal. The roller removes some of the surplus spent dye liquor from the fabric. In some winch machines, a pair of rollers are provided that exert a mild

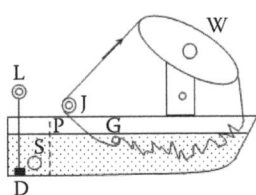

FIGURE 11.8
Swallow-draught winch.

squeezing action on the fabric. For dyeing delicate fabrics, it is advisable to fit the jockey near the winch reel.

Many shallow-draught winches are fitted with an adjustable horizontal bar, called mid-feather or gate (G) halfway back in the winch, and usually located about 15 cm below the top of the liquor. The bulk of the plaited fabric is kept submerged at the back of the dye-bath. The fabric slowly rises to touch the gate, which prevents bulk of the material from moving forward. From the gate onward, the fabric moves nearly in a horizontal sweep in crease-less condition up to the jockey. The formation of permanent lengthways creases is thus avoided.

A peg rail, fitted with numerous horizontal pegs, is provided in front (and little below) of the jockey. The pegs are usually 12 cm apart and act as separators for the ropes to avoid entanglement. The pegs may be fixed or they may rotate with the movement of the fabric, minimizing tension and chafing.

A perforated sheet (P) of stainless steel is fitted vertically across the full width of the machine and about 30 cm away from the front wall of the machine. The separate chamber, thus formed, is the additional chamber in which the concentrated solutions of dyes and auxiliaries are added and stirred thoroughly to make uniform dilute solutions before they enter the actual dyebath. The box also provides a house for the water inlet pipe, outlet (D), and open horizontal steam pipe (S), as liquor can easily diffuse into and out of the dyebath with minimum disturbance to the fabric ropes at the bottom of the vessel. Medium capacity winch machines are provided with a single drain valve operated by a lift rod (L), whereas large capacity machines have two 12-cm outlets, one at each side of the additional chamber. Direct steam injection into the dyebath is usually carried out by means of a perforated pipe mounted horizontally at the front and near the bottom of the additional chamber. This simple arrangement results in rapid heating and vigorous agitation—steam bubbles are forced into the liquor at high pressure. Most winch machines are totally enclosed, but it is essential to have some interlocking device to open a vent to allow the steam to escape before the front is opened. The steam loss is thus minimized and a good uniformity of temperature is maintained. To provide good accessibility to front and back, toughened glass doors are fitted to lift vertically.

In winch, the material-to-liquor ratio is usually around 1:25. For better-designed machines, it is around 1:15, whereas for small winches it may be as high as 1:40. The winch dyeing machine does not give very efficient movement of the fabric through the liquor as a large portion of it always remains at the bottom of the dyebath. Moreover, the fabric is irregularly piled and uneven dyeing may occur unless suitable dyes and leveling agents are used. Due to high liquor ratio, dye exhaustion is poor and considerable amount of dye remains in the bath after dyeing, especially in the case of low substantive dyes.

A fault with winch machines is the tendency for the temperature to vary in different parts of the dyebath, which may cause variation in color depth of different ropes, especially with temperature-sensitive dyes.

11.6.3 Beam Dyeing Machine

In beam dyeing, a length of fabric is wound on a perforated carrier and processed in a cylindrical autoclave. The system is fully enclosed and is normally pressurised to provide processing at temperatures up to 130°C (40Mpa, 4 atm.). Pressure beams came into use in late 1950s in order to dye lightweight, delicate warp-knitted nylon and acetate fabrics and later for dyeing of polyester fabrics at high temperature. The making of a good beam is an essential prerequisite for uniform dyeing. The perforated beam must be sufficiently robust to withstand the weight of the full roll of wet fabric. After invention of jet dyeing machines, fabric beam dyeing machines are obsolete now.

11.6.4 Jet Dyeing Machines

Until 1960s, the machines used for dyeing fabrics in batches were the winch, jigger, and beam. When polyester was first introduced as a textile material, conventional pressure-free winch beck equipped with a hood was mainly used for dyeing using carriers. Soon it became evident that dyeing above 100°C would be necessary to eliminate the use of carriers and accelerate the diffusion of the dyes for cutting down the dyeing time. This prompted machine manufacturers to construct high-temperature or high-pressure modifications of the existing winch dyeing machines. It was impossible to solve all the problems of polyester dyeing by mere modification of HT winch beck. It was increasingly felt that dyeing time was too long in these machines. All the disadvantages of HT winch becks had been overcome by the jet dyeing machine, based on a Venturi tube in which the circular movement of the liquor carries the fabric wound with it in a totally enclosed tubular chamber, annular in shape. The Venturi tube is a constriction in the annular passage through which the flow rate of the liquor must be increased, thus causing suction, which imparts movement of the fabric. In all true jet dyeing machines, the fabric rope is threaded through a ring or Venturi, to which the processing liquor is supplied. The jet of liquor issuing from the ring serves both to transport the fabric and to bring the liquor into intimate contact with it. The primary flow is given by a centrifugal pump, but it is usual to incorporate a few inclined steam jets also to boost the movement of both fabric and liquor. The first jet dyeing machine was commercialized by Gaston County Dyeing Machine Co. in Atlantic City in 1961.

Most jets of this type have two or three tubes mounted side-by-side within the same pressure vessel, but machines with up to 10 tubes have been made. When dyeing double-jersey fabric, weighting 600 g/m², loads of up to 150 kg per tube can be accommodated, and the entire rope circulating in about 1 min. Lightweight fabrics usually require a smaller diameter jet and each machine is supplied with a range of jets of different sizes—nowadays adjustable jets are also available.

In first generation jet dyeing machines, longitudinal creases are formed when the fabric is lifted from the dyebath. To minimize creasing and damage to the delicate material, a new type called fully flooded jet dyeing machine has been developed in which the fabric remains completely immersed throughout the dyeing cycle. Foaming problem is thus avoided. The partially flooded systems have been virtually replaced by fully flooded designs, often referred to as soft-flow machines. In order to reduce tension, the dye vessel is either horizontal or only slightly inclined. Although the liquor is circulated by a pump and extra motion is imparted to the fabric by an accelerator, there is usually a driven reel to assist movement. A good control on the fabric speed is provided by the driven circular reel; any jamming of the fabric increases the load on the driving motor and this can be used to initiate an alarm. The circular reel creates much less strain in the fabric than the conventional winch reel with its elliptical cross section.

In these machines, the transportation tube is filled with the dye-liquor using an overflow system and both liquor and fabric flow along the tube under gravity. The reduced speed of the fabric (100–200 m/min as compared to 250–400 m/min) and consequently gentler action on the fabric, led to this type of fabric movement system being called *soft flow*. As the

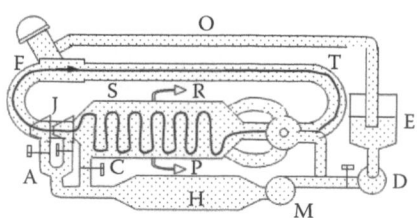

FIGURE 11.9
The jet dyeing machine.

transportation tube is filled by overflow, only a low pressure is to be exerted by the pump; hence, a less powerful pump (e.g., 5 kW) is required. In HT models, a secondary pump is usually available to generate adequate internal pressure.

The machine designed by them has a lower height convenient to operate, but occupies much more floor space (Figure 11.9).

The fabric passes through a high-speed jet (J) feeding fresh dye-liquor to the fabric and is carried along a tube (T) to the entrance of a much wider section chamber used as a storage compartment (S). In this section both liquor and fabric move very slowly until reaching the end when the fabric is drawn off through a narrow tube and is fed to the jet. By regulating the flow of liquor through the storage compartment, the speed of the fabric in this critical area can be controlled. The dye liquor is drawn off from both the ends of the storage compartment and passed through a heat exchanger (H) by the main pump (M) before being fed back to the jet. Heat exchangers should be regularly cleaned (mechanically or with caustic soda and sodium hydrosulfite) as dyes and oligomers are deposited on the heated surface causing slough off.

Other components of the machine shown in Figure 11.9 are

F—Fabric entry.

A—Jet controlling valve.

C—Connection for counter current flow.

E—Addition or expansion tank.

D—Secondary pump for dyebath.

O—Overflow pipe.

P—Dyebath discharge.

R—Overflow rinsing.

Fully flooded machines are particularly good at minimizing creasing in delicate fabrics, but are difficult to operate. The control on fabric in the storage compartment is limited. Lightweight fabrics cover a small portion of the compartment creating possibilities for entanglement. The material to liquor ratio is 1:15 onward.

The recent developments in jets have the following objectives:

1. Minimizing liquor usage
2. Minimum fabric distortion, pilling, and creasing
3. Improving rinse efficiency

11.6.5 Semicontinuous and Continuous Dyeing Methods

Continuous dyeing is aimed at maximum productivity, but considerations of resource conservation and environment protection are now overtaken in importance. Conservation measures include the recovery of heat, water, chemicals from waste process liquor, lower liquor retention in padding, slick rollers and loop transfer methods of application, and more efficient hydroextraction to achieve minimum cost of drying.

The padding mangle is perhaps the most familiar and universal of all textile dyeing and finishing machinery. It gained popularity for application of dyes and chemicals in continuous and semicontinuous dyeing and finishing processes. The continuous dyeing process is advantageous when a large quantity of fabric is to be dyed in a single shade. The uniform application of dye-liquor to the fabric is the most critical part of a continuous dyeing process and satisfactory performance of the padding mangle is absolutely essential for success of such processes.

The padding process is suitable for applying dyes and chemicals having low or no affinity for the fiber. Low-substantive naphthols, reactive, solubilized vat dyes, vat dyes by pigment padding or vat acid process, aniline black, mineral khaki, and so on can be applied by padding process, as the original dyes have little or no affinity for the textile materials. If the dyes used in the padding mangle have high affinity, their concentration in the trough decreases with time, resulting in a tailing effect. The effect can be minimized by using feeding liquor having higher concentration than the solution initially fed to the trough. The padding mangle is not a complete dyeing machine by itself—other machines are necessary for fixation of dyes in the padded fabric.

The padding operation consists of two steps. First, the fabric, usually woven, is immersed in the dye-liquor to achieve through impregnation, and second, the fabric is passed between two rollers to squeeze out the air and to force dye-liquor inside the material, the excess liquor being sent back along the fabric. The former step is known as *dip*, whereas the contact between the squeeze rollers and passing between the rollers is known as *nip*. For one dip and one nip padding process, the fabric is immersed in the trough once and passed through the nip once. For dyeing most fabrics at speeds of up to 50 m/min, single dip and single squeeze are sufficient. For higher speed (say 120 m/min) or for heavy fabrics, double-dip and double-squeeze process using a three-bowl padding mangle is preferred. The liquor retained in the fabric after padding is expressed by weight as percentage of the weight of dry fabric and is known as % pick-up or % expression. Thus, 80% expression means the fabric weight is increased by 80% of dry weight after padding. The expression less than 60% is difficult to obtain in padding mangle. An increased pressure at the nip results in a lower percentage expression, but at the same time, the penetration inside the fabric is better. The fabric is subsequently passed through various machines for fixation and postfixation processes. The padding mangle is also termed as dye pads or padders. The expression *mangle* is often confined to less critical machines for squeezing water from fabric, usually prior to dyeing.

In old mangles, one of the rollers is rubber-covered, whereas in the other one, the driving roller is made of stainless steel, brass, or ebonite. However, such types of mangle may give face to back difference on the fabric. Now both rollers of the mangle are made of rubber of similar hardness.

The chemical composition of the rubber surface should be chosen carefully so that it is unaffected by the liquor containing dyes and textile auxiliaries. In most of the two-bowl padding mangles, the lower bowl is in fixed bearings and the upper bowl is separated by about 2 cm, when the mangle is not operating.

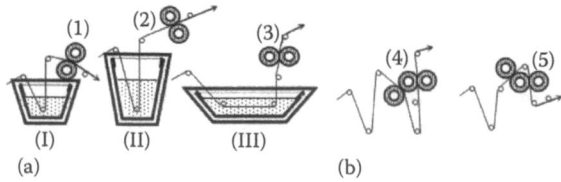

FIGURE 11.10
Padding mangles: (a) two bowl and (b) three bowl.

Situated below or at the front of the squeeze rollers is the padding box or trough, having a free-running roller at the bottom which guides the fabric through the trough. The box may be V-shaped, deep immersion type, or shallow type (shown in Figure 11.10a as (I), (II) and (III), respectively) having a width slightly more than maximum width of the fabric to be processed. The volume of the pad box should be as small as practicable (about 2 gallons for a two-bowl padding mangle with 60″ bowl width), so that the wastage of dye-liquor is minimum at the end of the run.

The mangles may be arranged as vertically opposed in ascending or descending order or as horizontally opposed position. The said three arrangements with Artos two-bowl padder are shown in Figure 11.10a as (1), (2), and (3), respectively. For three-bowl padders, all three may be inclined with a steady central bowl, whereas the top and bottom bowls press themselves against the central steady bowl with the action of the levers and pneumatic pressure. There may also be a combination of ascending and horizontal with two-dip (same or different bath)–two-nip (4) or descending and horizontal with one-dip–two-nip (5) arrangements as shown in Figure 11.10b. The ascending-bowl gives better observation of the nip, whereas in descending-bowl system, the fabric path does not change much. With horizontally opposed mangle, the fabric rises vertically from the padding box directly to the nip and straight through a dryer mounted above the nip. This arrangement is widely used for *Thermosol* dyeing process for polyester and its blends. In three-bowl arrangement, maximum impregnation occurs at the first dip-squeeze and maximum uniformity at the second squeeze.

The alternate sequences semicontinuous methods are as follows:

1. Pad → cold batch → wash
2. Pad → preheating → hot batch → wash
3. Pad → dry → jigger development

In cold-batch process, the fabric is padded with the dye and necessary auxiliaries, and is batched on a roll. The rolls are subsequently stored for 2–24 h depending on the time required for fixation. Covering the rolls with polyethylene sheets during storage is advantageous. During storage, the rolls are to be rotated slowly to prevent drainage of padding liquor. The method is very popular for application of cold brand as well as Remazol brand of reactive dyes.

Pad-hot batch system was first developed in Sweden in 1954 for application of direct dyes on cellulosic materials and acid dyes on nylon and then extended for application of metal-complex dyes to warp-knitted nylon automobile fabrics. The method may also be used for application of hot-brand or Remazol brand of reactive dyes. The fabric is padded with the dye and then preheated with infrared (IR) heater for only a couple of seconds before the fabric is wound on a large roll. The roll is kept in an atmosphere of saturated steam in a cottage or chamber for several hours for dye fixation to take place. The roll is slowly rotated during the period to prevent the liquor from settling. The heat input should be carefully regulated to raise the temperature of the wet fabric to near 100°C without drying and without excessive condensation. Very few dyers achieved this objective successively and consistently and now it is rarely used for coloration.

In another popular sequence, the fabric is padded with finely dispersed nonsubstantive dyes, dried, and then the fabric roll is transferred to a jigger in which the development or fixation is carried out, followed by washing.

The most common sequence for continuous dyeing is padding followed by drying, steaming, and washing in continuous machines.

In many cases, an additional padding step is necessary between drying and steaming for application of fixation chemicals just before steaming. Therefore, the most popular sequence for continuous dyeing is as follows:

Pad → dry → pad → steam → wash

The classical pad–steam route was developed by DuPont for the production of military uniforms with vat dyes, and was later adopted for other dye classes. It is still the major method used by continuous dyers to apply vat, reactive, and sulfur dyes to cellulose in a fully continuous process. However, the dramatic growth in the leisure and sportswear sector has had an impact on the industry. Frequent fashion changes have led to a requirement for more rapid response and lower stock inventories, which in turn have resulted in a dramatic reduction in lot size. The reduction in lot sizes forced continuous dyers to develop other methods apart from pad-steam. Some of the alternate routes rely on different ways of generating heat energy.

Pad–dry–bake method is well established for dichlorotriazine (DCT) and monochlorotriazine (MCT) reactive dyes and currently for bifunctional MCT reactive dyes. However, the use of urea in the process results in obnoxious fumes in the factory.

11.7 Dyeing of Cellulosic Materials

As mentioned in Chapter 6, cellulosic materials can be dyed with direct, azoic, sulfur, vat, solubilized vat, reactive, ingrain, mineral, pigment, and oxidation colors (limited use). All of the dye classes have their own merits and demerits.

Azoic, solubilized vat, ingrain, mineral, and oxidation colors have very restricted use. Pigment colors are water insoluble and can only be used by padding along with a film-forming polymer called binder. This can be simultaneously applied with finishing agents.

Direct, sulfur, vat, and reactive dyes are practically used in coloration of cotton fabrics. The application of sulfur and vat dyes are restricted on jute and flax fabric.

11.7.1 Dyeing with Direct Dyes

The most attractive feature of direct dyes is the simplicity of the dyeing process and lower cost for achieving high depth of shades. Direct dyes have varying fastness to washing, light, perspiration, and other wet fastness properties, and varying staining properties on wool, silk, polyester, and acrylic fibers. Most direct dyes have limited wet fastness in medium to full shades unless they are after treated, but some are better than others.

Resin finishing after dyeing produces a notable improvement in wet fastness, especially on regenerated cellulosic fibers.

Direct dyes are used in low-priced viscose or blended curtain fabrics, furnishings, and carpets. Cheap cotton apparels, casual wear, and bedspreads, low-quality discharge-printed materials, which are not washed frequently, are dyed with direct dyes.

Unfortunately many important disazo direct dyes are withdrawn from manufacture because of the carcinogenicity (cancer producing) of benzidine and some of its derivatives. The most toxic ones are found among the disazo and cationic classes. Nonbenzidine direct dyes derived from less hazardous diamine derivatives have replaced those prohibited—but those are costly and have much poorer built-up properties than the corresponding benzidine derivatives.

The direct dyes are pasted with cold water, if necessary, with an anionic or nonionic wetting agent. Sufficient boiling water is then added with constant stirring to bring it into solution. It is then added to the dyebath containing a large volume of cold water through a strainer. Soft water is preferred, but is not essential. However, some dyes are precipitated by hard water and this can be prevented by adding a sequestering agent such as sodium hexametaphosphate (not with metal–complex direct dye) or 1%–3% (o.w.m.) (0.5–2 g/L) of soda ash. Common salt amounting 5%, 10%, or 20% (o.w.m.) for pale, medium, or heavy shades, respectively (or about 10–15 g/L) is then added to the liquor. Pastel hues are preferably dyed without the addition of salt.

Wet cellulosic material, scoured and optionally bleached, is then entered into the dye-bath containing dye at 40°C–50°C. Alternately, the material is first run in water in the dyebath and then dye and salt solutions are successively added. The liquor is then heated slowly and the temperature is raised to boiling over a period of 30–40 min, after which dyeing is continued for 45–60 min. If the dye distribution is not uniform throughout the fabric, further boiling may be carried out. Improved yields can be achieved when applying full depths by cooling to 80°C at the end of the period at boil, adding a further 5 g/L salt and rising to the boil again. The exhaust dye liquor is drained. The material is washed with cold water, hot water, and lightly soaped water (optionally).

11.7.2 Dyeing with Sulfur Dyes

Sulfur dyes now constitute the largest class of dyes in terms of quantity with an estimated worldwide production of more than 80,000 tons per annum, even though a different picture emerges if the calculations are made on the basis of price (Shore, 1995). They are the most economical (cheapest) dye class and for most of them the chemical structures are unknown. They are normally distinguished in terms of the organic intermediates from which they are derived and the process of sulfurization used in their manufacture. The characteristic feature of the dyes of this class is that they all contain sulfur linkages within the molecules. There is no black dye for cellulose that can be compared with the attractive shade and depth possible with C.I. Sulfur Black 1. It is one of the most widely used dyes in the world.

Sulfur dyes are important for black, navy, brown, olive, and green colors in medium to heavy depths. The liquid brands are ideally suited for continuous dyeing in long runs.

Sulfur dyes are used widely on cellulosic fibers and their blends, especially with polyester, and also with nylon and acrylic. Cotton and polyester–cotton drill and corduroy are dyed continuously or on the jig, whereas cotton–nylon and cotton–acrylic knitted and pile fabrics are winch-dyed and jet-dyed. Since 1985, sulfur black finds increasing use on denim (Shore, 1995).

Sulfur dyes differ from vat dyes in being easier to reduce, but more difficult to reoxidize, different oxidizing agents produce variations in hue and fastness properties.

The dyestuff is pasted with equal quantities of T.R. Oil or suitable wetting agent and soda ash. Sodium sulfide is also added, the quantity varies between half (for browns, blues) and double (for yellows) of the weight of the dye, in most cases 1½ times. Boiling water (about 10 times) is then added to the paste until the dye is completely dissolved. The dye solution is strained to remove the insoluble matters. For blacks and heavy shades with large quantities of dyes, straining may not be possible, and in such cases the color must be dissolved in 1 or 2 feet of water at the bottom of the machine.

The dyebath is made up with 5% o.w.m. soda ash, 2% o.w.m. sodium sulfide, the dye solution, and 5%–25% o.w.m. common salt depending on the depth of shade. The salt may be added in several portions in the bath at the commencement or after raising the temperature to 100°C. Surface-active penetrating agent may be added. The dyeing is continued to boil for 30 min. The steam is then turned off and the dyeing continued for a further ½ h in the cooling bath. Some sulfur dyes are best exhausted at 70°C–75°C.

The sulfur dyes do not exhaust well in deep shades and dyeing is to be carried out with a low liquor ratio. A standing bath is usually used for sulfur black. After each dyeing, between ½ and ¾ of the original quantities of dye, sodium sulfide and soda ash are added. When the specific gravity of the bath exceeds 1.05 (10°Tw), the bath should be discarded or diluted. When free sulfur appears in the bath, sodium sulfite should be added till the whole free sulfite is converted to thiosulfate.

$$Na_2SO_3 + S \rightarrow Na_2S_2O_3 \tag{11.1}$$

It is important that the retained liquor should be rinsed out immediately after dyeing, to prevent deposition of the insoluble product of oxidation on the surface of the fibers.

Worldwide the standard oxidant for sulfur dyes has long been sodium bichromate, known simply as chrome and acetic acid. It oxidizes all reduced sulfur dyes rapidly and completely. For batchwise methods, treatment in 0.25–1.0 g/L sodium dichromate and 0.8–1.2g/L acetic acid for 15–20 min at 60°C is usually recommended. For continuous dyeing, the recommended quantities are 5 g/L and 6 g/L acetic acid, respectively and dwell time is 20–40 s at 60°C–70°C. For optimal oxidation, it is essential to maintain pH 4.5–5.5. However, the presence of heavy metal salts such as chrome in effluent has long been recognized as environmentally undesirable.

The economical per-compounds (e.g., H_2O_2) are widely used for oxidation of sulfur dyes. The use of 1 ml/L hydrogen peroxide (35%) or 1 g/L sodium perborate at pH 9 and 40°C for 10–20 min to give a brighter color, especially with sulfur blues, is well-known. However, these compounds barely oxidize the sulfur red–browns and may overoxidize sulfur blues.

11.7.3 Dyeing with Vat Dyeing

In spite of the highest all-round fastness, vat dyes are not much used for fabric dyeing due to high cost and poor leveling properties. The vat dyes are first passed with little water and reduced with sodium dithionite or $Na_2S_2O_4$ (sodium hydrosulfite), commonly known as hydros and caustic soda. The fabric is first run in blank dyebath having hydros and caustic soda and the concentrated reduced dye is slowly added in the bath. The concentrations of hydros and caustic soda, and temperature of the dyebath should be strictly maintained. The presence of the chemicals should be constantly monitored and added as and when required. Excess chemicals may cause overreduction. Closed dyebath is preferred to prevent premature oxidation. After dyeing, the dyes in the fabric are to be oxidized in air or with oxidizing agent followed by washing and soaping.

The steps to be followed in jigger dyeing are

- The fabric is loaded evenly onto the jig in a crease-free condition, with selvedge laid on selvedge to avoid introducing dyeing fault caused by premature oxidation.
- A blank dyebath is prepared with the required quantity of caustic soda and sodium hydrosulfite at the required dyeing temperature and is run one end to remove oxygen from the fabric.
- The reduced stock vat is added over two ends.
- Dyeing is carried out for 6–8 ends (45–60 min) at constant temperature with regular checks with Vat Yellow paper, and necessary additions are made.
- The dyebath is completely drained and the fabric is rinsed with overflowing cold water for 2–4 ends.
- Oxidized, soaped, and rinsed.

With the introduction of vat dyes in the form of finely dispersed particles, it is possible to have a uniform deposition of the dye particles in a nonsubstantive form either in yarn package (without the danger of filtration, as in the case of normal powder vat dyes) or on the fabric using a padding mangle. This *pigmentation process* is followed by vatting process. In a modification of the process, the circulating liquor containing the unreduced microdisperse vat dyes, the total quantities of alkali and hydrosulfite, along with a leveling agent (if necessary) is maintained at room temperature (30°C) and is circulated for 10–15 min. On account of lower temperature, the dye is not reduced completely during the period and is uniformly deposited on the fiber in nonsubstantive form. The temperature is then raised gradually when the dye is reduced slowly and absorbed by the material. The process is known as semipigmentation process.

The vat acid process is based on the principle that when a stock vat solution is treated with acetic acid to neutralize the alkali, in the presence of an acid-stable dispersing agent, the vat acid, that is, the leuco compound is formed as an extremely fine dispersion. The vat acid has no affinity for cellulose and it can be applied evenly throughout the material without the complications introduced by affinity. The vat acid dispersion is much finer than the finely dispersed pigments; hence, the penetration is much higher. Once evenly applied, it can be easily converted to the substantive form by the treatment with sodium hydroxide.

Similar to circulating machines, prepigmentation method may be carried out in jigger for better penetration of heavy fabrics. The bath is filled with 70% of the final volume of water and heated to 80°C. The pigment dispersion is added over two ends. Cold water is added and the temperature is lowered to 60°C. Both caustic soda and hydrosulfite are added over the two ends and the dyeing is continued as before.

11.7.4 Dyeing with Reactive Dyes

Reactive dyes are the youngest and the most important of the dye classes for cellulosic materials and 50% cellulosics are dyed with these dyes. Share of reactive dyes among all textile dyes is 29%, which is next to the disperse dyes (Wasif, 2010).

These dyes react with hydroxyl or amino groups of textile fibers forming covalent bonds. These dyes are fast so long as the covalent bonds are intact. As the fastness does not depend on the molecular size of the dye, dyes of low molecular weight can be used that usually provide brighter shades. These dyes were originally developed for cellulosic materials, but special reactive dyes for wool and nylon are also available. Depending on the reactive group, these dyes are classified into a number of subgroups: cold brand (M-brand), Remazol (vinyl sulphone), high exhaust (HE-brand), and so on.

The estimated application wise consumption (in percentage) of the reactive dyes is as follows:

Exhaust dyeing	48%
Pad-batch dyeing	14%
Continuous dyeing	10%
Printing	28%

11.7.4.1 Batchwise Dyeing

The batches of light delicate and heavy structurally stable fabrics can be dyed in winches and jiggers, respectively. The general principle in exhaust dyeing of reactive dyes is to exhaust as much dye as possible onto the fiber using salt under neutral or very slightly acidic conditions. Under the conditions in which the dye does not react or react negligibly, leveling takes place. The pH of the bath is then raised by the addition of alkali and the reaction is allowed to undergo for a particular time at a particular temperature dictated by the reactive group present in the reactive dyes. Dyes having different reactive groups are advisable not to be mixed as the reaction conditions differ.

11.7.4.2 Semicontinuous Processes

On account of their low substantivity during the neutral adsorption and the ease with which fast dyeing is really obtained in the subsequent alkali fixation, the reactive dyes have proved eminently suitable for continuous and semicontinuous dyeing processes. From the mechanical aspect, the simplest are the semicontinuous methods, that is, pad-batch methods.

For cold-brand reactive dyes, padding liquor of short batching process (2 h) contains the following:

Dye + 2–5 g/L wetting agent + soda ash equal to the weight of the dye (minimum 3 g/L and maximum 30 g/L).

The Remazol or vinylsulfone dyes are well suited for the short-time (2 h) pad-batch method. The padding liquor contains the following:

x g/L Remazol dye.

10–30 mL/L sodium hydroxide solution (32.5%) depending on the depth of shade.

30 g/L common salt or Glauber's salt.

5 g/L (or as required) wetting agent.

After padding, the fabric is batched up without selvedge overlap, and then the roll of fabric is wrapped with polyethylene sheet. Dye fixation is affected at room temperature and it takes 2–6 h.

This is followed by washing in an open-width washing machine.

11.7.4.3 Continuous Processes

In continuous dyeing, several variations are possible such as

1. Pad (with alkali)–dry.
2. Pad (with alkali)–dry–bake.
3. Pad (with alkali)–dry–steam.
4. Pad (dye)–dry–pad (alkali)–steam.

11.8 Dyeing Woolen, Silk, and Nylon Fabric

The breakdown of traditional wool coloration processes is estimated (Lewis, 1992) to be as follows:

Loose stock dyeing	16%
Top dyeing	16%
Yarn dyeing (hank and package)	40%
Piece and garment dyeing	28%

So, it is clear that woolen as well as silk and nylon fabrics are dyed in batch method by exhaust dyeing system and the methods of dyeing in yarn and fabric form (Chapter 6) are more or less similar—the difference is in the respective dyeing machines and corresponding liquor ratios.

11.9 Polyester Fabric Dyeing

Polyester fabrics are mostly dyed in jet dyeing machine using disperse dyes. The conventional exhaust method of dyeing is very similar for both yarn and fabric forms.

To reduce the dyeing time further, quick-fixable rapid dyes may be selected, so that satisfactory diffusion and fixation may be achieved within 10–20 min in rapid dyeing machine as compared to much higher time (even 50 min) by the conventional dyes in conventional jet machines. The total dyeing cycle may be reduced to 50–70 min as compared to 180 min in conventional process.

In the early 1950s, DuPont developed *Thermosol method*, a continuous method of dyeing polyester or polyester-blended fabric in which the fabric is padded with aqueous dispersion of dye, dried and baked at 180°C–220°C for a period of 30–60 s. On account of the hydrophobic characters of fiber and dye and high temperature of treatment, rapid diffusion of dye takes place into the fiber in the absence of carrier or pressurized equipment.

Dry-heat method became popular for woven polyester–cotton blends from about 1960s onward in the United States, the United Kingdom, and other parts of the world where continuous cotton-piece dyeing was already established. Numerous process sequences are possible, but for polyester component, the most popular sequence is

Pad with an aqueous dye dispersion → dry → bake → wash off

During baking, disperse dyes are transferred from the cellulosic component to the polyester, essentially by vapor transfer rather than contact migration. Migration in fabric thickness, that is, from back to face or from the fabric interior to exterior–will always occur to some extent, even under most properly balanced drying conditions possible in commercial operations. The commercial antimigrants are essentially polymeric electrolytes.

The material is padded cold using 50%–60% pickup with padding liquor consisting of

Disperse dyes	x parts
Migration inhibitor	5–20 parts
Sodium alginate	2 parts
Sequestering agent (if necessary)	1 part
Water to make	1000 parts

Tartaric acid is added to set pH at 5 (about 5 g/L).

After padding, the material is passed through a predrying unit, which may be a hot flue chamber, heated can, or an infrared predryer. It is then thermofixed at the desirable temperature between 180°C and 220°C for 30–60 s. In some countries, preheaters are hardly available and instead of thermosoling unit, pin stenters are used for fixation.

11.10 Blend Dyeing

The term blending is used by the yarn manufacturers to describe specifically the sequence of processes required to convert two or more kinds of staple fibers into a single yarn composed of an intimate mixture of the component fibers. To the dyer, however, the important type of staple fiber blend is that in which components are two different fibrous polymers, each with its own characteristic dyeing properties.

The kinds of colored effects achieved by dyeing a blend of two fibers can be categorized as

1. Solid shades, that is, both fibers are dyed as closely as possible to the same hue, depth, and brightness. This is most popular as the blends. Solid shades are difficult to obtain on blends of fibers having similar chemical composition but varying dyeability—the same dye class gives different build-ups on such varieties of fibers depending on their relative affinity.

2. Reserved effects, that is, only one fiber is dyed, whereas the other is retained white or undyed.

3. Shadow effect or tone in tone is obtained when two fibers are dyed with the same hue and brightness but of different depths. It may be considered as an intermediate stage between solid and reserve effects. The effect may be obtained by using two substrates (e.g., nylon–wool or acetate–polyester) or differential dyeing variants of a substrate having similar dyeability but with varying affinities.

4. Contrast or two-color effects or cross-dye, that is, the two fibers are dyed in contrast hues. The contrast may be sharp or subtle.

In order to dye large varieties of blends, it is necessary to classify the textile fibers according to the dye classes, which can exhaust on them in full depths (Nunn, 1979):

1. Dyeable with disperse dye, for example, acetates and polyester (D-class fibers).
2. Dyeable with anionic dyes, for example, cellulosic fibers, nylon, wool, and silk (A-class fibers).
3. Dyeable with basic dyes, for example, acrylic and modacrylic fibers, basic-dyeable nylon, basic-dyeable polyester (B-class fibers).

Most important binary blends are as follows:

D-blends: Acetate–polyester.
A-blends: Wool–silk, wool–cellulose, silk–cellulose, nylon–wool, nylon–cellulose, normal and deep-dyeing nylons.
DA-blends: Polyester–cellulose, polyester–wool, polyester–nylon, acetate–nylon, and acetate–wool.
DB-blends: Normal and basic-dyeable polyesters, polyester–acrylic, and acetate–acrylic.
AB-blends: Wool–acrylic, cellulose–acrylic, nylon–acrylic, basic-dyeable polyester–nylon (or wool), deep-dyeable, and basic-dyeable nylons.
Except a few AB-blends, ternary blends have not been accepted commercially.
AB-blends: Normal, deep, and basic-dyeable nylons and acrylic–modacrylic–wool (or cellulose).

11.10.1 Dyeing of D-Blends

Though all synthetic-polymer fibers are readily dyeable with disperse dyes, the dyeability of secondary acetate, triacetate, and polyester differs greatly at a given temperature. The rates of dyeing on acetates are much higher and it is difficult to obtain solid shades in acetate–polyester blends.

11.10.2 Dyeing of A-Blends

These blends are usually developed because of a desirable balance of physical characteristics, and the attainment of solid effects is, therefore, necessary.

Cotton-viscose blends are made to produce cheaper fabrics, but with higher luster. Cotton-viscose blends require careful treatment, as rayon fibers are more susceptible to chemical attack than cotton.

Both wool and silk can be dyed simultaneously in a single dyebath using milling acid dyes and 1:2 metal–complex dyes. From the technical standpoint, the main problem in dyeing such blends is the distribution of the dye between the two fibers. Silk fiber lacks the impervious structure of wool and the dyes are more readily adsorbed and desorbed by silk. At low temperatures, dye is taken up preferentially by silk and later migrates to the wool-hair at higher temperatures. Under normal wool-dyeing conditions, more acid dye is adsorbed by wool than by silk. Hence, the dyeing temperature is a decisive parameter in the control of dye distribution.

For wool-cellulosic blends, direct dyes alone or admixed with milling and/or 1:2 metal-complex dyes can be used. A slightly acidic condition facilitates exhaustion of wool dyes. If the dyeing with direct dyes is started under neutral or slightly alkaline condition, the pH may be lowered by adding acid. Nylon–wool blended materials may be dyed in solid shades using leveling and milling type of acid dyes and metal–complex dyes.

11.10.3 Dyeing of DA-Blends

The blends of polyester with natural fibers are most successful examples of achieving the benefits of the blend. The dyeing of DA-blends is usually directed toward solidity rather than differential effects, but unfortunately the most troublesome problem is the staining of wool with the disperse dyes. Cellulosic fibers are less sensitive to cross-staining by these dyes.

The biggest problem of dyeing polyester–wool blends is the staining of wool with the disperse dyes. The stain on wool is dull and exhibits poor wash and light fastness. Though the terms stain or cross-stain is frequently used, the wool is actually dyed by disperse dyes but with low levels of fastness.

In one-bath process, disperse and acid dyes are added at 50°C–60°C with 3%–5% ammonium acetate or sulfate and acetic acid to give pH 5–6, 0.5–1 g/L sodium dinaphthylmethane disulfonate and 3–5 g/L carrier. The temperature is increased to boiling or 105°C slowly and dyeing is continued for 60–120 min depending on the depth. If dyeing is done above 110°C and outside isoelectric range of wool (pH 4.5–5.2), a fiber-protecting agent, or suitable auxiliary should be added.

After one-bath dyeing, the stain on wool may be cleared by scouring with 1–2 g/L nonionic detergent and 0.5–1 ml/L acetic acid (pH 4–5) for 20–30 min at 50°C–70°C. In one-bath process, staining will be more if the affinity of disperse dye for polyester is lower (as in the case of C.I. Disperse Violet 27) than that when the affinity is higher (as in the case of C.I. Disperse Orange 30 or Blue 73). Full depths are usually dyed by a two-bath process. On polyester–cellulose, the degree of staining is less, prolonged boiling favors migration of disperse dyes to polyester without severe attack on cellulose and the stain can be removed by reduction clearing. Reactive dyes give negligible staining on polyester. Direct dyes give good reserve sequence.

Many of the hot-dyeing reactive dyes (e.g., HE-brand) are sufficiently stable to withstand the conditions of high-temperature dyeing and they can be dyed by one-bath two-step sequence.

The reactive and disperse dyes are added at 60°C with 0.5 g/L sodium dihydrogen phosphate (pH 5–6), 1 g/L dispersing agent, 5 g/L sodium-m-nitrobenzene sulfonate (to prevent reduction of reactive dyes at high temperature). The temperature is raised quickly to 95°C and then over 30–40 min to 130°C. The dyeing is continued for 45–60 min. The temperature is reduced to 80°C and required quantity of common salt (50–80 g/L) is added. After 15–20 min, soda ash (15–20 g/L) (pH 11) is added and the fixation is completed in 40–60 min at 80°C.

11.10.4 Dyeing of AB-Blends

Wool–acrylic blends can also be dyed with selected anionic and basic dyes, either at pH 2–2.5 for 1:1 metal–complex dyes or at higher pH for neutral-dyeing acid and 1:2 metal–complex dyes. 1:1 metal–complex dyes are more compatible with basic dyes due to their zwitterionic character at low pH.

The staining of wool by basic dyes is favored at the early stage of dyeing below 80°C. As the boiling is approached, transfer of basic dyes from wool to acrylic proceeds, thereby wool exerts a retarding action. The degree of wool staining depends on dyeing conditions and is minimized by dyeing for at least 1 h and at pH not higher than 5. Wool–acrylic blends may be dyed in pale shades by setting dyebath at 50°C with 2% acetic acid (pH 4.5), 5% salt, 1%–2% anionic retarder and 0.5%–1% alkanol polyoxyethylene. The anionic and basic dyes are added at short interval. The temperature is raised to 90°C over 45 min, held for 15 min, again raised to boiling and continued for 60–90 min. Better compatibility may be attained in medium depth by adding anionic dyes in the starting bath and treating for 15–30 min at 80°C. This is followed by addition of basic dye, retarder, acid (pH 4.5). The temperature is raised to boiling. After completion of dyeing, the fabric is cooled to 70°C and scoured with nonionic detergent.

In the two-bath process, the acrylic is dyed first by adding basic dyes, sufficient acetic acid to bring the pH to 4.5–5 and a cationic retarding agent. The temperature is raised to boiling, taking care in the critical range of 80°C–100°C. If the wool is heavily stained, it can be cleared in a bath containing 3 g/L sodium dithionite at 60°C followed by rinsing. Fresh liquor is then made up; the wool is dyed in the normal manner with the appropriate dyes, and after cooling to below 70°C it is taken out.

11.11 Textile Printing

The process of making a pattern or design on the fabric surface by using some kinds of dyes or pigments, which causes localized dyeing, is called textile printing. In other words, textile printing is the process of applying color to selected portions of the fabric as per definite patterns or designs. In printed fabrics the color should be physically or chemically bonded with the textile fiber, so that the colors can withstand washing and rubbing. The mechanism of textile printing is similar to dyeing but, in dyeing the target is uniform penetration or spreading of a single color in all directions of the fabric, whereas in printing one or more colors are applied in selected parts of the fabric in sharply defined patterns.

In printing, wooden blocks, stencils, engraved plates, rollers, or silk screens are used to restrict the flow of color paste as per design. More or less same colorants are used in dyeing and printing. However, dyeing is carried out by dipping textile materials in large volume of dilute colorant solution, whereas in printing low volume of concentrated and thickened colorant solutions are applied through any of the printing machine or tools. Thickening agents prevent the color from spreading by capillary attraction beyond the limits of the pattern or design.

There are several methods for producing colored patterns on cloth. Among them commercially most important are

1. Hand block printing
2. Engraved roller printing
3. Flat or table screen printing
4. Rotary screen printing
5. Transfer printing
6. Digital textile printing

11.11.1 Hand Block Printing

This technique for printing text, images, or patterns was used as a method of printing on textiles and later paper widely throughout East Asia and probably originated in China in antiquity. In this process, a design is drawn on, or transferred to, a prepared wooden block. A separate block is required for each distinct color in the design. Woodblock printing on textiles is the process of printing patterns on textiles, usually of linen, cotton, or silk, by means of incised wooden blocks. It is the earliest, simplest, and slowest of all methods of textile printing. Block printing by hand is a slow process. It is, however, capable of yielding highly artistic results, some of which are unobtainable by any other method.

Block making required patience and skill. A fairly hard wood was required, such as pear wood, and four or five layers were usually glued together with the grain running in different directions. The design was traced on to the surface and a fine chisel used to cut away the nonprinting areas to a depth of perhaps 1 cm (Figure 11.11a). Finally, each block required corner *pitch pins*, which printed small dots; these allowed the succeeding blocks to be correctly positioned by accurately locating the pitch pins above the already printed dots.

In addition to the engraved block, a printing table and color sieve are required. The color sieve consists of a tub (known as the swimming tub) half filled with starch paste, on the surface of which floats a frame covered at the bottom with a tightly stretched piece of mackintosh or oiled calico. On this color sieve, a frame similar to the last but covered with fine woolen cloth is placed, and forms when a position is sort consisting of elastic color trough over the bottom of which the color is spread evenly with a brush.

Figure 11.11 shows that a man is printing with a wooden block. The printer commences by drawing a length of cloth from the roll over the table, and marks it with a piece of colored chalk and a ruler to indicate where the first impression of the block has to be applied.

He then applies this block in two different directions to the color on the sieve and finally presses it firmly and steadily on the cloth, ensuring a good impression by striking it smartly on the back with a wooden mallet. The second impression is made in the same way with the printer and taking care that it fits exactly to the first, a point at which he can make sure by means of the pins with which the blocks are provided at each corner and are arranged in such a way that when they are at the right side or at the top of the block they fall upon those at the left side or the bottom of the previous impression, the two printings join up exactly and continue the pattern without a break. Each succeeding impression is made in precisely the same manner until the length of cloth on the table is fully printed. When this is done it is wound over the drying rollers, thus bringing forward a fresh length to be treated similarly.

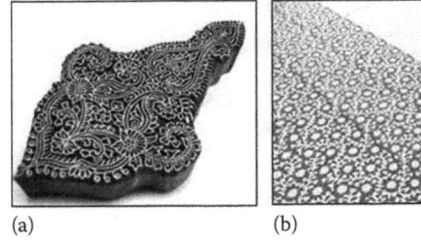 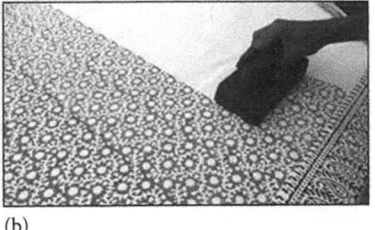

(a)　　　　　　　　(b)

FIGURE 11.11
Hand block printing: (a) wooden block and (b) block printing method.

11.11.2 Roller Printing

Roller printing, also called cylinder printing or machine printing, is a textile printing process patented by Thomas Bell of Scotland in 1783 in an attempt to reduce the cost of the earlier copperplate printing. This method was used in Lancashire fabric mills to produce cotton dress fabrics from the 1790s, most often reproducing small monochrome patterns characterized by striped motifs and tiny dotted patterns called *machine grounds*.

In its simplest form, the roller-printing machine consists of a strong cast iron pressure bowl (A) mounted in adjustable bearings capable of sliding up and down the slots in the sides of the rigid iron framework. Each engraved roller (F) mounted on a steel mandrel (L) is forced against the fabric being printed (E) as it travels around a pressure bowl (A) with a resilient covering (B). As the pressure bowl rotates, a furnishing roller (G) transfers print paste (color) from a color box (H) to the engraved roller (F), filling the engraved surface and smearing the whole surface. This surface color is almost immediately removed by the steel blade known as the color doctor (J). This doctor must be precisely ground, sharpened, and set, at the optimum angle and tension, to leave the surface perfectly clean. Engraved areas retain the color in parallel grooves and the doctor blade is *carried* on the crests between the grooves. The fabric is then forced into the engraved roller and most of the paste is transferred. The cushion between the pressure bowl and the engraved roller clearly plays a critical role in the uniform transfer of color across the fabric width and along its length.

To protect lapping from the color printed beyond the fabric edges or those that are forced through the fabric, an endless printing blanket (C) must be used and washed and dried continuously before returning to the point of printing. In addition, a back–gray (D) is used to absorb color and give greater resilience, unless the blanket provides enough resilience and is able to hold the excess color satisfactorily. In the printing of lightweight or knitted fabrics, the woven back-gray is often *combined* with the fabric to be printed, using small amounts of a suitable adhesive. This helps to maintain dimensional stability under the tension applied during printing (Figure 11.12).

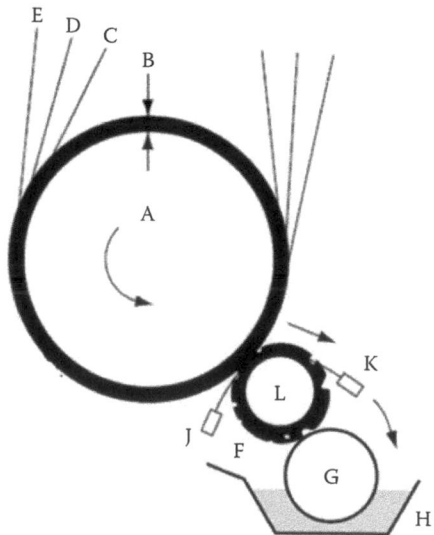

FIGURE 11.12
Schematic diagram of a roller printing machine.

After transferring its color, the engraved roller (F) is cleaned by a brass blade known as the lint doctor (K). This removes any loose fibers that may have stuck to the surface of the roller and could become trapped under the color doctor, where they could cause a color streak to be printed. The printed fabric is then separated from the back–gray and blanket and carried on to the drying section of the machine, avoiding any contact of the print face until some drying has been achieved.

On a multicolor machine the printing rollers, with color boxes and other auxiliary equipment, are arranged around the larger pressure bowl (A) with a minimum separation of the rollers. Setting and maintaining the correct registration (i.e., fitting each color of the design relative to the others) requires an arrangement for separately rotating each roller a small distance while the drive to all the rollers is engaged. The original box-wheel device was elegantly simple but introduced a serious hazard to the printer's hands as it required the insertion of a rod, or Tommy key, into a hole in a wheel rotating near the meshing drive gears (Miles, 2003).

11.11.3 Screen Printing

Screen printing is a form of stenciling that first appeared in a recognizable form in China during the Song dynasty (960–1279 AD). The main benefit of screen printing is that it is cost-effective and large-scale production is possible with it.

The advantages of screen printing became increasingly appreciable, especially in the fashion houses. Designs are relatively easy to transfer to screens and the frame size that can be readily varied. The designer, freed from the restrictions of copper rollers, thus had far greater freedom to choose repeat sizes. In addition, the pressure applied in screen printing is much lower than in roller printing with the result that surface prints with an improved *bloom* or color strength are obtained, and textured surfaces are not crushed.

In stenciling the pattern is cut out of a sheet of stout paper or thin metal with a sharp pointed knife, and the uncut portions representing the part that is to be reserved or left uncolored. The sheet is then laid on the material to be decorated and the color is brushed through its interstices.

Screen printing is a method of printing in which a design is imposed on a screen of silk, nylon, polyester, or other fine mesh on a rigid wooden or metal frame, with nondesign areas coated with an impermeable substance, generally, with photosensitive lacquers, which harden when exposed to light, especially ultraviolet light. The screen containing open mesh in the design area is pressed against the substrate. A fill blade or squeegee is moved across the screen stencil, forcing or pumping ink into the mesh openings for transfer by capillary action during the squeegee stroke. The open areas of mesh transfer ink or other printable materials that can be pressed through the mesh as a sharp-edged image onto a substrate. It is basically the process of using a stencil to apply ink onto another material whether it is open-width fabric, T-shirts, posters, stickers, vinyl, wood, or any material that can keep the image onto its surface (Figure 11.13). As silk fabric was originally used, it is also known as silkscreen, serigraphy, and serigraph printing. A number of screens can be used to produce a multicolored image or design.

Printing is carried out on a flat, solid table covered with a layer of resilient felt and a washable blanket (usually coated with neoprene rubber). Heat for drying the printed fabric may be provided either under the blanket or by hot air fans above the table.

Fabric movement or shrinkage must be avoided during printing in order to maintain registration of the pattern. The fabric to be printed is laid on the table and stuck to the blanket directly, using either a water-soluble adhesive or a semipermanent adhesive; alternatively, it

Squeegee

FIGURE 11.13
Screen printing.

is *combined* with a back–gray. In the latter instance an absorbent fabric is stuck to the blanket and the fabric to be printed is pinned down on top of it. Sometimes fabric and back–gray are combined before fixing to the table using an adhesive and a specially adapted pad mangle. Combining is most suitable for printing lightweight fabrics in which there is a danger of smudging or loss of adhesion caused by the presence of excessive print paste. It can also be advantageous for knitted fabrics.

The manual process has been semiautomated by mounting the screen in a carriage and driving the squeegee mechanically across the screen. Long tables, typically 20–60 m long, are used, and some provision is usually made for drying the printed fabric. Semiautomated flat-screen printing is still very popular in which the design size is very big, which is not possible to print manually (as in the case of bed sheet), the scale of production is not large, or capital investment is limited. In semiautomatic machine, the fabric is temporarily stuck to an endless rubber blanket by hot melt adhesive. All screens are arranged sequentially above the blanket keeping one repeat gap between each of them. The blanket stops moving, all screens brought down over the fabric glued to the blanket. The squeeze of each screen is moved manually for the required number of times. The screens are lifted and blanket moves for one repeat distance. The printed fabric is detached from the blanket and is sent to the dryer fitted above the printing machine. In fully automatic screen machines, the movement of squeeze is done automatically.

11.11.4 Rotary Screen Printing

Fully continuous printing is best achieved using cylindrical (rotary) screens and many attempts were made to form flat wire mesh screens into cylinders, despite the necessity of a soldered seam. When printing through a cylindrical screen with a seam, a line will show across the fabric once every cylinder circumference, unless the seam can be hidden within the design. The invention of seamless screens of electrodeposited nickel was a really significant step, which heralded the rapid expansion of rotary-screen printing. Peter Zimmer (Austria) introduced the galvano screen in 1961 and Stork (Holland) introduced the lacquer screen in 1963.

Lacquer screens have uniformly spaced hexagonal holes arranged in lines parallel with the axis of rotation of the cylinder and offset in alternate lines, as in a honeycomb, for maximum strength. The walls of the holes through the thickness of the screen are sloping, so that the holes are larger on the outside of the wall than on the inside. In the screens used for printing textiles in the open area, measured on the inside of the screen, varies between 9% and 13% of the total. Lacquer screens are manufactured by electrodeposition of nickel

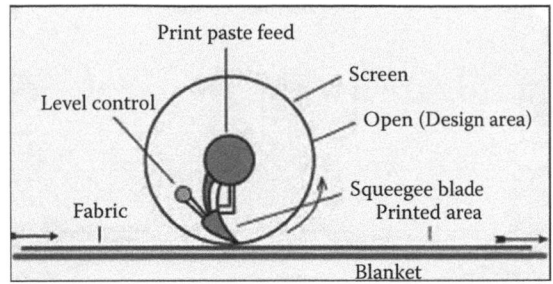

FIGURE 11.14
Schematic diagram of a rotary screen printing machine.

on to a mill-engraved mandrel. The method of introducing the design on to the screens is similar to that used for flat screens, except that the shape of the screens necessitates modifications in details. The screen is coated with a photosensitive lacquer and is exposed to light keeping a positive film of the design in between. The design portion is marked with black ink and remains unexposed, whereas nondesign portion is exposed to light and thereby the lacquer in those portions gets hardened. During washing, lacquer from the design portion is washed out as it is still water soluble. The lacquer in the nondesign portion is further hardened with a suitable varnish.

The galvano screen by Peter Zimmer is manufactured quite differently. The design is introduced at the same time as the nickel is electrodeposited. This means that the nondesign areas are solid nickel instead of a uniform mesh filled with a thin layer of lacquer as in the case of lacquer screen. They are therefore stronger and less susceptible to pinholes.

In rotary-screen printing, continuous rotation of a cylindrical screen while in contact with the fabric ensures genuinely continuous printing. Print paste is fed into the inside of the screen, and during printing it is forced out through the design areas with the aid of a stationary squeegee. Figure 11.14 shows a schematic diagram—only a single-screen assembly is shown. About 8–12 rotary screens for different colors are placed horizontally on the moving endless blanket. The screens are positioned along the top of the blanket. The blanket washing and drying are effected underneath during the return passage of the endless driven blanket. Provision for the use of a thermoplastic adhesive is common in rotary machines, with a curved-surface heating plate to heat the fabric before it is pressed on to the blanket. The cylindrical screens can be much closer together than is possible with flat screens and so the blanket is shorter (for a given number of colors). The fabric dryer, however, must be long enough to enable the printed fabric to be adequately dried at higher running speeds. Typically, speeds of 30–70 meter/min are used depending on the design and the fabric quality. It is quite possible to run the machine faster than this, the limitations often being the length and efficiency of the cloth and blanket dryers and the difficulty of observing printing faults at high running speeds.

The print paste is pumped into the screen through a flexible pipe from a container at the side of the machine; inside the screen, the paste pipe has a rigid structure as it also acts as a support for the squeegee. Holes in the pipe allow the paste to run down into the bottom of the screen; as the paste is pumped in from one end, the holes need to be larger at the end that is furthest from the pump to achieve an even spread across the full width of the screen. A sensor (level control) actuates the pump when the paste level falls below a preset height.

| Conventional | Airflow | Magnetic |
| squeegee | squeegee | squeegee |

FIGURE 11.15
Three types of squeezes used in rotary screen printing.

The first squeegees used by Stork were of the traditional rubber type, but excessive wear of the rubber due to the continuous movement and the drag on the inside of the screens, which caused screen distortion, soon led to their replacement with flexible, stainless steel blades (conventional squeeze). For controlling squeeze pressure independently the stainless steel blades are supported by air sacks in Stork's Airflow squeeze. A completely different approach was adopted by Zimmer, who invented a rolling-rod squeegee moved by an electromagnet, driven under the blanket. The three types of squeezes (conventional, airflow, and magnetic rod) are shown in Figure 11.15.

11.11.5 Transfer Printing

In transfer printing the design is first printed on to a flexible nontextile substrate (e.g., paper) and later transferred by a separate process to a textile material. The advantages are

1. Designs may be printed and stored on a relatively cheap and nonbulky substrate such as paper, and printed on to the more expensive textile as and when required.
2. The production of short-run repeat orders is much easier by transfer processes than it is by direct printing.
3. The method requires lower skill input and rejection rates are lower.
4. Storage costs are lower as the designs are held on a paper which is cheaper than textiles.
5. Many complex designs can be produced more easily and accurately on paper than on textiles.
6. Most transfer-printing processes enable textile printing to be carried out using simple, relatively inexpensive equipment with modest space requirements. No need for washing-off and hence, no effluent production.

The sublimation transfer method depends on the use of a volatile dye in the printed design. When the paper is heated the dye is preferentially adsorbed from the vapor phase by the textile material with which the heated paper is held in contact. In this method, paper is printed with disperse dyes of low sublimation fastness (B and C class as per ICI disperse dye classification). Methyl cellulose or other suitable thickener is used for making print paste. The printed paper is dried and stored. When required, the printed paper and the polyester fabric or garment are placed one on the other and passed through transfer machine consisting of two stainless steel plates heated at around 180°C–200°C.

The paper-fabric/garment sandwich is kept between the plates under pressure for about a minute. The plates are separated and the fabric is removed. No washing is required.

For transfer printing on cotton and other substrate, a special coated paper is printed in the normal inkjet printer and again kept in the heat transfer machine in Sandwich with cotton fabric for a minute. This method is now quite popular for T-shirt printing.

11.11.6 Digital Printing

It is a fact in history that the first digitized manufacturing process involved weaving of textiles by a card with punched holes and was the invention of Joseph-Marie Jacquard in 1801. Digital technologies have developed since then. Digital information can now be stored and processed electronically and mechanical storage devices are no longer required (Tyler, 2005). Keeling (1981) was able to identify the following key characteristics of digital printing technologies:

- The input data are not constrained by size, nor does it need to be repeated, because the data are drawn from a computer file that is, in principle, unrestricted in size.
- The ink is dropped onto the substrate (noncontact), making it possible to print on a flat or curved, smooth or rough, and delicate or hard substrates.
- In most cases, inks can be developed that are compatible with any chosen surface; hence, they are versatile.
- Based on the cyan–magenta–yellow–black color gamut, thousands of colors can be printed without the requirement for a color kitchen.
- The printing rates are high, but dependent on resolution, the type of printing required, the head technologies, and so on.
- No moving parts. Printer motions are limited to oscillators within heads and a system to move the heads in relation to the substrate. It is the ink that moves during printing, not a mechanical device. Consequently, ink jet printers are inherently reliable.

Conventional paper printing is now a big user of digital technologies, but has the advantage of working with relatively standardized substrates—typically smooth and flat. Textile digital printing has to address all the issues of conventional digital printing, but also new issues are raised by the media to be printed: surface texture, endues performance, color physics, and so on.

Inks are composed of a colorant, a carrier base, and various additives. The carrier base (or vehicle) may be water, solvent, oil, phase change fluid (hot melt), or UV curable fluid. However, to date, nearly all the inks for textile printing are based on water because they are designed for heads that require water-based inks.

Colorants for textile substrates have to be fast for washing, light, and rubbing. Functionality issues are just as important as putting color on the textile. There are four main categories of colorant:

1. Reactive dyes (for cellulosic fibers)
2. Disperse dyes (for polyester)
3. Acid dyes (for protein fibers and nylon)
4. Pigments (for all substrates)

Digital printing technologies are considerably cleaner than conventional ways of applying color to textiles. The environmental aspects of digital printing are

- Digital printing machinery do not waste ink or dye. After conventional printing, surplus dyestuff is dumped. This has implications for both effluent and cost. All this is avoided with digital systems using the CMYK color system.
- The digital process has a higher fixation rate. In conventional printing, the amount of dye placed on the substrate is not metered. It is extruded in sufficient quantities to do the job, allowing for some waste. Thus, if reactive dyes are used, a fixation rate of 65%–70% would not be unusual. The 30% waste is because there are not enough sites on the substrate to bind with the dyes. With digital technologies, the amount of dye put down is precisely controlled; and made just enough to do the job. The fixation rates are found to be higher, above 90%. With disperse dyes, the situation is very similar. Transfer is efficient with the appropriate dwell time and transfer temperatures.
- Conventional printing makes use of carriers that are later washed off—both reactive and disperse dyes have thickeners and carriers for the dye paste. After use, they are washed off as waste. These materials are not incorporated in the inks used for digital printing. With reactives, fabrics for digital printing require a *prepare for print* process, which puts a precisely monitored mix of materials on to the fabric to help achieve print integrity (Tyler, 2005).

11.12 Printing Methods

Printing is first carried out with a thick viscous paste containing dyes, auxiliaries (e.g., humectants, pH controller, buffer) and a thickening agent, for example, gums to increase viscosity, so that the paste does not flow outside the printing or design area. After printing, the fabric is immediately dried and then the print is fixed by steam or dry heat. The printed fabric is thoroughly washed with cold and hot water followed by thorough soaping and rewashing.

Textile printing can be made in three styles:

1. Direct style of printing
2. Resist style of printing
3. Discharge style of printing

In direct or normal style of printing, the colors or printing pastes are directly applied on the fabric in the design portion. The term *direct* originally indicated that no prior step of mordanting or the following step of dyeing was required. Textile fabrics are mostly printed by this style.

In resist style of printing, the design portion in the fabric is blocked with some substances, which physically or chemically resist penetration of colors in those areas. After resisting selected portion of the fabric, it is quickly passed through dye liquor and the color is fixed by a suitable method. The resisted portion may be colorless (white resist) or colored (color resist) if suitable colors are applied along with the resisting agent. An example of resist printing is batik printing in which wax is used as a resisting agent.

In discharge printing, the fabric is first dyed overall with selected dischargeable dyes by normal dyeing procedure. The dyed fabric is subsequently washed, dried, and printed with a dischargeable chemical (a strong reducing agent, e.g., sodium formaldehyde sulfoxylate sold as Formosul or Rongalite C) followed by steaming. During steaming, the dyes in the design portion are reduced and become colorless (white discharge). Alternately, suitable dyes that are stable to the reducing agent may be added in the printing paste, which gets fixed during steaming in the discharged areas (color discharge).

Some direct style of printing methods is discussed in Section 11.12.1.

11.12.1 Printing on Cotton

In earlier days cotton fabrics were mostly printed with direct and vat dyes. Direct dyes were used where fastness was not of much concern and the printed fabrics were rarely washed. On the other hand, vat dyes were used where a high degree of fastness was required. But due to their high cost and difficulty of application method their use at present is limited. Reactive dyes and pigment colors are presently predominantly used in cotton printing.

11.12.1.1 Reactive Printing

The full gamut of colors in the reactive class of dyes, which is one of its significant advantages, is obtained by employing a wide range of chromophores. Alginates are the only natural thickeners suitable for use in printing with reactive dyes. All other carbohydrates react with the dye and this result in low color yields or unsatisfactory fabric handling due to insolubilization of the thickener. Sodium alginate also contains hydroxyl groups but it reacts very little, presumably because the ionized carboxyl groups on every ring of the polymer chain repel the dye anions. At first a stock thickening paste is prepared following the recipe as under:

Alginate thickener (3–12%)	400–500 g
Urea	100–200 g
Sodium bicarbonate	20 g
Sodium m-nitrobenzenesulfonate	10 g
Water	470–270 g
Total	1000 g

Sodium alginate powder is sprinkled in hot water under stirring till the paste is quite thick (viscous). The paste is then cooled. A typical print paste is prepared by sprinkling the required amount of reactive dye into the stock paste. The solubility of most reactive dyes is sufficient for this sprinkling method, followed by high-speed stirring, to give perfectly smooth prints.

The dye can alternatively be predissolved, using the urea to increase the solubility, in a small volume of hot water. Sodium bicarbonate should only be added after the mixture of dye solution and thickener has cooled to room temperature.

On account of the relatively high cost and limited supply of alginates, attention has recently been paid to finding alternatives. Synthetic thickeners with anionic charges show great potential. Poly (acrylic acid) does not react at all with typical reactive dyes and color yields are higher than with alginates. However, the washing off is difficult and the handling of the fabric may be impaired. Emulsions of both o/w and w/o types are also suitable and *half emulsions* have been widely used.

The fixation of most reactive dyes is effected by saturated steam at 100°C–103°C within 10 min. Reactive dyes with high reactivity may require only 1 min. Faster fixation is obtained in superheated (high-temperature) steam at temperatures of 130°C–160°C, 30–60 s only being required. Such short reaction times allow the use of smaller steamers (flash agers).

11.12.2 Pigment Printing

In pigment printing, insoluble pigments, which have no affinity for the fiber, are fixed on to the textile with binding agents in the pattern required. Since around 1960 these have become the largest colorant group for textile prints. More than 50% of all textile prints are printed by this method, mainly because it is the cheapest and simplest printing method. After drying and fixation, these prints meet the requirements of the market. The washing process, carried out on classical prints to remove unfixed dye, thickening agents, and auxiliaries, is not normally necessary when using the pigment printing technique.

Most of the pigments used in textile printing are synthetic organic materials, except for carbon black, titanium dioxide, and some other metallic compounds. The chosen pigments are treated in a disintegrator or grinding mill in the presence of suitable surfactants until they have been reduced to the optimum particle size—in the region of 0.03–0.5 μm. The binder is a film-forming substance made up of long-chain macromolecules, which, when applied to the textile together with the pigment, produce a three-dimensional linked network.

At first a reduction thickener is prepared. Earlier emulsion thickener was prepared by adding kerosene or white spirit (about 1:7 by weight) slowly into a water containing binder and urea under stirring. But kerosene and white spirit are now no longer used due to their fire hazard and health hazard properties. Following is the all-aqueous recipe for reduction binder (Miles, 2003):

Water	x g
Flow moderator	0–5 g
Softener	5–20 g
Defoamer	2–3 g
Emulsifier	2–5 g

Stir in and mix well

Synthetic thickeners	25–30 g
Acrylic- and/or butadiene-based binder	80 g
Cross-linking agent	0–10 g
Total	1000 g

Just before printing, the pigment emulsion is mixed with a reduction binder and catalyst (diammonium phosphate) under slow speed stirring

Pigment emulsion	$x\%$
Catalyst	1%–2%
Reduction binder	rest to make 100

After printing the fabric is dried (below 100°C) and cured at 140°C–150°C. No after-wash is required.

11.13 Printing of Nylon

Dyes are selected from the ranges of acid, metal–complex acid, and direct dyes, according to solubility, print paste stability, washing-off properties, and fastness properties. The use of both thiodiethylene glycol, to improve dye solubility, and thiourea, which acts as both a fiber-swelling agent and a dye-solubilizing agent, is recommended. The amount of swelling agent, or carrier, required for printing nylon 6 needs to be only half of that for nylon 6.6. For high color yield it is essential to include an acid or acid donor, and both ammonium sulfate and citric acid have proved to be suitable.

A typical recipe for knitted fabrics of bulked nylon 6.6 is as follows:

Acid or metal–complex dye	20–40 g
Thiodiethylene glycol	50
Water	x g
Thickening (8%–14%)	500 g
Containing urea	50 g
Coacervating agent	0–10 g
Antifoam	5 g
Ammonium sulfate (1:2)	30–60 g
Total	1000 g

The coacervating agent has a beneficial effect on the levelness and surface appearance of carpets and bulked-yarn fabrics. Ensuring the absence of uncolored surface fiber gives an increase in the apparent color yield. Modified natural products based on guar, locust bean, and karaya gums have been widely used in nylon printing.

Saturated steam is used for print fixation of nylon fabric. By careful selection of dyes the steaming time required can be shortened, but 20–30 min is recommended at 100°C–103°C. With many dyes, improved yields may be obtained by using a pressure steamer at 35–105 kPa. High-temperature steaming, or thermofixation, may also be used.

Great care is taken during washing off; otherwise, unprinted white ground areas and pale printed areas may be stained. Nylons, especially nylon 6, have high affinity and high rate of sorption of acid dyes.

Final rinsing in cold water may be preceded by a *back tanning* after treatment with a synthetic tanning agent (Syntan), to improve the fastness to washing at temperatures up to 80°C.

11.14 Printing of Wool and Silk

Acid, basic, and direct dyes are all theoretically suitable for printing wool and silk; however, application is mainly limited to acid dyes for both direct and to discharge printing. Basic dyes can provide extraordinary brilliance in some cases, but basic and direct dyes do not show adequate fastness.

Wool has a well-known tendency to felting. Chlorination, however, changes the scaly layer of the wool fiber in such a way that the material wets out and swells more easily. The scales lie flat on the cortical layer; the wool no longer felts readily and has increased luster. Unfortunately, it is difficult to carry out this process uniformly and can result in irregular uptake of dye. The two approaches have been used to improve the uniformity of wet chlorination. First, nitrogenous auxiliary agents may be used in the chlorination bath, which temporarily bind the free chlorine in the form of chloramines; these then react slowly with the wool (an example of this technique is the Melafix method, CGY). Alternatively, the sodium salt of dichloroisocyanuric acid is added (such as Basolan DC, BASF) and is slowly hydrolyzed, liberating hypochlorous acid.

The less soluble acid dyes need urea or thiourea to assist solution along with auxiliary solvents, such as thiodiethyleneglycol and hot water. When printing wool, glycerol may be used to reduce the adverse effects of superheat in the steam during fixation; with silk, however, the addition of glycerol can cause *flushing* problems.

Locust bean or guar derivatives are used as thickening agents, either on their own or in mixtures with cold water-soluble British gum. Crystal gum is used for printing silk, rather than tragacanth and mixtures of tragacanth with British gum or gum Arabic, which were formerly the main thickening agents. Thickeners of high solid content are used for fine effects and outlines, whereas products of low solid content are preferred for larger areas because of their better leveling effect and the reduced possibility of crack marks occurring after printing.

The printing pastes contain an acid donor for fixation of the dye. This may be ammonium sulfate, ammonium tartrate, oxalate, or even in some cases acetic acid or glycolic acid. Small amounts of sodium chlorate are added to counter the reductive effect of wool, and possibly the thickeners, during steaming. Defoamers and printing oils are also usually necessary for smooth prints with sharp outlines.

Relatively long steaming times of 30–60 min are usually needed to fix acid dyes on wool or silk. The most brilliant and fast prints can only be obtained in saturated steam fixation at 100°C–102°C. For this reason it is most important to avoid overdrying the goods after printing. Wool is even *spray damped* sometimes before steaming, or put into artificially moistened steam. The final wash of the printed goods usually takes place on the winch or washing machines designed for knitted goods. Wool and silk should be treated very carefully, with as little mechanical strain as possible. To prevent the staining of unprinted areas, as well as to counter the risk of bleeding, it is recommended to use polycondensation products of aromatic sulfonic acids, such as Mesitol HWS (BAY) or Erional NWS (CGY). These products considerably improve the wet fastness on wool. To obtain the best results, 5%–6% of the mass of the goods should be used at 60°C in an acetic acid or formic acid bath. Treatment for approximately 20 min in the final rinsing bath is required.

Both 1:1 and 1:2 metal complexes offer higher fastness than acid dyes, particularly light fastness, but produce somewhat duller colors. The way in which metal–complex dyes are applied hardly differs at all from the method used for acid dyes, except that these dyes do not require the use of an acid or acid donor. At low pH, not only is the stability of the paste diminished but there is also a tendency for the dyes to aggregate and to give rise to specks. At the same time acid donors can adversely affect the levelness of prints, and these high-affinity dyes often give leveling problems in blotch printing anyway.

Otherwise both the dissolving process and the choice of printing thickener, as well as fixing and after-washing are the same as with acid dyes. It is possible, therefore, to mix acid and metal–complex dyes, but printing should be without acids or acid donors. The recipe of the printing paste is presented in the following table:

	Acid (g)	Metal–Complex (g)
Dye	x	x
Urea	50	50
Thioethylene glycol	50	50
Water	y	y
Thickener (approx.)	500	500
Ammonium sulfate 1:2	60	x
Sodium chlorate 1:2	15	15
Total	1000	1000

11.15 Printing of Polyester Fabric

Heat setting, preferably after scouring, is normally required to ensure fabric stability in subsequent processing and use.

On unmodified polyester, only disperse dyes are used. Two other important criteria in selecting dyes are the fastness to sublimation and the behavior during after-washing. As dye fixation takes place in high-temperature steam at temperatures up to 180°C or in dry air up to 210°C, it is necessary to reject dyes that sublime on to the adjacent white fiber under these conditions. Similarly, it is necessary to avoid staining of the ground by unfixed dye during washing at high temperatures by selecting either dyes of low affinity or those that can be destroyed by a reduction clearing treatment.

High-solid-content thickeners, such as crystal gum or British gum, give optimum sharpness of outlines, but form brittle films that crack and scatter dye by *dusting off*. The low-solid-content thickeners, such as alginates and locust bean ethers, form elastic films and are easily washed out.

A typical recipe is given in the following table:

Disperse dye	1–20 g
Water	x g
Acid donor	5 g
Fixation accelerator or carrier	0–20 g
Oxidizing agent	0–5 g
Thickener	500 g
Total	1000 g

The acid donor is important because many disperse dyes are affected by hot alkaline fixation conditions. Sodium dihydrogen phosphate is recommended because, unlike some organic acids, it has no corrosive effect on nickel screens and is compatible with alginate thickeners.

Destruction of dye by reduction, especially in steaming under pressure, is prevented by adding an oxidizing agent, either sodium chlorate or sodium nitrobenzenesulfonate.

Depending on the dyes being printed and their concentrations and the fixation conditions, the use of a carrier or fixation accelerator can have a substantial effect on dye sorption.

As the development of high-temperature festoon steamers and high production speeds are compatible with steaming times of 5–20 min and this has become the most important of the fixation methods. Temperatures of 160°C–185°C can be used, depending on the dyes and times of treatment chosen. Steam condenses on to the cold fabric, raising its temperature to 100°C and swelling the thickener film. The condensed water is again largely evaporated during the exposure to superheated steam, but the thickeners are not *burnt in* to the fabric as in dry fixation and the subsequent handling of the fabric is softer.

The thermofixation process also has some significance, the printed fabric being passed through stenters or specially built equipment. The best results are obtained at 200°C–220°C, at temperatures where the sublimation characteristics of the disperse dyes used are particularly critical. The time of treatment required may be 40–50 s, but in the absence of a carrier fixation is only in the range of 50%–70%. The use of 20–30 g/kg of fixation accelerator in the print paste considerably increases the depth of color obtained.

11.16 Visual Color Measurement

The accurate description of color is essential for communication and for accurate reproduction of color across a wide range of products. The color of any object is commonly registered or recorded in two ways:

- Preserving colored physical samples
- Recording in terms of common color names

Physical samples of dyed/printed fabrics, yarns, or fiber paint panels, patches of printing inks, colored papers, and so on are frequently used in the trade. Collections of such color samples are very useful as examples of colored product if the number of colors required is fairly limited. A good example of such collections is the dye manufacturers' *shade cards*. Shade cards carry numerous colored objects on specific substrates (e.g., pieces of paper or various textile materials) along with procedures and names of the colorants to be used. Paint and printing ink manufacturers also publish shade cards for their products (colors) with names that are very specific to the concerned industry. However, the exemplifications are very limited. They are restricted to the specific type of colorant or substrate and cannot be used for general reference.

When we deal with a reasonable number of specimen, say a few thousands, to cover the whole range of possible colors (1 million or more), the specimen must be selected according to a system or plan. The color naming systems were popular for a long time, but they were not very systematic—hence, the accuracy of the specification was limited. It is utmost necessary to arrange the colors in a systematic manner to tackle with the enormous number of colors we can perceive.

It is well-known that the colors are three-dimensional. However, the dimensions of color are expressed in various ways in different fields. For systematic arrangements, the dimensions should be independent of each other. The question is, therefore, what dimensions are to be chosen to arrange colors in a three-dimensional space?

Wright (1984) identified two sets of visual attributes:

1. Group A attributes are lightness, hue, and chroma
2. Group B attributes are whiteness, blackness, and chromaticness

A color order system is a systematic and rational method of arranging all possible colors or subsets by means of material samples. Once the colors are arranged systematically they are named according to some descriptive terms and/or are numbered (Graham, 1986). A technical committee of the International Organization for Standardization, ISO/TC187 (Color Notations), has defined a color order system as a set of principles for the ordering and denotation of colors, usually according to the defined scales (Slideshare, 2013).

A color order system is usually exemplified by a set of physical samples, sometimes known as a color atlas. This facilitates the communication of color, but is not a prerequisite for defining a color order system.

There are a large number of color order systems and a large number of color atlases globally. There is little scope to discuss all these in details. Two popular systems, that is, Pantone and Munsell color order systems are discussed here.

11.16.1 Pantone Matching System

Pantone Inc. is a corporation headquartered in Carlstadt, New Jersey. It is basically a color mixture system. It is not a color order system, because it does not include a continuous scale. It is more appropriately considered a color naming system.

The Pantone matching system (PMS) (www.pantone.co.uk) began life in 1963 in the United States for defining colors for printers, but expanded into other fields later, for example, textiles in 1984, plastics in 1993, and architecture and interiors (1925 colors) in

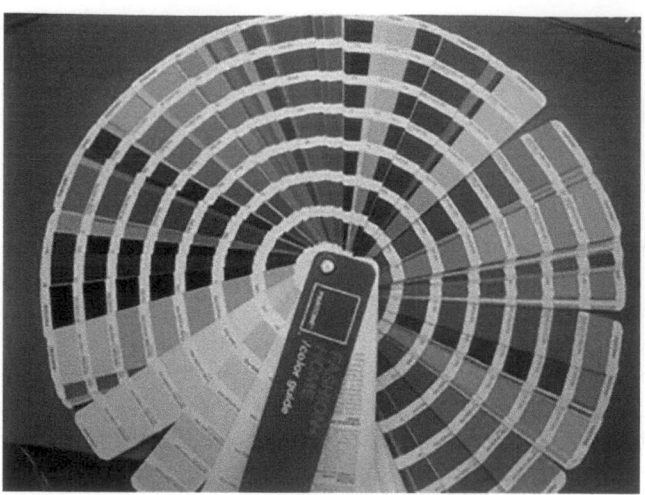

FIGURE 11.16
Pantone color guide.

2002, each of which has a six-digit numerical notation (e.g., # 19–1764) and an *inspirational* color name. A six-digit number is assigned to each color:

- The first pair (19) refers to lightness (light or dark), which has nine levels designated from 11 through 19.
- The second pair of numbers (17) specifies the hue. The hue circle is divided into 64 sectors, with 01 containing yellow–green and so on through the spectrum of sector 64, which contains green–yellow. The 64 sectors cover all the pure colors. 00 represents the neutral point.
- The third pair of numbers (64) describes the chroma level of the color.

It is divided into 65 steps, starting 00 a neutral and ending with 64 being the maximum chroma of the color.

Pantone's primary products include the Pantone guides (Figure 11.16), which consist of a large number of small (15 × 5 cm) thin cardboard sheets, printed on one side with a series of related color swatches and then bound into a small *fan deck*.

11.16.2 Munsell System

Professor Munsell (1905), an artist, wanted to create a *rational way to describe color* that would use decimal notation instead of color names, and developed the oldest and by far the most popular color order system to fill the gap between art and science. The Munsell atlas was released in 1915 and got commercialized in 1929.

The system consists of the following three independent attributes represented cylindrically in three dimensions as an irregular color solid (Plate VIII [see color section]):

1. Hue(H), measured along circumference of the horizontal circles
2. Chroma (C) or purity of color, measured radially outward from the neutral (gray) vertical axis
3. Value (V), measured vertically from 0 (black) to 10 (white).

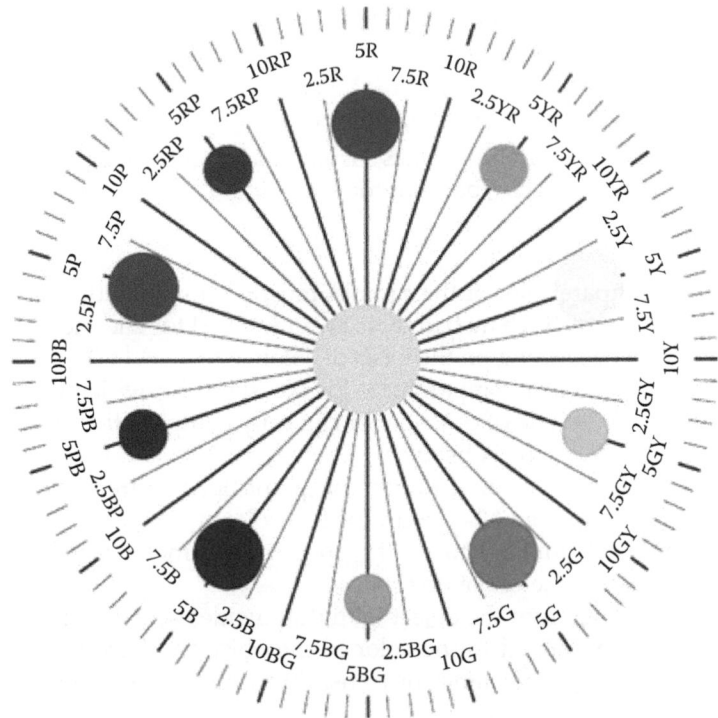

FIGURE 11.17
Munsell color order system.

The Munsell system divides each horizontal hue circle into five unique or principal *hues*: (1) Red (5R), (2) Yellow (5Y), (3) Green (5G), (4) Blue (5B), and (5) Purple (5P), along with five intermediate hues (5YR, 5GY, 5BG, 5PB, 5RP) halfway between the adjacent principal hues. In the original Munsell book, each hue sector (H) is divided further into four finer categories, 2.5H, 5H, 7.5H, and 10H (0 for the next H) (Munsell originally sampled only 20 hues, later 40) as shown in Figure 11.17. Each of the 10 steps may also be broken into further 10 substeps, so that 100 hues are obtained with integer values.

11.17 Instrumental Color Measurement

Color Measuring Instruments are of three types:

1. Colorimeter
2. Spectrophotometer
3. Spectroradiometer

They are available in the market with varying degrees of sophistication and specialization.
Although the spectroradiometer measures in illuminant mode (light mode), the other two generally measure in the object mode.

Colorimeters can be classified into two types:

1. Visual
   a. Absorption meters or color comparators
   b. True colorimeter that emphasizes visual equivalence or psychophysical esti-
      mation, for example, Lovibond tintometer
2. Photoelectric

Color comparators compare color of the sample, mostly liquids, with that of the standard
and find a match of the two. Such instruments are employed for chemical analysis, concentration determination, or grading on the basis of color.

The oldest and simplest color comparator is the Nessler tube, which was developed into
Duboscq colorimeter. This type of colorimeter can only compare the optical properties of
solutions of a particular coloring substance.

True colorimeter measures colors in terms of quantities of three color primaries (red,
green, and blue) when mixed matched with the test color.

The most popular instrument, that is the spectrophotometer, measures the spectral
reflectance or transmission factors of objects.

It compares light leaving from the object with that incident on it at each wavelength.

The data are primarily related to the color of the object and are usually presented as
curves in which percentage reflectance or transmittance is plotted against wavelength at
regular intervals of 1, 5, 10, or 20 nm.

The colorimetric property of an object is characterized by its reflectance and/or its transmittance. Majority of the colored objects encountered by us are either opaque or translu-
cent for which reflectance property is most important.

The perceived color of a material is determined by the relative amounts of light reflected
throughout the visible range of light, that is, 380–750 nm or for practical purposes
400–700 nm.

11.18 Numerical Description of Color

The spectrophotometric data provide us the scope of describing colors by real numbers.
Our color sensation is not analytical in nature. We cannot judge the simultaneous existence
of lights of different wavelengths. We get the sensation from cumulative effect. As this
cumulative quantity can be matched by mixing three primary lights, it is proved that our
eyes have three types of color-detecting cones; the stimuli generated by them are mixed
before reaching to the brain. Various other phenomena of color have also led to the conclu-
sion that our eyes have only three types of cones. Each color object is sensed by each type
of cone separately and each type sends a stimulus to the brain. So, for each object color, the
brain receives three separate stimuli.

11.18.1 Tristimulus Values

The spectral sensitivity of the three color-detecting cones has been measured and named
as color matching data $(\bar{r}(\lambda), \bar{g}(\lambda), \bar{b}(\lambda))$ (bar stands for statistical average data of a number

of color normal observers) and subsequently transformed into more usable CIE standard observer functions, $\bar{x}(\lambda), \bar{y}(\lambda), \bar{z}(\lambda)$. The area under the functions signifies the amounts of three stimuli to be transmitted to the brain by the incidence of light having one unit of energy at each of visible wavelength.

The three stimuli reaching our brain depend on the three sets of parameters:

1. Spectral energy distribution of the illuminating light source, $E(\lambda)$, which is being modified by
2. Reflectance characteristics of the opaque object, $R(\lambda)$, the modified light falls on our eyes and is sensed by the three types color-detecting organs resulting three stimuli depending on
3. Their spectral sensitivities represented by standard observer functions, $\bar{x}(\lambda), \bar{y}(\lambda), \bar{z}(\lambda)$

An object color can, therefore, be represented by three numbers (X, Y, Z) called tristimulus values. In the first stage, $E(\lambda)$ is multiplied by $R(\lambda)$ at each wavelength interval between 400 and 700 nm to obtain the energy at each wavelength of reflected light. In the second stage, the quantity $E(\lambda) \times R(\lambda)$ is multiplied by the spectral sensitivity functions $\bar{x}(\lambda), \bar{y}(\lambda), \bar{z}(\lambda)$ at each wavelength interval and are summed for all wavelength intervals to get the area under each curve. Objects are always seen in relation to its surroundings and not in terms of the absolute level of reflected light. Therefore, CIE recommends the specification of tristimulus values as relative to the luminosity of a perfect diffuser having reflectance of 1.0 at each wavelength. In the third stage, the luminosity of a perfect diffuser is calculated as shown in Equation 11.2.

$$K = \sum_{\lambda=400}^{700} E(\lambda)\bar{y}(\lambda) \tag{11.2}$$

By definition, Y value for the perfect diffuser is always 100.

Therefore, in the fourth stage, the X, Y, Z values are normalized by multiplying the summations of the second stage with $100/K$.

The calculation of CIE tristimulus values may be expressed mathematically as follows (Equation 11.3):

$$X = \frac{100}{K} \sum_{\lambda=400}^{700} R(\lambda)E(\lambda)\bar{x}(\lambda)$$

$$Y = \frac{100}{K} \sum_{\lambda=400}^{700} R(\lambda)E(\lambda)\bar{y}(\lambda) \tag{11.3}$$

$$Z = \frac{100}{K} \sum_{\lambda=400}^{700} R(\lambda)E(\lambda)\bar{z}(\lambda)$$

where $E(\lambda)$ is the relative spectral energy distribution of the illuminant, $R(\lambda)$ is the spectral reflectance factor of the object, and $\bar{x}(\lambda), \bar{y}(\lambda), \bar{z}(\lambda)$ are the color matching functions of the CIE standard observer.

K is the luminosity function of a perfect reflectance diffuser (Equation 11.1) and is a normalizing constant.

When the calculations are made for a wavelength interval of 10 or 20 nm, the weights are so adjusted that the value of K is 100.

11.18.2 Chromaticity Coordinates

In order to identify chromaticity independent of lightness, CIE proposed chromaticity coordinates or trichromatic coefficients x, y, and z as in Equation 11.3.

$$x = \frac{X}{X+Y+Z}, y = \frac{Y}{X+Y+Z}, z = \frac{Z}{X+Y+Z} \tag{11.4}$$

As $x + y + z = 1$, only two chromaticity coordinates x and y have been recommended by CIE to specify chromaticity.

However, none of the chromaticity coordinates is correlated with any meaningful visual attribute of color.

11.18.3 Uniform Color Scales

CIE tristimulus values or chromaticity coordinates are not very convenient for identifying color of the objects, because these were designed for color stimuli of different modes. None of the values are directly correlated with any visual attributes of color.

Only Y value has high correlation with luminance and object lightness. The spacing of colors in chromaticity diagram is not visible uniform.

The relative size of differences between colors in different portions of Y, x, y color space does not correspond to the sizes of the differences as seen by the human eye.

A number of uniform color scales are, therefore, developed, which can represent colors with equal visual spacing and are directly related to meaningful attributes of color appearance. The color scales have been formulated by linear or nonlinear mathematical transformation of CIE tristimulus values.

Most of the scales are based on opponent color theory, that is, the chromaticity is represented by a red–green and a blue–yellow attributes.

A number of uniform color scales have been developed by different color scientist.

11.18.4 CIELUV and CIELAB Color Spaces

Before 1976, CIE could not recommend a single color space and a single color-difference formula to satisfy all color measuring industries.

Colorant industries were in favor of a formula similar to Adam–Nickerson (AN40) formula. The CIELAB formula was acceptable as color-difference values were about 1.1 times those produced by AN40 formula.

On one hand, television industries preferred a color space associated with a chromaticity diagram because of its simple way of presentation of additive mixture, which also occurs in television and other display devices.

No simple relation exists between the two color scales.

Both CIELUV and CIELAB formulae are plotted on rectangular coordinates. Lightness L^* function is same for both color spaces and is represented by the formula

$$L^* = 116 \left(\frac{Y}{Y_n} \right)^{1/3} - 16, \quad \text{if } \frac{Y}{Y_n} \ll 0.008856 \tag{11.5}$$

For object colors, CIELAB formula is used where:

$$\text{Red–green attribute, } a^* = 500\left[\left(\frac{X}{X_n}\right)^{1/3} - \left(\frac{Y}{Y_n}\right)^{1/3}\right] \tag{11.6}$$

$$\text{Yellow–blue attribute, } b^* = 200\left[\left(\frac{Y}{Y_n}\right)^{1/3} - \left(\frac{Z}{Z_n}\right)^{1/3}\right] \tag{11.7}$$

The tristimulus values for the two standard daylight illuminants for both 2° and 10° standard observer are as follows:

	Illuminant/Observer Combinations			
	$D_{65}/2°$	$D_{65}/10°$	$C/2°$	$C/10°$
X_n	95.047	94.811	97.074	97.285
Y_n	100.000	100.000	100.000	100.000
Z_n	108.883	107.304	118.232	116.145

11.18.4.1 CIE Psychometric Attributes

In recent years efforts have been made to define CIE correlates for perceptual (visual) attributes such as lightness, chroma, and hue.

The quantity L^* serves as a correlate of lightness.

$$h_{uv} = \tan^{-1}\left(\frac{v^*}{u^*}\right), \quad h_{ab} = \tan^{-1}\left(\frac{b^*}{a^*}\right) \tag{11.8}$$

The psychometric correlate for hue is called hue angle, which may be defined as follows: The hue angles are expressed in 0°–360° scale. The relation between the signs of a^* and b^* and the range of hue angles are presented in the below-mentioned table:

a* Value	b* Value	Hue Angle Range
Positive	Positive	0°–90°
Negative	Positive	90°–180°
Negative	Negative	180°–270°
Positive	Negative	270°–360°

11.19 Conclusion

Fabric dyeing was known since the beginning of civilization. Initially, people used color available from nature (vegetable, animal, and mineral sources). From the beginning of the twentieth century, synthetic dyes of various properties came in the market and the use of natural colors became insignificant. In recent years, the environmental pollution created by synthetic dyes slowly turning us back to natural colors. For each type of textile fiber (cellulosic, proten, and synthetic) a number of dye classes are available. Proper dye is to be selected considering easy availability, end use, fastness requirement, available machine, and so on. Tremendous development has been in the design of dyeing machinery. From batch machines such as kier, winch, jigger, and so on, the production is more and more shifted toward continuous machines such as combi steamer, continuous bleaching and dyeing ranges, and so on. Automation has been incorporated at all the stages of textile coloration from dye weighing to autodispensing of dyes, automatic process control, and so on. Communication about color or the shade desired can now be easily made with the help of color atlas such as Pantone or Munsell. Disputes arising from the color discrepancy between samples ordered and delivered are very common and difficult to resolve as visual judgment is observer specific. Instrumental assessment may be binding to accept for both buyers and sellers.

References

Duckworth, C. (Ed.). (1983). *Engineering in Textile Coloration*, Bradford, UK: SDC.
Graham L.A., (1986). *Colour Technology in the Textile Industry*, Ed. by G. Celikiz and R.G. Kuehni., (North Carolina, USA: AATCC).
Karmakar, S.R. (1999). *Chemical Technology in the Pre-treatment Processes of Textiles*, Amsterdam, the Netherlands: Elsevier.
Keeling, M.R. (1981). Ink-Jet printing, *Physics in Technology*, 12, 196–203.
Mahapatro, B. et al. (1991). Orientation programme in chemical processing for senior executives, Book of papers, The Textile Association, Bombay, India.
Miles, L.C.W. (2003). *Textile Printing*, 2nd ed., Bradford, UK: SDC.
Mock G.N. (1985). Bleaching, In: *Encyclopedia of Polymer Science and Engineering*, Vol. 2, 2nd ed. (New York: John Wiley & Sons) pp. 310–323.
Nunn, D.M. (Eds.). (1979). *The Dyeing of Synthetic-polymer and Acetate Fibres*, Bradford, UK: SDC.
Lewis D.M. (Ed.). 1992, *Wool Dyeing* (Bradford, UK: SDC).
Peters, R.H. (1967). *Textile Chemistry*, Vol. 2, Amsterdam, the Netherlands: Elsevier.
Rao, A.L.N. and Gandhi, R.S. (1991). Orientation programme in chemical processing for senior executives, Book of papers, The Textile Association (India), Bombay, India, p. 58.
Shore, J. (Ed.) (1995). *Cellulosics Dyeing*, Bradford, UK: SDC.
Slideshare (2013), http://www.slideshare.net/Tanveer_ned/color-order-system,accessed on 30.1.13.
Trotman, E.R. (1968). *Textile Scouring and Bleaching*, London, UK: Charles Griffin.
Tyler, D.J. (2005). Textile digital printing technologies, *Textile Progress*, 37(4), 1–65.
Wasif, A.I. (2010). Reactive dyes: Past and future, *Indian Textile Journal*, 120(5): 38–41.
Wright, W.D. (1984). The basic concepts and attributes of colour order systems, *Color Research Application*, 9, 229–233.

12

Color Knowledge

Tracy Cassidy

CONTENTS

12.1 Introduction

In this chapter, the importance of color as a design, marketing, and sales tool for industry is outlined to substantiate the need for a strong knowledge of color and the need to work confidently and proficiently with color. This chapter is designed in a manner that promotes knowledge development and inspiration, and for the more proactive reader, a rich experience of color can be gained by working through the chapter activities, initially at a computer and later with pigments (paints and dyes). Practicing these activities is highly encouraged and recommended. This workshop approach enables you to experiment with color mixing and color schemes to better appreciate color and its application, without which the theories become just a concept; there really is little substitute for experimentation. A short history of color that focuses on the building of the scientific underpinning of the most predominant theories of color used today in relation to textile-based products (fashion) is explained. This chapter also gives an insight into color psychology in relation to creating colorways, color ranges, and color themes and to color forecasting. Color terminology is given and discussed throughout this chapter, and the following fundamental theories and color characteristics are also discussed:

- The principles of additive color mixing and subtractive color mixing
- The 6-hue color wheel and the 12-hue color wheel
- Color schemes and color combinations
- Metamerism
- Color deficiency (color blindness)
- Color measurement

12.2 The Importance of Color

Color is considered one of the most important elements of design to assist or even influence decisions to purchase, thus promoting sales opportunities. For some, color is the essential ingredient to *get right*, particularly for fashion-related products, where color relates directly to consumer preferences, style-image perception, and other psychological and emotional reasons. While getting the color of a product *right* in relation to individual consumer preferences or desires *en mass*, or for a substantially large number of consumers, a target market is fraught with problems. Nevertheless, color has become an important design, marketing, and sales tool for mass production and mass retailing. The concept is relatively simple. Retailers stock staple product lines, which allow for style features, including color to be changed, without the need to totally design or invent a follow-on product for continual sales potential. The consistency in product type enables retailers to develop a strong brand identity. For the mass manufacture of such products, the production methods used require little change, which is cost-effective, as the factory layout can remain the same and machine operatives do not need to continually learn new construction processes. Making subtle changes to a product to make it look different, which is essentially still the same product, employs a process known as *product development*; for example, a lady's fitted t-shirt could be of a certain length in yellow, with a scooped neckline and three-quarter-length sleeves for a spring season and of the same length in red, with a v-neckline and the same sleeve type for the subsequent summer season. Product development facilitates planned obsolescence. *Planned obsolescence* is a process where a particular product model is superseded by a new version. This process is used in several industries, in particular, the electronics industry, and many other consumer product industries, including fashion. The purpose of planned obsolescence is to render products out-of-date to encourage new sales. In the fashion industry, seasonal product ranges are used to promote sales. More recently, retailers have adopted shorter lead times on fashion changes, bringing new product ranges into stores, typically every 6 to 8 weeks. Some retailers endeavor to bring in new ranges even more frequently, so that potentially each time a consumer visits the store, it looks fresh and different, thus encouraging regular sales.

Style and color changes for fashion-related products are highly influenced and supported by the trend-forecasting sector. Here, the process begins with the development of *color stories* as part of the *color-forecasting* process. Through a systematic approach, color stories are developed as *color themes,* based on inspirations and the principles of color psychology, using mood boards or color story boards. The color stories are marketed throughout the industry to build a color consensus, usually on a seasonal basis, enabling the color coordination of products to create a *look*. The color stories feed into a broader trend-forecasting process and sector, where style and design features are included. Designers and retailers interpret the color and style direction into their own product ranges. For a comprehensive understanding of color forecasting and its importance in the fashion and textiles industries and in the consumer product industry, you are urged to read *Color Forecasting* (Diane and Cassidy).

In principle, a product will be available to the consumer in a small number of color choices—a *color range*. The color range will usually be extended across a range of complimentary products, in order for consumers to create a complete outfit. For the retailer, this is an opportunity to sell more than one product at a time to any one consumer. Examples of suggested *looks* are displayed on mannequins, both in store and in window displays. When a designer or a product developer is deciding on the color range, a wider range of the colors are considered. These are usually reduced to six or eight, and from this, a set of *colorways* are created. When fabrics are created using more than one color, the designer will experiment with different color combinations on the same pattern structure. Such experiments may be made in a software package before working with fabrics. Decisions will be made from this experimentation process on the final colorways to be used.

To create convincing and meaningful colorways and color ranges, the principles of color psychology can be particularly helpful. Essentially, color is a sensation. Color in the most basic form of color psychology is experienced as having temperature, being warm or cool. This is related to color undertone, which is discussed in Section 12.10. While individual hues have different cultural meanings within societies, different qualities of color can be used to communicate a feeling or a range of feelings and thus meaning. With the help of imagery, the psychological inference of color is used to create color stories, colorways, and color ranges through the development of themes. Strong themes can be realized by consumers, for example, current predominant themes in both the fashion and home industries are vintage and retro. The colors and color combinations are distinctive and quite different; vintage is generally floral and pretty and utilize soft greens and pale blues with pinks, cream, and red; retro uses more muted colors with strong bright colors, such as sage green and orange, and employs large abstract shapes. The choice of color and the color combinations used within a pattern or in a product range reinforces the theme, which, if made popular by consumer engagement, translates into a trend. In the 1990s, minimalism was a popular trend, which was mostly conveyed through simplistic styling and the feel of masculinity. *Achromatics*, black, grays, and white were strongly associated with this theme and trend because of the psychological meanings of these noncolors. Color psychology is supported by a relatively large body of literature, and it is recommended that you explore such texts to more fully develop your understanding of the meaning of color.

12.3 Basic Terminology

Color is a general term that we use. It may infer the source such as light and pigment (dyes, paints, inks, and so on) or the quality, which could be how it appears to the viewer or its psychological attributes. When we refer to colors such as those of the rainbow (red, orange, yellow, green, blue, indigo, and violet), we are actually using another term, hue.

Hue is simply the name given to a color, for example, red, yellow, and blue. It is neither a description of the color nor an indication of its quality; rather, it is the family name to which the color belongs. This will become very clear later in Section 12.19 when we work with the color wheel and with color-mixing principles. Other *color names* that we generally use include pink, brown, magenta, and turquoise. It is important to note that such color names do not exist in all languages and that individuals have different concepts of what these actually look like. Each, in reality, belongs to a hue, a *color family*. For example, pink is essentially red with added white and therefore belongs to the red family. Warm browns are orange with added black and therefore belong to the orange family. Cool browns are neutrals. We will discuss neutrals later in Section 12.11.

Knowledge of color is required to help us determine how a hue is made, or its color composition. Without a good understanding of color, coloration problems are inevitable when discussing and working with color within industry, as miscommunication will be time-consuming and costly. *Color names* are highly important to the fashion and textile industry as a marketing tool. This is linked very much with the trend-forecasting industry and, in particular, the color-forecasting sector.

Color theory comprises two fundamental approaches to color mixing in order to produce a gamut of color, depending on the color agent you use. To produce color on textiles and on other product materials where pigments such as dyes and paints are used, *subtractive color principles* apply. When working with colored light, *additive color principles* apply, which are relevant to the use of computer-aided design (CAD) packages. In simple terms, the more the color added to a composition by including more colored light, the lighter and brighter the resulting colors are, and the more the color added by including more pigment, the darker and more muted the resulting colors become. We discuss and apply the principles of both additive and subtractive color theory throughout this chapter.

Color is principally categorized into, and classified as, three fundamental hue types: primary, secondary, and tertiary colors. *Primary colors* are those that cannot be created through color mixing; they exist in nature. In each system, additive and subtractive color mixings, there are three primary colors; however, some colorists and theorists also include black and white as primary colors. Here, we keep it simple by using black and white to change the value of color, which is discussed in Section 12.11. To distinguish *pure* primary colors from those that contain any amount of impurity, which can be another primary color, we make reference to *true* color. Depending on the source of primary colors, some impurities may exist; the impurities will determine the quality of the hue and will affect the quality of the resulting colors from their composite colors. We expand on this point later when we discuss undertones in the Further Terminology section and again when we work with the principle of true color while developing our color wheel.

Secondary colours are a result of mixing two primary colours. There are only three possible secondary colours for both additive and subtractive systems. True secondary colors are an equal mix of their two composite colors. *Tertiary colors* are a mix of one primary color and one secondary color. In reality, there are a wide range of tertiary colors, depending on the percentages of composite colors used in the mix, but in order to simplify color knowledge, theoretically, we consider a true tertiary color as an equal mix of its composite colors. Some colorists refer to tertiary colors as intermediates to cover the range of possible qualities of color that sit between the color representatives on the standard color wheel. As we will be using the 12-hue color wheel later in Section 12.9, we consider true tertiaries as the mid-point color between their two composite colors and the remaining possibilities as intermediaries or color range. This will become very clear as we develop our color wheel.

The quality of a hue or color can be described as it is seen in terms of color composition and also by using sensory descriptions. To do the former, a strong knowledge of color mixing is required. For the latter, we draw on our life experiences and on color psychology theories. *Color temperature* is a contentious issue, as some who work in the field of color dismiss this as a concept while others fully endorse it. For those who are receptive of color temperature, it is generally accepted that reds, oranges, and yellows are *warm* hues and blues, greens, and violets are *cool* hues. Further color terminology is given later in Section 12.11, as we now consider the more theoretical aspect of color, employing a brief historical context.

12.4 The Underpinning Theories of Additive Color

Additive color mixing is essentially creating color by mixing different amounts of colored light. This principle is used, for example, on television and computer screens. The principle can be applied to any software package with a color facility. Before we get into the technicalities of additive color mixing, let us first understand how this all came about. There are numerous textbooks readily available that fully describe the history of color theory discoveries and discuss those theories at length; here, we keep it simple to put additive mixing principle into context.

Sir Isaac Newton (presumed to have being knighted by Queen Anne in 1705) contributed substantially to the scientific knowledge of color when he discovered the principles of light refraction. During the mid-1660s, the British physicist, mathematician, and philosopher found white light (daylight) to be a composition of a spectrum of colors while projecting rays of light onto a prism of solid glass. Newton observed the separation of the different light energies (frequencies), which, due to the formation of the prism, had created a band of individual colors. He eventually declared seven colors in total: red, orange, yellow, green, blue, indigo, and violet. Through further experiments, Newton discovered that the same separated colors could also be reflected back through a second reversed prism to reproduce the original beam of white light. From this, Newton substantiated his claim that white light is made up of (seven) different colored energies. We now recognize three of those colors, red, blue, and green, as being the primary colors of light and the remaining four, orange, yellow, indigo, and violet, as a result of mixing colored lights. Before Newton's discovery, some 2,000 years earlier, philosophers worked with differing theories. Pythagoras believed that the objects themselves emit color, and Plato believed that the eye emitted light, which then reflected color back from the object to the viewer. In other words, Pythagoras and Plato believed that light traveled in particles. However, it was Aristotle who first theorized that light travels in waves. Furthermore, Newton concluded through further investigations that each of the individual colors he observed in his band of colored light had a different specific wavelength and frequency. He found that red has the smallest angle of refraction and hence the longest wavelength, lowest frequency, and least energy, whereas violet, at the opposite end of the spectrum, has the largest angle of refraction and hence the shortest wavelength, highest frequency, and most energy. It can therefore be concluded through Newton's work that light is made up of energy vibrations of differing wavelengths and that each is perceived by the eye as a color; hence, color is an energy vibration and each color oscillates at a different rate or travels at a different speed.

Later, in 1887, the physicist Phillip Lenard theorized that an electric current exists, as electrons emit from objects by absorbing light. This theory was expanded upon by the

German physicist Max Planck in 1900. He concluded that energy is released and absorbed in small quantities or packets, which we now call quanta; Albert Einstein's quantum theory is also based on Planck's theory, further substantiating that light is quanta or photons that move in waves. *Wavelength* is the term given to the distance between successive waves. Colors are simply photons of various but specific wavelengths, where the longer the wavelength, the more spaced out the photons are, which result in less energy content. Wavelength is measured in meters (m). *Frequency* is the term given to the number of times a wave oscillates per second. Frequency can be measured in cycles per second (s^{-1}); however, hertz (Hz) is more commonly used.

The rule is—the longer the wavelength, the lower the frequency.

We measure wavelength in nanometers (nm), where 1 nm equals one millionth of a millimeter. Humans can discern only a small range of the electromagnetic spectrum, known as the *visible spectrum*, shown in Figure 12.1.

Another term used for the measurement of color is *amplitude*. Simply put, this is the height of the wave. Amplitude relates to the intensity of a color. *Intensity* is the quality that determines how bright a color is.

The rule is—the larger the amplitude, the brighter the color of a particular wavelength.

Bright colors have high amplitude, and muted colors have low amplitude. We discuss intensity later in the additive color mixing activities, where we can appreciate the brightness of color and what we mean by muted colors in a practical sense.

How we see color was further theorized by the English physician Thomas Young in 1802. From this, we can summarize that the retina of the human eye receives light waves, transmitting them through optic nerves to the brain to be decoded. The retina is believed to be composed of around 12 million light-sensitive rods and around 6 million cones. The rods aid our night vision and help us differentiate between white, grays, and black (values). The cones are sensitive to color and assist our daylight vision. German physicist Herman Helmholz continued Young's work in 1867, and in 1871, Scottish physician James Maxwell

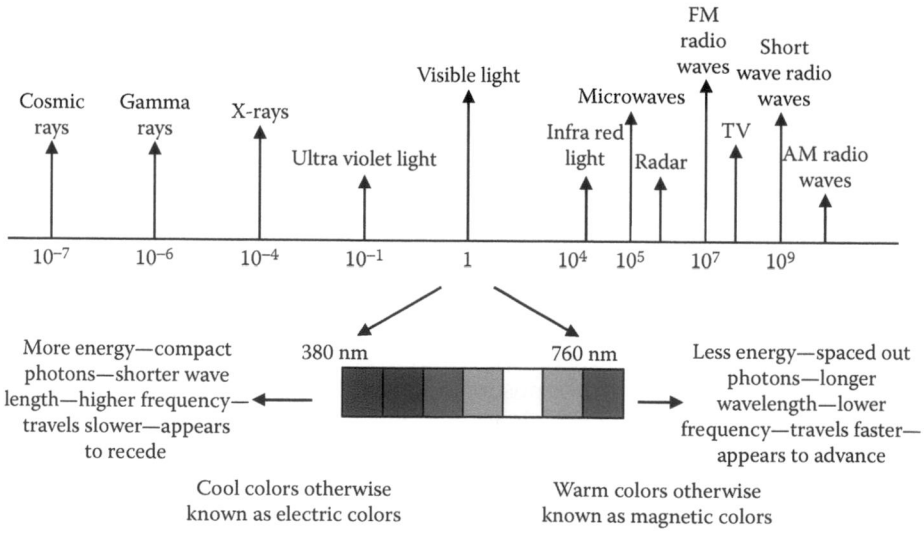

FIGURE 12.1
The Visible Spectrum.

continued with this body of work. The three sets of work compiled by Young, Helmholz, and Maxwell are collectively known as the Tristimulus theory of color perception. This body of work concludes that the human eye comprises three different types of cone, each capable of receiving just one of the three primary colors of light, which, as we previously stated, are red, green, and blue. The visible spectrum of the human eye ranges between 760 nm and 380 nm, as shown in Figure 12.1.

12.5 Color Deficiencies

Eye disorders can affect the way in which individuals perceive color. The cones and rods of the human eye contain slightly different pigments known as photopigments, which absorb the various light wavelengths. One of the three cone types (S) is sensitive to the shorter wavelengths; through these, we experience the blue–violet end of the visual spectrum. The cone type (M) is sensitive to the mid-range wavelengths, through which we experience green, and the final type (L) is sensitive to the longer wavelengths that sense red. The range of wavelength sensitivity of each of the three cone types is somewhat overlapping, which enables us to see the wide range of colors in the spectrum. This overlapping is known as the trichromacy theory, which can be used to explain color deficiency. Color deficiency occurs if one of the cone types is defected or missing. If the defect occurred from birth, the individual will mostly likely not be aware, unless tested. Color deficiency is believed to affect around 8% of the global male population due to a faulty gene inherited from their mothers, while women are largely unaffected by this defected gene. The most common deficiency is of red/green, which occurs when either the M-cone (which affects around 5% of the male population) or the L-cone (approximately 1%) is faulty, as in both cases, the response curve between the two cones reduces, resulting in a lack of distinction between the wavelengths experienced as red and green. If the S-cone is missing or faulty, the deficiency is blue/yellow; this is extremely rare. If two cone types are missing, the individual cannot differentiate between colors at all, therefore resulting in total color blindness, which is incredibly rare. The most common test for red/green color deficiency is the Ishihara test, shown in Figure 12.2. Having a color deficiency will restrict a person's ability to hold a post in the industry that involves color quality control such as color matching.

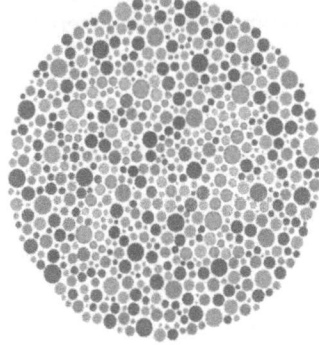

FIGURE 12.2
An Ishihara test image. (Courtesy of Wikimedia Commons (2017). *Ishihara 9*. Available at: https://commons.wikimedia.org/wiki/File:Ishihara_9.png [Accessed February 6, 2017].)

FIGURE 12.3
Light box: The image on the left-hand side has the light set to daylight, on the middle, the light is set to shop, and on the right-hand image, it is set to fluorescent. The difference in the color of the fabric can be clearly seen.

12.6 Metamerism

Another phenomenon of light is a condition known as metamerism. Under different lighting conditions, the absorption and reflection of light from an object will affect the quality of the color perceived. The quality of natural daylight changes at different times of day and as a consequence of different weather and climate conditions; similarly, artificial light differs. The variations in the quality of light can change the colors perceived quite dramatically. In the textile industry, a light box is commonly used to view the metameric change of color under controlled lighting conditions; see Figure 12.3.

12.7 Additive (Light) Color Theory in Practice

As we have established earlier, the primary colors of light relate to the three cone types of the human eye, red, green, and blue. Color facilities in software packages use a range of color value systems, including RGB (red, green, and blue) and HSL (hue, saturation, and luminosity). For the purpose of our activities and explanations of color light theory, we have used the RGB system, as this relates directly to the primary colors. The lowest value that can be entered into any one of the color values is 0, meaning that none of that color is represented in the mix. The highest value is 255; this indicates a full strength of that particular colored light. Black and white are thought by some as additional primary colors. In additive color mixing, this makes some sense, as when adding more of a colored light, the result is a lighter color, whereas less light results in darker colors. Moreover, full-strength red, green, and blue light, or an RGB value of 255/255/255, results in white, and no colored light, or an RGB value of 0/0/0, results in black.

The basic principles of additive color theory therefore are as follows:

- The primary colors of light are red, green, and blue.
- When mixing more than one of the colored lights, more light is added, the resulting colors are lighter.
- When mixing the three primary colors of light together in equal amounts, the result is white light.
- The secondary colors of light are produced by mixing two primary colors; in additive mixing, these are commonly known as magenta, yellow, and cyan.

In theory:

- Magenta is a 50/50 mix of red and blue.
- Yellow is a 50/50 mix of red and green.
- Cyan is a 50/50 mix of green and blue.

ACTIVITY

To further develop your understanding of how additive color mixing works, it is recommended that you switch between systems in your color facility to see the relationships between hue, saturation, and luminosity. For example, while referring to Figure 12.4, begin with one hue and note the RBG and HSL values. Systematically change one value in one system and view the change in the other system. Work with different RGB and HSL values, noting the changes both in the system values and visually to develop a more thorough understanding of additive color theory in practice.

In order to better understand additive color mixing principles in practice, we will apply color to the color wheel by using the computer in the next section. We should point out at this stage that the red, green, blue primary colors of light differ slightly from the red, green, and blue colors of subtractive color mixing using pigments, which we will discuss later. The primary light color red has a more orange quality and therefore a slightly warmer appearance; green has a slightly yellow quality; and blue has more of a purple quality or an indigo quality than is more commonly associated with the primary pigment color blue. It is important to become visually aware of such subtle differences, which comes from experimenting with different mediums, working with both additive and subtractive theory principles.

While computers work with additive color theory, printers use inks rather than light and therefore use subtractive color theory. Printer inks are the bright secondary colors of

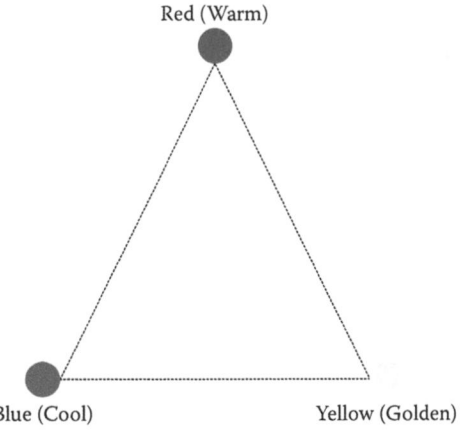

FIGURE 12.4
Primary colors and their associated color temperatures.

light—magenta, yellow, and cyan. Further colors are produced on paper by layering the inks; the transparency of the ink works like a filter, reducing the brightness. The secondary colors of the printing process are red, green, and blue, the primary light colors. In reality, there are a number of issues with color consistency between the colors that we see on the computer screen and those printed on paper, due to the two processes applying different color mixing principles; this is an area that we do not intend to cover in any depth in this chapter but this is introduced just to make you aware of the compatibility implications when printing designs from computer packages.

The basic principles of printer ink mixing are as follows:

- The primary colors are magenta, yellow, and cyan.
- When mixing equal amounts of the primary colors, black results.
- The secondary colors are red, green, and blue.

In theory:

- Red is a 50/50 mix of magenta and yellow.
- Blue is a 50/50 mix of magenta and cyan.
- Green is a 50/50 mix of cyan and yellow.

Commonly, printers use a pure black ink to save on colored inks, as large amounts of text are usually printed and a more pure black can be obtained. The white of the paper is utilized by the print process to produce lighter colors by applying thin layers of color. If printing on colored paper, the quality of the resulting colors will be affected, depending on the color of the paper.

We will now work with the principles of additive color mixing to develop the color tools. We will manipulate the colors to best represent pigment colors on screen; however, they may slightly differ if printed. In doing so, we will also refer to subtractive color theory principles. It is therefore useful to set out this theory before developing our color wheel.

12.8 Subtractive Color Theory

Fibers, yarns, and fabrics are mostly colored by using pigment dyes. Designers and product developers also use different types of pigment when experimenting with colorways and color stories. Among others, these are likely to be paints, which allow for the precise mixing of color variations. Pigments are found in natural sources, as previously stated, and when mixed, they effectively begin to cancel out reflected light, which results in darker color combinations. This implies that subtractive color mixing principles are at play. We will discuss this further in practice later in Section 12.12. For now, we need to be aware of the basics.

The basic principles of subtractive color theory are as follows:

- The primary (pigment) colors are red, yellow, and blue.
- When mixing more than one pigment color, light is reduced; the resulting colors are darker.

- When mixing the three pigment primary colors together in equal amounts, the result is black.
- The secondary pigment colors are produced by mixing two primary colors.

In theory:

- Orange is a 50/50 mix of red and yellow.
- Green is a 50/50 mix of blue and yellow.
- Violet is a 50/50 mix of red and blue.

ACTIVITY

To further develop your understanding of how additive color mixing works and the theory of subtractive color mixing, it is recommended that you work through the development of the color wheel by using your computer, as it is quick and easy to manipulate the color values in a software package. Here, we are using Microsoft Word for added ease; however, you may wish to use another package familiar to you. Additional benefits of beginning with the computer include no mess and no cost. It is also good practice to develop a further color wheel by using paints to better appreciate the differences in color mixing.

12.9 The Color Wheel

The *color wheel* is a simple tool that can be used for understanding the basic principles of color mixing and color relationships and for producing ranges of color combinations. The most basic color wheel has only six colors, three of which are known as the primary colors and the remaining three are called secondary colors. The primary colors are red, yellow, and blue; they are called primary colors because they cannot be created by mixing other pigment colors. We can plot the three primary colors on an equilateral triangle, as shown in Figure 12.4. As previously noted, we often talk about color in terms of temperature. In particular, designers sense and see color temperature as a quality of a hue. Generally, true reds are considered very warm colors; yellows are also warm but have a more golden quality, for example, sunshine. Blues are considered to have a cool quality.

ACTIVITY

Reproduce Figure 12.4 following the discussion in the following. It is recommended that you do this on a computer first. Then, to repeat the process with pigments (paints), particular artists' colors for this purpose are recommended later in Section 12.12.

When using a computer to fill a shape, such as the red circle in Figure 12.4, the RGB value to use is 255/0/0; printing this color value should give you a good representative true red. To produce true yellow, the RGB value 255/255/0 is recommended. This is because additive mixing of red and green creates yellow; therefore, a full-strength red light and a full-strength green light will produce a full-strength yellow light. However, as previously stated, yellow pigments are obtained directly from natural sources and are not created through a mixing process. To produce true blue, 0/0/255 is recommended, which is a full-strength blue.

In theory, the primary pigment colors are as follows:

- True red (100% red)
- True yellow (100% yellow)
- True blue (100% blue)

To complete the six-hue color wheel, we produce three secondary colors by mixing two of the three primary colors. In theory, we mix two primary colors in equal quantity (50/50 mix) to produce colors that are midway between to the two mixed colors; the resulting colors are to be visually recognized as true secondary colors. In practice, the mix is not 50/50, as you will realize when you produce a color wheel by using pigments.

ACTIVITY

Reproduce Figure 12.5 following the discussion in Figure 12.5.

The three secondary colors are green, orange, and violet; see Figure 12.5. Some color practitioners, designers, and authors refer to violet as purple, which is a mix of red, a warm color, and blue, a cool color; for this reason, here, we use the term *equilibrium* to describe the

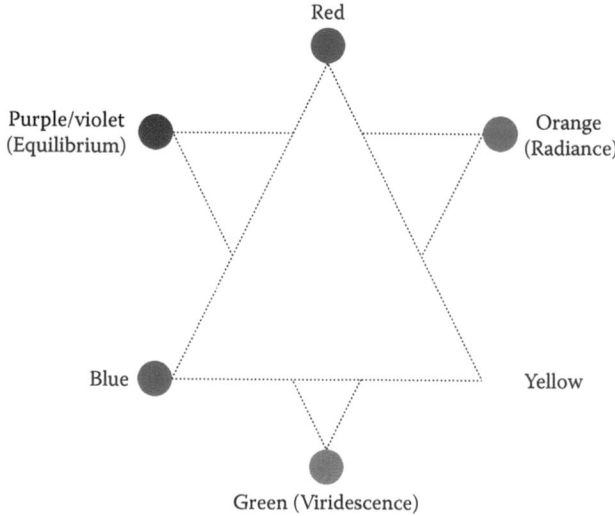

FIGURE 12.5
Secondary colors and their associated color temperatures.

psychological feel of purple/violet, as it is a balance of these two extremes. To produce true violet color on the computer, it is easy to assume that an RGB value of 255/0/255 would suffice, as this represents a full-strength red and a full-strength blue, but in reality, this will give a magenta color, which is, of course, a result of the additive mix of red and blue lights. To produce a color that represents the pigment violet color, red and blue need to be reduced in strength, preferably half strength, hence the RGB value 128/0/128. Similarly, it is easy to assume that 0/255/0 would produce a true green on the computer, but in reality, this is the color of green light, which is a bright light (pale)-colored green. Again, by reducing the strength of the green value by half, we can produce a better true pigment green, representative color on the computer, hence the RGB value 0/128/0. Here, we use the term viridescence, or having the quality of being viridescent, to describe the psychological character of greens. Producing a true orange on the computer is a matter of experimentation and a personal evaluation. An RGB value of 255/255/0 represents true yellow. To produce an orange color, we need to reduce the strength of the green. However, reducing the green value by half, 255/128/0, produces a light yellow–orange that would be better placed between true orange and true yellow. An RGB value of 255/102/0 gives a good true orange representation. As orange has the quality of being radiant, the term *radiance* is used here as the psychological description.

Therefore, in theory, our secondary pigment colors are as follows:

- True violet (50% red + 50% blue)
- True green (50% blue + 50% yellow)
- True orange (50% yellow + 50% red)

While the six-hue color wheel provides us with a useful color tool, it is very rudimentary, whereas the 12-hue color wheel affords the color practitioner with a wider color choice and provides us with a much greater appreciation of the variety of color variations at our disposal, but the model itself still remains simplistic. We use the same color mixing principles, the 50/50 mix of neighboring colors on the color wheel, to produce representative colors, known as tertiary colors. Each tertiary color is a mix of a primary color and a secondary color, making six tertiary colors in total, each sitting at equal distance between its component colors.

In theory, our tertiary pigment colors are as follows:

- True red–orange (50% red + 50% orange)
- True yellow–orange (50% yellow + 50% orange)
- True yellow–green (50% yellow + 50% green)
- True blue–green (50% blue + 50% yellow)
- True blue–violet (50% blue + 50% violet)
- True red–violet (50% red + 50% violet)

Color names tend to differ at this stage of color mixing, depending on the actual quality of the color and on trends in fashion. The most common color names are given in Figure 12.6, with the psychological feel of each. The primary and secondary colors are plotted on a hexagon.

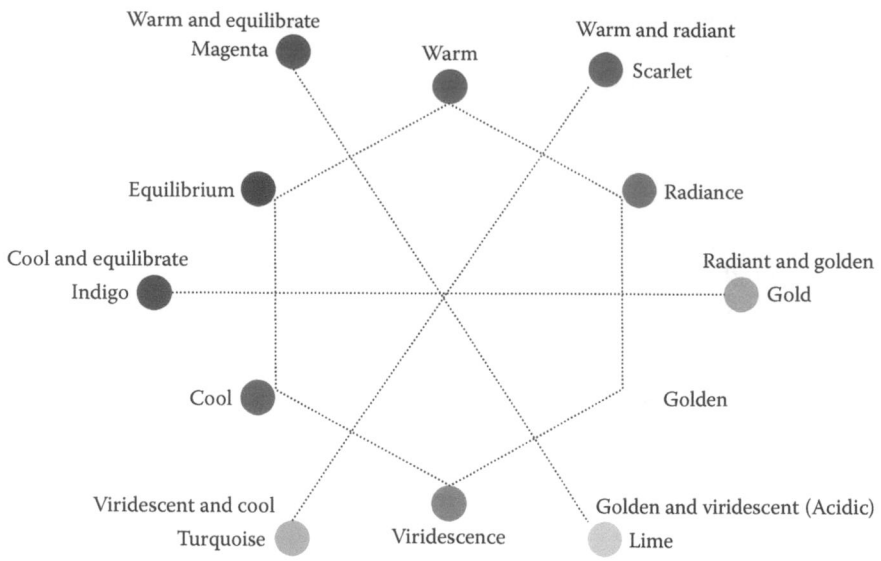

FIGURE 12.6
Tertiary colors with their associated color temperatures and suggested color psychology terms.

ACTIVITY

Reproduce Figure 12.6 following the discussion in the following.

To produce true red–orange, a mid-way point between red (255/0/0) and orange (255/102/0) is logical (255/51/00); however, by adjusting the yellow content slightly to 255/49/0, a more visually true mid-way red–orange is achieved. To produce yellow–orange a mid-point between yellow (255/255/0) and orange (255/102/0), 255/153/0 gives a visually good true yellow–orange. True yellow–green is quite difficult to judge as a mid-way representative between yellow and green, as the colors quickly become too muddy or too bright. The RGB values for yellow and green are 255/255/0 and 0/128/0, respectively. Equal values of red and green (153/153/0), for example, are very muddy, whereas a higher level of green than that of red, such as 128/204/0, produces a green that is too bright for pigment mixing, giving an artificial-looking light green. The RGB value of 128/192/0 works well. True blue–green is equally difficult to produce, as resulting colors are again too artificially bright, for example, 0/255/255. An RGB value of 0/153/153 is too blue and also too dark, whereas 0/255/128 is a reasonably good turquoise color but again too bright. An RGB value of 0/172/128 is recommended. True blue–violet sits between blue (0/0/255) and violet (128/0/128), again a difficult color to judge, and therefore, 45/0/128 is recommended. True red–violet, which sits between 255/0/0 (red) and 128/0/128 (violet) requires less blue than the result of 255/0/255, and therefore, the RGB value of 255/0/102 is recommended.

The purpose of the tertiary colors on the 12-hue color wheel is to represent mid-way colors between each primary and neighboring secondary color. By adjusting the percentage quantities of each mix variations of tertiary colors, intermediaries can be achieved, which will move the resulting color closer to the primary or secondary color having the larger percentage in the mix. When adjusting percentages by large amounts, the difference in color is obvious. In reality, many of the tertiary colors, when produced on the computer, benefit from a small amount of the third colored light component.

ACTIVITY

Remember that due to personal color perception, the color representatives selected are determined by the colorist. When working on the computer, it is very easy to make adjustments to the RGB values to see how the colors subtly change. However, working solely on the computer may reduce an appreciation of subtractive color mixing, as colors become naturally darker than when working with additive mixing principles on the computer. It is recommended that you experiment with pigments too, in order to reduce any tendencies to err toward the brighter representations that occur when working with colored light.

Before we explore the complexities of subtractive mixing while working with pigment, let us explore the benefits of using the color wheel to create color combinations.

12.10 Color Schemes

For some, working creatively and successfully with color comes natural, and for others, using theoretical color schemes aids this process. With practice color selections for colorways will eventually become more intuitive. Working with color schemes also helps students better understand color. The idea behind successful, or pleasing, color schemes is harmony. Harmonious color combinations can be created in many ways. Here, we keep it simple, referring to the true colors of the 12-hue color wheel and the more basic six-hue color wheel. In Section 12.11, we will consider the incorporation of black, white, and gray while discussing value.

Analogous color schemes are linked by a common color and are therefore considered to be harmonious. Using the 12-hue color wheel, analogous colors sit next to, or very close to, each other. For example, referring to Figure 12.6, red–orange and red–violet sit either side of red and are therefore the analogous colors for red. When using the six-hue color wheel, the analogous colors of red are orange and violet (refer to Figure 12.5), and therefore, the color relationship is less pronounced. In reality, any quality of red–orange or red–violet will produce an analogous color scheme with red. In addition, yellow–orange and blue–violet are also considered to be analogous colors of red, as both of them contain an amount of red.

The rule is—the relationship through one color, and they need not appear directly next to each other on the color wheel.

Indeed, the further apart the composite colors are on the color wheel, the more distinctive the color difference is, which will create more diversity in the appearance. Analogous color schemes comprising two colors are known as *dyads*. Similarly, three color combinations are known as *triads*, four colors *tetrads*, five colors *pentads*, six colors *hexads*, and so on. Greater contrasts are achievable with colors of different color temperature. Colors next to each

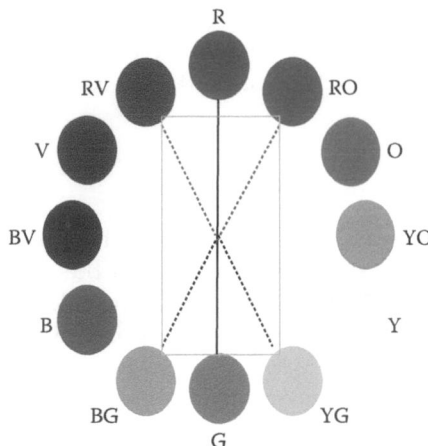

FIGURE 12.7
Color scheme examples.

other on the six-hue color wheel naturally have a greater contrast than those on the 12-hue color wheel.

In contrast to analogous schemes, *complementary colors* are created when using colors that sit opposite to each other on the color wheel. Complementary color schemes are harmonious through contrast. The contrast is greater when using the six-hue color wheel, which contains the primary and secondary colors of the 12-hue color wheel. The tertiary colors of the 12-hue color wheel have a degree of color relationship. For example, both red–violet and its complementary color yellow–green contain blue. The contrast between tertiary complementary colors is therefore less than that of primary and secondary color complementaries, as there is no color relationship. In Figure 12.7, complementary hues red and green are shown by the unbroken black line.

A variation of the complementary color scheme is the *split complementary color scheme*. This combination type creates a triad by using the two hues on either side of the complementary color. On the six-hue color wheel, two split complementary combinations are possible, the first contains all three primary colors, red, yellow, and blue, and the second contains all of the secondary colors, orange, green, and violet. Greater contrasts are created using the six-hue color wheel than those achieved using the 12-hue color wheel, where there is an inherent color relationship between at least two of the colors within the color combination. In Figure 12.7, the two black broken lines indicate the two split complementary colors of red; BG and YG used with red make a triad.

A variation of the split complementary color scheme is *double split complementary*. These color combinations are tetrads, which use two pairs of split complementaries. This combination can be achieved in two ways. The first method is where hues on either side of the first of the complementary pair are used with hues on either side of the secondary complementary hue. This is usually identified using an oblong within the hub of the color wheel. In Figure 12.7, a double split complementary is shown using the split complementaries of red and green, BG and YG, plus RV and RO. The split complementaries of green are shown by the red dotted lines. Alternatively, and more simply, the oblong shown in blue can be used.

The second *double split complementary* method comprises two pairs of complementaries of equal distance on the 12-hue color wheel. Begin by selecting one hue; the second selection will be the next third color, the third selection will again be the next third color round

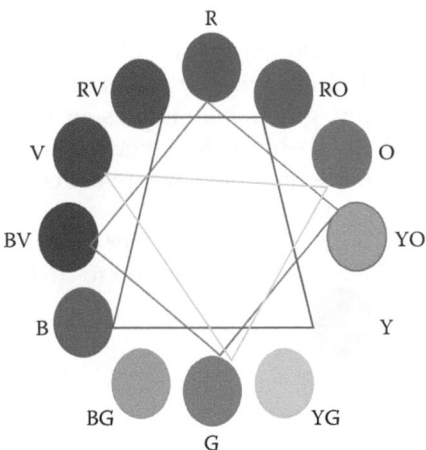

FIGURE 12.8
Further color scheme examples.

the wheel, and the fourth will be the next third color. For example, if red is our first color, the second is yellow–orange, the third is green, and the fourth is blue–violet. A square can be drawn in the hub of the color wheel to easily select colors of equal distance, as shown in blue in Figure 12.8. This is not possible when using the six-hue color wheel.

There is another variation of the *double split,* where a trapezoid shape is used instead of a square. First, we select a color, then we determine this hue's two split complementaries, for example, we select green, then select green's complementary, which is red, and then select the split complementary colors of red, which are red–violet and red–orange, which are now the first two colors of the combination. We now use one of the split complementaries of each of these hues, in this case, those of red, red–violet, and red–orange. The complementary of red–violet is green–yellow, and the split complementary colors are green and yellow. To form the trapezoid shape, we need to select the yellow. Likewise, the complementary color of red–orange is blue–green, and the split complementaries of blue–green are blue and green. To complete the trapezoid, we select blue. This explains the principle behind the combination. However, this does not have to be memorized; simply use the trapezoid shape, shown in red in Figure 12.8. Alternatively, select a hue on the color wheel, miss one hue and select the next hue, miss two hues and select the next one, miss three hues and select the next one, and there will be two hues between the last and first selected hues. This method does not incorporate any complementary pairs at all in the color combination; it just uses them as a basis for the selection. There is some degree of color relationship between the four resulting colors.

There is one more harmonious triad. Using an equilateral triangle in the hub of the color wheel to select three colors of equal distance, see Figure 12.8, shown in yellow–green. On the six-hue color wheel, there are only two combinations—red, yellow, and blue, the primary colors, and orange, green, and violet, the secondary colors. There are an additional two combinations when using the 12-hue color wheel—red–violet, blue–green, and orange–yellow and red–orange, yellow–green, and blue–violet.

A *clash* combination is an example of discord. *Discordant* is considered the opposite of harmonious but need not be unpleasant. A *clash combination* is a variation of a split complementary color scheme, where instead of using both split complementary colors, only one is used. For example, yellow–green is one of the split complementary pairs of red; when used with red, it is discordant because it is considered not to be harmonious without the balancing influence of blue–green. On the 12-hue color wheel, the clash color is related to the complementary color and all three primary colors are present in the combination, as with our example, red and yellow-green, blue is present in the yellow-green. The clash contrast is greater when using the six-hue color wheel, as the combinations will be either the three primary colors, red, yellow, and blue, where the contrast is greatest, or the three secondary colors, orange, green, and violet.

So far, we have only worked with true hues; further color combinations are possible by changing the quality of color. Here, we introduce you to further color terminology and to more color scheme possibilities.

12.11 Further Color Terminology and Color Schemes

As previously stated, *intensity* refers to the strength or brightness of a color, which may also be expressed as a color's saturation. A pure or true hue can be fully saturated or less saturated without altering its strength. It is important to note that different fibers, yarns, and fabrics have different inherent light reflectance. Fibers with a natural high luster, such as silk, and many of the man-made fibers that imitate the characteristics of silk will often give the perception of a fully saturated color, whereas color applied to opaque fibers, such as wool, generally appear less saturated. Developments and advancements in the dyeing and finishing industry have largely overcome this to obtain stronger, brighter colors on opaque fibers such as wool. Without the ability to achieve this, certain fibers would become unpopular during times of certain color trends.

ACTIVITY

It is interesting to experiment with intensity on the computer to realize how the quality of color changes when the value of intensity is changed. With reference to Figure 12.9, change between the RGB and HSL values to engage with the following activity.

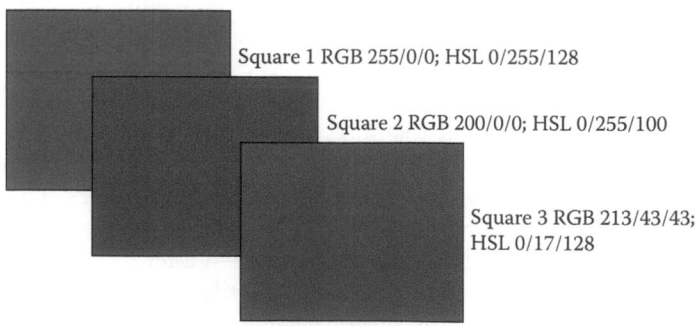

Square 1 RGB 255/0/0; HSL 0/255/128

Square 2 RGB 200/0/0; HSL 0/255/100

Square 3 RGB 213/43/43; HSL 0/17/128

FIGURE 12.9
Changing intensity.

Create a square and fill in with an RGB value of 255/0/0. Change the color model in the color facility to HSL, and note that the value is now 0/255/128. As there is no green or blue light, the color is fully saturated and bright. Refer to square 1 at the back of the image in Figure 12.9. Square 2 in the middle of the image is darker but still bright. By changing the RGB value to 200/0/0, the HSL value automatically changes to 0/255/100, reducing the luminosity from full (128) to 100. While square 3 in the front of the image is also darker than square 1, this color change is achieved by reducing the saturation instead of the luminosity, which gives a darker but muted variation. It is recommended that you continue to experiment in a similar manner with other colors and variations to further develop your color knowledge.

The *value* of a color refers to its clarity or purity. In its pure form, a color has no added white, black, or grey. Pure colors may be referred to as being bright (brights). Colors with added white are known as *tints* or may also be referred to as pastels. Colors with added black are known as *shades*; the quality of the resulting colors is rich and dark. Colors with added gray are known as *tones* and are often referred to as being muted, as the brightness is reduced. When adding gray, the resulting colors have a dusty appearance. Many variations of a hue can be achieved by altering the value. Values alone can be used as colors with no added hue, known as *achromatics*; the generic names of achromatics are white, gray, and black. Gray is made by mixing white and black; color names or fashion names such as slate and silver are often given. The gray scale is shown at the bottom of the image in Figure 12.10. Tones may be light, moderate, or dark, depending on the amounts of black and white used in the mix; larger quantities of white to black produce lighter tones than tones that contain more black than white. Equal amounts of white and black produce a moderate tone. In the English language, there are a few generic color names for tints, tones, and shades, including navy (blue shade), burgundy (red shade), pink (red tint), peach (orange tint), and khaki (yellow tone). Tints create lighter-weight colors than tones and shades. Light tones appear misty; dark tones also appear misty but are heavier and more muted than light tones. Shades appear heavier than the other values but are not muted; they are sometimes referred to as darks, or dark colors. These colors are often rich in appearance and may also be referred to as such.

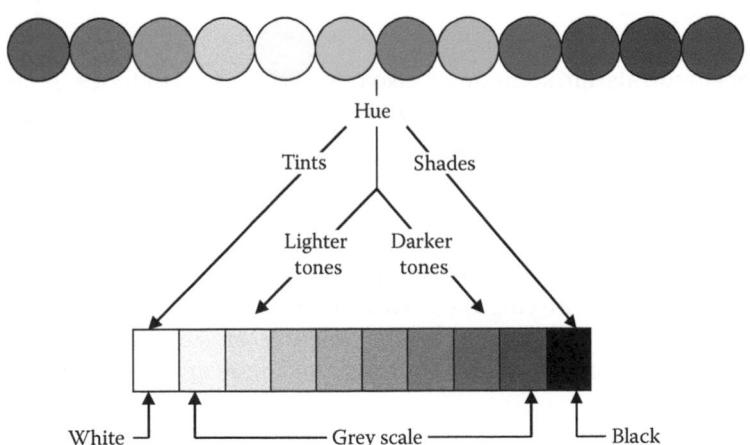

FIGURE 12.10
Adding value.

Monochromatic color schemes, while comprising two or more colors, are created using only one hue. The composite colors are simply variations of that one hue. As the colors are closely related, they are said to be harmonious but may also be considered to be limited in appeal, particularly when incorporated into fabric patterns. When working with the 12-hue color wheel, the variations are more subtle than when using the six-hue color wheel. When using black, white, and grays, the color scheme is known as *achromatic*.

Achromatics are color combinations based on value and include black, white, and gray; strong contrasts can be achieved using color combinations of shades, tints, and tones, respectively. More subtle combinations can be created with different saturations of hue, known as intensity, discussed previously. Any of the color schemes and color combinations can be enhanced by changing the tonal value of one or more of the composite colors.

Discordant color schemes, in theory, are the opposite of being harmonious; however, a discord in itself can be considered as being harmonious, or at least pleasant. In addition to a clash combination produced through a split complementary, as previously described, a discord can also be created by breaking the *natural order for color*, the natural order being elated to the inherent value of a hue. For example, yellow is naturally lighter than its complementary violet. If white is added to the violet and black to the yellow, the natural order is changed and is considered discordant. Another example of discord is where two colors do not share the same value, for example, pink and brown. Pink is red and white, and brown is orange and black. Again, the results may not necessarily be undesirable, depending on the actual qualities of the composite colors. Discordant combinations may also be made more pleasant by adding a further color or color variation to the combination.

ACTIVITY

To further develop confidence in working with color combinations, it is recommended that you experiment putting color combinations together. This can be achieved by working on a computer or in practice by using any medium such as paints or simply pieces of colored fabrics, yarns, and so on.

Neutral colors (neutrals) are sometimes also referred to as earth tones. A *neutral* is a combination of all three primary colors of any percentage mix. Neutrals are often muddy in appearance due to the principles of subtractive color mixing, as complementaries cancel out each other. Theoretically, this cancellation should produce black, but because colors are not 100% pure, which means that they contain some impurities, the results are not true black. For instance, a neutral may be made by mixing complementary colors orange and blue, as orange is made up of the two primary colors red and yellow and blue is the third primary color. The result of mixing these two colors would be a neutral of some kind, but the actual resulting color will depend on the amounts of each primary color in the mix. Brown is often thought of as a neutral but may be the result of mixing an orange with black, an orange shade.

As with pure hues, achromatic values can also be added to a neutral to produce further variations of a color. However, as neutrals are naturally muddy in appearance and grays and white have a misty appearance. The addition of white and gray may not produce very appealing colors, particularly for fashion apparel.

An *undertone* is the underlying primary color of any hue. The undertone determines the color temperature bias of a hue. A blue undertone gives a hue a cool bias, whereas a red undertone gives a hue a warm bias. Hues have either a warm or a cool undertone, regardless of the family to which the hue belongs. Therefore, blue, while being a cool color in general, can have a warm bias if the undertone is red. Similarly, red, which is generally considered to be a warm hue, can have a cooler appearance if the undertone is blue. Therefore, each primary hue has a warm and a cool variation. If two primary colors with a warm bias are combined, a warm-biased secondary color will result. Similarly, if two primary colors with a cool bias are mixed, a secondary color with a cool bias will result. If colors with a different color temperature bias are mixed, the resulting colors will have a more temperate appearance, though the undertones will still be identifiable. This theory forms the basis of the following section, where we concentrate on subtractive color mixing.

12.12 Subtractive Color Mixing in Practice

While the color theory presented earlier in Section 12.9 through the development of the 12-hue color wheel, it is very rudimentary, as when mixing pigments, the results may not be as predictable as you would have thought. This is because, as stated earlier, the undertone of any color when using pigments will affect the resulting color of a color mix. For more accurate color mixing prediction, it is essential to consider six primary colors rather than three—two reds, two yellows, and two blues, each with a different undertone or *color bias*. Table 12.1 demonstrates this more realistic subtractive color mixing theory when working with pigments.

Using the six primary colors, there are now four possible secondary color qualities for each of the basic three secondary colors, as shown in Table 12.2.

For each tertiary color, we now have eight color options, each having a different quality due to the undertones of each component color in the mix. Table 12.3 shows the variations for red–orange.

TABLE 12.1

Six Primary Colors of Subtractive Color Mixing

Bias 1	Basic Primary Color	Bias 2
Blue bias (blue–red) Red 1	Red	Yellow bias (yellow–red) Red 2
Blue bias (blue–yellow) yellow 1	Yellow	Red bias (red–yellow) Yellow 2
Red bias (red–blue) Blue 1	Blue	Yellow bias (yellow–blue) Blue 2

TABLE 12.2

The Twelve Secondary Colors of Subtractive Color Mixing

Red + yellow = Orange	Red 2 (yellow bias) + yellow 2 (red bias) = Orange 1 (yellow/red–orange) – orange bias	Red 1 (blue bias) + yellow 1 (blue bias) = Orange 2 (blue/blue–orange) – blue bias
	Red 1 (blue bias) + yellow 2 (red bias) = Orange 3 (blue/red–orange) – violet bias	Red 2 (yellow bias) + yellow 1 (blue bias) = Orange 4 (yellow/blue–orange) – green bias
Yellow + blue = Green	Yellow 1 (blue bias) + blue 2 (yellow bias) = Green 1 (blue/yellow–green) – green bias	Yellow 2 (red bias) + blue 1 (red bias) = Green 2 (red/red–green) – red bias
	Yellow 2 (red bias) + blue 2 (yellow bias) = Green 3 (red/yellow–orange) – orange bias	Yellow 1 (blue bias) + blue 1 (red bias) = Green 4 (blue/red–green) – purple bias
Blue + red = Purple	Blue 1 (red bias) + red 1 (blue bias) = Purple 1 (red/blue–purple) – violet bias	Blue 2 (yellow bias) + red 2 (yellow bias) = Purple 2 (yellow/yellow–purple) – yellow bias
	Blue 2 (yellow bias) + red 1 (blue bias) = Purple 3 (yellow/blue–purple) – green bias	Blue 1 (red bias) + red 2 (yellow bias) = Purple 4 (red/yellow–purple) – orange bias

TABLE 12.3

The Eight Red–Orange Tertiary Colors of Subtractive Color Mixing

Red–orange 1 Red 1 (blue bias) + orange 1 (orange bias) Blue–orange bias	Red–orange 5 Red 2 (yellow bias) + orange 1 (orange bias) Yellow–orange bias
Red–orange 2 Red 1 (blue bias) + orange 2 (blue bias) Blue bias	Red–orange 6 Red 2 (yellow bias) + orange 2 (blue bias) Green bias
Red–orange 3 Red 1 (blue bias) + orange 3 (purple bias) Blue–purple bias	Red–orange 7 Red 2 (yellow bias) + orange 3 (purple bias) Yellow–purple bias
Red–orange 4 Red 1 (blue bias) + orange 4 (green bias) Blue–green bias	Red–orange 8 Red 2 (yellow bias) + orange 4 (green bias) Yellow–green bias

ACTIVITY

To further enhance color knowledge, it is recommended that you experiment mixing color pigments to sensitize to the variations of color possibilities. To do this, we recommend the following six primary colors (artist's paints):

- Red with a yellow bias (cadmium red)
- Red with a blue bias (quinacridone violet)
- Yellow with a red bias (cadmium yellow)
- Yellow with a blue bias (Hansa yellow)
- Blue with a red bias (ultramarine blue)
- Blue with a yellow bias (cerulean blue)

12.13 Color Measurement

To verify the actual color composition of pigment colors, scientific instruments known as *spectrometers* are used. Such instruments enable colors to be analyzed and accurately described for repeatability. A range of sophisticated devices are available, most of which work with computer software. Handheld devices are also available, offering designers, product developers, color forecasters, and other color specialists a portable tool to record desired color qualities on spec while undertaking product research. Spectrometers measure the reflection or absorption of light (wavelengths) quantitatively from all manner of objects, including fabrics. The measurement values can be plotted graphically by using appropriate software packages. The measured values can also be entered into CAD packages for garment design. Color codes for precise color repeatability are also provided by the color company Pantone for their color chip color guides, RGB values being one of the systems that they use. Pantone color guides have become probably the most widely used color measurement products in the textile and fashion industry. Pantone color references are also commonly used by color and trend forecasters to assist color consistency and accuracy in color reproduction for trend-following practice.

12.14 Conclusion

In this chapter, we have continually highlighted the importance of color as a design, marketing, and sales tool for industry while assisting and encouraging you to develop and further enhance your color knowledge through discussions and suggested activities. This chapter gave a brief history of color theory to provide a basic scientific underpinning for the important aspects of color application in relation to the textile and fashion industries. A balance of creativity and methodological approaches to the delivery of the principal concepts has deliberately been used in order to appeal to a range of readers and disciplinary requirements. Insight into areas such as color psychology for the development of colorways, color ranges, color themes, and color forecasting is given to help readers better understand why a sound knowledge of color is necessary. Color terminology is given and discussed throughout this chapter, and the fundamental theories and color characteristics are also discussed. In addition to color mixing principles to aid readers' ability to recognize and visually analyze color, the scientific measurements of color, color deficiency, and metamerism are included to synthesize the discussion points as a comprehensive body of work in a condensed and informative format. The following bibliography serves as indicative reading to further develop your knowledge.

Bibliography and Recommended Reading

Chijiiwa, H. (2000). *Color Harmony: A Guide to Creative Color Combinations*, Edition Olms AG: Zurich.
Diane, T. and Cassidy, T. (2005). *Color Forecasting*, Blackwell Publishers: Oxford, UK.
Feisner, E. A. (2006). *Color Studies*, Fairchilds Publications: New York.

Holtzschue, L. (2011). *Understanding Color: An Introduction for Designers*, John Wiley & Sons: Hoboken, NJ.
Lloyd, D. (2007). *The Color Book,* Craft Print International, Ltd: Singapore.
McLannahan, H. (Ed.) (2008). *Visual Impairment: A Global View,* Oxford University Press: Oxford, UK.
Whelan, B. M. (1994). *Color Harmony,* Rockport Publishers: Beverly, MA.
Wilcox, M. (2009). *Blue and Yellow Don't Make Green,* Imago Productions (FE): Singapore.

13

Textile Finishing

Andrew J Hebden and Parikshit Goswami

CONTENTS

13.1 Introduction

Textile finishing refers to any process, mechanical or chemical, post dyeing, which leads to an improvement in the look, handle, or performance of the fabric, be it a woven, knitted, or nonwoven material. Generally, these processes are carried out on the textile in fabric form, but some can also be applied to fibers or yarns.

In one school of thought, *textile finishing* could be defined as any process after the fabric is formed and so could include fabric preparation steps such as scouring and bleaching or, indeed, coloration. A second school of thought considers finishing as *the final step* and thus considers finishing as preparing the fabric or fiber for the consumer and excludes fabric preparation and coloration. Thus, in this chapter, we will adopt the second school of thought and focus primarily on finishing steps, which occur post coloration.

Textile finishes generally fall into two classes: dry or mechanical finishes, such as calendering or mangling, and wet or chemical finishes, such as fluorochemical or flame retardancy. There is room for crossover, however, with some finishing techniques involving both mechanical and chemical finishes.

In addition, due to consumer's desire for performance, there is a scope for multiple finishes to be applied to textiles, each with a distinct function to add functionality or, indeed, one finish to provide multiple functionalities.

13.2 Drying and Setting

Wet textile fabrics can hold more than their own weight in water, and thus, it is important to sufficiently dry fabric as part of the finishing process. When we consider removing water from fabrics, there are two methods by which this can be achieved, either physically removing the water (wringing, mangling, centrifugation, etc.) or the usage of heat to thermally dry the fabric. Obviously, mechanical removal has a much lower relative energy cost compared with thermal drying, because of water's high latent heat of evaporation, and thus, it may be the case that water is removed in two stages, initially mechanically and then later by heat.

The easiest way to remove water from a garment is by wringing it out, a mechanical method that has been in use since textiles existed. In a washing context, however, this depends on the strength of the person washing the garment and is likely to lead to a relatively high amount of water being left in the fabric. Many years ago, this process was mechanized with the introduction of the mangle, which squeezes the fabric between two rollers, increasing the pressure on the material and thus increasing the amount of water removed. The exertion of such forces can, however, lead to abrasion of the fabric, which is undesirable in an industrial setting.

Water can be successfully removed from textiles by centrifugation, in which essentially industrially sized tumble driers are present. Owing to the large forces involved with a large load of fabric rotating in excess of 1000 revolutions per minute, it is important that the load is balanced and the machine securely mounted.[1]

When it comes to removing water from fabric with heat, there are three possible heat-transfer mechanisms: conduction, convection, and radiation. To remove water via conduction, the wet textile is placed in direct contact with a heated surface. This is the most efficient heat-transfer method, but it offers no control over fabric width. Tenter frames allow control of the fabric dimensions during drying, allowing water to be removed via convection; in this case, hot air comes in contact with the wet textile. Radiant heaters dry via infrared and are often used as a predryer to reduce the moisture content before the textile enters the actual drying process. It is important that the fabric is not overdried, as this may produce a harsh handle, and thus, drying must be carefully controlled.

Heat setting involves achieving dimensional stability of a fabric (or yarn) that has a thermoplastic component. During the fabric-manufacturing process, the textile is exposed to processes such as drawing, spinning, knitting, or, indeed, one or more of the finishing treatments outlined in this chapter. As a result, significant stress is created within the fibers of the fabric, and thus, when the material is subjected to subsequent heat treatment (e.g., ironing) or laundering, relaxation of the fibers takes place, and this can be observed as shrinkage or a change of shape in the garment.[2,3] It is thus desirable to heat set the textile in advance to reduce this effect for the consumer. During such a treatment, the melting point of the polymer must not be exceeded, but sufficient energy must be supplied to break

TABLE 13.1

Setting Temperatures of Common Textiles

Fabric	Setting Temperature (°C)
Nylon 6,6 filament (woven)	200–210
Nylon 6,6 filament (knitted)	220–225
Nylon 6 filament	180–190
Polyester yarn	150
Blended wool	170
Acrylics	140

interchain bonds, and this leads to an observable decrease in tensile strength. The typical setting temperatures for some common textiles are included in Table 13.1.

Pin stenters can be used for fabric setting, allowing both the width and length of the fabric to be controlled. Obviously, the thickness of the fabric will determine the heating time, and when that temperature approaches 200°C, there is the potential for fiber oxidation in air, leading to undesirable yellowing of the fabric. It should be noted that setting can have a major effect on the fabric handle—in some cases, a detrimental effect. In addition, setting of the fabric after dyeing can lead to a significant change in perceived color.

Heat setting is important in the carpet industry, as heat-set yarn is more durable, attractive, and, most importantly, comfortable for the user. When a yarn is cut, typically the ends will fray, and as a result, the carpet will have a reduced life cycle and poorer appearance. By heat setting the carpet's yarn ends will take on a shape known as *pinpoint tip definition*, which protects the ends of the yarn and improves the carpet's life span. This process is particularly relevant for carpet's pile yarns made from acrylic, nylon, polypropylene, and polyester materials.[4]

13.3 Application of Mechanical Finishes

Mechanical finishes use heat, pressure, and rollers to impart a particular finish to the textile, with the aim of improving either its appearance or its handle. Mechanical finishes include calendering, embossing, and napping. A comprehensive overview of mechanical finishes is beyond the scope of this chapter, but key finishes are discussed.

13.3.1 Calendering

Calendering uses high temperatures and pressures to change the nature of the fabric; this will obviously depend on the initial state of the fabric. Soft fibers, or open-weave fabrics, are obviously much more readily affected as compared with fabrics with a tight weave or those composed of hard fibers.[5]

13.3.2 Frictional Calendering

In frictional calendering, there is a speed differential between the pair of rollers, such that a smooth, shiny appearance is created on the textile material. One of the rollers is made from smooth metal and may be heated; this rotates faster than a second softer roller, which effectively polishes the fabric.[6] In addition to the friction created, which produces a glossy finish to the fabric, the process also reduces the gap between the warp and weft threads. Using this process finishes, such as Chintz can be produced. In this case, a padding machine applies the finishing solution to the fabric, which is then partially dried and friction calendered. Utilization of a resin-based solution leads to a permanent finish, whereas with starch or wax, the effect is temporary.

13.3.3 Embossing

This technique uses heat and pressure to impart an elegant aesthetic appeal to a fabric or garment. The fabric is pressed between an engraved heated roller and a second softer roller, which causes the embossed image to be raised when compared to the background. In a textile context, embossing is typically used in nonwovens such as nappies, napkins, and tissue papers, where one may wish to create a logo or a pattern. In addition, embossing can be seen on home furnishings such as curtains and cushion covers. It may also be used on fashionable evening gowns worn by celebrities attending the Oscars.

13.3.4 Napping

The napping process can be traced back to Roman times, when dried teasel pods were used as part of the process, and indeed, for woolen clothes this technique persisted until relatively recently. Napping involves raising the ends of fibers out of the fabric (Figure 13.1) and is performed on both woolen and cotton fabrics, with flannelette being an example of the cotton fabric that has undergone this process. The nap is typically brushed in one direction in fabrics such as corduroy and velvet such that light will reflect off the surface in a particular way. Thus, when making garments from pieces of napped fabric, it is important that they are all laid in the same direction; otherwise, the finished garment will look like it is made of fabrics of different colors.

To effectively raise the nap in woolen fabrics, they must be damped and then subsequently dried. Modern techniques use metal needles with 45-degree hooks on the ends to pull the ends of the yarn from the fabric. The rollers usually alternate with one roller, with hooks directed toward the fabric feed direction, followed by one with hooks counter to the fabric feed direction. Rotating brushes counter to the rollers cleans the napped fibers from the hooks.

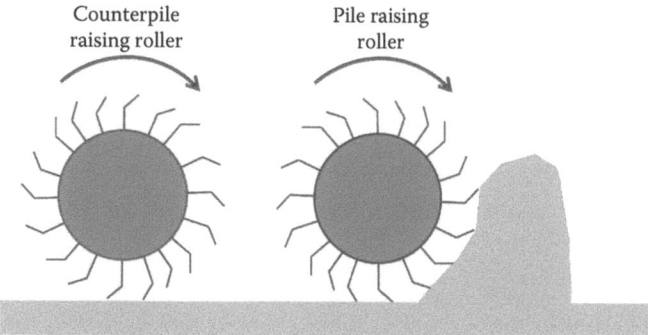

FIGURE 13.1
Nap-raising rollers.

13.4 Application of Chemical Finishes

There are many finishing treatments and finishing agents. A thorough review of all those agents, however, is beyond the scope of this chapter, and thus, only the main finishes are discussed here. The application and formulation of chemical finishes depend on several factors such as the compatibility of different chemical finishes, the nature of the material to be treated, and the chemistry of the functional chemical. It is also important to consider the environmental and sustainable credentials of the process.

If the desired functional chemical displays a high affinity toward the substrate, then exhaust (immersion) application methods may be suitable, and thus, any of the textile-dyeing machines typically employed for batchwise dyeing can thus be used for this type of textile finishing. However, should the applied functional chemical have low or limited substantivity, a continuous application method should be employed. One of the most widely used machines for the application of finishes in a uniform manner to fabrics is the pad mangle. Using the pad–dry–cure method, the fabric is first immersed in a solution containing the functional chemical(s), followed by partial drying and lastly *curing* to permanently fix the finish to the fabric (Figure 13.2).

13.4.1 Sanforizing

Mainly used on cotton fabrics and patented by Sanford Lockwood Cluett in 1930, sanforizing is a controlled mechanical shrinkage that does not require chemicals.[7] If fabric is not sanforized, shrinkage rates up to 10% can be observed, but this can be reduced to less than 1% following treatment. To treat the fabric, it is moistened with steam or water and then passed between a series of rubber belts and cylinders, before been compacted to its finished size. It should be the last treatment applied to the fabric.

13.4.2 Water Repellency

This is often referred to in the literature, particularly historically, as waterproofing. In fact, when it is raining outside, people in Britain will often say "Think you will need your waterproofs today" rather than "Best wear your water-repellent jacket." However, what does waterproof mean? Waterproof in a textile sense generally means a material that prevents the penetration and absorption of water, thus providing a barrier to water.

Thus, it is perhaps more appropriate to talk of water repellency, but again, this term has some shortcomings, as a liquid in contact with a solid will always experience some form of attraction, however small. The easiest method to produce a waterproof fabric is by

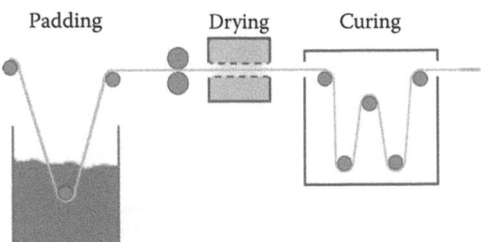

FIGURE 13.2
Schematic of typical padding machine.

coating it with a polymer coating of polyvinyl chloride or polyurethane. This means that such materials could be used for tarpaulins, as the coating provides a solid barrier, which prevents the penetration of water (or indeed other liquids). However, such a solid barrier also prevents the passage of both water vapor and air, meaning that if it were to be used in a garment context, the wearer would quickly become uncomfortable. In this scenario, where a wearer is likely to perspire, a garment must allow the water vapor to pass through the fabric to maintain wearer's comfort while also maintaining its waterproof properties.

To achieve a water-repellent surface, the free energy at the fiber's surface must be decreased. In doing this, the adhesive interactions between the drop of water on the garment and the textile surface are less than the internal cohesive interactions of the drop. As a result, the drop will not spread but will remain in a bead form. The most popular way to achieve this in recent times has been through fluorocarbon finishes, which also offer good oil-repellent properties.[8] However, environmental concerns surrounding the prevalence of perfluoro octanoic acid (PFOA) and the classification of polyfluorinated compounds as persistent organic pollutants call into question their long-term future. W. L. Gore, who make the market-leading Gore-Tex® fabric, eliminated PFOA from all their weatherproof fabrics in 2013. Fluorine-free alternatives, such as silicones and microporous polyurethane coatings, are available in the market already, but non currently match the oil-repelling abilities of fluorine-based coatings in addition to offering water repellancy.[8]

13.4.3 Flame Retardancy

Textile materials that are flame-retardant are intended to resist fire but may still catch fire. To create fire, three elements are required: fuel, oxygen, and heat (or a source of ignition). Flame retardants aim to disrupt the fire triangle by depriving or reducing one of these three essential components (Figure 13.3).[9]

Since the Second World War research into this area developed at pace, and certainly, within the domestic market, it led to a sharp downturn in the number of deaths associated with house fires. It is predicted that the Furniture and Furnishing Regulations,[10] which required all fabric and polyurethane foams used in the manufacture of furniture to be fire-resistant, have saved at least 710 lives between 1988 and 1997 and have contributed, along with other factors such as smoke alarms and a reduction in the number of smokers, to a reduction in the number of house fire deaths in recent years, which have been at their lowest recorded levels. While great strides have been made, there is still room for improvement. This was highlighted by the recent case of Strictly Come Dancing's host Claudia

FIGURE 13.3
Three components that lead to fire.

Winkleman's daughter Matilda Thykier.[11] During Halloween, her fancy-dress outfit caught fire when it came into contact with a tea light in a decorative pumpkin. At the time of writing, such outfits are classified as toys and are thus not subject to the much stricter flame retardancy regulations that clothing such as children's nightwear must meet. Such fancy-dress outfits are typically made of fabrics that are not only flammable but are often thermoplastic, such that when heated, they will melt and then stick to the wearer, increasing the amount of injury that the victim suffers. Hopefully, improvements to the flame retardancy of the materials used in these types of outfits will occur soon, in addition to a stricter regulatory regime, such that fancy-dress outfits fall under clothing regulations rather than toys.

Cellulosic textiles, including regenerated fibers such as viscose, account for more than 50% of fiber consumption per annum.[12] Owing to the flammable nature of these materials, there is a large market in flame-retardant cellulosics, which comprise more than three quarters of the total flame-retardant market.

To understand how flame retardants can offer a benefit to textile materials, we first need to understand the process by which textile combustion occurs; initially, an ignition source, such as a small flame, perhaps from a match or a cigarette, or heat, perhaps from an electric heater, provides energy to the system. This ignition source causes the material to ignite and burn, leading to pyrolysis and the release of flammable gases. In the case of some materials, they will remain in the solid phase and slowly smoulder, sometimes even self-extinguishing if a char barrier is formed—this is a carbonated barrier between the ignition source and the unburnt bulk material, which leads to a breakdown of the fire triangle.[13] If flammable gases are released from the material, they will mix with oxygen from the air and burn; this will cause more of the textile material to burn, leading to more flammable gases being released, which will cause further propagation of the burning process.

Flame retardants must act to disrupt at least one aspect of the fire triangle, in order to retard if not inhibit fire. Three main ways to achieve this are as follows:

- Promotion of a char layer to insulate the available fuel within the textile material, providing a physical barrier to further burning
- Emission of water, nitrogen, or an inert gas to reduce the oxygen concentration and dilute the levels of flammable gases, inhibiting flame formation
- Delay or inhibition of *flashover*, which makes it harder for occupants to escape from a burning room, by disrupting the combustion stage of the fire cycle

If you consider a typical sitting room in the average British household, it is likely to contain multiple soft furnishings such as settees, cushions, and curtains. While the usage of smoke alarms within households has led to the early detection of fires, thus increasing the time available to escape, at the same time, improvements in flame retardancy have meant that when treated soft textiles catch fire, they burn much more slowly, thus increasing the amount of time before a flashover-type event occurs and hence the likelihood of the occupants escaping.

Work by Sun et al. have focused on creating a flame-retardant coating, which additionally offers self-healing superhydrophobicity functionality.[14] This is an example of a multifunctional finishing treatment. Cotton fabric is coated with a trilayer of branched polyethylenimine, ammonium phosphate, and fluorinated-decyl polyhedral oligomeric silsesquioxane via sequential deposition. On exposure to flame, the coating exhibits an intumescent effect, as the porous char layer is formed.

13.4.4 Ultraviolet Protection

Since the 1980s, Australia has run a succession of health campaigns based around Slip, Slop, Slap!—slip on a shirt, slop on sunscreen, and slap on a hat.[15] Obviously, applying a textile covering to the skin is going to increase the amount of time taken for the skin to burn compared with bare skin, but ultraviolet (UV) radiation is still able to penetrate the textile, and thus, it can be desirable to add UV protection finishes to apparel to increase protection for the wearer. Thus, not only the textiles that are likely to be worn in the sun, such as t-shirts, sun hats, swimwear, but also items such as tents and caravan awnings are given a UV protection finish.

UV radiation refers to light with a wavelength between 280 nm and 400 nm, but it is the radiation in the UV-B region (280–320 nm), particularly 300–310 nm, that poses the greatest danger to the skin. Thus, for a textile to be effective at protecting the wearer from UV radiation, it must offer effective protection between 300 nm and 320 nm. To quantify the protective effect offered by a garment, the solar protection factor (SPF) is determined spectroscopically; the higher the SPF, the greater the protection to UV-B offered by the fabric.[16]

For a molecule to be suitable for application to textile apparel, it must not add color to the garment, be easy to apply, and have the ability to quickly transform the UV energy absorbed efficiently into vibrational energy. The structures of several common UV absorbers for synthetic and natural fibers are shown in Figure 13.4 and Figure 13.5, respectively. Varying the ring substituents allows the UV absorption wavelength to be carefully controlled.[17] Such molecules can be applied at the same time that the material is dyed, and thus, they often contain chemical groups similar to those found in typical dyes to aid dissolution and fixing to the fiber. Ultraviolet absorbers, in much the same way as dyes, must be wash- and light-fast; this can be evaluated by standard laundry trials.

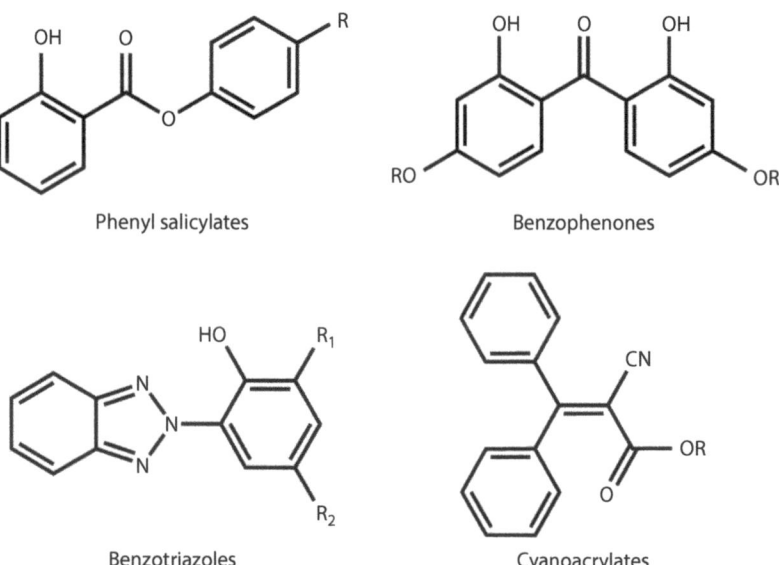

FIGURE 13.4
Structure of UV absorbers for synthetic fibers.

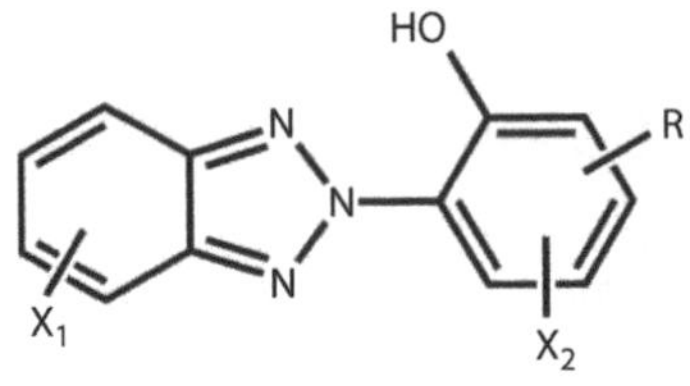

R = alkyl, alkoxy, sulfonate
X_1, X_2 = H, halide, sulfonate

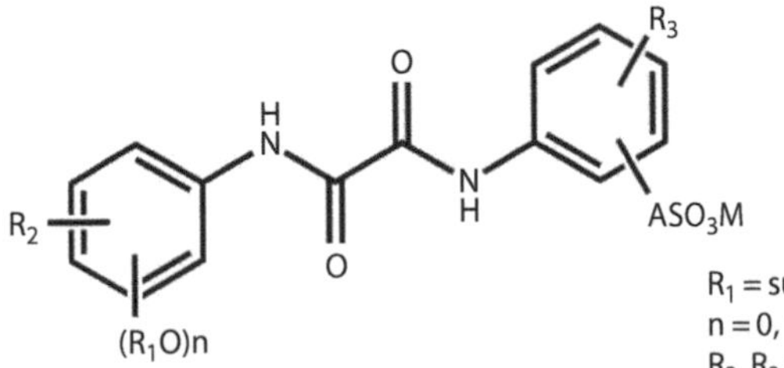

Benzotriazoles

R_1 = substituted or unsubstituted alkylbenzyl
n = 0, 1, 2
R_2, R_3 = H, halide, alky
A = direct bond or alkylene linkage
M = H or alkali metal

Oxalic acid dianilide derivatives

FIGURE 13.5
Structure of UV absorbers for natural fibers.

13.4.5 Antimicrobial

There are two main reasons for antimicrobial finishes: to protect the user from pathogenic or odor-causing microorganisms and to protect the fabric from conditions under which mold, mildew, and similar organisms would proliferate. This is achieved by either inhibiting the growth of microorganisms or destroying them. While some natural fibers such as wool (bacterial attack) and cotton (fungal attack) are susceptible to attack by microorganisms, it is by no means limited to this class, with synthetic fibers such as polyurethane also being able to be damaged by microorganisms.[18] In recent years, consumers have seen a plethora of silver-containing materials entering the market, for example, Freshfeet™ socks produced by Marks and Spencer. However, this itself offers problems, with the development of silver-resistant bacteria and the subsequent leaching of silver from textiles into wastewater during the laundering process.

Octadecylaminodimethyl trimethoxysiyl propyl ammonium chloride (DMOAP) has been successfully used as a bound antimicrobial, with a curing step required to form a siloxane polymer on the fiber surface. Polyhexanide, also known as polyhexamethylene biguanide (PHMB), has long been used as an antimicrobial in both cosmetics and swimming pools. It can be applied to textiles by either exhaust or pad method. The structures of both molecules are shown in Figure 13.6.

Modern health care provides a wealth of opportunities for both preventative methods to halt the spread of infection, be that antimicrobial textile surfaces, specialized wound dressings, or textile-based implantable material such as sutures, in which case biocompatibility is important. Healthcare-associated infections are a growing issue within modern medicine, and textiles can offer novel solutions to maintain cleanliness. Surface attachment and subsequent biofilm formation are a key survival mechanism for bacteria such as *Escherichia coli*, *Staphylococcus aureus*, and *Enterococcus faecalis*. The usage of nonwoven

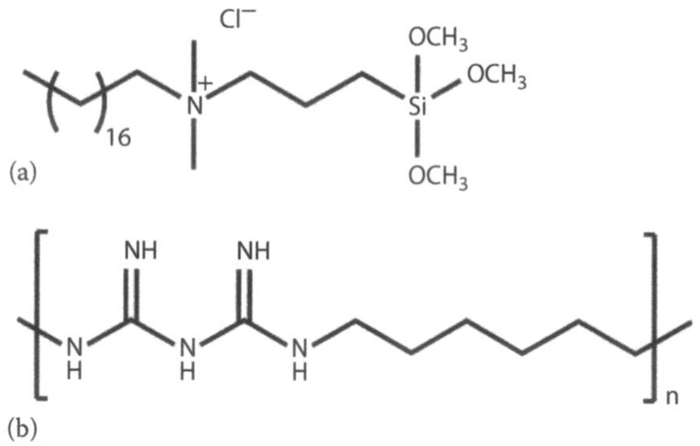

FIGURE 13.6
Structure of (a) DMOAP and (b) PHMB.

textile wipes offers a cheap and effective mechanism to remove bacterium from such surfaces. Plasma treatment of the nonwoven fibers with hexafluroethane (C_2F_6) leads to a reduction in the surface energy of the fibers, which increases removal efficiency of *E. coli*.[19]

13.4.6 Easy-Care Finishing

The apparel industry is dominated by cellulosic fibers; for example, cotton is a strong, breathable garment to be manufactured, which unfortunately has a natural tendency to crease when worn and shrink during the laundering process. Easy-care clothing is a technology designed to allow garments to be washed with minimal shrinkage or creasing, while requiring little, if any, subsequent ironing to restore the original garment appearance.

Cellulose fibers are composed of a bundle of polymer chains consisting of β-glucose molecules and contain crystalline and amorphous regions.[20] Moisture can be readily absorbed within the amorphous regions of these polymers, breaking hydrogen bonds between the chains and creating stress within the fibers. The polymer chains subsequently reorient to a lower-energy confirmation, at which point the hydrogen bonds reform, causing a wrinkle or a crease in the garment. Ironing, by the application of heat and mechanical force, provides sufficient energy to overcome the internal forces within the fiber and remove the crease, reorienting the polymer chains to their original position.

Easy care provides a mechanism to prevent the reorientation of the polymer chains within these amorphous regions, effectively locking them in a *crease-free* confirmation. This is achieved by the reaction of multifunctional crosslinking agents with hydroxyl groups of the cellulose fibers. It should be noted that this crosslinking reduces the elasticity and flexibility of the cellulosic fibers.

It was the mid-1920s that the first crosslinking agent that provided dimensional stability to cellulose was disclosed, but generally, they all work on the same principle; crosslinking is achieved by using amide-formaldehyde crosslinkers in an acid-catalyzed reaction with the cellulosic substrate.[5]

Easy-care finishes are often applied by a pad–dry–cure procedure, where the crosslinking agent, catalyst, softener, and any other components are dried onto the fabric before the curing step, which causes the crosslinking reaction to occur. It may be possible to add additional finishes such as stain repellency or fire retardancy to apparels at this stage if they are compatible. When this curing step takes place depends on the application of the final garment. For formal trousers or pleated skirts, which need to retain their creases, the curing step takes place after garment manufacture, whereas for casual trousers or sheets, the fabric is cured before garment manufacture.

Formaldehyde is intimately linked to easy-care fabrics, and concerns have been expressed regarding possible health risks associated with the usage of formaldehyde. It is known that at concentrations ~1%, formaldehyde can irritate both the eyes and respiratory tract, and in susceptible individuals, it may cause a skin irritation.[21] This has led to the establishment of maximum permitted *free* formaldehyde levels to be established within textiles, with different countries adopting different standards. In addition, there has been a drive within the industry to move toward low-formaldehyde products, which lead to a reduction in free formaldehyde.

Consumers are increasingly becoming more environmentally aware, and thus, the importance of recyclability and lessening the impact on natural resources becomes more important. Research into repurposing textile waste has increased with companies looking to create *closed-loop* systems. Haule et al. have investigated the removal of dimethylol dihydroxyethylene urea (DMDHEU) from cotton fabrics in order to form cellulosic pulp, which could be combined with wood pulp to provide a commercial feedstock for lyocell production.[22]

13.4.7 Softening Agents

The handle, or feel, of a textile material is an important property for the consumer but is often a difficult parameter to measure scientifically and so can be subjective. Technical textiles can often become brittle or feel rough to the consumer due to the processes required to impart their technical functionality, and thus, softening is often a key part of obtaining an acceptable finish for the end user.[5] Softeners are widely used, and in addition to improving the handle of the material, they can enhance the perceived quality of the finished fabric. They can also counteract the harshness imparted by other finishes such as easy care. The majority of softeners used are not covalently bound to the substrate and thus are removed from the material over time or by laundering. Thus, they need replenishing; as a result, softeners are used extensively in domestic washing formulations.

Each individual type of fiber has a specific softness value, which will depend on both its chemical composition and its physical structure. So, for example, the finer the wool used to make a woolen jumper, the softer the final garment will feel to the wearer. In addition, how a fabric is woven will affect the handle of the finished article; a looser-weave garment will feel softer than a closely woven garment. Moreover, a greater thread count will reduce the fabric softness due to an increase in stiffness. The increased usage of microfibers within textiles has led to a subsequent improvement in perceived softness.

If the softness characteristics of the textile fall outside the desired range and there is no scope, due to technical performance restrictions, to alter the physical properties in one of the ways outlined earlier, then an alternative approach is required, and that is where softening agents can play a role.

Early softeners were based on waxes and oils, in attempts to mimic natural softeners such as lanolin, which is secreted by wool-bearing animals such as sheep to soften their wool.[23] The main categories of softeners are anionic, cationic, pseudo-ionic, nonionic, amphoteric, and reactive. Softeners are one of the most important classes of textile finish that are used to improve the handle of the fabric, which is a very important property for the consumer. It is important that softeners are easy to handle (e.g., liquid, pumpable, and stable to dilution) for ease of application; ideally, they should be stable with respect to

TABLE 13.2

Charge of Common Fibers and Typical Softener Used

Fiber	Electric Charge	Preferred Ionogenity of the Softener
Cotton	Negative	Cationic
Polyamide (neutral)	Slightly negative	Cationic
Polyamide (acid)	Slightly positive	Anionic
Wool (neutral)	Slightly negative	Cationic
Wool (acid)	Slightly positive	Anionic

other chemicals, such that several types of finishes can be added simultaneously (e.g. flame retardant added at same time as softener). Softeners should not impart color shade changes on application and should not yellow with age. Ideally, they should be dermatologically harmless, nontoxic, and easily biodegradable. A summary of the typical softener used with common fibers is included in Table 13.2.

Anionic softeners were one of the first groups of softeners to be exploited commercially. They are negatively charged and thus are repelled from negatively charged fibers, such as cellulosic, leading to higher hydrophobicity. They include long-chain alkyl sulfates, diesters of phosphoric acid, sulfonates, and succinates (Figure 13.7) and were one of the first groups to be exploited commercially. The majority of fluorescent brightening agents are anionic and thus compatible with anionic softeners and anionic surfactants used in laundry products.

Cationics (Figure 13.8) are the most utilized group of softeners, both industrially and domestically, exhibiting a good level of softness and durability to laundering. Parts of the

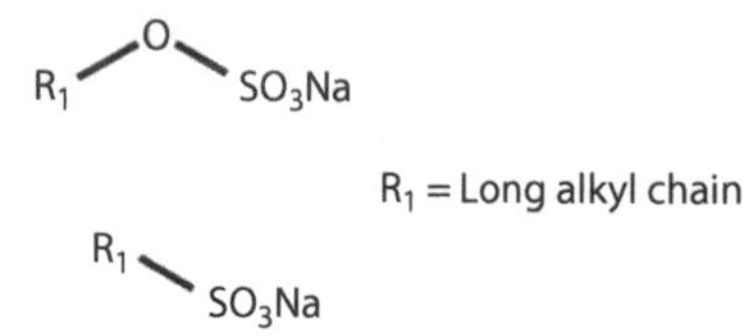

$R_1 = $ Long alkyl chain

FIGURE 13.7
Structure of typical anionic softeners.

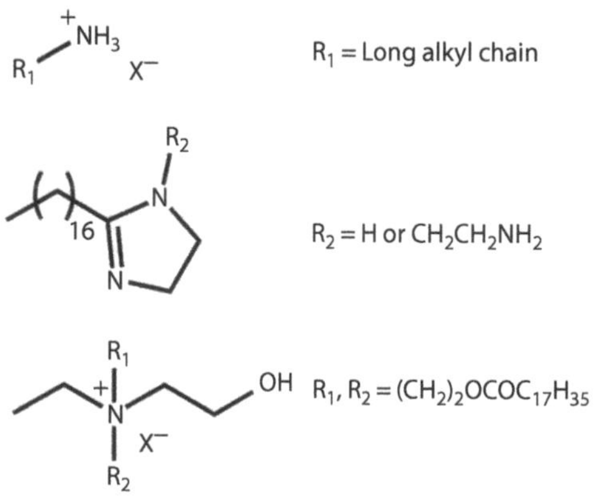

$R_1 = $ Long alkyl chain

$R_2 = $ H or $CH_2CH_2NH_2$

$R_1, R_2 = (CH_2)_2OCOC_{17}H_{35}$

FIGURE 13.8
Structure of typical cationic softeners.

popularity of these molecules are their ease of application and the low levels required to give a soft handle. Orientation of the positively charged ends of the softeners toward the negatively charged fibers creates a surface of hydrophobic carbon chains on the surface, which leads to the softening effect end user's feel. Owing to the charge on these softeners, application can be by exhaustive means (e.g., winch, jet, beam, and package). Rate of exhaustion generally increases with an increase in temperature; however, some wax-containing products become unstable at higher temperatures. Good agitation helps improve the rate and evenness of softener deposition with a bath temperature between 40°C and 50°C typically used. It is also desirable to apply the softener at a pH 5–6; below pH 4, yellowing can occur if anionic brighteners are present in the fabric. However, there are drawbacks to the usage of cationic softeners, as they increase the soiling propensity of fibers and inhibit soil removal. It is also important to ensure that cationic softeners are not discharged into the waste stream, as this will disrupt the biological treatment of sewage.

Distearyldimethyl ammonium chloride (DHTDMAC) was introduced as a fabric softener in the 1950s to counteract the harsh feel of cotton nappies after machine washing. This has subsequently been discontinued because of the slow rate of biodegradation and has been replaced with mono- and di-*esterquats*, which show a much faster rate of hydrolysis.

In addition, the use of silicone-based cationics has grown steadily in recent years. These provide a high degree of softness and impart a good handle to the fabric. It should be noted that depending on the method of manufacture, some silicon-based softeners may contain significant amounts of siloxane oligomers, which may cause air pollution.[24] Weakly cationic amino-functional siloxanes have received some attention, and although they can be applied under a mild alkaline solution, durability is improved by application under mildly acidic conditions (pH 4).

Nonionic softeners are able to perform functions such as stabilization, emulsification, and lubrication, in addition to their softening abilities (Figure 13.9).[5] Paraffin waxes and similar substances are the simplest type of nonionic softeners. More recently, low-weight polysiloxanes have become an important class of nonionic softener, in particular amino-functional polysiloxanes. Nonionics show good compatibility with cationic and silicon products, which may be present in the finishing formulation. Under extreme temperatures, nonionics show low levels of yellowing and have a minimal effect on any fluorescent brighteners that may be present on the fabric. These softeners give a temporary soft feel and show a low incidence of skin irritation.

Reactive softeners contain functional groups capable of forming bonds with particular groups within the fiber. Creating a covalent bond between the softener and the substrate leads to a permanent finish, which shows excellent wash fastness. Cellulosic fibers, with their proliferation of hydroxyl groups, are ideal for usage with this type of softeners.

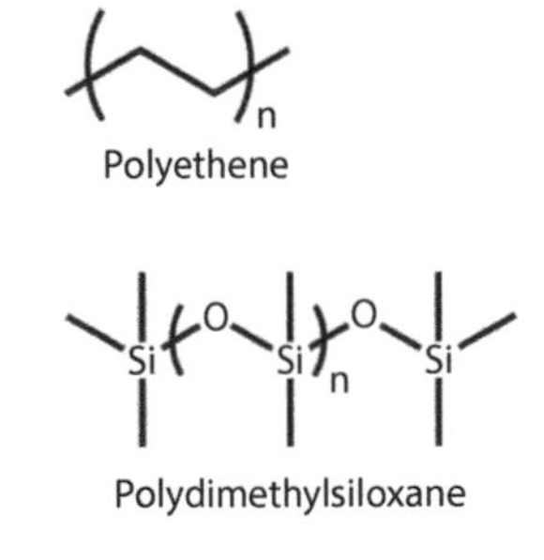

Polyethene

Polydimethylsiloxane

FIGURE 13.9
Structure of typical nonionic softeners.

A pad–dry–calender method can be used for application due to their solubility in water. N-methylol derivatives of stearic acid amides and urea-substituted compounds have been successfully utilized within this area (Figure 13.10). Reactive softeners are often used less often than cationic and nonionic softeners, as they are relatively more expensive.

Amphoteric softeners are more often seen in personal care products due to their low levels of toxicity, making them ideal for shampoos. Examples of amphoteric softeners are shown in Figure 13.11. They show good compatibility with easy-care and hydrophilic finishes when applied in a pad bath. Amphoterics contain potentially anionic and cationic groups within the same molecule, depending on the conditions to which the molecule is

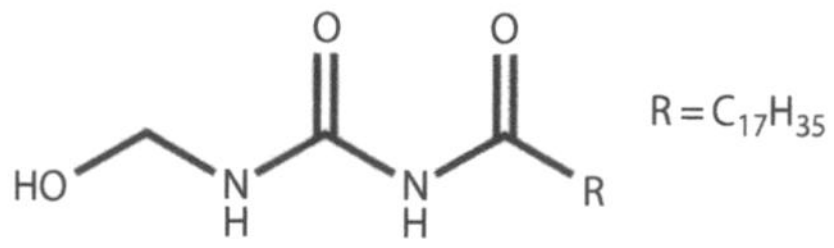

$R = C_{17}H_{35}$

FIGURE 13.10
Structure of reactive softener N-methylol urea.

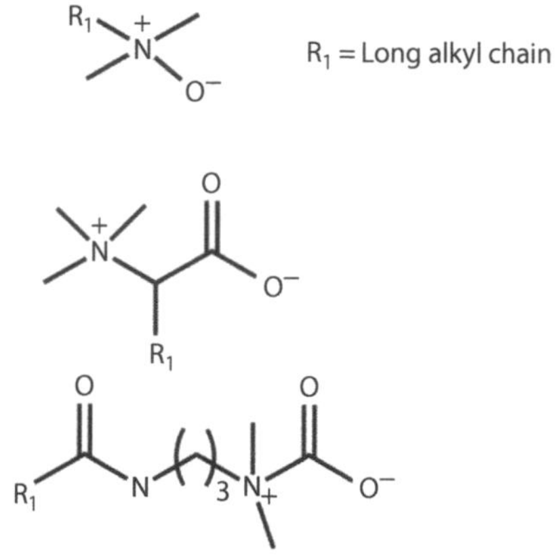

R_1 = Long alkyl chain

FIGURE 13.11
Structures of typical amphoteric softeners.

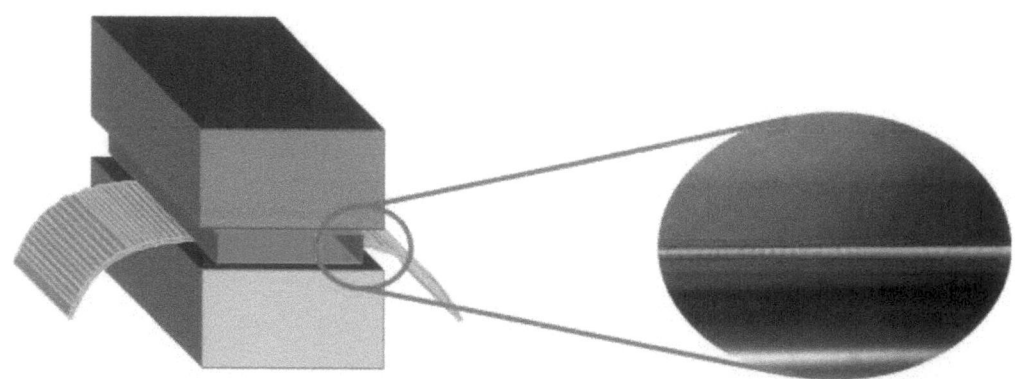

FIGURE 13.12
Continuous plasma machine showing helium plasma generated between a pair of parallel plates.

exposed. Thus, at low pH, protonation will occur and the softener will be cationic, and at high pH, an anionic species will be formed. Between these two extremes, there will be an isoelectric point, at which point the molecule is uncharged.

13.5 Environmental Plasma Treatment

Plasma is considered the fourth fundamental state of matter and is created by heating a gas or exposing it to a strong electromagnetic field to yield a cloud of charged ions. Such technology allows the modification of the textile surface without changing the bulk properties of the material. This can be either by the chemical modification of the surface of the textile under a particular gaseous atmosphere and/or the bonding of additional chemical moieties to the textile or by the usage of plasma to modify the topology of the textile. As plasma only treats the surface of the fiber, there is no significant loss in the strength of the material. Plasma treatment consumes minimal chemicals compared with a conventional water-based finishing method. As it is a *dry* method, no effluent is produced, which may need to be cleaned or disposed of. There is also the additional benefit that when plasma is used, a post-treatment drying step is not required. It is also worth noting that a side effect of the plasma treatment is that the surface of the fiber is *cleaned*. Thus, any sizing agents, lubricants, and so on are removed from the surface of the fiber. It is possible to plasma treat the textile at the fiber or yarn stage and thus improve the wettability of the fiber, such that it is easier to dye the resultant fiber, thus increasing the durability of the dye on the fiber.[25] Plasma treatment can be achieved under vacuum conditions in either a batch or a semi-continuous method. It can also be achieved at atmospheric pressure via corona discharge. Durability of the finish and/or effect can be a disadvantage of plasma treatment.

Oxygen plasma can be used to increase the number of functional groups on the surface of the fiber, resulting in an increased polarity at the surface, which increases wettability. This means that hydrophilicity can be introduced to hydrophobic fibers such as polyethylene and polypropylene. In contrast, the hydrophilic nature of cotton is well known, and it can thus be desirable to impart a hydrophobic coating to the fiber/fabric. To this end, the usage of a plasma-containing siloxane (or fluorine-containing gases, e.g., C_2F_6) leads to a fabric from which water will run off, while retaining its moisture and water vapor permeability. Common plasma applications, the types of material that can be treated, and the nature of the treatment are listed in Table 13.3.

TABLE 13.3

Typical Plasma Treatments Used and the Types of Material to Which They Are Applied

Application	Material	Treatment
Hydrophilic finish	Polypropylene, polyester, polyethylene	Oxygen plasma, air plasma[26]
Hydrophobic finish	Cotton, P-C blend	Siloxane plasma[27]
Antistatic finish	Nylon, polyester	Helium plasma, air plasma[28]
Reduced felting	Wool	Oxygen plasma[29]
Improved dyeing	Polyester	$SiCl_4$ plasma[30]
Bleaching	Wool	Oxygen plasma[31]
UV protection	Polyester	TiO_2 nanoparticles with plasma[32]
Flame retardancy	Cotton, rayon	Phosphorus-containing plasma[33]

13.6 Summary

As sustainability and environmental impact continue to be watch words within the textile industry, efforts will continue to advance the area of textile finishing. Industrially, technologies that require low or no water, for example, plasma, are becoming more popular. In addition, multifunctional finishes, or those finishes that can be applied simultaneously to reduce the number of treatment steps required, are beginning to receive more attention.

For a more in-depth look at textile finishing, Heywood's *Textile Finishing* and Hauser's *Chemical Finishing of Textiles* expand on the ideas covered in this chapter.

References

1. W. D. Schindler and P. J. Hauser, *Chemical Finishing of Textiles*, Oxford, UK: Woodhead Publishing, 2004.
2. V. B. Gupta and S. Kumar, *Journal of Applied Polymer Science*, 1981, 26, 1865–1876.
3. D. Phillips, J. Suesat, M. Wilding, D. Farrington, S. Sandukas, J. Bone and S. Dervan, *Coloration Technology*, 2003, 119, 128–133.
4. M. Dadgar, A. A. Merati and S. M. Hosseini Varkiyani, *The Journal of The Textile Institute*, 2016, 107, 107–115.
5. D. Heywood, *Textile Finishing*, Bradford, UK: Society of Dyers and Colourists, 2003.
6. D. R. Jackman, M. K. Dixon and J. Condra, *The Guide to Textiles for Interiors*, Winnipeg, MB: Portage & Main Press, 2003.
7. H. B. Sturtevant, *Journal of Chemical Education*, 1944, 21, 308.
8. H. Holmquist, S. Schellenberger, I. van der Veen, G. M. Peters, P. E. G. Leonards and I. T. Cousins, *Environment International*, 2016, 91, 251–264.
9. D. Lilley, in *39th Aerospace Sciences Meeting and Exhibit*, American Institute of Aeronautics and Astronautics, 2001.
10. U.K. Government, in *No. 1324*, HMSO, 1988.
11. M. Oppenheim, in *Independent*, 2016.
12. S. Chang, R. P. Slopek, B. Condon and J. C. Grunlan, *Industrial & Engineering Chemistry Research*, 2014, 53, 3805–3812.
13. S. Bourbigot and S. Duquesne, *Journal of Materials Chemistry*, 2007, 17, 2283.
14. S. Chen, X. Li, Y. Li and J. Sun, *ACS Nano*, 2015, 9, 4070–4076.
15. R. Marks, *Australasian Journal of Dermatology*, 1990, 31, 1–4.
16. J. Xin, W. Daoud and Y. Kong, *Textile Research Journal*, 2004, 74, 97–100.
17. J. Keck, H. E. A. Kramer, H. Port, T. Hirsch, P. Fischer and G. Rytz, *The Journal of Physical Chemistry*, 1996, 100, 14468–14475.
18. K. Kleman-Leyer, E. Agosin, A. H. Conner and T. K. Kirk, *Applied and Environmental Microbiology*, 1992, 58, 1266–1270.
19. N. W. M. Edwards, E. L. Best, S. D. Connell, P. Goswami, C. M. Carr, M. H. Wilcox and S. J. Russell, *Science and Technology of Advanced Materials*, 2017, doi:10.1080/14686996.2017.1288543, 1–35.
20. A. C. O'Sullivan, *Cellulose*, 1997, 4, 173–207.
21. M. H. Garrett, M. A. Hooper, B. M. Hooper, P. R. Rayment and M. J. Abramson, *Allergy*, 1999, 54, 330–337.
22. L. V. Haule, C. M. Carr and M. Rigout, *Cellulose*, 2014, 21, 2147–2156.
23. M. Delacorte, E. V. Sayre and N. Indictor, *Studies in Conservation*, 1971, 16, 9–17.

24. D. Graiver, K. W. Farminer and R. Narayan, *Journal of Polymers and the Environment*, 2003, 11, 129–136.
25. M. Lehocký and A. Mráček, *Czechoslovak Journal of Physics*, 2006, 56, B1277–B1282.
26. A. Vesel, I. Junkar, U. Cvelbar, J. Kovac and M. Mozetic, *Surface and Interface Analysis*, 2008, 40, 1444–1453.
27. J. Zhang, P. France, A. Radomyselskiy, S. Datta, J. Zhao and W. Van Ooij, *Journal of Applied Polymer Science*, 2003, 88, 1473–1481.
28. K. K. Samanta, M. Jassal and A. K. Agrawal, *Fibers and Polymers*, 2010, 11, 431–437.
29. S. Shahidi, A. Rashidi, M. Ghoranneviss, A. Anvari and J. Wiener, *Surface and Coatings Technology*, 2010, 205, Supplement 1, S349–S354.
30. M. Sarmadi, A. R. Denes and F. Denes, *Textile Chemist and Colorist*, 1996, 28, 17.
31. K. S. Gregorski and A. E. Pavlath, *Textile Research Journal*, 1980, 50, 42–46.
32. D. Mihailović, Z. Šaponjić, M. Radoičić, R. Molina, T. Radetić, P. Jovančić, J. Nedeljković and M. Radetić, *Polymers for Advanced Technologies*, 2011, 22, 703–709.
33. B. Edwards, A. El-Shafei, P. Hauser and P. Malshe, *Surface and Coatings Technology*, 2012, 209, 73–79.

14

Clothing Technology

Aileen Jefferson

CONTENTS

14.1 Introduction

The textile and clothing industry is one of the largest and most intricate supply chains in manufacturing. Advances in technology have facilitated its globalization and enabled companies to span geographical borders, adding to its complexity. It has been suggested by Teng and Jaramillo (2006) that the process can be split into a six-point structure: cotton growing (raw material), fiber yarns (yarn producers), textile manufacturers, apparel (clothing) manufacturers, retailers, and customers. The life cycle of a garment, which resides in Teng and Jaramillo's (2006) latter three points, is said to encompass stages from initial research and design through product development, manufacture, distribution, retail, and, in recent years, disposal. Previous chapters have covered in great depth the textile element, whereas this chapter will focus on clothing technology utilized predominantly within the product development area. Within a typical life cycle of a garment, Rodriguez (2001) stated that product development can be as high as 70%; thus, time saved in this area can greatly reduce time to market and therefore increase cost savings.

This chapter will discuss the product development process from an industrial perspective, commencing with an overview in Section 14.2.1, followed by pattern design and incorporating seams and opening, costings, and measurements in Sections 14.2.2–14.2.5. Grading is discussed in Section 14.2.6, whereas Sections 14.2.7 and 14.2.8 provide a brief overview of marker making and cutting. This chapter then moves to reviewing technology within product development (Section 14.3), whereas Section 14.4 and its subsections discuss the digital product development process and the role that computer-aided design/computer-aided manufacture (CAD/CAM) can play. This leads to Section 14.5 and the three-dimensional (3D) process, where advances in this technology are examined. Section 14.6 witnesses a brief section on 3D printing, and Section 14.7 presents applications and examples of how the technology has been utilized. Finally, to enhance the readers' knowledge, questions can be undertaken in Section 14.9, and further reading can be explored by referring to information in the Bibliography (and sources of further information).

14.2 The Product Development Process

14.2.1 A Brief Overview

The product development cycle begins when a conceptual garment design is received. Following this, two-dimensional (2D) patterns are created, which represent the design; the patterns are cut out and constructed into a 3D prototype; and the prototype is reviewed and revised until a sample is approved. Progressing onto manufacturing, grading (the increasing and decreasing of the size of pattern pieces to create differently sized garments), which can be carried out by the product development team, is undertaken before the pattern pieces are placed in a lay plan or marker that will be cut out in bulk cloth, and the garment parts are sent to be manufactured.

14.2.2 Pattern Design

A design of a garment is received. This is followed by the development of a 2D pattern, which represents the design sketch and enables a garment to be cut out in cloth and constructed into a 3D prototype; a time-consuming interactive process occurs, where the prototype is reviewed until the desired shape, style, and fit are approved. A combination of skills is required for the highly specialized task of pattern design. These skills include creativity, technical pattern-making knowhow, and knowledge of fabric properties, as all of the following need to be considered during the creation of an initial pattern for a prototype garment: aesthetic qualities, technical information, function, fit, and handling of the selected fabric. Such skills of the pattern maker can greatly reduce the number of prototypes samples being made. There are generally two recognized pattern-generation methods in industry: flat pattern cutting and draping. Furthermore, flat pattern cutting has two approaches: first, a basic pattern, called a block or block pattern, is used as a starting point for the desired garment. The pattern maker interprets design shape and features, and block patterns are manipulated and details added to achieve a final set of patterns, which reflects the design sketch and allows for fabric handling. In the second approach, the process of drafting directly onto paper from a set of measurements can be used. Measurements are taken from the required positions on the body; an allowance (or ease) of varying amounts is added to specific measurements, which will allow the prototype garment to fit correctly against the areas of the body and the body to move within the garment. Referring to these measurements and considering the style features of the design and the selected fabric, a pattern is drafted directly.

Draping or modeling, the second recognized pattern-generation method, can be used to create a pattern for a prototype garment or part of a garment. Fabric is draped, pinned, and modeled on a mannequin or dress stand to achieve the aesthetic qualities of the sketch and fit of the garment. This tactile process aids in determining the silhouette and proportions, the drape, and suitability of the selected fabric or allows prints, plaids, and stripes of the fabric to be positioned for visual effect. Specific points are marked on the draped garment to facilitate the joining of the parts. The fabric is carefully removed from the stand and laid flat on a piece of paper; the shape is traced out to form the pattern, noting any style features and specific points identified, and marked during the draping process.

In addition, if a constructed garment has a silhouette similar to the desired shape, then a copy of such a garment can be made either by deconstructing the garment and tracing round the parts or by manipulating the garment and copying the parts while the garment is constructed. This could form a starting point for the new set of patterns, or indeed, patterns of a previously created prototype could be utilized. There are no hard and fast rules when creating patterns for an initial prototype garment, and a combination of techniques can be used to achieve the desired effect. The processes of reviewing the prototype, amending the pattern, and generating additional prototypes continue until an approved sample is established.

14.2.3 Seams and Openings

During pattern development of an approved sample, makeup and joining methods are incorporated within the pattern to facilitate the manufacturing process and produce a well-constructed final garment. End use of the garment, market level, and fabric properties are considered throughout. Relevant seam allowances (an additional amount added to the perimeter of the pattern to enable the garment to be joined together) are incorporated, depending on the elected joining method. Joining methods include traditional sewn seams by using lockstitch machines; three, four, and five thread overlockers; etc. or non-sew technologies such as ultrasonic welding and bonding machines. In addition, openings, such as button openings, zips, etc. are considered and integrated within the pattern. Edge finishes could include binding, trims, or facings, and again, a relevant seam allowance is incorporated.

14.2.4 Costing

Simultaneously, vital information is required by the costing department to ascertain a final selling price and ensure that a profit can be made on production of the garments. Given that around 50% of a garment cost can typically be the fabric cost, an estimate of the amount of fabric required for the garment is essential. This is achieved by the creation of a costing lay plan, a rectangle that represents the measured width (the weft) of the fabric by the length of the lay plan (the warp of the fabric). Pattern pieces generated for the prototype sample are positioned within the width of the fabric in the most efficient manner to achieve the shortest length and thus best utilization of the fabric. It is vital that the straight grain of the pattern is positioned parallel to the length of the lay plan, or warp of the represented fabric. The length of the lay plan is noted. This process is repeated for all the fabrics necessary to construct the garment, including contrast fabric(s), lining, interlining, etc. Additional information, such as trims (buttons, zips, applique, etc.), to complete the garment is also relayed. The costing department will incorporate labor, overheads, distribution, profit margin, etc. in their costings to attain a final selling price.

14.2.5 Measurements

The approved sample is created to represent the base or sample size of the required size range, for example, in womenswear, a typical size range could be 10 through to 16, with a sample size 12 being utilized for the prototype; each company will have its own specific size range and body fit measurements on which they work. Thus, accurate body measurements are crucial for creating the initial block and approved prototype. Traditionally, a tape measure was used to attain body measurements; however, body scanners have been employed in recent years (see Section 14.4.5 for body scanning and Table 14.1 for body measurements). Ease (additional measurements to ensure that the garment fits comfortably on the body) is added to the body measurements. From the body measurements, a size chart for the range of garment sizes can be created; measurements in the size chart will be for the specific finished garment (see Table 14.2). However, should the garment be made from stretch fabric, which needs to be tight over the body, garment measurements

TABLE 14.1

Simple Example of Body Measurements Derived from Sizing Survey Information

	Sizes				
Area	10	12	14	16	Increment between Sizes
Bust	82 cm	87 cm	92 cm	97 cm	5 cm
Waist	60 cm	65 cm	70 cm	75 cm	5 cm
Hip	87 cm	92 cm	97 cm	102 cm	5 cm

TABLE 14.2

Simple Example of a Garment Size Chart (In This Case, 5-cm Ease Has Been Added to the Body Measurements)

	Sizes				
Area	10	12	14	16	Grading Increment
Bust	87 cm	92 cm	97 cm	102 cm	5 cm
Waist	65 cm	70 cm	75 cm	80 cm	5 cm
Hip	92 cm	97 cm	102 cm	107 cm	5 cm

will be decreased in accordance with the stretch properties of the fabric and the desired look or function of the final garment.

14.2.6 Grading

Once a sale has been agreed, grading is undertaken. Grading, the art of increasing or decreasing the patterns pieces to ensure the shape, aesthetics, and fit established in the prototype sample size, is retained throughout the size range. This task can be performed by a pattern designer or a personnel working within the team. In addition, graded sizes, particularly within childrenswear, will need to be considered during the initial stages of designing placement prints or appliques, particularly when costing of the garment is being undertaken. Each pattern piece within the set of patterns, depending on its position within the garment, will have its own amount of grading applied.

There are two broad systems used for grading: two dimensional and three dimensional. Two dimensional grading grades a pattern only in the girth and height, darts are not graded. In other words, suppression in the form of a dart is neither increased nor decreased; this will therefore limit the range of sizes and type of garment graded. Three dimensional grading not only increases or decreases the pattern for size but also increases or decreases suppression, for example, around the bust area. This is the optimum form of grading and is essential when grading over a wide range of sizes, however, a good working knowledge of pattern cutting is required to use this method.

While reviewing the size chart for the style in question, grading increments are determined. A grading increment is the difference between the measurements of the sizes in the range, either the whole measurement or a proportion of this measurement, depending on where the pattern piece will fit within the garment.

Table 14.3 illustrates a simple size chart for a skirt, and, for example, if the waistband consists of one pattern piece that goes all the way round the body, then the increment, and thus the amount the pattern piece needs to increase or decrease, will be the whole

TABLE 14.3

Simple Size Chart for a Skirt

	Sizes				
Area	10	12	14	16	Grading Increment
Waist	65 cm	70 cm	75 cm	80 cm	5 cm
Hip	92 cm	97 cm	102 cm	107 cm	5 cm
Length	60 cm	60 cm	60 cm	60 cm	0 cm

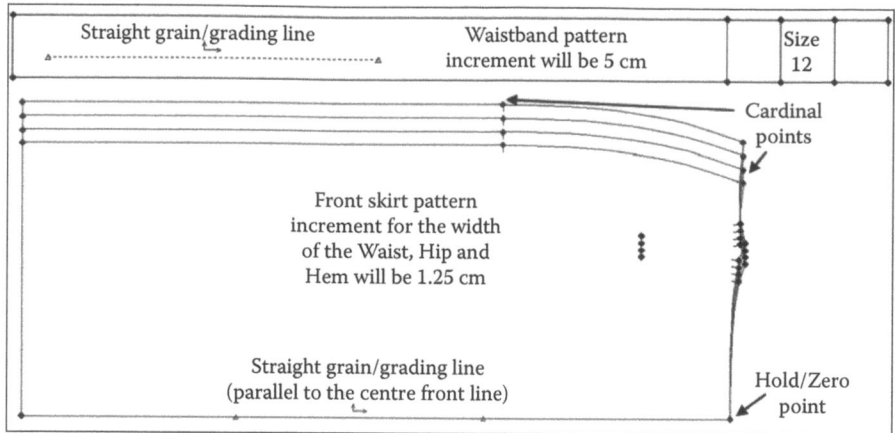

FIGURE 14.1
Illustration of grading a waistband and front skirt.

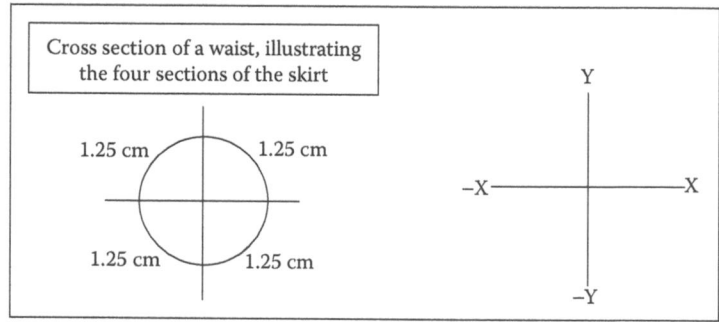

FIGURE 14.2
Illustration of cross section of garment and grading movement.

incremental measurement from the size chart. Whereas if the front and back skirts of the garment are being considered, these will usually have a left side and a right side. Thus, four pattern pieces will make the whole of the garment, and therefore, the whole of the waist (and the hip) incremental measurement will need to be divided in four ways, illustrated in Figures 14.1 and 14.2.

Grading is generally applied through X (horizontal) and Y (vertical) movements. The garment size chart is studied, and key points (called cardinal points) are identified on the pattern to ensure that each size of the finished garments corresponds to the measurements on the size chart. In the example of this simple skirt, where the waist, hip, and length are key, a horizontal grading line is established, usually the center front or center back of a garment, and a hold point is identified. The increments are calculated, and the cardinal points move accordingly, with the lines between them being drawn to achieve a smooth shape. This process is repeated until a full set of patterns are created, which produces a full range of required sizes.

14.2.7 Production Lay Plan/Marker

A series of production lay plans (sometimes called markers) is planned and created. These are similar to the costing lay plans but will incorporate different sizes to ensure that the correct quantities of each size and color (should several colors of the garment have been ordered) are able to be cut and ready for production and that the optimum utilization of the fabric is achieved.

14.2.8 Cutting

The quantities, planned during the creation of the lay plans, are reviewed; fabric is spread to the required length of the lay plan, with further lengths of fabric being laid on top, until the correct amounts of plies of fabric are achieved. This enables the precise quantities and sizes of the production garments to be cut. The cut work is then sent to be manufactured.

14.3 Technology within Product Development

The following sections will detail technologies incorporated in the product development process described previously.

There are many drawing and design packages adopted by the clothing industry and utilized across numerous industries. However, during the late 1970s and early 1980s, CAD/CAM software, written specifically for the clothing industry, was introduced. It is this bespoke software that will be discussed in this chapter. Since its introduction, such software has become ubiquitous within the industry, and there is no doubt that 2D CAD/CAM technology is an enormous success. It enhances many aspects of the process; the benefits of time saving, and therefore cost savings, are well known. In an early study comparing traditional methods with CAD/CAM methods, Neil Wainwright (1992, cited in Gray, 1992) gathered the information illustrated in Figure 14.3.

The 2D CAD/CAM systems that were introduced in the 1970s and 1980s have continued to develop and update their functionality, as hardware has become more powerful. There are numerous companies writing such software, however, the initial players continue to lead the industry: Assyst-Bulmer, now part of Human Solutions Group, Gerber Technology Ltd, and Lectra. Their software incorporates pattern input methods, extensive pattern cutting and direct drafting facilities that emulate manual pattern cutting (pattern design system [PDS]), marker making (including check, stripe, and pattern-matching functionality), and made-to-measure (MTM) software that also links with body scanners. These companies

Time saved making a seven piece ladies blouse		
	Traditional methods	CAD/CAM methods
Concept cost	4 hours	1 hours
First pattern confirmation	8 hours	2 hours
Grading	12 hours	2 hours
Production marker	6 hours	0.5 hours
Total	30 hours	5.5 hours

FIGURE 14.3
Time saved making a seven-piece ladies blouse. (Courtesy of Wainwright 1992, cited in Gray, 1992, p 51.)

also produce hardware such as plotters that enable patterns and markers (lay plans) to be plotted (printed) out, and plotters that can plot or cut individual pattern pieces. In addition, they produce automatic fabric cutters that are driven by the data generated from the software. Data can be communicated through varying formats that are used extensively and have facilitated the fast-paced global industry of today.

During the latter part of the twentieth century, 3D CAD visualisation software was introduced to the industry. Initial offerings were aimed at the online retail market, where customers could input their personal measurements, colouring, and hair style to create a virtual model, which could be dressed in available styles and viewed before purchase. However, in the last decade or so, a shift has been observed toward 3D offerings being employed and developed within the design and product development arena, where 3D links with 2D PDS software to shorten the process. Commercial 3D CAD software has two approaches: 2D to 3D and 3D to 2D. First, 2D to 3D enables 2D patterns to be assembled into a virtual garment on a 3D human form called an avatar; the garments can be rotated and viewed from full 360 degrees and any zoom level. In addition, 3D garment data can be returned to the 2D format, presenting amendments undertaken within the 3D environment. Second, 3D to 2D enables garment shapes to be created on a 3D avatar and flattened into 2D data, however, currently, this is for close-fitting garments only, and some human interaction is required to produce good quality pattern shapes.

14.4 The Digital Product Development Process

14.4.1 2D CAD Pattern Design

Pattern designers receive the design sketch, which could be in a digital format produced on one of the many design and drawing packages. The pattern makers, utilizing the PDS and their extensive skills, create a set of patterns for the prototype sample and save these into the computer system. This again can be a combination of the methods, as described earlier. Block patterns and previous styles are stored digitally; they can be retrieved and viewed on screen, where manipulation or creation of a new style can be undertaken. Sophisticated and extensive commands, along with basic functionality, are available to the PDS user. Furthermore, a complete set of sizes, or graded sizes, can be stored on the block pattern and patterns of previous styles, when modifications are made, the new pattern will automatically have the full size range available. In addition, if the grading needs modification or additional sizes are required, this can be adjusted quickly. If combinations of pattern generation methods are used, full scale paper patterns can be inputted into the computer via a digitizer or digital photography. A digitizer is a special table with a series of wires beneath the surface; pattern pieces are secured to the table, a cursor is used to position its cross hairs over specific points, which will register the point position in the software (Figure 14.4). Digital photography encompasses the capture of a bitmap image of the pattern pieces (Figure 14.5); this image is converted into a vector-based pattern shape recognized by the PDS. To emulate the tracing of the shape within the draping method to achieve a pattern, once fabric has been perfected on a mannequin or dress stand, it can be removed and pinned to a special design table connected to the computer, which can also act as a digitizing table. A cordless stylus is used to trace the outline of the piece of fabric and add any internal markings; the shape is then displayed on the computer screen (Figure 14.6).

FIGURE 14.4
Digitizing. (Courtesy of Gerber Technology Ltd., www.gerbertechnology.com.)

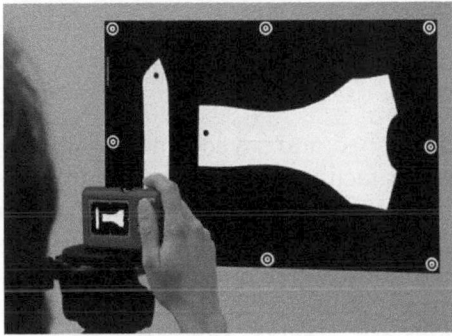

FIGURE 14.5
Digital photography. (Courtesy of Gerber Technology Ltd., www.gerbertechnology.com.)

FIGURE 14.6
Gerber's silhouette table. (Courtesy of Gerber Technology Ltd., www.gerbertechnology.com.)

This hardware, known as a silhouette table, facilitates the interaction of designers with the PDS, as it also enables them to work on full scale patterns by using conventional pattern cutting tools, as opposed to working on pattern shapes scaled on the screen. In addition, patterns can be drafted directly on the silhouette table with conventional tools, and again, results are presented visually on the screen. Furthermore, patterns can be drafted directly on the computer screen by using the drafting functionality within the PDS software.

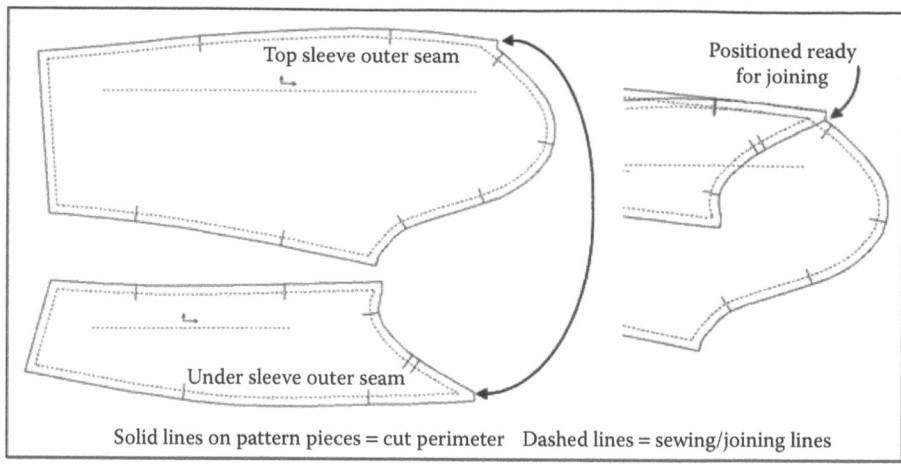

Solid lines on pattern pieces = cut perimeter   Dashed lines = sewing/joining lines

FIGURE 14.7
Square corners generated at the outer seam and patterns positioned ready for joining.

Once the patterns are established and saved in the computer, seam allowances can be added; in addition, corners to facilitate make-up (or joining of the garment parts) can be applied. The software will consider the width of seam allowance, the angle of the area of the pattern where the corner is to be applied, and the type of corner. Furthermore, if the patterns are graded, these considerations will be undertaken on each size in the range, meaning that the corner shape could change on different grade sizes, all within one pattern piece. Moreover, corners can consider and match-up with corners on separate patterns pieces, should one corner change, the other corner will adjust to reflect the change. For example, a two-piece sleeve typically has a square corner at the sleeve head, enabling the two pieces to be placed in the correct location in preparation for joining (Figure 14.7).

14.4.2 2D CAM Grading

As Wainwright (1992) identified, time can be greatly reduced using CAM methods for grading. Within CAD/CAM systems, grade rule tables can be created and saved in the computer. These inform the system of the base size and size range. X and Y increments can be input for each specific movement or rule, and each rule is given a unique number. These rules can be used time and time again. When paper patterns are being input via a digitizer, different rule numbers can be assigned to the cardinal points during the digitizing process; thus, when the pattern piece is viewed on the PDS screen, the piece will be fully graded (Figure 14.8). Within PDS, there is extensive grading functionality which emulates manual grading; should grading need to be adjusted or applied to pattern pieces that have not been graded, these grade rules and any of the extensive grade functions can be applied with great speed and accuracy. In addition, as pattern pieces can be stored with grading embedded within them, during pattern manipulation to create new styles, the system will automatically adjust the grading according to the modifications made to the pieces. Many functions in PDS can be applied to countless pattern pieces simultaneously; should pieces have grading applied, the modifications will not only be made to the base size but to all sizes in the range. This demonstrates the advantage of speed over manual methods.

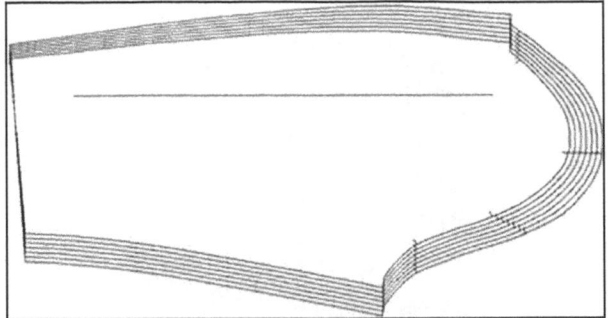

FIGURE 14.8
Graded top sleeve (hidden seam allowance for clarity).

14.4.3 2D CAM Costings

Incorporated within PDS is the ability to automatically generate a costing lay plan. All the pieces to make a garment are grouped together, different fabric codes that represent all the fabrics required to make the garment are assigned and the number pieces of each pattern are specified. The group of patterns is then opened in PDS. Sizes and the width of the fabric are indicated along with the code for the specific fabric and pieces for the particular fabric are positioned in the lay plan in a matter of seconds. This process is repeated for each fabric required for the garment. The lay plan(s) can be saved in the system for future reference (see Figure 14.10 for a lay plan discussed in Section 14.4.6).

14.4.4 Alterations

In addition, there is the ability to generate altered patterns, such as short, regular, and long versions of a style. Special point numbers are assigned to key points of the pattern, which could be undertaken during digitizing or in PDS. Editable forms within the system are created to move the points a specific percentage and saved, meaning that the actual amount for the alterations can simply be input according to the adjustment required at the time. Pattern pieces will adjust automatically. Furthermore, MTM software is also available, similar to alterations, but with greater movements and options. Typically, this software lends itself to men's tailoring. Customers measurements are input, rules, previously created in tables/editable forms, identify the closest size in the range of graded patterns to the customer's measurements; and distribution of the adjustments automatically occurs. Measurements can be input manually or extracted from body scan data, and a lay plan and cut data can automatically be generated. Brooks Brothers in New York has been providing this service for its customers for many years; they have an in-house body scanner and utilise Gerber Technology's MTM software. The first human intervention would be to spread the fabric to the length of the lay plan to be cut.

14.4.5 Body Scanning

The body scanner is a piece of equipment that captures the shape of a human by using lights or lasers and cameras. Since its introduction in the 1990s, advances in technology, their accuracy of collecting data, and lower financial outlay have led to increased availability. The scanning system consists of a booth or scanning assembly, a motor controller, and a personal computer. Generally, it requires a subject to enter the booth,

FIGURE 14.9
Body scanner. (Courtesy of Size Stream, www.sizestream.com.)

dressed in close-fitting undergarments. However, developments in scanning technology are being introduced where a subject can be scanned without removing clothes. The subject is required to stand or sit at a specific location within the booth (Figure 14.9). A series of lights or lasers, located at the corners of the booth, are projected, while cameras detect and capture in a matter of seconds the X, Y, and Z points from the surface of the body shape. The computer triangulates these points, which are generally spaced 1–2 mm apart, to form a 3D point cloud that represents the 3D human body. This raw data are then processed, and predefined body measurements can be extracted. Body scanners can output a number of file formats that can be imported into numerous CAD applications such as the aforementioned MTM CAD systems and 3D CAD for the creation of an avatar (see Section 14.5). To extract body measurements, body scanners are able to automatically identify landmarks on the body, or user defined landmarks can be positioned before scanning. In addition, to enhance the point cloud data, a virtual surface (a skin) can be added and virtual lighting used to present a visual image of the subject.

The clothing industry had been working with body measurements collected in the mid-twentieth century to create their block patterns and garments. However, due to the speed in collecting sizing data through body scanners, in 2001 the first ever UK National Sizing Survey was undertaken, the project was known as SizeUK. SizeUK captured the shape of 11,000 male and female subjects between 16 and 95 years of age. Size and shape analysis was conducted with data being available to the industry. From this data tailor's dummies have been produced that replicate not only average measurements but also the changing shape of the body within different age ranges. Since 2001, SizeUK has formed the basis for numerous sizing surveys in other countries.

14.4.6 2D CAM Production Lay Plan/Marker

Creating lay plans on a computer far outweighs manual methods speed, particularly as lay plans can be several meters long. The manual method was to draw round each pattern piece individually on a length of paper; very often, only one pattern piece would be produced where a pair is required, this one piece would have to be flipped over before being drawn around again. In contrast, the computer operator will have every pattern and size available on screen and can quickly slide these into position. Again, there is extensive functionality within this software and manual operations emulated. Several patterns can be positioned with one movement of the mouse or patterns grouped to ensure that they are positioned together. In addition, with a few clicks of the mouse, the width of the fabric can be adjusted and patterns repositioned, and lay plans can easily be attached together or split. Extensive check, stripe, and pattern matching can be undertaken, this entails special indications being added to the patterns in PDS or incorporated while digitizing the patterns into the system. Checks, stripes, or patterns are set up on the lay plan, to which the patterns pieces will automatically adhere. Moreover, should several hard copies of a lay plan be required, further copies can be plotted out within minutes. Again, accuracy and speed are advantages of the system. Figure 14.10 illustrates a lay plan of a jacket.

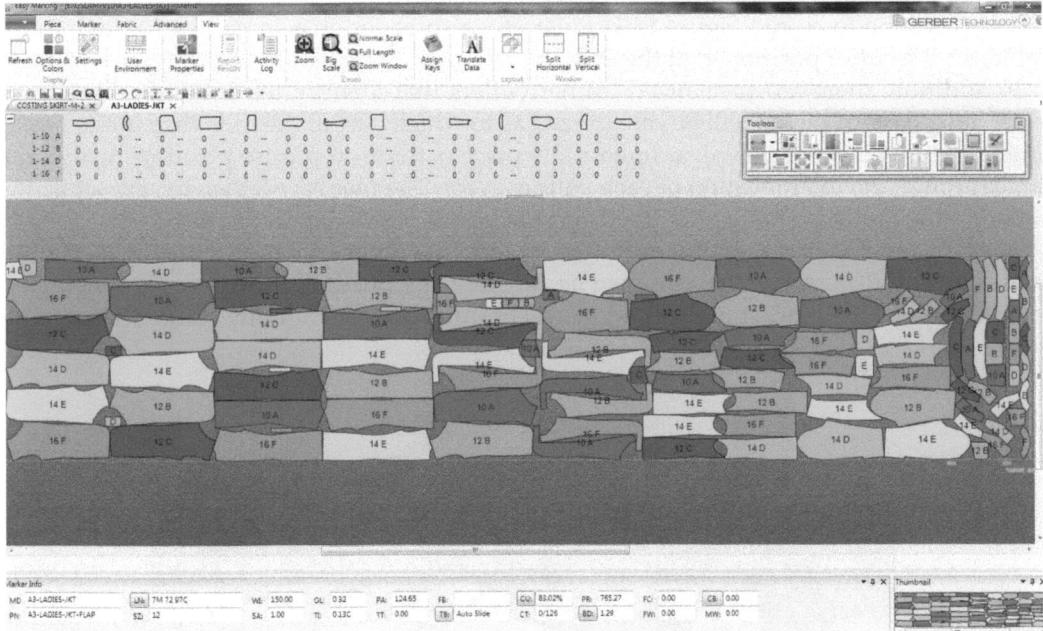

FIGURE 14.10
Lay plan of a jacket.

14.4.7 2D CAM Spreading and Cutting

As described earlier for production, fabric is spread to the length of the lay plan to be cut, with further lengths spread on top. A printed copy of a lay plan can be positioned on top of the final ply and the pieces cut ready for production. However, pulling fabric from a roll for several meters can stretch and distort it; therefore, special spreaders are utilized to feed the fabric from the roll at a controlled speed while the spreader travels the length of the table. The spreader continues to travel, applying further plies until the desired amount is achieved. Stretch fabric is particularly challenging. When spreading, care is paramount, and it will need to be allowed to relax for several hours before it is cut, ensuring that cut parts will be the correct size for manufacture.

Many companies have automatic bulk cutters; once the lay plans have been produced, the data can be sent directly to these cutters. There are many cutters on the market, ranging from single-ply cutters to multi-ply cutters, and from cutters that will cut delicate fabric for lingerie through to heavy-duty cutters for fabric such as denim and kevlar®. Single-ply cutters come in different sizes; numerous design rooms utilize smaller ones to cut parts for their prototype samples, whereas multi-ply cutters can accommodate not only the specific characteristics of the fabrics but also differing height of plies.

Companies can have several tables to layout the fabric before cutting, called spreading or laying-up tables. The cutter can be positioned at the end of these tables on dedicated tracks and moved into position at the end of each spreading table. The fabric (or lay), which has been laid on top of special perforated paper, is moved onto the cutter bed; the cutters have a vacuum to hold the fabric in place during the cutting process, and thus, a plastic sheet is positioned on top of the lay to provide suction for the vacuum. The lay is cut in sections, the cut pieces removed and the lay is conveyed forward, facilitated by a cushion of air, for the next section to be cut. Figure 14.11 illustrates a spreader on a spreading (or laying-up) table, with a cutter positioned at the end.

In addition, there are specialized leather cutters that identify the individual faults on each hide; pattern pieces will be categorized as to which part of the hide they are to be cut from. Integrated software will automatically place patterns in precise positions within the hide through the identification of each pattern piece. The pieces will then be cut, while the second hide faults are being identified and pieces are positioned in the software.

Further developments within software include the ability to apply images to pattern pieces within PDS. A lay plan can then be generated by utilising the extensive functionality available to the marker maker and then sent directly to be digitally printed.

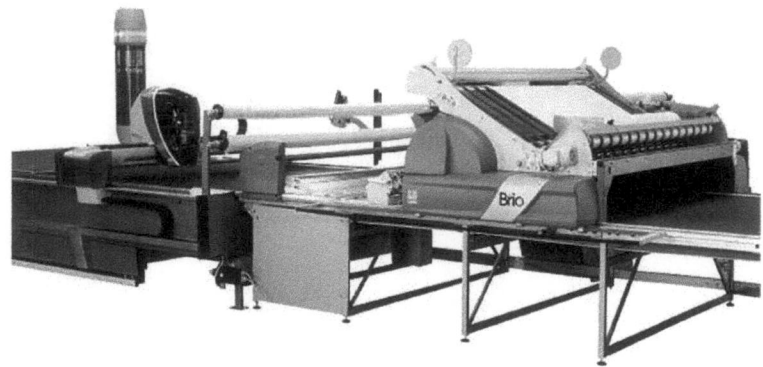

FIGURE 14.11
Spreader and cutter. (Courtesy of Lectra, www.lectra.com.)

14.5 3D CAD Garment Visualization

The 2D CAD software has become ubiquitous within the clothing industry, with many innovations since its introduction, whereas 3D software, utilized in film and game animation for many years, has been slow to be adopted within the clothing industry. However, unlike these industries, the clothing industry requires the garment to flow and react in accordance with the fabric properties being simulated, and garment simulation is one of the most technically demanding areas, with fabric modeling being highly complex due to its anisotropic nature. These complex algorithms and visual data simulation require extensive computing power. Advances in computing technology however, have led to the availability of 3D software within the design and product development arena.

The 3D CAD interacts with many technologies that are constantly emerging and developing. Garment visualization was initially performed on a virtual mannequin but now takes place on an avatar. The avatar can be created by modifying the extensive set of measurements on one of the avatars provided, or data captured through body scanners can be imported into the 3D software to recreate an exact shape of a human being.

Fabric selection is a key area within the design of a garment, and thus, 2D patterns are assembled and fabric images applied to specific patterns. Fabric properties of mass, thickness, bend, and stretch are associated with each fabric image to reproduce a realistic look of a garment and produce fit analysis. Figure 14.12 illustrates fabric test result being input.

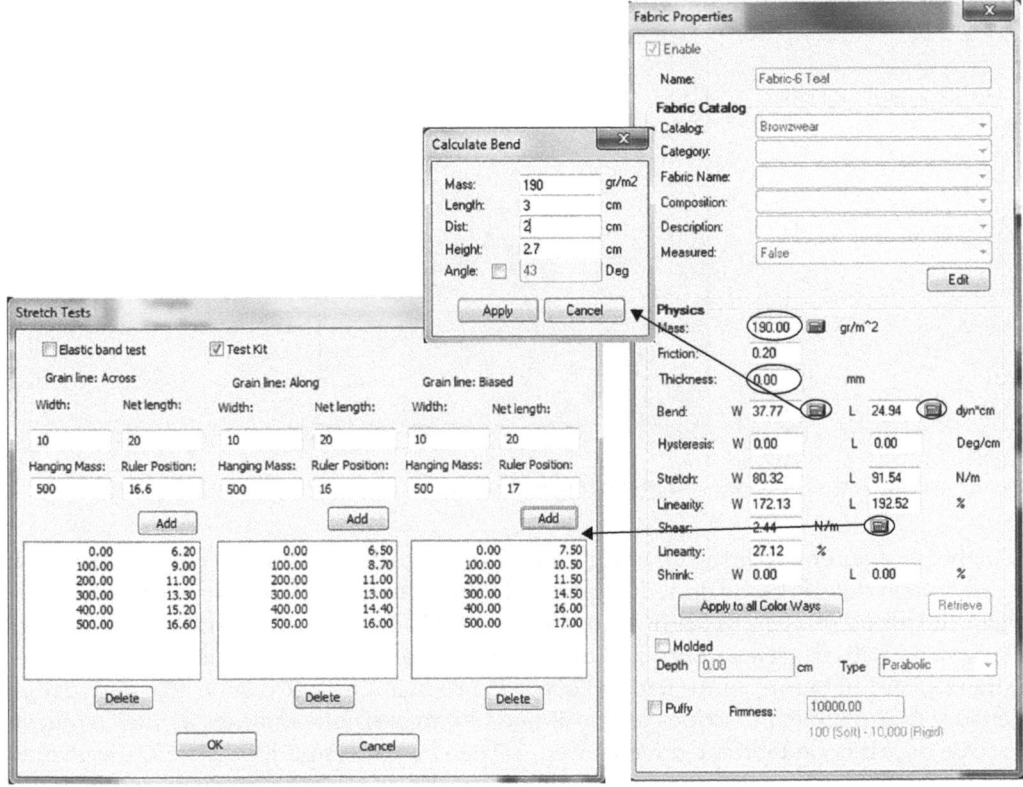

FIGURE 14.12
Fabric test results being input into Browzwear software.

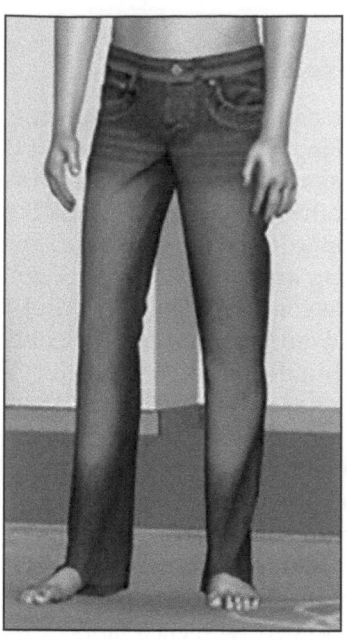

FIGURE 14.13
Jean details.

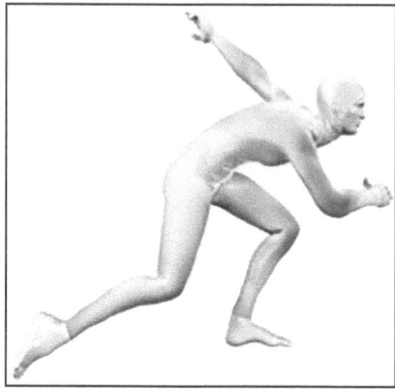

FIGURE 14.14
Pressure map. (Courtesy of Browzwear, www.browzwear.com.)

Further images enhance the virtual garment with logos; trims such as buttons, studs, and etc. and stitch details being added. Figure 14.13 illustrates a pair of jeans with fabric, stud, rivets, and stitch images, providing an authentic look. The visual display enables personnel involved with design and product development the opportunity to assess proportions, aesthetics, and fit before a tangible prototype is produced. Fabric can be adjusted visually to ensure that plaids and stripes are positioned in an aesthetic manner, while changes in the scale of prints on fabrics can be viewed, 2D can be patterns adjusted, and the garment can be amended virtually. Pressure mapping can be applied to garments, a color scale identifies the pressure of the garment on areas of the body (Figure 14.14 illustrates an avatar in an athletic pose, with pressure mapping applied). All of which allows early decisions

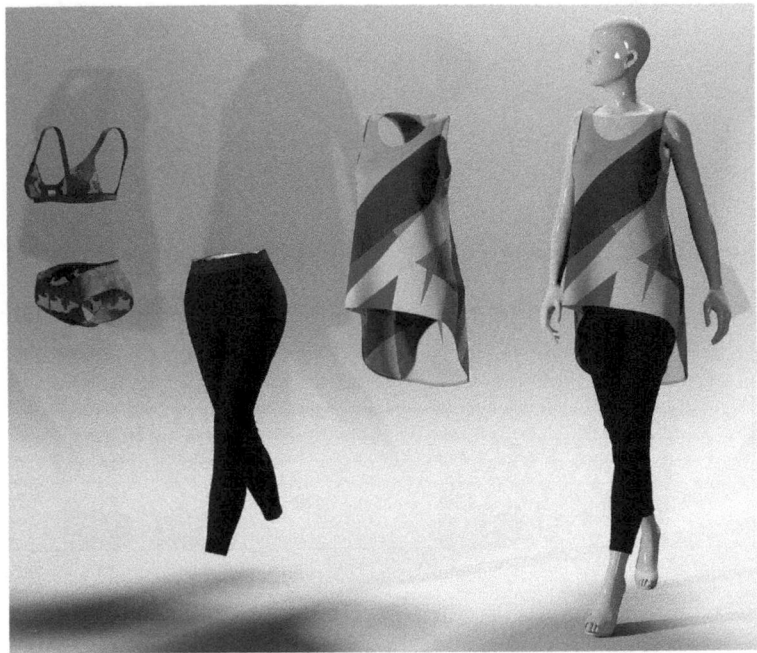

FIGURE 14.15
Image of 3D. (Courtesy of Human Solutions, www.human-solutions.com.)

to be made in the product development process. Furthermore, benefits of buyers placing orders against garments viewed on screen are now being realized. Moreover, 3D catwalk effects can be created and utilized as marketing tools.

Links with body scanners and sizing surveys have enabled companies to generate their own virtual fit models, enabling them to visualize garments on these avatars. Benefits are that many virtual fit models can be created, their shape will not change, and they are available 24/7. In addition, personal MTM clothing can be realized with further linking of body scanners, 2D software, and 3D software. The 2D pattern data and body landmarks are noted and saved in 3D format on an avatar. When a new body scan is generated with corresponding landmarks, these can morph the saved avatar, including the 3D pattern shapes, which can be returned to a 2D format, from which cloth can be cut and constructed into the personalized garment. A further image of 3D software can be viewed in Figure 14.15.

The 3D software is an area that will continue to grow as greater numbers of students are being introduced to the software, more companies are embracing it, and computers continue to increase in power.

14.6 3D Printing

The 3D printing is another emerging technology being investigated by clothing personnel. New materials and the reduced costs of printers are making this technology more accessible. Peleg graduated in 2015, where she 3D printed a collection for her final degree

FIGURE 14.16
Top and skirt from Peleg's collection. (Courtesy of Danit Peleg, www.danitpeleg.com.)

FIGURE 14.17
Peleg's dress and jacket. (Courtesy of Danit Peleg, www.danitpeleg.com.)

(Figures 14.16 and 14.17). Subsequently, she was invited to create the 3D printed dress worn in the Rio 2016 Paralympic Opening Ceremony. To ensure accuracy and fit of this garment, Peleg captured the athlete's 3D body shape by using an app called Nettelo and utilized Gerber Technology's 2D and 3D solutions.

14.7 Applications

This section provides examples of utilising CAD/CAM technology available today. It discusses creating and moving 2D pattern data through to cut data and the results, followed by an example of digital printing.

14.7.1 Example of Using the Gerber's Accumark Software and Single-Ply Cutter

York selected a red leatherette fabric for her top and skirt (Figure 14.18); she wanted the perimeter and the intricate internal shapes to have clean-cut edges and employed Gerber Technology's Accumark 2D software and their single-ply cutter.

First, York experimented with paper to develop her desired look (Figure 14.19). Then, she digitized her paper patterns into the 2D CAD system and created a lay plan (as described in Sections 14.4.1 and 14.4.3). She utilized the functionality within the software to generate data that would drive the cutter. As the blade on the single-ply cutter is circular, trial cuts were conducted, and the internal lines within the pattern were adjusted in PDS to ensure that a clean cut was achieved, with no over- or undercuts. Figure 14.20 illustrates York's data for her top in PDS—the solid line is the perimeter, and dashed lines are the intricate internal lines.

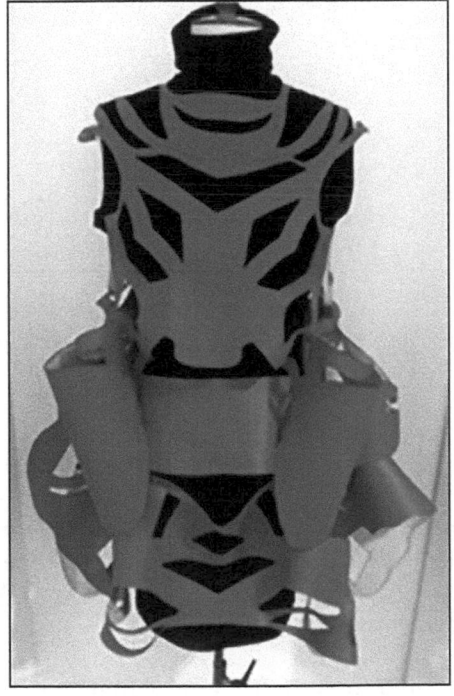

FIGURE 14.18
Top and skirt. (Courtesy of Amy York, Manchester Metropolitan University graduate.)

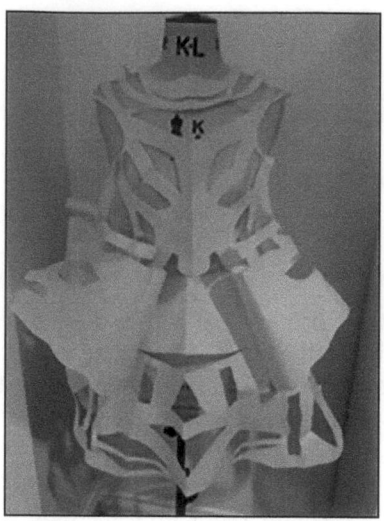

FIGURE 14.19
Paper experimentation. (Courtesy of Amy York, Manchester Metropolitan University graduate.)

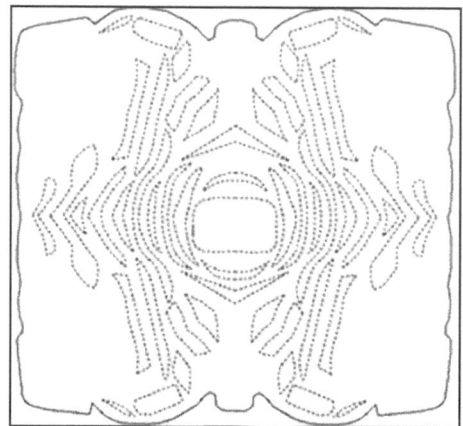

FIGURE 14.20
PDS data of top. (Courtesy of Amy York, Manchester Metropolitan University graduate.)

14.7.2 Example of Using a Laser Cutter 1

Humberstone's skirt has strategically positioned seams and laser-cut detail repeated around the hemline to form a continuous boarder effect (Figure 14.21). CorelDraw® is the software that drives the laser cutter; however, Humberstone created her lace detail in Adobe Illustrator® and her skirt pattern in Gerber Technology's Accumark software. Thus, the process of exporting and importing data through the various software packages was required.

The skirt pattern was drafted directly from measurements in Gerber's Accumark PDS (as described in Section 14.2.2). The seams on the skirt were planned to accommodate the laser-cut lace detail before seam allowances were added.

FIGURE 14.21
Complete outfit. (Courtesy of Lucy Humberstone, Manchester Metropolitan University graduate.)

Humberstone had created her cut detail in her preferred drawing package, Adobe Illustrator®; therefore, the skirt panel was exported via the functionality in the Accumark software as a .dxf file (Drawing Exchange Format) and imported into the drawing package for the laser-cut detail to be repeated and positioned at the hemline of the panel. The skirt panel was then exported from the drawing package, again as a .dxf file, and imported into CorelDraw®, the software that drives the laser cutter, where the cut data were created. The final CorelDraw® file was a .cdr file.

As there was a limited bed size on the laser cutter, a review of the fabric revealed that the panel could be cut on the fold. Therefore, data were created to ensure that the waist, panel side seam, and lace detail at the hem would be laser cut, but not a fold line of the center of each panel. Cut data are presented in Figure 14.22, indicating the lines to be cut.

As the panel side seam was laser cut, this sealed the edges and eliminated the need to neaten the seam allowance during construction, producing a very clean finish to the inside of the garment.

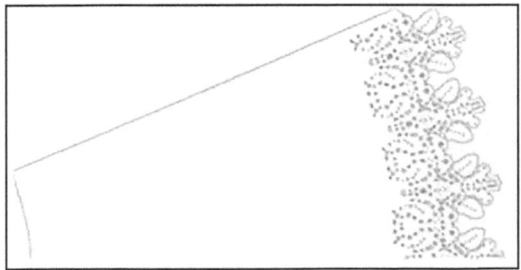

FIGURE 14.22
Cut data. (Courtesy of Lucy Humberstone, Manchester Metropolitan University graduate.)

14.7.3 Example of Using a Laser Cutter 2

Fellows was creating her leather and fur coat (coat back detail illustrated in Figure 14.23). The silver leather panels were etched to give further detail. Figure 14.24 illustrates details of the leather. Again, Fellows employed Gerber Technology's Accumark software, Adobe Illustrator®, and CorelDraw®.

Once Fellows had established the desired shape and fit, the patterns were digitized into Gerber's 2D CAD system (as described in Section 14.4.1); they were exported from the

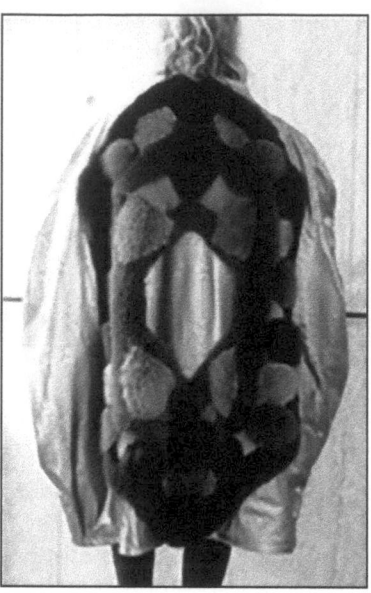

FIGURE 14.23
Back view of coat. (Courtesy of Maizie Fellows, Manchester Metropolitan University graduate.)

FIGURE 14.24
Detail of the leather etching. (Courtesy of Maizie Fellows, Manchester Metropolitan University graduate.)

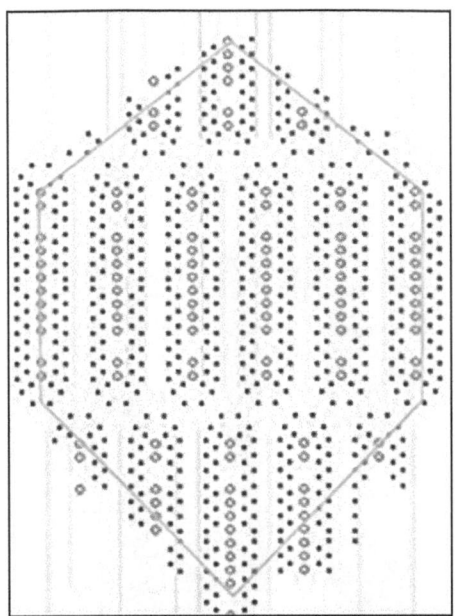

FIGURE 14.25
Data of etching detail in Adobe Illustrator. (Courtesy of Maizie Fellows, Manchester Metropolitan University graduate.)

Accumark software via its functionality as a .dxf file and imported into Adobe Illustrator®, where Fellows positioned the detail on the leather panels (Figure 14.25). The panels were exported, again as a .dxf file and then imported into CorelDraw®, the software that would drive the laser cutter to produce the etched pattern. Again, the final CorelDraw® file was a .cdr file.

14.7.4 Example of Digital Printing

Digital imagery on garments is becoming increasingly popular. Zapp wanted to portray the images she that captured around the world and display them on garments (Figure 14.26). Gerber Technology's Accumark and Adobe Photoshop® software were employed.

Working with Zapp and liaising with the digital printers, patterns were developed and graded sizes were created in Gerber's PDS. In addition, an allowance to accommodate a bleed of color during the printing process was added round the perimeter of the patterns. Each pattern's piece and size, including the 2-cm bleed allowance, were exported from PDS as a .dxf file and imported into Adobe Photoshop®, where Zapp's images were strategically positioned. Figure 14.27 illustrates the initial image, a graded front pattern, a front pattern with bleed allowance, and a front pattern with the image positioned.

Utilizing the functionality within Adobe Photoshop®, a lay plan was created and the resulting .psd file was sent to be digitally printed (Figure 14.28).

FIGURE 14.26
Image of final garment. (Courtesy of Andrea Zapp, www.andreazapp.com.)

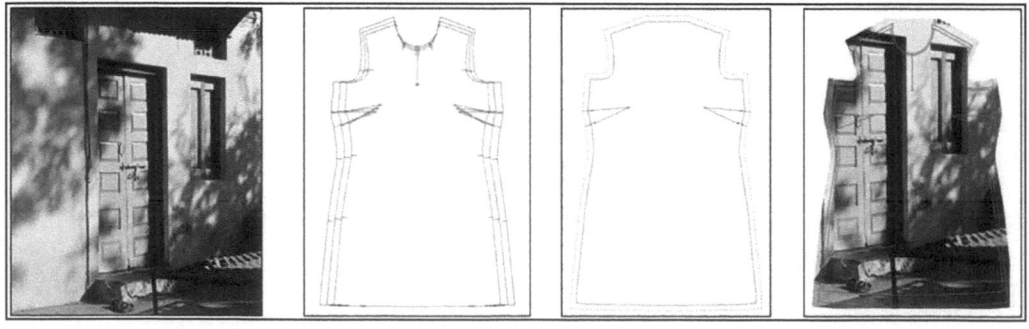

FIGURE 14.27
From initial image to a front pattern with positioned image. (Courtesy of Andrea Zapp, www.andreazapp.com.)

FIGURE 14.28
Lay plan created in Photoshop.® (Courtesy of Andrea Zapp, www.andreazapp.com.)

14.8 Summary

This chapter has described briefly some of the technologies available to personnel in product development and the start of manufacturing areas. Such technologies are constantly evolving and, as computing power increases, these boundaries will continue to be pushed. Further information can be gleaned by investigating published and online material, and a starting point can be found in bibliography (and sources of further information) of this chapter.

14.9 Try to Answer the following Questions

- What stages are encompassed within the product development area?
- In a typical life cycle of a garment, what is the percentage of the product development area?
- What are the methods for creating a set of pattern pieces for a prototype garment?
- How many methods of grading are there, and what are they called?
- When grading a pattern, what are the key points called?
- What is the other name for a marker?
- What is the typical percentage of fabric within the cost of a garment?

- What does CAM stand for?
- What does PDS stand for?
- What does .dxf stand for?
- What are spreading or laying-up tables?
- How is fabric prepared for bulk cutting?
- In the example given here, which software drives the laser cutter?

Bibliography

Gray, S. (1992) *The Benefits of Computer-Aided Design and Manufacture,* pp. 1, 3, 51. London: The Design Council.

Soni, G. and Kodali, R. (2010) International benchmarking for assessment of supply chain performance, *Benchmarking: An International Journal,* 17(1), 44–76.

Taylor, P. J. and Shoben, M. M. (1990) *Grading for the Fashion Industry.* (2nd ed.) Cheltenham, UK: Stanley Thornes (Publishers) Ltd.

Teng, S. G. and Jaramillo, H. (2006) Integrating the US textile and apparel supply chain with small companies in South America, *Supply Chain Management: An International Journal,* 11(1), 44–55.

Web Site Resources

Andrea Zapp (digital printing), www.andreazapp.com.

Browzwear, www.browzwear.com.

Danit Peleg (3D printing), www.danitpeleg.com.

Defra. (2011) Sustainable clothing roadmap progress report 2011 [online], https://www.gov.uk/government/uploads/system/uploads/attachment_data/file/69299/pb13461-clothing-actionplan-110518.pdf (accessed July 12, 2016).

Gerber Technology Ltd, www.gerbertechnology.com.

Human Solutions, www.human-solutions.com.

Lectra, (www.lectra.com).

Nettelo (body scanning), www.nettelo.com/

Optitex, www.optitex.com.

Rodriguez, A. (2001) Are you lean in product development? *Apparel Sourcing* [online], www.themagazineapparelsourcing.com/magazine/are-you-lean-in-product-development (accessed June 2, 2016).

Size Stream (body scanning), www.sizestream.com.

Tc[²] (body scanning), www.tc2.com.

Tukatech, www.tukatech.com.

Vitronic (body scanning), www.vitronic.com/industrial-and-logistics-automation/sectors/3d-body-scanner.html.

15

Stitches and Seams

John McLoughlin

CONTENTS

15.1 Introduction

For the past century, the clothing industry has been dominated by one simple and long-lived tool, the basic sewing machine. Its importance can be described as not only facilitating entry into the industry but also virtually dictating its structure. The past two decades have seen the revolution of the microprocessor (the silicon chip), on which much of our existing technology depends. This small component has been largely responsible for developing the sewing machine into one of the most versatile and intricate pieces of engineering used in industry today. As the sewing machine has continued to prosper, other types of equipment have evolved along with it. Most clothing companies today will have within their walls a combination of progressive equipment of latest specification and design. Methods of production have advanced to meet the ever-increasing needs of the consumer. Factories have to be lean, mean, and, above all, disciplined in order to survive in an ever-increasingly competitive world. Companies must be able to change and adapt to the market place quicker and more efficiently and have quicker response times if they are to survive.

To have a successful clothing industry, you also need the resources necessary for the industry to survive. This involves not only technology that the business needs but also the most important resource of all, which is the workforce. Without a well-disciplined and well-trained workforce, you deprive the business of the skills and expertise needed to make the business successful.

15.2 The History of Sewing

The use of clothing dates back to the history of mankind, the main purpose being the protection against the elements, and also to the development of civilization, adornment, and fashion. Early methods of construction utilized methods of lashing/joining pieces of skin together, which can be compared to stitching used today. Primitive sewing tools such as awls and needles made from bone, fishbone, and spines from plants, wood, and bronze give evidence of sewing techniques of prehistoric times.

It was not until the middle ages that sewing needles were made from steel and the first efforts to sew with a machine are reputed to be in 1750 by an Englishman named Charles Wiesenthal.

In 1790, a man named Thomas Saint applied for a patent to build a wooden sewing machine that produced a chain stitch.

In 1810, a German hosier named Balthasar Krems devised a machine capable of sewing the edges of caps by using a chain stitch. In 1830, a Frenchman named Barthelmy Thimmonnier built 80 wooden sewing machines; apparently, all of them were capable of producing efficient stitches. However, his success was brief. Tailors, worried about their livelihood, completely demolished his shop and tools. He barely escaped with his life. This was the start of the production revolution.

Wooden machines started to be replaced by metal ones, and in 1850, there were approximately 104,000 machines in the United States alone. The principles of the sewing machine have hardly changed to this day, but the technology has changed significantly (Figure 15.1).

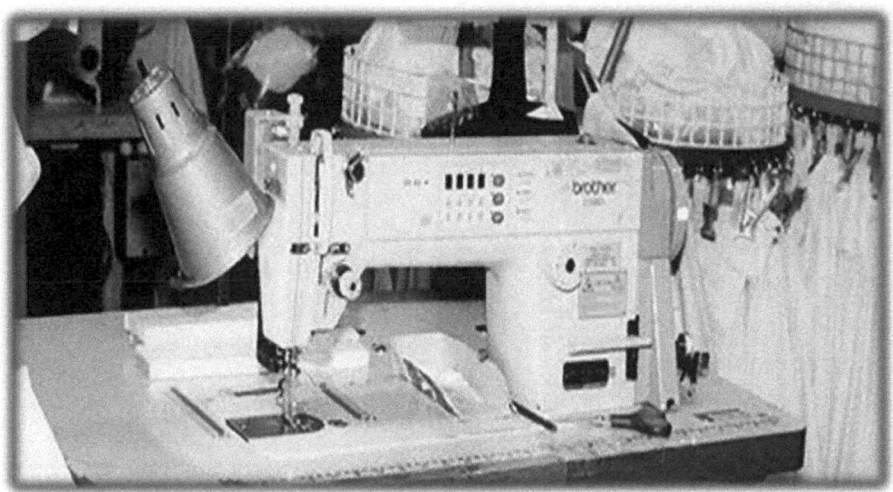

FIGURE 15.1
A modern-day industrial sewing machine.

The machine illustrated in Figure 15.1 is commonly known as an *integrated stitching/ sewing unit (ISU)* and is called so because it has many different functions that make life easier for the machinist.

15.3 Examples of Sewn Products

For a moment, let us consider how diverse the sewn product industry is. In the automotive industry, we make cars. In the maritime industry, we make boats, yachts, and ships; in the military, we make warships, armored vehicles, and fighter aircraft; and in the fashion industry, we make clothes, underwear, shoes, and so on. All of these industries use textiles in one form or another. All of these industries use some form of joining these materials together, the most common form being stitching by using sewing threads. Welding and bonding have also become more common in joining materials together, particularly synthetic fabrics.

15.4 Sewing Machines and Categories

Basically, the sewing machine is categorized into four main types, very often called divisions. These are as follows:

- The basic sewing machine—ISU
- Mechanized stitching units—long and short cycles
- Semiautomatic machines
- Automatic transfer lines

15.4.1 The Basic Sewing Machine

The basic sewing machine (Figures 15.2 and 15.3) will consist of the following:

- The sewing head
- The machine bed
- The work top
- The stand
- The drive motor
- The treadle

The ISU is so called because it provides many useful purposes that help reduce operator fatigue, while at the same time improving the efficiency of production. These machines (Figure 15.8) are called basic units, as described earlier, but there is nothing basic about them. Most have built-in computers that produce a variety of time-saving cost-effective

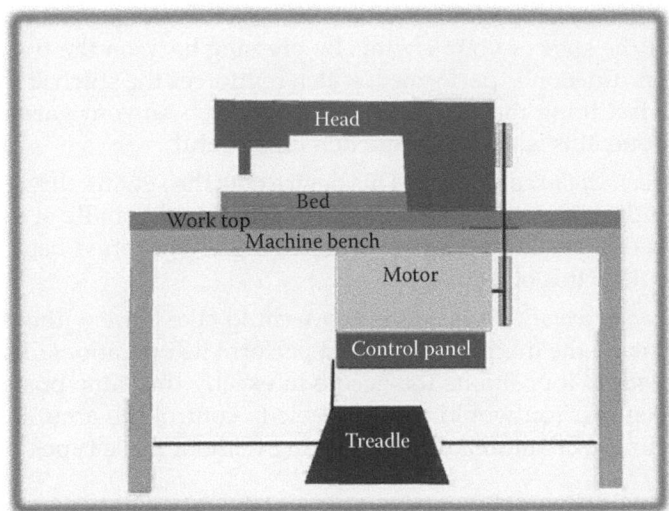

FIGURE 15.2
Schematic diagram of machine.

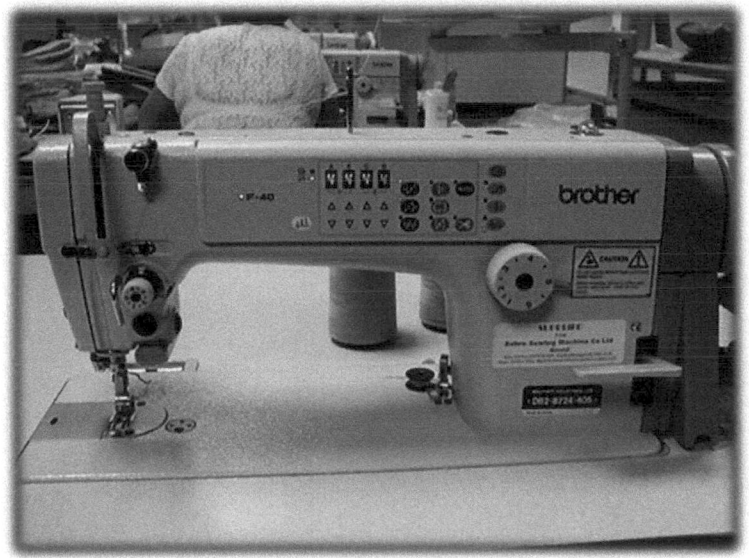

FIGURE 15.3
Computer-controlled sewing machine.

functions. Figure 15.9 shows the various components that the electronics in the machine can control. These include the following:

- *Automatic foot lift*: Pressing back with your heal on the treadle causes the presser's foot to rise automatically.
- *Automatic back tack*: You can set the machine, so that when you press the foot treadle to start the machine cycle, the machine will feed forward a set number of

stitches and then reverse a set number of stitches to reinforce the stitch area at the beginning of the start of your sewing. By pressing back on the treadle with your heal, a similar function is performed, which reinforces the stitch at the end of your sewing. The last thing that you want is for the stitch on your garment to unravel or come undone; this is why this function is so useful.

- *Automatic underbed thread trimmer*: This device cuts the sewing threads without the need to use scissors. Again, this function is performed usually at the end of your sewing operation on the seam you have just sewn. You press back firmly on the treadle to achieve this objective.

- *Needle-positioning motors*: None of the above could take place without these special motors that make the machine move and perform its operations. These machines use a *synchronizer*. It positions the needle in exactly the same position each time the machine stops. You would never be able to control the accurate trimming of the sewing thread or automated back tacking without these types of motors.

Figure 15.4 shows the vital components for this machine to produce a lockstitch.

Check out the diagram below and try to memorize some of the components, as this will help later on.

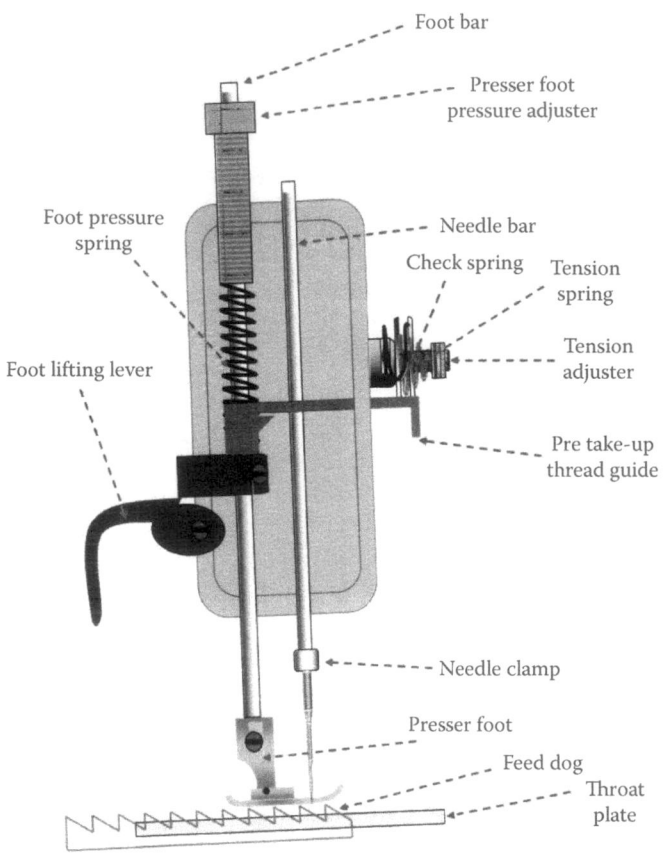

FIGURE 15.4
Stitching components.

15.4.2 Mechanized Sewing Machines

Mechanized sewing machines are machines that sew a predetermined stitch line or pattern but still require the machinist to place parts underneath the machine. Examples of these machine operations are as follows:

- *Bartacking (Figure 15.5)*: For example, belt loops on jeans
- *Button sewing (Figure 15.6)*: Trousers, skirts, and so on
- *Button holes (Figure 15.7)*: Trousers, jackets, skirts, blouses, shirts, and so on
- *Profile stitching (Figures 15.8)*: Pocket flabs, designs on the pockets of denim jeans, and so on

FIGURE 15.5
Bartack machine.

FIGURE 15.6
Button sewing machine.

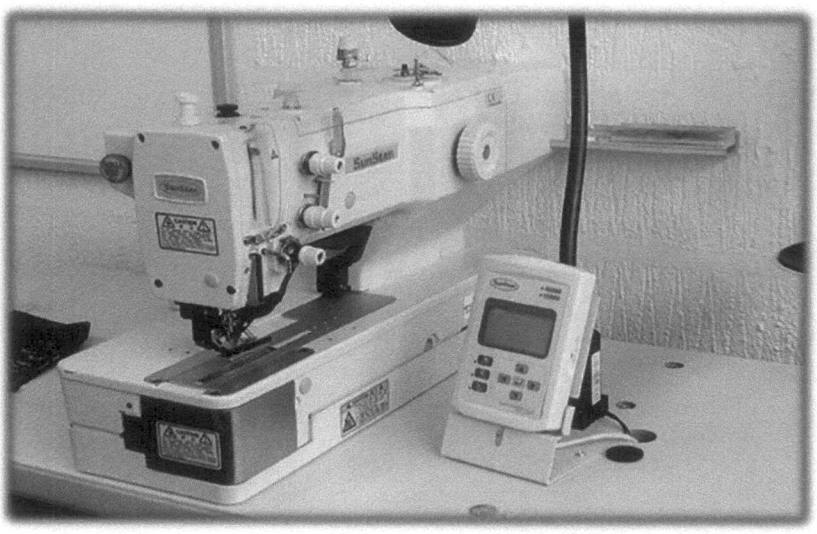

FIGURE 15.7
Button hole machine.

FIGURE 15.8
Profile stitching machine.

15.4.3 Semiautomatic Machines

These offer many features similar to mechanized machines, but they are electronically controlled and are more flexible than their mechanized counterparts. Parts are still entered by a human operator and sewing takes place using an x-y stepper motor arrangement, which can be programmed to give different stitch profiles, similar to the profile stitch machine mentioned previously. However, component parts of the garment can be much larger, and there is usually less human intervention than with mechanized machines. Examples of these machine operations are as follows:

- Embroidery machines (Figure 15.9)
- Semiautomatic leg seamers (Figure 15.10)

If we consider the fact that electronics and automation have to some extent taken over our lives and influence our world, particularly in the twenty-first century, technology has enabled the speeding up of production in the apparel industry. We rely on fashionable clothing to make us look good. However, this comes at a price. Companies are investing in automated technology to speed up production, product quality, and profitability. Sewing machine automation is often described as a process by which machine processing of a product is achieved with little or no human intervention. Other descriptions include a product that is machine driven. Despite these definitions described previously, we are still not able to fully automate clothing production to the same extent as the production in the automobile and engineering industries. The reasons for this are given later in this chapter.

FIGURE 15.9
Embroidery machine—flat design.

FIGURE 15.10
Semiautomatic leg seamer.

Figure 15.9 shows a 12-headed embroidery machine producing the same pattern along the length of the fabric. The more heads on the embroidery machine, the more production you can produce, and hence, the more cost-effective the product.

15.4.4 Automatic Transfer Lines

Machine-building companies have, for years, been trying to move toward fully automated environment, trying to develop complete automated production systems. Robotic systems for manufacture have been developed, but these have proved very difficult to implement. Why? A major reason why they have not been successful is because it is very difficult for robots to handle limp materials such as apparel fabrics. Moreover, robotic fingers do not have the same dexterity as human fingers.

15.5 The Machine Bed

Sewing machines are classified by three distinct factors that differentiate the different types of machines for many types of seaming operations. These three factors are as follows:

- Machine bed type
- The feed system used
- The stitch type

For example, you can have a machine with a flat bed and a top and bottom feed system with a lockstitch. Alternatively, you can have a machine with a raised bed and a drop feed system with a chain stitch. Here, you would have the following options:

Machine 1

1. Flat bed
2. Top and bottom feed
3. Lockstitch

Machine 2

1. Raised bed
2. Drop feed
3. Chain stitch

The machine bed usually offers only one stitch type, the exemption being a combination stitch machines, where more than one stitch type are produced on the same machine. An example of this would be a five-thread overlock machine, where you have the overlocking stitch number 504 and one row of chain stitch number 401. These will be discussed later.

The machine bed is also determined by the volume of fabric that passes beneath and around the needle and the amount of movement required on this fabric (Figure 15.11).

It is primarily used where large open parts are to be joined and when more than just the seam allowance must pass to the right of the needle; it is easy to mount guides and markers on this type of bed (Figure 15.12).

It is used when parts are small or curved and when the operation is being performed on partially closed components (e.g., hemming a trouser leg) (Figure 15.13).

It is used when the machine operator needs to get close to the stitching area on small intricate items such as women's bra cups and so on (Figure 15.14).

By its very nature, the overedge machine requires very little clearance to the right of the needle, because excess fabric is trimmed by reciprocating blades as stitching takes place. The bed may be raised from the worktop in order to allow better handling around the work area for sewing smaller pieces, for example, pockets, underwear, and so on.

FIGURE 15.11
The flat bed.

FIGURE 15.12
The cylinder bed.

FIGURE 15.13
The post bed.

FIGURE 15.14
The overedge/overlocking machine.

15.6 Seam Definition and Properties

A seam joins more than one piece of material together, and the type of seam used depends on the product to be sewn. Seams in clothing have different levels of complexity, depending on the design of the product. Seams in other sewn products have similar levels of complexity.

If we take into consideration the different requirements of a seam on the outside leg of a pair of jeans compared with a seam used in joining the perimeters of an air bag, there may be:

- More consistent stress loading of a seam on an air bag due to an even air pressure.
- An uneven pressure on a jean due to leg movements.
- Stress/pressure on a seam of an air bag due to rapid inflation of the bag.
- A gradual build-up of stress loading on the seam of the jean.
- A difference in the fabric ply angles, which could alter the strengths of the seams. Accordingly, the angle of the ply means that it is the direction in which the fabric is cut.

There is one factor that both of these and other seams have in common; it is that the seams must combine the required standards of appearance and performance, while ensuring economy of production.

The aesthetic characteristics of a seam are usually influenced by the accuracy of the stitching, the visibility of threads of differing color to each other, and the surface of the fabric. Seam performance relates to the strength, extensibility, durability, security, and comfort and to the maintenance of specialist fabric properties.

Generally, the seam must exhibit similar properties to the fabric into which it is being introduced. In addition, depending on the physical properties of the area within a product, occasionally modify that product area.

The strength of a stitch line is assessed both longitudinally in the direction of the seam and transversely in perpendicular direction to the seam. It is assessed in terms of peak tensile load at break. The seam extensibility can be quantified simultaneously while applying the load longitudinally, and the seam *grin* can be measured while a seam is under transverse load.

Factors influencing the strength and extensibility of a stitch line seam include the following:

- Stitch density (stitches per centimeter)
- Thread tension (static thread tension setting on the sewing machine)
- Thread properties (strength and elasticity, determined by fiber type)
- Fabric integrity (cover or tightness factor)

The seam performance relates to the following:

- Strength
- Aesthetic appeal

- Extensibility
- Durability
- Ease of assembly
- Security
- Comfort in wear

The only British standard test in existence is the BS 3320:1998, "method for determination of slippage resistance of yarns in woven fabrics: seam method," often referred to as the grab test. Seam strength is often referred to as seam efficiency, which is an expression of seam strength as a function of fabric strength.

$$\text{Seam Efficiency} = \frac{\text{Un-seamed Fabric Strength} \times 100\%}{\text{Seamed Fabric Strength}}$$

The requirements of seam extension can vary dramatically with the product application; Comfort stretch (up to 30%) for casual garments and performance wear and support garments can exhibit extensions of more than 100%—power stretch.

The durability and security of the seam are directly related to these properties and the integrity of the sewing thread under normal use and laundering conditions.

Many types of seams can be constructed in ways involving different raw materials, stitch types, and machinery. An optimum combination of these, along with minimized thread consumption, will provide an economical approach to manufacturing.

There are hundreds of different seam types, so only the most common are discussed here. They are basically categorized into eight simple categories, which are described in the following subsections.

15.6.1 Class 1 Superimposed Seams

Class 1 superimposed seams are generally two or more plies of fabric laid on top of each other, superimposed in the same orientation. They are stitched near the edge with one or more rows of stitching. The rows of stitching may be sewn simultaneously or consecutively (Figures 15.15 through 15.18).

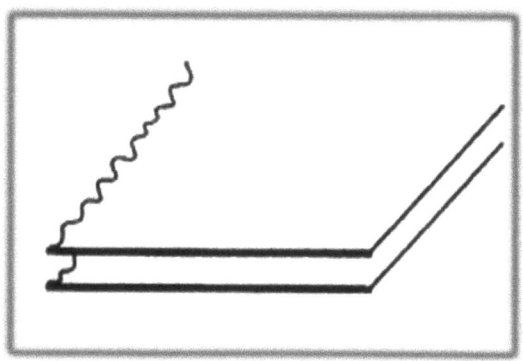

FIGURE 15.15
Superimposed seam.

FIGURE 15.16
Superimposed seam cross section.

FIGURE 15.17
Double-lapped seam.

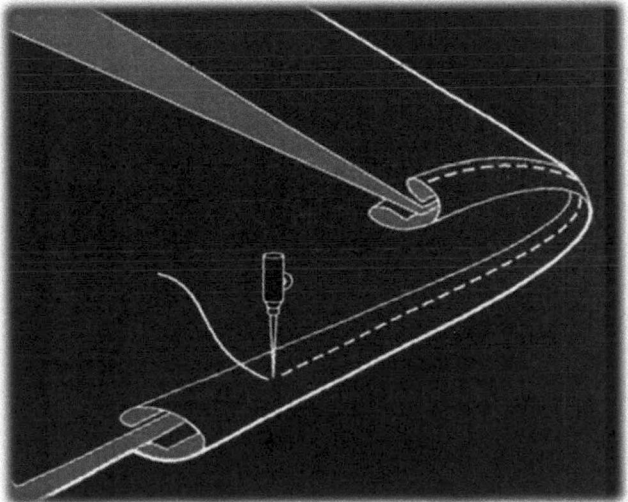

FIGURE 15.18
Bound seam.

15.6.2 Class 2 Double-Lapped Seams

In class 2 double-lapped seams (Figure 15.17):

- Two more plies are lapped together, overlaid, plain, or folded.
- These are secured with one or more rows of stitching.

- One of the most common is the double-lap felled seam, with two or more rows of stitching.
- This provides a strong seam with fabric edge protection.
- These are often used on the side seam of jeans, other denim products, and tents.

15.6.3 Class 3 Bound Seams

Class 3 bound seams (Figure 15.18) are formed by:

- Folding a binding strip around the raw fabric edge(s) and securing with one or more rows of stitching.
- This produces a secured neat seam on an edge often exposed to view or wear.
- These are often used on luggage (Figure 15.19).

15.6.4 Class 4 Flat Seams

Class 4 flat seams (Figure 15.19) are:

- Produced with a minimum of two plies of fabric butted together at the raw edge on the same level.
- A minimum of two rows of stitching are introduced simultaneously, which are joined on one or both surfaces.
- These are often used in knitwear to reduce bulk around the join, especially in underwear.

15.6.5 Class 5 Decorative Seams

In class 5 decorative seams (Figure 15.20):

- One or more rows of stitching are introduced to one or more folded or plain plies of fabric.
- These are often used to create effects such as pin tucks and channel seams.

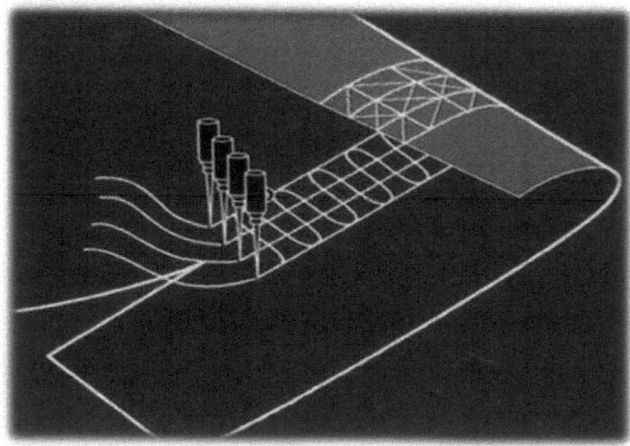

FIGURE 15.19
Flat seam.

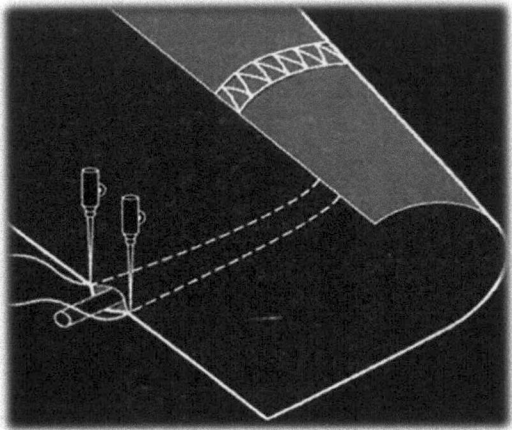

FIGURE 15.20
Decorative seam.

15.6.6 Class 6 Edge Finishing

- Line of stitches is formed at or over the edge of a plain or folded fabric ply(s).
- The simplest of these is serging, where the raw fabric edge is finished with a 500 class stitch.

15.6.7 Class 7

- These seams are constructed with one piece of base material in a similar manner to lapped seams but include an added component, which is limited on both sides (a braided tape or a lace trim).

15.6.8 Class 8

- These are classified as seams used on one ply of fabric, which is limited on both sides; this includes things such as belt loops.

15.7 Fabric Feed Systems

The principle of the feeding mechanism is to move the material from one stitch position to the next over the prescribed distance, given by the length of the stitch. In doing so, the fabric must be controlled precisely under minimum pressure. The feed system is made up of the following components (Figure 15.21):

- Throat plate
- Feed dog (feeder)
- Presser foot

These are often described as the fittings, because they fit together.

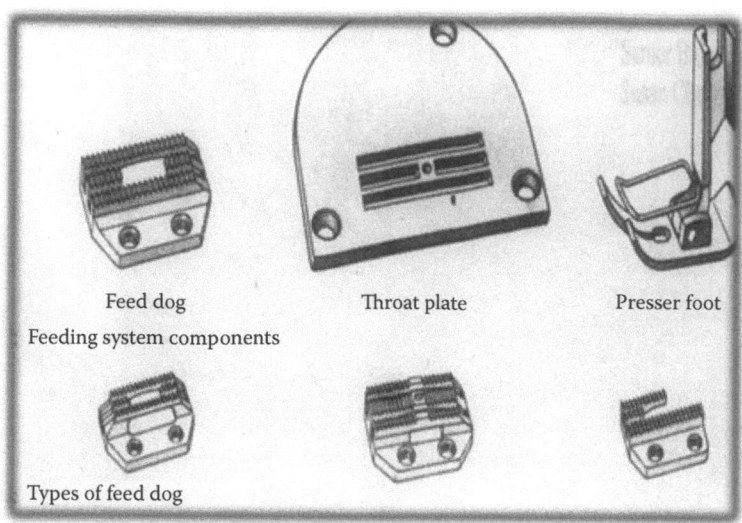

FIGURE 15.21
Fittings of the sewing machine.

15.7.1 The Throat Plate

The throat plates are designed to support the material being stitched and allow the easy passage of the material over its surface. The throat plate is manufactured with a needle hole—to allow the needle to penetrate the fabric.

- The needle hole must be large enough for the needle and sewing thread.
- There are slots in the throat plate that allow the feed dog to rise and engage the underside of the fabric.
- The slots should match the width of the rows of teeth on the feed dog, without allowing contact.

15.7.2 The Presser Foot

The primary functions of the presser foot are as follows:

- To hold the fabric with the required pressure against the throat plate.
- To enable the feeder to advance the material through the machine during feeding.
- To provide stability around the needle point for correct loop formation of the sewing stitch.

There are many types of presser foot. Some of them, among others, are as follows:

- Hinged presser foot
- Compensating foot

- Piping foot
- Narrow hemming foot

15.7.3 The Feeder (Feed Dog)

- A solid mass of hardened steel, which takes the form of multi-rows of teeth
- The number of rows effects the level of support offered to the fabric during feeding
- The number of the teeth per centimeter on the feed dog—determined by the fabric, weave/knit, and its thickness
- A feeder with fine teeth (high tooth density) used to feed delicate light-weight fabrics
- A feeder with course teeth used to feed heavier fabrics
- Sometimes, necessary to round off the teeth to prevent damage to certain types of fabrics

15.7.3.1 Feed Settings

- Feed setting is usual to allow the full depth of one tooth to be above the throat plate at the point of maximum lift (Figure 15.22).
- The feed dog can be tilted to the front or the back, which can help prevent fabric slippage or extension (Figure 15.23).
- Feed raised at the rear to stretch the material (Figure 15.24).

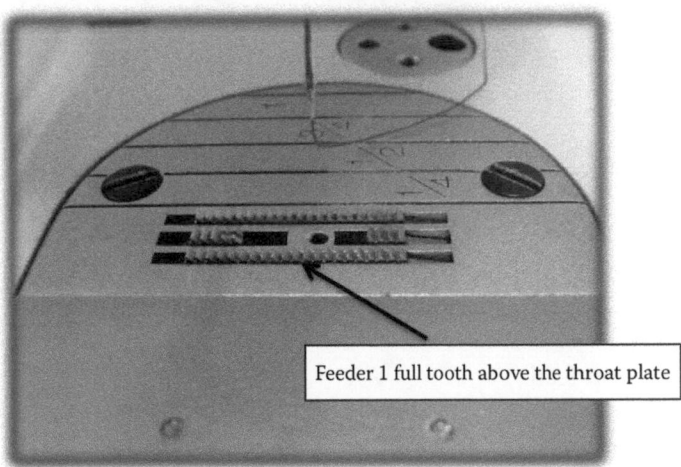

Feeder 1 full tooth above the throat plate

FIGURE 15.22
Feeder setting, one full tooth above the throat plate.

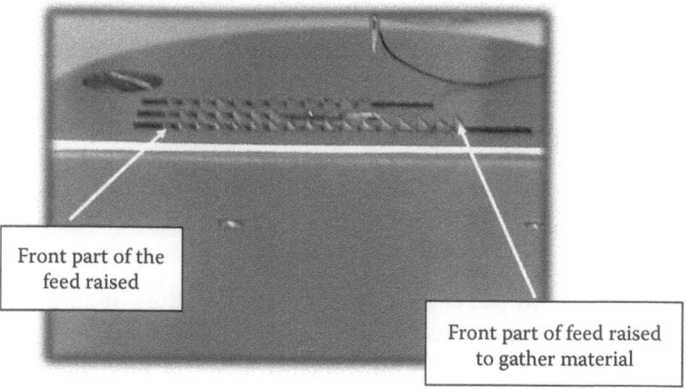

FIGURE 15.23
Feeder raised at the front to gather the material.

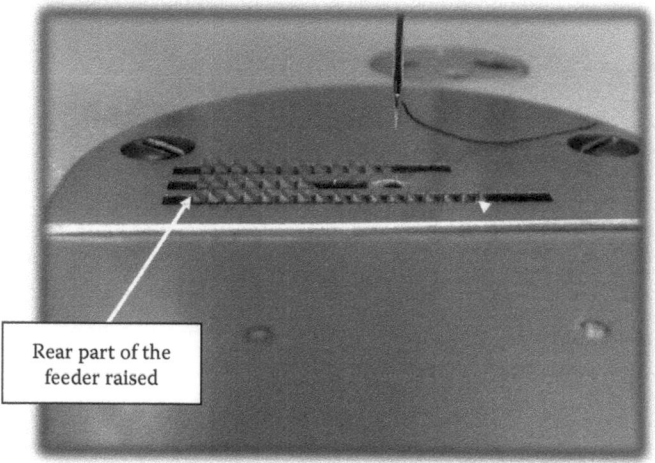

FIGURE 15.24
Feeder raised at the rear to stretch the material.

15.8 Feed Mechanisms

To correctly feed the material, as required for effective stitch formation, several different feed systems have been adopted. These are given as follows:

1. Four-motion drop feed
2. Differential drop feed
3. Needle feed
4. Compound feed
5. Feeding foot
6. Variable top and bottom feed

7. Alternating foot
8. Unison feed
9. Puller feed
10. Wheel feed
11. Cup feed
12. Manual feed

Many of the feed systems are used in specialist areas of production; therefore, only the most common ones are illustrated here.

15.8.1 Four-Motion Drop Feed

The four-motion drop feed (Figure 15.25) is the most common of all the feed systems.

- The feed dog engages the underside of the bottom fabric ply intermittently.
- It is set up (timed) to engage the fabric as late as possible after the needle has risen clear of the top ply.
- This ensuring that the fabric movement has finished before the needle redescends into the fabric.

15.8.2 Differential Feed

Differential feed (Figure 15.26) has the following features:

- By feeding more with the back feed dog than the front feed dog (in the same stitch cycle), it is possible to stretch the bottom fabric ply.
- Conversely, by feeding more with the front feed dog than the back feed dog, it is possible to introduce fullness into the bottom ply.

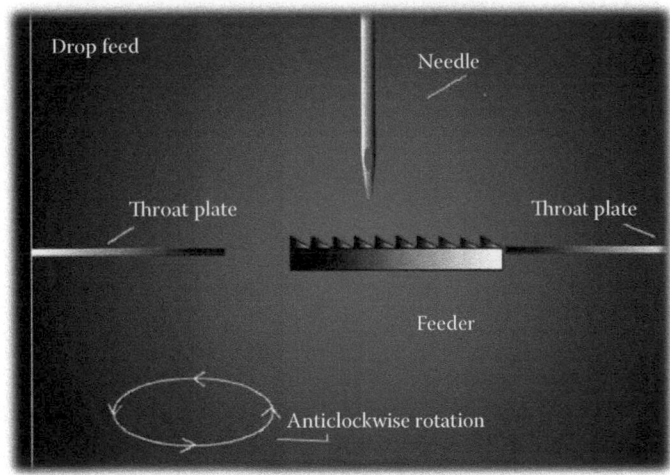

FIGURE 15.25
Drop feed.

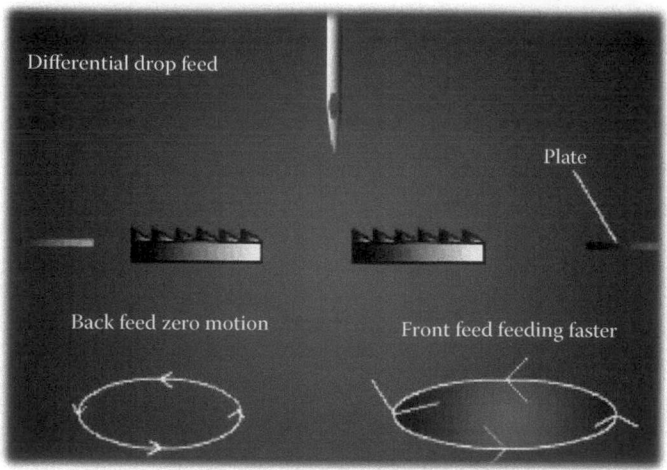

FIGURE 15.26
Differential feed.

- By feeding more with the back feed dog than the front feed dog (in the same stitch cycle), it is possible to stretch the bottom fabric ply.
- Conversely, by feeding more with the front feed dog than the back feed dog, it is possible to introduce fullness into the bottom ply.

15.8.3 Compound Feed

Compound feed (Figure 15.27) combines needle feed with the four-motion drop feed system.

- The feed timing is such that the feed dog rises and engages the bottom fabric ply at the same time as the needle descends into the fabric.

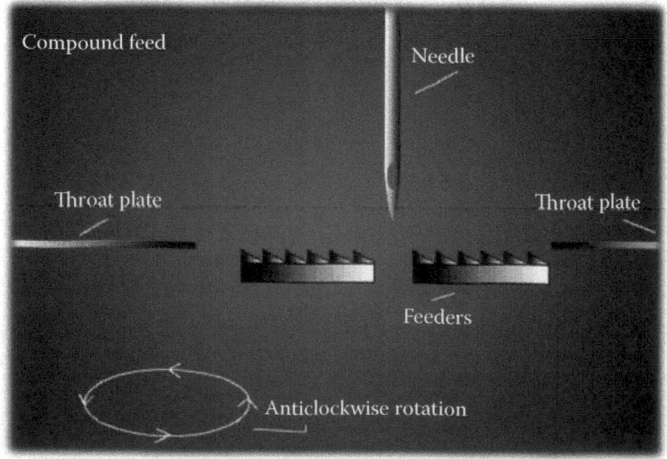

FIGURE 15.27
Compound feed.

- Both the feed dog and needle move the fabric through the prescribed stitch length.
- This can help reduce feed pucker and ply slippage during the stitching, because the fabric plies are pinned together during stitching.

15.8.4 Variable Top and Bottom Feed

- Variable top and bottom feed (Figure 15.28) is a combination of the feeding foot, synchronized with a four-motion drop feed system.
- It is often used for the sewing of high-friction materials such as simulated leather and composites, where the use of a static presser foot is unsuitable.

15.8.5 Pulley Feed

Pulley feed (Figure 15.29) is an auxiliary feed (additional to the main feed system) that takes the form of a continuously or intermittently turning weighted roller positioned at the rear of the needle.

It is mainly used in the seaming of heavy fabrics or to maintain tension in the lighter-weight materials.

15.8.6 Other Feed Systems

15.8.6.1 Feeding Foot

- No feed dog is employed in this system.
- The action of the feed dog is mimicked by the movement of the presser foot descending and engaging the fabric, moving it through the machine, and then disengaging and rising before traveling forward for the next stitch.

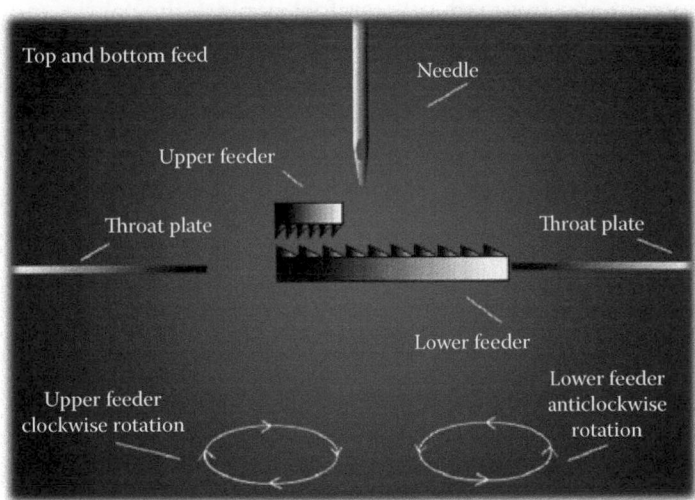

FIGURE 15.28
Top and bottom feed.

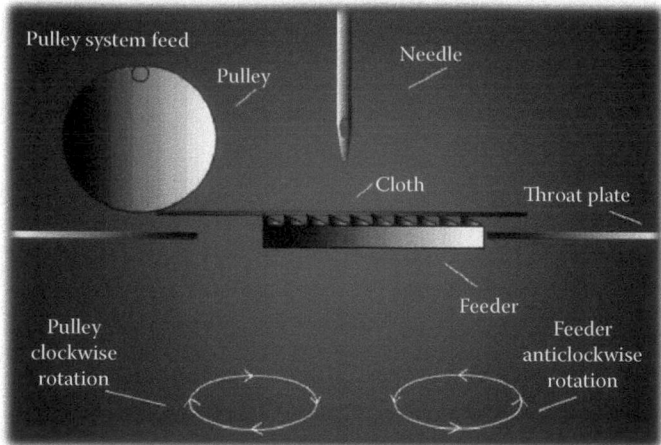

FIGURE 15.29
Pulley feed.

15.8.6.2 Alternating Feed

- It comprises a feeding foot and a lifting foot in conjunction with a feed dog.
- The feed dog and the feeding foot transport the fabric while the lifting foot is clear of the work piece.
- The lifting foot descends to hold the fabric while the other two components return for the next stitch.
- It is excellent for very bulky seams.

15.8.6.3 Unison Feed

- It is similar to alternating feed but provides more positive feed again.
- Both sections of the two-part presser foot contact the fabric and transport it through the machine sequentially.
- This means that one of the presser feet is in contact with the fabric at all times.

15.8.6.4 Cup Feed

- In cup feed (Figure 15.30), two horizontally mounted feed wheels are driven simultaneously, in order to move the fabric from right to left through machines with horizontal needles.
- It is most often used in the seaming of knitted goods and skins.
- The positioning of the feed system allows greater visibility of both fabric plies.

15.8.6.5 Wheel Feed

- In this system, a driven roller, wheel, or belt is employed above, below, or in both positions to move the fabric, instead of using a feed dog and presser foot.
- The surface of the wheel is often knurled or coated to provide increased friction for feeding leather or plastics.

FIGURE 15.30
Cup feed.

15.9 Stitch Formations

15.9.1 Definition

One or more separate thread supplies being combined in the following manner form the stitches within a seam.

15.9.1.1 Intralooping

The passing of a loop of thread through another loop of the same thread supply. Example: 101 single-thread chain stitch.

15.9.1.2 Interlooping

The passing of one loop of thread through a loop formed by a separate thread supply. Example: 401 double-lock chain stitch.

15.9.1.3 Interlacing

The passing of a thread around, or over, a separate thread supply or a loop of that supply. Example: 301 lockstitch.

British Standard (BS) 3870 or International Organization for Standardization (ISO) 4915 categorizes the different stitches available under the following classes. Each stitch formation has its own unique number.

Class 100	Chain stitches
Class 200	Stitches originating as hand stitches
Class 300	Lockstitches
Class 400	Multithread chain stitches
Class 500	Overedge chain stitches
Class 600	Covering chain stitches

15.9.2 Class 100 Chain Stitches

These are formed by the intralooping of a needle thread supply on or around the fabric.

Single-thread chain stitch seams are often used for temporary applications due to their ease of removal. This is because each successive loop is dependent on the previous loop for security.

15.9.2.1 *101 Single-Thread Chain Stitch*

- In 101 single-thread chain stitch (Figure 15.31), basting, temporary holding of fabric pieces before final securing stitch, is introduced.
- Sack closure and ease of removal allow use for securing industrial sacking openings.
- Coinelli, up to 65 needles can be employed to produce decorative stitch with the chain on the face fabric.
- It can be used to produce button holes and to attach buttons.
- End of seam must be secured in all cases to avoid *running back* of seam.

15.9.2.2 *Stitch Type 103 Blind Hemming or Felling*

- This stitch type (Figure 15.32) is formed with one needle thread (1), a loop of which is passed into the material from the needle side, through portions of the material, emerging on the needle side, where it is intralooped at the next point of needle penetration.
- A minimum of two stitches describes this stitch type.
- Partial penetration of surface fabric by curved needle means that the hemline cannot be seen on the face fabric. Seam must be tacked to secure stitch line.

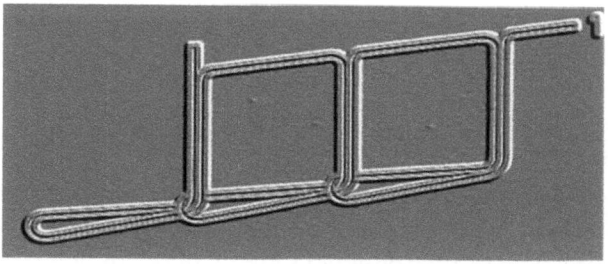

FIGURE 15.31
Single-thread chain stitch.

FIGURE 15.32
Single-thread chain stitch for blind-stitching operations.

15.9.3 Class 200 Stitches Originating as Hand Stitches

These are formed by a single thread being passed from one side of the fabric to the other on successive needle penetrations. It is not widely used; however, stitch type 209 is used to produce *pick stitching* around the edges on some suit jackets.

15.9.4 Lockstitches (Interlaced)

Often referred to as double-lock stitch (Figure 15.33), interlacing a needle thread supply with the bobbin thread supply underneath forms this stitch type.

These stitches are very secure, as a break in one stitch will not cause the seam to completely unravel; however, it will compromise the overall seam performance. A disadvantage is that the machine needs constant bobbin changes, and if the bobbin runs out on a seam, the stitch has to be unpicked, which can leave needle holes in the fabric, damaging the garment. This stitch is the only stitch to be used in parachutes. There are also more than 50 derivations of this stitch.

This stitch type is formed with two threads: one needle thread (1) and one bobbin thread (a). A loop of thread 1 is passed through the material from the needle side and is interlaced with thread a on the other side. Thread 1 is pulled back, so that the interlacing comes midway between the surfaces of the material being sewn.

It is sometimes produced from a single thread, in which case the first stitch differs from subsequent stitches. A minimum of two stitches describes this stitch type. This stitch is the most commonly used of all the stitches used in the production of apparel.

All stitches need to be balanced correctly, and this means that the tensions on the machine must be set properly; otherwise, the aesthetics of the seam will be poor and a weakened seam will result. Examples are given in Figures 15.34 through 15.36.

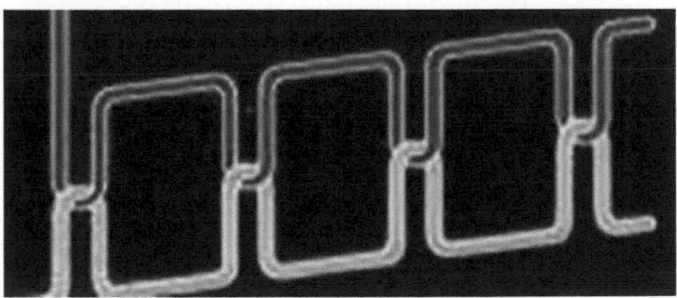

FIGURE 15.33
Single-needle lockstitch number 301.

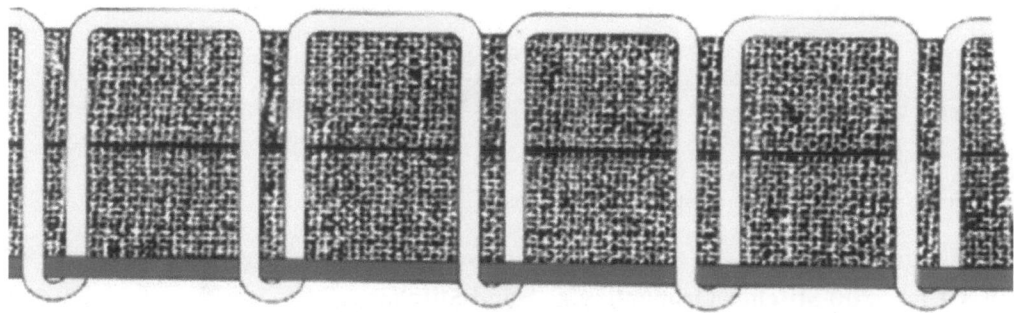

FIGURE 15.34
Stitch slack underneath as a result of a slack needle tension.

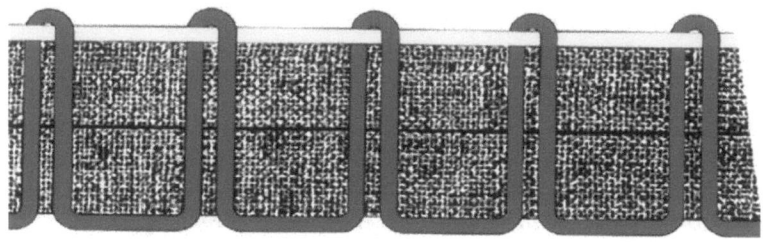

FIGURE 15.35
Stitch slack on top as a result of a too slack bobbin tension.

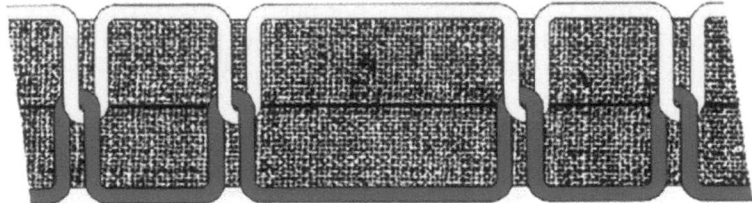

FIGURE 15.36
Tension balanced correctly.

15.9.5 Class 400 (Interlooped Stitches)

These stitch types are formed by the interlooping of a needle thread with a separate looper thread on the underside of the fabric. These are often referred to as double-locked stitches, because each needle thread loop is interconnected with two loops of the same single under-thread. The most common of these is chain stitch number 401 and is given in Figure 15.37.

This stitch type is formed with two threads: one needle thread (1) and one looper thread (a). A loop of thread 1 is passed through the material from the needle side and through one loop of thread a on the other side. It is then interlooped with a second loop of thread a. The interloopings are drawn against the material.

A minimum of two stitches describes this stitch type.

Appearance of lock-stitch (301) on surface with a double chain underneath.

It exhibits good strength and increased extension/recovery properties due to lower static thread tension and interlooped threads.

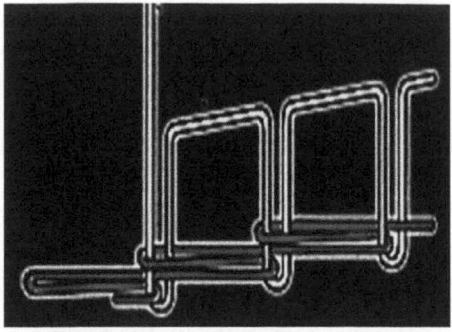

FIGURE 15.37
Chain stitch number 401.

It is less prone to pucker, again due to lower static thread tension and interlooped threads on underside and is excellent for long seams because of continuous thread supplies. These stitches also have many derivations from the basic 401 to 410 and beyond. All of these stitch types can be found in the British standards for stitch formations, as mentioned earlier.

15.9.6 Class 500 Overedge Chain Stitches

This stitch type is formed with two threads: one needle thread (1) and one looper thread (a). A loop of thread 1 is passed through a loop of thread (a), already laid across the needle side of the material and through the material. The loop of thread 1 is interlooped with a second loop of the thread a at the point of emergence on the other side of the material. The loop of thread is brought around the edge of the material and extended to the next point of needle penetration (Figure 15.38).

15.9.7 Coverstitches

Coverstitches are those that give full cover on the face of the fabric and the underneath of the fabric. They provide the highest extensibility of all stitch types and are commonly used particularly on weft knitwear fabrics, where the loop structure gives a high level of stretch. Therefore, you need a stitch formation that can cope with the high extensibility of the material.

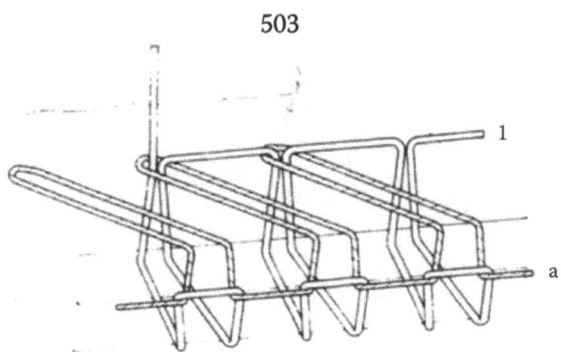

FIGURE 15.38
Three-thread overlock, stitch number 503.

If we consider the fact for a single-needle lockstitch that the extensibility of this stitch is approximately 30% and for zigzag lockstitches these values are higher so these will be discussed later.

For overlocking stitches, the values are higher still, but for coverstitches, the value can be 100% extensibility, and when you look at the stitch, you can see why. It almost looks like a spring.

Let us put it this way. If you take a hem on a knitted garment or a woven garment with a high level of stretch, you would not use a lock stitch, because when the seam was pulled or stretched, the stitch would crack and the seam would be compromised. Similarly, you would not use a two-thread chain stitch, as described previously, because even at 60%, the seam could break under the stresses demanded by the seam.

There is another very important reason, and it is that before certain machines were invented that produced these stitch formations, the traditional way of stitching a hem, for example, was to overlock the edge of the hem, in order to prevent the fabric from fraying and then stitching the hem using a 401 two-thread chain stitch. This meant that two operations were needed: two machines and two operators for two operations. Now, this has been reduced to one operation and one operator, as there is no need to overlock the edge of the fabric. The coverstitch seals the edge of the fabric with its cover, providing the higher extensibility, as described. This can be achieved in one operation. One of the most common stitch types used in this way is stitch type/number 605, which is described in Figure 15.39.

15.9.8 Stitch Type 605

This stitch type (Figure 15.40) is formed with five threads: three needle threads (1, 2, and 3), one looper thread (a), and one cover thread (z). Loops of threads 1, 2, and 3 are passed through loops of thread z already laid across the needle side of the material and then through the material and through three separate loops of thread a. They are then interlooped with a further loop of thread a, and the interloopings are drawn against the material. A minimum of two stitches describes this stitch type.

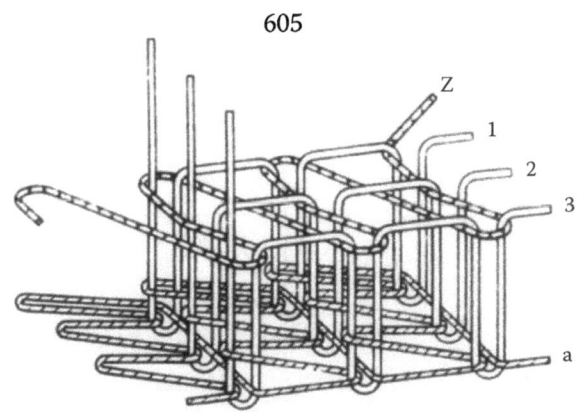

605

FIGURE 15.39
Coverstitch 605.

FIGURE 15.40
Hem of a pair of jeans stitched by using a 301 lockstitch.

15.9.9 Summary of Stitch and Seam Formations

There are hundreds of stitch and seam formations, so it is impossible to describe them all in this chapter. Therefore, only the most common and widely used stitches and seams have been described here. As mentioned previously, the British Standards Institute lists all the stitch and seam formations, and this is widely available from their website (BSI). Moreover, all of the seam definitions are contained. Another very useful website is American and Efird, which has lots of information in their technical bulletins on stitching many types of fabrics.

As a useful addition to this chapter, I considered it necessary to provide some images of certain operations on garments that contain some of these stitch formations. These are as follows.

Lockstitches are used on many other operations, attaching trims, for example. These are also used extensively in outdoor wear, rain jackets, and tents due to the locked formation of the stitch. A chain stitch would be too extensible and could unravel, so the last thing you want is to get wet in the night (Figures 15.41 and 15.42).

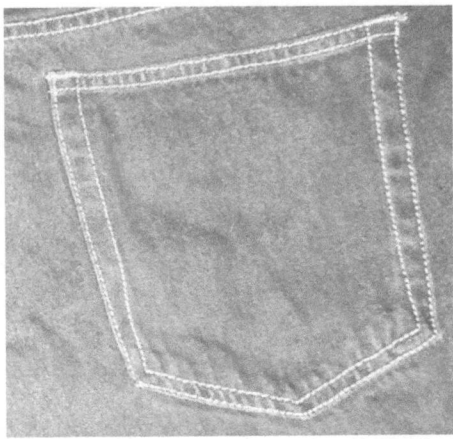

FIGURE 15.41
Back pocket on a pair of jeans by using a 301 lockstitch.

FIGURE 15.42
Two-thread chain stitch number 401 by using a double-lapped seam.

Two-thread chain stitch 401 is used extensively on side seams on jeans, waist bands, and other product areas where you need some extensibility and durability on the seam. If you are into hiking, for example, you may be walking up hills or climbing turnstiles; therefore, if you are lifting your leg high and as the fabric stretches, you want the seam to stretch with it. In addition, the same applies to the waist band when you are fastening the jean, as it needs to expand to fit around your waist (Figures 15.43 and 15.44).

Now, this is an interesting stitch, as it has two rows of stitching produced on the same machine. In America, they call this one row of 504 and one row of 401. By international European standards, it is known as stitch number 516 and is commonly used for denim jeans and other heavier fabrics. It combines neatening on the edge of the seam and durability with the 401 chain stitch (Figure 15.45).

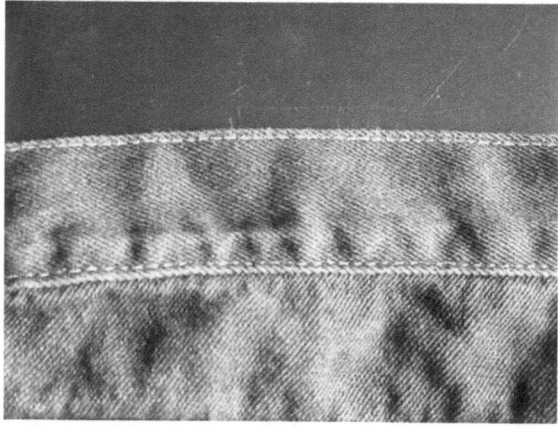

FIGURE 15.43
Waistband attached and stitched with a 401 two-thread chain stitch and a bound seam.

FIGURE 15.44
This is the machine that produces the bound seam. You can see a white tape that runs through a binding folder. The tape starts from 3 cm and is folded into the bound seam at 1 cm. The fabric would be slotted in between the two pieces and stitched together.

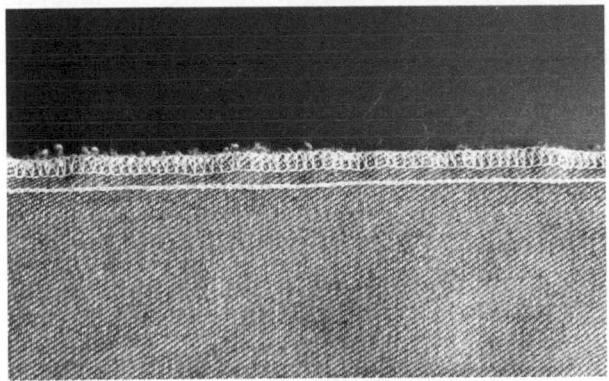

FIGURE 15.45
516 stitch.

15.10 The Sewing Needle

The sewing machine needle is the most important component of the lockstitch sewing machine because it is the carrier and deliverer of the sewing thread to the sewing mechanism. If it is not changed regularly, it can be responsible for major quality problems. It is also subject to the most abuse of all the machine parts, as it penetrates the material at speeds of 5000–6000 times per minute for lockstitches and 8000–10,000 times per minute for chainstitches. The friction caused by the penetration of the needle into the fabric causes extreme needle heating, with temperatures in excess of 250°C. Therefore, care needs to be taken not to burn your fingers if you are changing the needle after you have just used the machine.

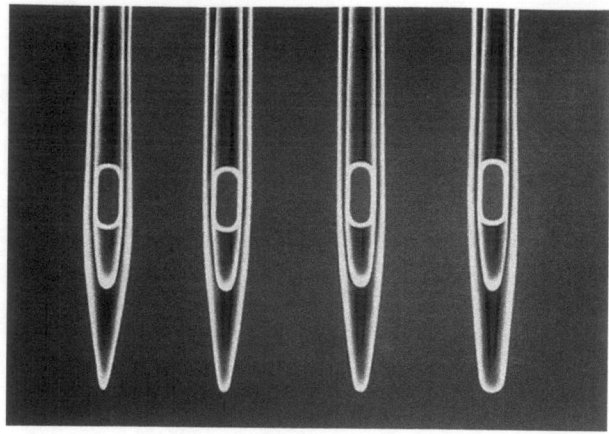

FIGURE 15.46
Examples of needle point types from right to left—heavy ball point, light ball point, round point, and acute round point.

Sewing needles come in a wide range of point types that are used for sewing many different types and properties of the material.

The needle is a slender pointed instrument usually made of tempered steel. It contains an eye to carry a thread through into a fabric for sewing. The main function is to pierce a hole in the material, so that the thread can form a stitch in the fabric.

Factors to consider when choosing a needle are as follows:

- Fabric type
- Fabric density
- Fabric composition
- The type of machine
- The type of sewing thread
- Fabric thickness

The most important aspect of needle design is the needlepoint, because it has to penetrate the fabric, without causing any damage to the material. It is also the most diverse part of the needle due to the many different type of points used. These needle points are designed for sewing on many different fabric types and seams. An example of needlepoint designs and the component parts of the needle can be seen in Figure 15.46.

15.11 The Designer and Garment Costs—The Commercial Designer

A garment design does not exist in a vacuum but is the end product of a chain of activities that can be said to start with the production of textile fibers. Various authorities have estimated that the time span between fiber production and the garment sampling stage can be as long as between 6 and 8 months and as short as 6–8 weeks for *fast-fashion* items. For the clothing manufacturer, the internal chain of activities starts some time before the forthcoming season's materials are available, because the company has to have some firmed-up ideas of what it intends to do before selecting materials.

The internal chain usually starts with the marketing/sales department doing some formal evaluation of historical sales performance to evaluate what the market sector served by the company could be looking for and at what prices. Parallel to this, the design team has researched trends in both fabric and garment styles through such channels as online trend bureaux, trade shows, and street trends (along with some comparative shopping). Marketing and design put their heads together and start formulating the framework of the sample collection. Fabrics and trimmings are selected and preordered, and the designer starts to prepare the core designs, which will represent the central theme of the collection with a clear market orientation and brand identity, if needed. Core designs, when approved, will be the basis for developing planned groups of variations. The presales design room processes are shown in Figure 15.47.

In this context, garment design tends more toward a goal-directed planning process, because apart from developing the appeal factors of each design, the designer also has to take into account the many technical and commercial factors involved; this process is then often referred to as product development. So, when designs have been approved and materials delivered, the design team must become involved with the production of sample garments.

15.11.1 The Designer and Garment Costs

Under a free enterprise system, it is accepted by the business world that money is the name of the game, and the clothing industry is no exception. The success of designs produced by a manufacturer can only be judged by the color printed on the company's bank statement at the end of a season: red or black. There are many factors that can influence profitability, but in normal circumstances, profitability originates, to a large extent, in the design section.

15.11.1.1 The Framework

Where does it all start for the designer? The answer is a combination of two factors: market specialization and the average garment concept. This linking of these two factors provides the designer with a reasonably accurate basis for initial cost estimates.

15.11.1.2 Market Specialization

The clothing industry is divided into sectors according to garment types, and within each sector, there are subdivisions or sections, based primarily on price. For example, one sector could be women's separates, with sections having prices ranging from very cheap to highly expensive. The prices reflect not only the manufacturing costs and fashion content of the products but also the brand equity associated with them.

The majority of clothing manufacturers concentrate on serving and expanding their share of a specific section within a sector. As a result, they accumulate a great deal of expertise regarding the suitability of products, prices, and production demands. Knowing this, the company is able to break down its average ex-factory price into the main components, such as materials, labor, overheads, and profit. The results of this analysis provide the designer with an accurate indication of what can be invested in an average garment in terms of materials and labor. The proportions between these two cost factors can vary

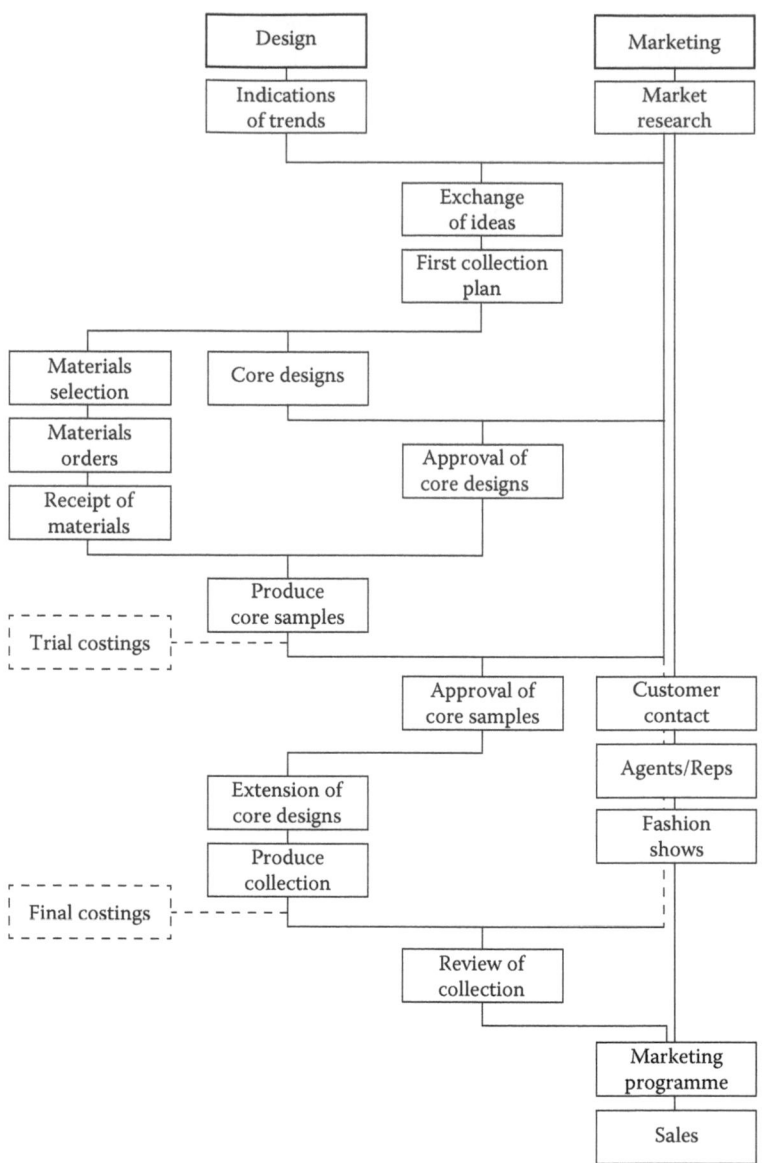

FIGURE 15.47
Presales design room processes.

from style to style, but their total has to be on, or very close to, the target, in order that a new sample will be able to slot into the correct price bracket.

15.11.2 The Average Garment Concept

Most production units, irrespective of the production system employed, are built around the average garment concept, where this term refers to a typical garment produced by the unit. This typical garment has an acceptable work content, and the balance between the various groups of operations is reflected in the staffing and equipment of the unit.

This type of factory would have the capability to handle a reasonable range of cloth and/or styling variations, without serious modifications regarding staff, machinery, and layout. However, in order to be more responsive to the market, factories can be arranged with several independent production lines, which can be modified to cope with higher degrees of change in fabric and style, with minimal disruption to the overall running of the factory.

The average garment concept is very widely used because of the production commonalities, which exist between garments of the same type. In practice, this means that regardless of individual styling, nearly every garment produced goes through the same standard operations. For example, the common operations for a unit producing skirts could be overlocking, dart sewing, closing side seams, zip setting, preparing and setting waistbands, top and underpressing operations, finishing, and inspection procedures.

It is important that production people communicate with the designer regarding the times for each group of operations in an average garment produced by the factory. With knowledge of these times and the average garment concept, the designer and technical staff can make amendments to the original garment design, without detracting from the original design concept. This ensures that operations required will fit into the production balance of the factory.

While nobody expects a fashion designer to be an expert in garment costing, designers must be aware of their influence on costs. Garments must be evaluated for costs at the sampling stage, because making samples without regard to price is often futile. So, when necessary, the designer and pattern cutter have to modify designs and patterns, so as to bring a new sample into the correct price framework.

15.11.3 The Garment Costing

Also known as the bill of materials, the garment costing details the costs of every item attributable to the production of a particular garment. The sum of these costs plus the profit margin is the selling price, which the company will quote to customers. Alternatively, the reverse is true, where a customer is only prepared to pay a certain amount for a product. The manufacturers must reverse engineer the product from here to ensure that they set production costs that allow them to achieve their desired profit margin. While each company has its own method of preparing costings, generally, the components of a costing are grouped under four headings: direct materials, direct labor, factory overhead, and general overhead.

15.11.3.1 Direct Materials

Direct materials are all the materials and trimmings that go into the construction and finish of the garment. Typically, these materials could include fabric, lining, interlining, buttons, zips, pads, tapes, labels, tickets, hangers, packaging materials, and so on.

15.11.3.2 Direct Labor

This covers the cost of all the labor directly involved in producing the garment and could include cutting, fusing, sewing operation, special machine operations, pressing, finishing, inspection, and packing. Labor of all types and grades has a direct overhead, which includes holiday pay, sick pay, fringe benefits, and so on, and the statutory payments made by the employer for each employee. This is usually expressed as a percentage of salary, and when this percentage is added to the employee's wage, it becomes the basis for calculating direct labor costs.

15.11.3.3 Factory Overhead

There are different methods of calculating the factory overhead, but most of them use a combination of the following three elements:

1. *Indirect labor*: This covers every person in the factory who does not directly perform a production operation, such as managers, supervisors, engineers, store personnel, clerks, maintenance staff, porters, canteen staff, security, cleaners, and so on.
2. *Expenses*: All fixed and variable expenses incurred in operating the factory, such as rent, rates, utilities, insurance, depreciation, and maintenance, and the various types of energy consumption/generation required by a clothing factory are included in this element.
3. *Indirect materials*: Also known as consumables, this element contains all the materials not directly connected to the make-up of a garment. Some of the typical items involved are office materials, spare parts, marker paper, and maintenance materials.

The total of these three elements is the factory overhead, and because it cannot be conveniently applied to specific cost units, it is generally expressed as a percentage of the direct labor costs. For example, using the arbitrary figures, if the costs for a given period are:

Direct labor	£38,000 (including direct overhead)
Factory overhead	£45,600

The factory overhead is 120% of the cost of direct labor. From this, it is simple to calculate the cost of 1 minute's work for every production operator:

Labor rate per hour	£5.93 (UK minimum wage for more than 21s as of October 2010)
Factory overhead at 120%	£7.12
Total cost	£13.05
Cost per work minute	£0.22

Therefore, the price of an operation is the rate per minute multiplied by the time allowed for the operation.

15.11.3.4 General Overhead

The general overhead comprises all the labor costs and expenses that are incurred in running the company, such as management, marketing, finance, insurance, warehousing, rent, and utilities. The design department costs are usually allocated to this component.

Again, because of the practical difficulties of apportioning this component to specific cost units, it is expressed as a percentage of the total for direct labor, factory overhead, and direct materials, as in this example, where all the costs are for the same period:

Direct materials	£114,000
Direct labor	£38,000
Factory overheads at 120%	£45,600
Total	£197,600
General overhead	£88,920

Therefore, conveniently, the general overhead is 45% of all the other costs. So, the framework of a garment costing would be the sum total of these four components.

While the method of computation, detail, terminology, and format can vary from company to company, the primary objective of the costing are always the same—how much does the garment cost to produce?

15.12 The Designer's Role

The preparation of a garment costing is usually the work of a costing clerk, who collates all the relevant information and calculates money values. Before the costing process starts, the design needs to be checked and approved as to the basic viability, within cost, for production by the design team and production/technical personnel. Skilled marker planners can reduce materials requirements, and production engineers can accurately analyze work content, but if the sample garment is carrying excessive costs of materials and/or labor, there is very little that these people can do to make the garment an acceptable proposition without the input and collaboration of the designer. So, the designer should never be designing in isolation and is a key member of the product development team.

15.13 Future Trends

Future trends in the joining of stitching fabrics will be dictated by the developments in the material itself, environmental drivers, and, of course, economic factors. More extensible materials are being developed, emulating the characteristics of knitted fabrics rather than woven materials. These materials may be impossible to match in performance with traditional sewn seams and could pave the way for increased use of bonded nonsewn technology to provide a softer feel and more extensible seam, but durability is still an issue to be met. The recent development of soluble sewing threads that can be easily *removed* from a garment, allowing for easier remanufacture of the material into a second life product, will undoubtedly have some impact on the sewing operations for both the product's first and second incarnations. For example, standard denim products, due to their consistency in design and the relatively stiff nature of the denim materials, lend themselves perfectly to semiautomatic operations, and this, combined with a drive for on (or near)-shoring production, could lead to further developments in this manufacturing technology as a way of minimizing direct labor costs in the manufacture of denim products. As the open-design wave propagates, it is possible that individuals could *build* their own denim products at (perhaps even at home)—the very nature of the material and the construction techniques lending themselves to personalization—which would lead to the development in joining techniques and technologies.

Bibliography

Carr, H. and Latham, B. *Clothing Technology for Fashion Designers*, 4th ed. Blackwell Publishing, Oxford, UK.

Cooklin, G., Hayes, S., and McLoughlin, J. (2006) *Introduction to Clothing Manufacture*, 2nd ed. Blackwell Publishing, Oxford, UK.

Fairhurst, C. Ed. (2008) *Advances in Apparel Production*. Woodhead Publishing, Cambridge, UK.

16

Knitwear Design Technology

Tracy Cassidy

CONTENTS

16.1 Introduction

In this chapter, we begin by rationalizing the importance of design and technological knowledge for knitwear specialists and briefly discuss the growth of the knitwear industry: an industry sector that was revolutionized by influential designers at particular times during the twentieth century and has continued to be an important sector of today's fashion and textiles industry.

We introduce the processes of machine-produced knitwear with a brief discussion of circular knit and flatbed fabric production, which leads to the characteristics of knitted fabrics and making-up technologies before outlining the four types of garment production:

- Fully cut or cut and sew
- Cut stitch shaped
- Fully fashioned
- Integral knitting and new full garment technology

We discuss and demonstrate how to recognize the key garment production methods integrating basic knit-garment styles and design features (hems, necklines, sleeve shapes, and fastenings including buttonholing) relating to those that are relevant to trends.

Garment specifications and sample garments (prototypes) are discussed, and further production techniques and technologies are given that are associated with garment finishing plus a mock factory layout is provided. Computer-aided design (CAD) is briefly made reference to in this chapter, and the stitch manipulation types that are used to create a range of knit fabrics are described. To conclude, this chapter highlights trend forecast sources specific to knitwear. Although knit fabrics can be used for a number of garment types such as T-shirts, undergarments, and popular accessories such as hats, scarves, gloves, and socks, this chapter focuses on the type of garments that are more generally associated with the term knitwear, which includes sweaters, cardigans, knitted dresses, skirts, and outerwear. Some retailers refer to cardigans and sweaters as jersey wear due to the basic stitch type of the fabric.

Traditionally, knit textbooks either focus only on design, or on technology, or they discuss these two as separate entities by making it difficult to appreciate how they work together and how they are dependent on each other. All knit specialists, be they designers or technologists, require a sound understanding of fibers, yarns, dyeing, and finishing that can be found in other chapters of this book. Such knowledge is essential for all fashion and textile professionals. In order to connect design and technology in the context of knitwear, the topics in this chapter combine the technicalities and the practicalities of knitwear production for a more fully integrated design technology approach to the understanding of this subject.

16.2 Growth of the Knitwear Industry

Until the 1920s, knit fabrics were largely used for underwear; Coco Chanel is renowned for using such interlock fabrics known as jersey for fashion apparel. The cost of the fabric production was relatively low, making the resulting garments affordable for many mass market consumers. Chanel's suits often comprised long cardigans with long knitted belts, skirts, and long tunic style tops; a completely knitted outfit. Meanwhile, Jean Patou met the demands of the growing sports leisure industry. He is renowned for the tennis sweater and cable knits. A decade later, Elsa Schiaparelli popularized knit sweaters with single motif designs at the time when hand knitting was still popular. Later in the 1970s, companies such as Missoni brought a wealth of color to machine-produced knitwear and again the popularity of hand-knit garments also gained importance. From the 1980s to the end of the twentieth century, the demand for hand knit and the hand-knit look declined but mass produced machine-knit garments flourished with fashion designers such as Sonia Rykiel, Kenzo, and Issey Miyake revolutionizing knit again as must-have fashion items. Throughout the 1990s, CAD supported the mass production of knit, and the technical yarns of the early 2000s helped to launch knit into areas such as performance sportswear. Knit continues to be an important sector of the fashion apparel industry today. At times when hand knit is on-trend, the look is reproduced in machine-made knitwear using large-gauge machines and heavier yarns to achieve a chunkier fabric construction.

In its humble beginnings, knitwear was handmade by a mass population for personal use and to clothe family members. The garments were shaped on the needle during the knitting process. Whereas the mass production of knitwear began by mimicking the manufacture of woven wear by cutting into knit fabrics to make into garments akin to the most basic of production methods used today; fully cut. This production method is discussed later in this chapter. Knitwear production has now come full circle with the aid of new technology in which whole garments are knit and mostly made up on the needle bed through a fully operational technique. It is therefore easy to appreciate how both design and technology have been equally responsible for the development of this industry sector. Today, the industry supports businesses of all sizes from designer-makers operating small-scale enterprises to large conglomerates operating on a global scale. Small-scale knitwear specialists may fully use industrial machinery or domestic versions that can be almost as electronically sophisticated and fully automated as industrial machines. The design technology approach to production is therefore ever more critical, as we shall see in the following section.

16.3 The Importance of Design Technology for the Knitwear Specialist

Knit is an exciting approach to fashion and garment production, largely because the process involves both fabric creation and garment realization within one complete design package. For this reason, there is a very distinct difference between the technicalities of producing knitwear and the practicalities of designing knitwear. The former requires knowledge of how the machines work, and how fibers, yarns, and knit stitch structures behave due to their constructions, and it is necessary to understand different fabric design capabilities; whereas the latter is essentially designing knitwear for a target market. In addition to knowledge of the characteristics of fibers and yarns and of fabric structures, design requires considerations to the aesthetic appeal of the garment in relation to the intended wearer, and toward trends, and its relevance to the retailer's brand. In microsize and many small-to-medium size enterprises (SMEs), the knitwear designer is often responsible for the entire design and production process. The knitwear designer therefore requires a sound understanding of all aspects of design and production including technological knowledge of materials and machinery. In larger businesses, the fabrics are often produced by technologists in relation to the designer's specifications that are produced as a result of the design process. The technological production and the creative design processes are more likely to be conducted in isolation of each other and by individuals from different educational training backgrounds. Technologists will have a sound technological knowledge of fibers, yarns, stitch structures, and fabric behavior but little or no knowledge of the design process. Designers with knitwear qualifications will have at least a basic working knowledge of the technologies but usually little experience of their application. Designers who are trained mostly in the design of woven fabrics or woven wear will have less specific technological understanding for knit but may still be employed in this specialist industry sector. However, designers and technologists must be able to work together to produce successful knitwear ranges and in order for a complete fusion between these two sets of professionals, it is essential that knitwear designers have a solid technological knowledge to produce garments that are fit for purpose and being desirable for consumers; and knitwear technologists must have an appreciation of aesthetic desirability and marketability of end products as an integral design factor of the functional product. Therefore, in this chapter we unify the technological knowledge and the rudiments of design specific to knitwear production to emphasize this necessary synthesis.

16.4 Basic Fabrics, Production Methods, and Machine Types

Knit fabrics can be produced in two basic forms in relation to the placement of the needles and needle beds on the machine. Single-bed fabrics are produced on machines that have only one bed of needles. All of the needles operate in the same direction. The most basic single-bed fabric is a plain fabric known as single jersey, which can be recognized by the characteristic *V*-shaped stitches on one side, known as the technical face of the fabric, or the knit face, and by the *U*-shaped stitches on the reverse side, known as the technical back, or the purl face. The fabric has the same appearance as hand knit plain fabric (Figure 16.1).

FIGURE 16.1
Single-bed knitting machine and single-jersey fabrics showing the *U*-shaped loops on the technical purl face and the *V*-shaped stitches on the technical knit face. Notice how single-bed fabric curls in at the selvedge prior to steaming or pressing, known as finishing.

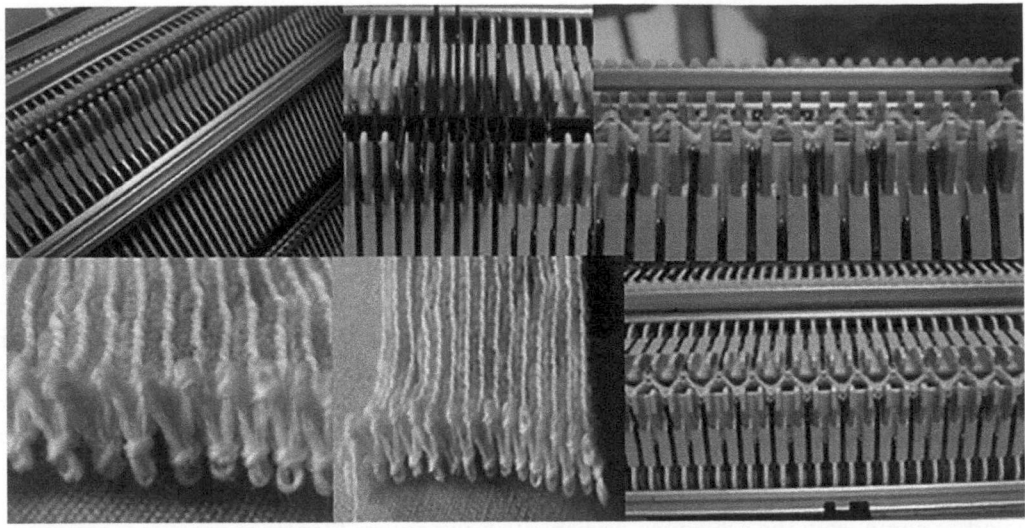

FIGURE 16.2
Two beds on a double-bed knitting machine and the needles are clearly shown to be operating in opposite directions. The double-bed fabric shows the V-shaped stitches on both sides of the fabric and the U-shaped loops sandwiched within the fabric core.

Double-bed fabrics are produced on machines that have two beds of needles, which operate in opposite directions. The most basic double-bed fabric is a plain fabric known as double jersey, which looks identical on both sides having the characteristic V-shaped stitches on the technical face and technical back, or reverse. The U-shaped stitches can be seen sandwiched in between the two technical sides. As double-jersey fabric is produced on a double-bed machine, technically it is known as a rib weft fabric (Figure 16.2).

Single jersey and double jersey provide the basis for all knit fabric types. The range of fabrics that can be achieved by manipulating the stitches on both single- and double-bed machines is discussed later in this chapter. The number of needles in a needle bed will depend on the size of the machine. In addition, different sized beds house different sized needles (needle gauge), which affect the stitch size appearance on the fabric.

In addition to the number of needle beds on a machine, there are two fundamental types of knitting machine relative to the shape of the beds: flatbed and circular. Both types can consist of one or two needle beds. Double-bed machines are commonly called rib machines because they are capable of producing fabrics that resemble the rib stitch of hand-knit fabric due to the two sets of needles that operate in opposite directions. In addition to producing double-bed or double-sided fabrics in which the two sides of the fabric can look identical, one bed can be used to produce single-bed fabrics. However for a single-bed fabric on a double-bed machine, it is more likely that the two beds will be used to produce a single thickness tubular fabric piece, mimicking the type of fabric that can be produced on a circular machine outlined next, rather than producing a flat fabric piece from only one bed due to cost-efficiency. Circular bed machines are more commonly used to produce fabrics for T-shirts rather than the type of knitwear that is covered in this chapter. However, it is worth noting that fabrics produced on flatbed machines are often more stable than those produced on circular machines due to the yarn feed. On flatbed machines, the yarn is fed via yard feeders on the carriage(s) from side-to-side across the bed. On circular machines, the yarn is fed in only one direction in a continuous circular motion. This circular motion can cause a condition of plain knit fabrics known as spirality, which is caused by twist stress that creates a distortion of the fabric. This is particularly evident on cheap T-skirts in which the side seams can clearly be seen to spiral off centre. Spirality can be minimized in production by using two-fold yarns (double yarn) though this will inevitably increase the cost of the garment.

The fabric pieces produced on knitting machines are known as blanks. Circular machines generally produce blanks quicker than flatbed machines and are therefore economical to use. The blanks produced by circular machines are tubular with a circumference relative to the circumference of the machine and needle gauge and also dependent on the type of stitch used. Depending on the circumference size, garments can be produced directly from the tubular blanks with no side seams. Alternatively, the blanks can be cut and opened out to produce a flat fabric. Flatbeds produce fabric blanks with sealed edges known as selvedges. Knitted fabrics will generally unravel easily if the stitches are not sealed in some way; the selvedges prevent such unroving. Blanks of various sizes can be produced on flatbeds up to the maximum width of the bed. Several separate blanks can be knit at the same time on one bed to increase productivity. The garment production type employed will invariably determine the type of machine most suitable for the task as we will see in the next section.

16.5 Characteristics of Knitted Fabrics and an Introduction to Making-Up (Assembly) Technologies

Knit fabrics can be made from natural and synthetic yarns both textured and untextured. Elastene yarns can also be incorporated to improve the stretch–recovery properties, think leggings! Knit fabrics generally have a soft handle and drape, and they tend not to crease easily due to the stitch formation. Knitwear is soft and comfortable lending it well to informal casual wear. However, knitted fabrics tend to be limited in their dimensional stability and therefore require careful handling both during the making-up processes and during aftercare such as washing and drying. Dimension stability can be improved through finishing, depending on the yarn type. The formation of the stitches determines the amount of stretch or extensibility of the fabric, which determines the most appropriate making-up processes to be used; knitted fabrics will unravel (unrove) easily if stitches are not sealed. Dropped stitches during the knit process can quickly unrove creating a ladder in the fabric. When cutting into knitted fabric, the raw edges need to be covered and neatened to prevent unroving or fraying; overlock machines are used for this purpose. Cup seam machines are used to assemble fully fashioned garments or shaped edges that have sealed fabric edges as they reduce the risk of overstretching the seams during the making-up process. Linking machines are commonly used to attach knitted neckbands though they can also be used to make up shoulder seams. It is important to use the correct type of sewing thread for knitwear; it should be the same or very similar to the yarn composition used in the knitted fabric. When linking, the same yarn used to knit the fabric is used on the linking machine. Depending on the yarn type, used knitwear has a tendency to pill. Pills are small bobbles of yarn on the surface of the fabric. These occur through the constant rubbing of the fabric against itself or by other fabrics. Under the armholes is the most commonly prone to pilling due to the constant movement of arms. Yarns made up of short staple fibers are more prone to pilling.

Before we move on to the different garment production types, the next section introduces you to basic garment styles, which will aid the recognition of the production methods.

16.6 Basic Garment Styles

As previously said, knit is an exciting medium as the designer can create both the garment and the fabric. Garment styles are largely subject to trends, and for mass production the most simplistic styles are largely preferred. The most common knit garments are the sweater and the cardigan. In their most basic form, they are essentially the same with the exception of the front of the garments. The sweater is made up of one garment front, one garment back, two sleeves, and most commonly a neckband. The cardigan is made up of two garment fronts, one garment back, two sleeves, and most commonly a neckband plus two button bands. Neckbands, button bands, and other band types are generally known as trimmings. Both garment types can be made without sleeves. A sleeveless sweater is generally known as a slipover, and a sleeveless cardigan is generally known as a waistcoat or gilet. Sleeveless garments commonly have trimmings around the armholes to neaten the edges.

The cardigan style is often used for heavier outerwear using thicker, warmer yarns to produce jackets and coats though they are rarely waterproofed. The length of sweaters and cardigans is subject to trends. The sweater can be extended well below hip level to form a long-length jumper popularly worn with skintight trousers or leggings or further extended to form a knitted dress. Depending on the fashion of that time, dresses may or may not be shaped at the waist. Long jumpers and knitted dresses gained particular popularity during the 1980s as part of the casual and sportswear fashion trends of that time. Similar to knitted dresses, knitted skirts appear in lesser frequent fashion trend cycles. Skirts are generally tubular with a small amount of hip to waist shaping to reduce the bulk around the waist. Alternatively, an A-line shaped skirt known as a sunray skirt periodically gains popularity subject to trends, which was last popular during the early 1990s. This style is knit sideways and incorporates pleating to achieve a good drape. Some sleeve types also benefit from sideway knitting, which are discussed in the following section. A waistband is common on knitted skirts. Depending on the extensibility of the fabric around the waist and the design of the skirt, a zip fastening may be necessary. The seating area of long sweaters and dresses often quickly distorts during wear, making them appear unflattering. Knit skirts can also quickly lose their shape and hang appeal and for the same reasons, though often not the case with the sunray design; and probably for the same reason, trousers are not popular as knitwear. The length of sleeves can be altered depending on the desired design style and sleeve length options as shown in Figure 16.3, and an example of basic garment blocks are shown in Figure 16.4 with a garment specification diagram.

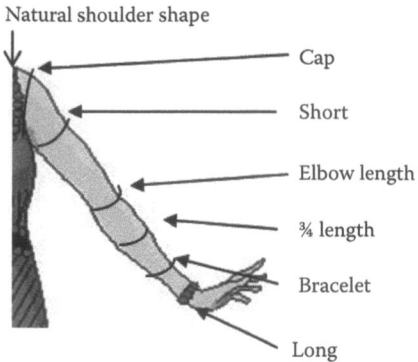

FIGURE 16.3
Popular sleeve lengths.

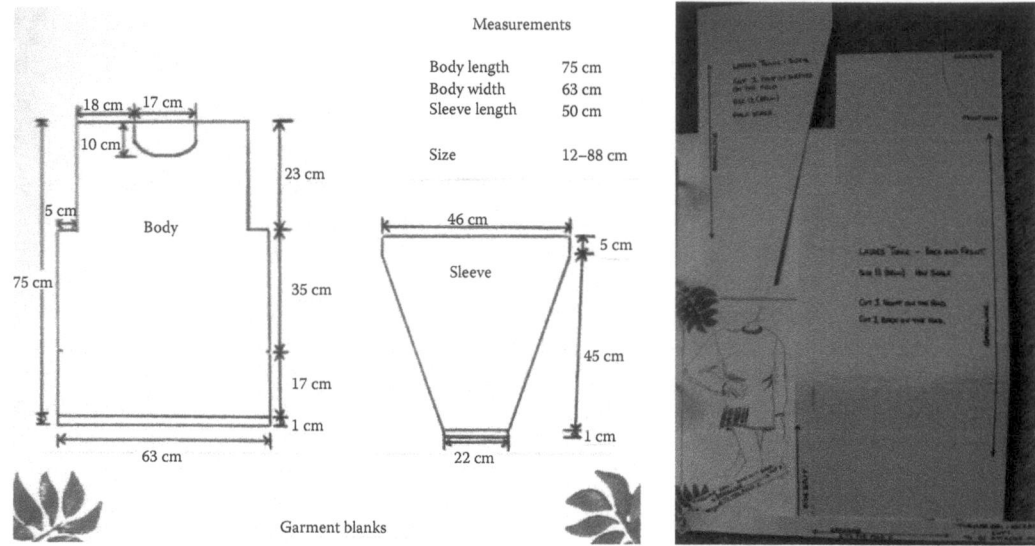

FIGURE 16.4
An example of a basic pattern block, showing a sweater body with a round neckline, no shoulder shaping, and a basic semi-set-in sleeve. The measurements are given for a standard UK size 12. The accompanying photograph shows the pattern pieces. The front and back blocks are to be cut with the center grain line on the fabric fold for a symmetrical cut. There is a hem trim for the two lower body side slits and a tubular hem neckband.

16.7 Garment Production Types, Popular Knit Design Styles, and How to Recognize Them

There are four garment production methods available for the production of knitwear:

- Fully cut
- Cut stitch shaped
- Fully fashioned
- Integral knitting and full garment knitting

Each production method requires a special making-up processes and machinery. The design style will indicate a preferred or even essential production method depending on whether or not edges; for example, around the neckline or the front edges of a cardigan/ jacket can be left untrimmed utilizing a sealed edge. Fully cut and cup stitch shaped methods require trims to seal and cover raw edges, whereas the fully fashioned method allows for the creation of sealed edges. The fully fashioned method also allows for additional style features to be included that can improve the aesthetic. In order to create even the most basic knitwear styles, an element of shaping is required. One of the first considerations of styling is the shoulder of the garment. A style known as drop shoulder results when no shaping is undertaken, such as the example in Figure 16.4. As the human form has some degree of shoulder shaping, even those who are considered to be very square shouldered, any garment to be worn draped over the shoulders will benefit from some shoulder shaping. The natural shoulder slope of the human form is shown on the sleeve length form in Figure 16.3. Cheap knitwear often has no shoulder shaping. No shoulder shaping will tend to slightly distort the drape of the garment's body and sleeve parts though this is often not really noticeable when worn. Shaping the shoulder using the fully cut or cut stitch shaped production method will not greatly incur additional cost as the assembly procedures for both shoulder types are the same; the front and back shoulders are cut with or without shoulder shaping and then stitched together. However, when using the fully fashioned and integral production methods, the knit productivity is reduced to undertake the additional shaping process that will incur higher costs for garment production per unit. Assessing shoulder shaping is simple, either there is a visible gradient or not. In addition to shoulder shaping, the sides of upper body garments can also be shaped. Knitwear generally utilizes straight side seams but subject to trends side shaping, or tapering, can be included. The trend for tapered knitwear, both machine-produced and handmade, was particularly popular during the 1930s to create a very feminine, shapely silhouette in contrast to the straight-bodied masculine silhouette of the 1920s. The style returned in the 1980s, supporting the power dressing trend where often the shoulders incorporated shoulder padding. The broad exaggerated shoulder line and side shaping to create a smaller waistline create an inverted triangular body shape synonymous with a strong masculine upper body shape. The style was popular for both male and female knitwear and for tailored jackets. Similar to shoulder shaping, shaping the side seams using the fully cut and cup stitch shaped production methods incurs little to no additional assembly cost, but additional cost will be incurred using the fully fashioned and integral knit methods as productivity is reduced due to the shaping process that is being undertaken on the machine. However, the fully fashioned method will not incur any fabric waste from the tapered side seams that result from the cut and sew method.

In order to create different styles of knitwear, shaping is required around the neckline and armholes. However, there are two basic styles that do not have any shaping: the bateau or boat neckline and the drop sleeve (Figure 16.5). The bateau neckline may be trimmed, or a sealed edge may be desired. The drop sleeve is commonly used with a drop shoulder giving a very basic garment style. The bateau neckline, however, is considered to be particularly flattering and sensual as the shoulders are fully exposed. The style is more sensual coupled with some side seam shaping and a set-in sleeve that accentuates the curves of the upper body silhouette. The set-in sleeve basically fits around an imaginary line from under the armhole following the contour of the body to the top of the edge of the shoulder and down the back of the arm to join under the armhole. To achieve a good fit, shaping is required both on the sleeve head and on the front and back body parts. Figure 16.5 shows the set-in sleeve style. A semi-set-in sleeve, as previously shown in Figure 16.4, allows for

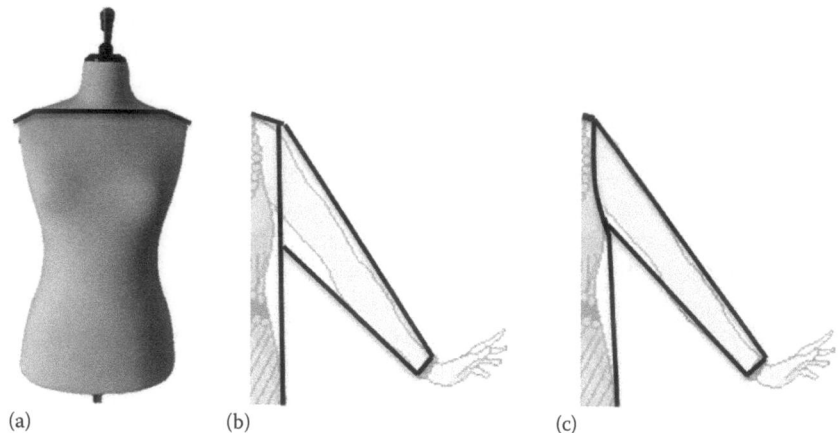

FIGURE 16.5
(a) Bateau/boat neckline shown on a dress form, (b) drop sleeve (without drape), and (c) set-in sleeve illustrated on a body form.

excess bulky fabric to be removed from under the armhole and to lessen the extent of a drop-shoulder line in which a set-in sleeve is fully shaped (curved) around the sleeve head; the semi-set-in sleeve has an unshaped sleeve head.

Further basic styles and the most popular design styles will be introduced in the following sections linked to one of the four production methods, which are discussed with their appropriate making-up methods and technologies. Although it is possible to produce almost all the styles through any of the four production methods, some processes work better than others and are more cost-effective.

16.8 Fully Cut, the Most Basic Production Method, and Basic Garment Styles

From a knit perspective, fully cut garment production is the most straightforward of the four production methods as the blanks are produced on the machine without any shaping. The garment-making process is similar to making garments from woven fabrics as the garment is cut out of the blank, or lay, which is why this process is also referred to as cut and sew; however, because knit fabric is generally very stretchy (extensible), darts are not required and little side-seam shaping is favored. The blanks are steamed if made of a man-made fiber such as acrylic, or pressed if made of wool or cotton. For mass production purposes, several layers of fabric are marked and cut through in one operation in the cutting room area of the factory. Although the fully cut method is akin to the woven-wear garment production process, lockstitch is unsuitable for knitwear production as the raw edges will fray and be unsightly; in addition, the stitch is not extensible enough for knit fabric. Three thread overlock machines are therefore used to seal all the raw edges of the garment. The overlock stitch joins the garment pieces, covers the edge with a chain stitch structure, and cuts the edge to neaten all in one operation. A four thread overlock machine is often used for heavier knitwear such as a jacket or coat. If a neckband is attached using the linking method (discussed in Section 16.10), it is usual to overlock the cut neckline first to prevent fraying; alternatively, the neckband can be attached using a mock-linking process. The fully cut method is the cheapest production method and is particularly suited to the more basic design styles with minimum shaping and fabrics that are produced on a circular knitting machine. This method is used more for garments such as T-shirts rather than the type of knitwear that we are concerned with in this chapter, which is more likely to be produced on flatbed machines, employing cup stitch shaped and fully fashioned assembly methods.

Knitwear is typically classified in relation to the shape characteristics. Sweaters, knitted dresses, slipovers, cardigans, jackets/coats, and waistcoats/gilets all require a neckline. The most basic neckline style is a round neck. As the fully cut method produces a raw cut edge around the neckline, overlocking is necessary to seal the edge and a neckband is required to cover the edge and thus neaten the garment. If a round neck is produced using the fully fashioned method with a sealed edge, a neckband is not required to neaten the edge but may be used as a design feature. The neckband needs to stretch sufficiently for the head to pass through the neckhole of a sweater, dress, or slipover and is therefore often constructed on a double bed using a rib stitch structure, which is more extensible than single-jersey fabric. Rib stitch fabrics are discussed later in this chapter. The neckband can be attached using a linking process, which is described in Section 16.10. Linking is a highly skilled process that adds cost to a garment. Such an expense would not be justified for a garment that is produced using a fully cut method, and therefore a mock-linking technique is more likely to be used. The edges of the neckband need to be sealed with a cast off edge or an overlocked edge because mock-linking does not secure open stitches as the linker machine method does. The principle of linking and mock linking is illustrated in Figure 16.6.

The fully cut production method allows for the most basic method of assembly. First, the shoulder seams are joined, thus joining the front and back pieces together. For a garment with no front fastening, such as a sweater, the neckband is joined to form a tube, which is then mock linked to the neckline either at this point in the assembly line or at the end of the process. For a garment with a front fastening, such as a cardigan, the neckband may be mock linked in place before attaching the front fastening bands or after attaching

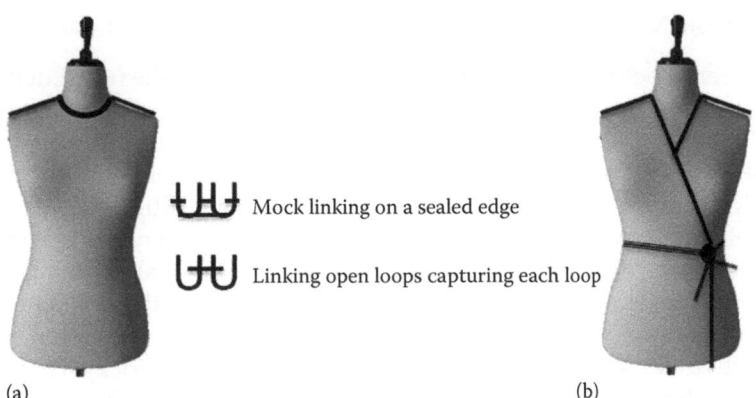

(a)   (b)

FIGURE 16.6
Round neckline shown on a dress form (a); mock linking and linking (illustrations) used to attach neckbands and other trims such as button bands; wrap and tie front shown on a dress form (b).

the front fastening bands, depending on the designer's specification. The sleeves are then positioned within the armhole shaping or between the armhole markings if a drop sleeve is used. The sleeves are then overlocked in place. The side seams are made up continuing through to join the sleeve seams as one operation on each side of the garment.

For a garment with a front fastening, the most basic style is a single buttoned front. This requires a button band on one side and a buttonhole band on the second front side. Buttonholes are made using a buttonhole machine similar to those used for woven wear. The ends of the holes are secured with a dense zigzag stitch known as a bar tack, and the buttons are attached using a special machine.

A button fastening can also be applied to one or both shoulders on a sweater. For adult wear, this would more likely be used as a design feature rather than being functional and would add cost to the garment. This style feature is however useful on baby wear and children's wear, particularly for children under the age of five, as babies' and young children's heads are proportionately larger than their body size; the shoulder opening makes it easier and less stressful for the infants when the garment is put on and taken off them.

The overlock stitch is somewhat bulky that would be uncomfortable in certain areas of the garment when worn if not flattened. A covering seam stitch is therefore used around the back neck and along the front fastening trims on cardigans and jackets, and so on. Once the garments are fully assembled, they are inspected for faults and undergo a final press.

Another popular basic cardigan style is the wrap and tie front (Figure 16.6). This style requires slightly wider front pieces to wrap one over the other and a long knit belt to fasten, thus not requiring buttons and buttonholes. The front garment pieces are akin to a double-breasted cardigan front fastening that normally has a button fastening along the outside edge and a matching column of nonfunctioning buttons down the other corresponding front aspect, and may have a round neckline. The length of the wrap and tie garment varies mostly from hip level to knee level but can be as long as ankle length, depending on the current trend. Belted wrapover cardigans were particularly popular during the 1970s.

16.9 Cut Stitch Shaped, the Most Economical Knitwear Production Method, and Further Basic Garment Styles

The main difference between the cut stitch shaped process and the fully cut method is that instead of cutting garment blanks from one large piece of fabric, the blanks are knit to the required width of the garment, plus a small seam allowance. The fabrics are generally produced on flatbed machines. The bottom of each blank is sealed either with a hem in which a flat, even edge is desired or a rib stitch is used when a snug fitting sleeve cuff or a figure hugging waistband is required such as at the bottom of a sweater; this is known as a welt. Most often, the blank begins with a rib stitch for the waistband and cuffs to incorporate a degree of shaping and to provide a snug fit at the extremities of the garment. The garment blanks are then cut to provide the correct garment length, armhole shaping, neck shaping, and possibly shoulder shaping. This process produces a less fabric/yarn waste than the fully cut method; however, cheaper acrylic yarns are more likely to be used than the more expensive wool or cotton yarns. The blanks are given a steam treatment to relax and stabilize the fabric and to control the shape and size. Wool and cotton fabrics are pressed rather than steamed. The blanks are then cut either by machine or by hand. The two sleeves and the two body blanks are usually cut together to ensure that they are the same to aid the making-up process. A process known as press-off shaping may be incorporated on a double-bed machine to shape the neckline without the need to cut. The loops are allowed to drop with the aid of pressers without unraveling. The neckline is then trimmed during the cutting process with minimum waste. Alternatively, the loops can be held (held-stitch method) on the needles until the shaping is complete; two or three rows of knit then complete the edging to secure the loops. This process is often used on sloped shoulders and sleeve heads.

The press-off shaping method or the held-stitch method can be used for all the basic neckline shapes including the round neck but is particularly beneficial for *V*-shaped (V neck) and square necklines. The production method is much easier to undertake, and the finishing is more superior to that which is achievable through a cut and a sew neckline. V necklines and square necklines require a neckband to cover the edges unless they are produced using the fully fashioned method in which a sealed edge can be achieved, as described in Section 16.10. The filled-in V neck is a popular variation on the V neck, the fill-in is often made from a contrasting colored fabric to give the appearance of a garment worn underneath the sweater. Popular variations on the round neck include the polo neck and the turtle neck; both styles are produced with a collar rather than a neckband. The polo neck collar is knit using a rib stitch that fits snuggly around the neck. A turtle neck is a loosely fitting collar that stands away from the neck; it may be constructed using a rib stitch or a jersey fabric. Polo and turtle necks when subject to trends tend to be considered as classical styles that can be popular even when they are not particularly *on-trend* (Figure 16.7). Another collar type that is more subject to trends is the sailor collar. This style gains more popularity within the children's market for girls and last popularized during the 1980s among women by Lady Diana, Princess of Wales. The collar itself sports a square shape, giving the appearance of a square neckline though the neckline itself is V shaped often with a tie feature at the base of the V on a sweater.

Round necks, polo necks, and turtle necks work particularly well with raglan sleeves and saddle shoulder styles (Figure 16.8). Sleeve heads may also be shaped on the machine, using the press-off shaping process. Raglan and saddle shoulder sleeve styles are the variations on the set in sleeve. Shoulder shaping is incorporated into the sleeve and body pieces, producing a comfortable and stylish fit. The assembly requires an additional seam, making it more expensive to produce than a set-in sleeve, and therefore these styles are more likely to be used with higher quality yarns to further justify the cost.

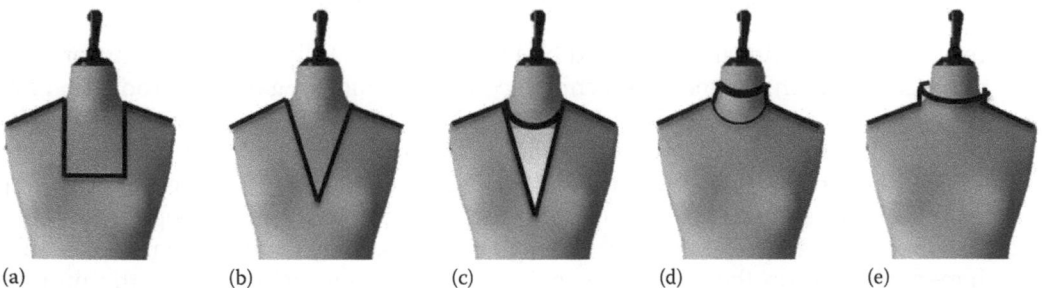

(a) (b) (c) (d) (e)

FIGURE 16.7
(a) Square neckline, (b) V neckline, (c) a filled-in V neckline, (d) Polo neck, and (e) Turtle neck.

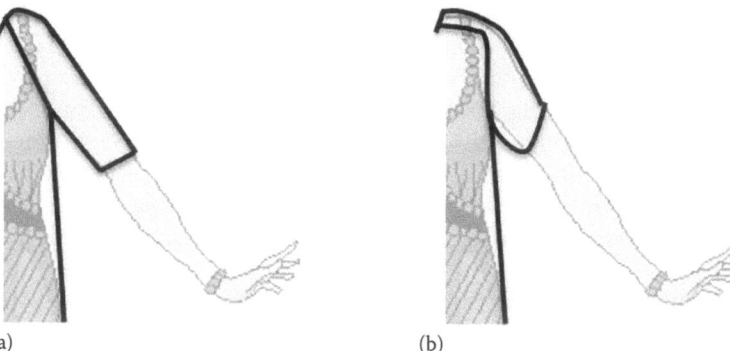

(a)   (b)

FIGURE 16.8
(a) Raglan sleeve and (b) saddle shoulder sleeve. The sleeve head of the raglan forms a larger part of the neckline than the head of the saddle shoulder, which is essentially a strip that extends from the head of an otherwise set-in sleeve that sits along the top of the shoulder and forms a small part of the neckline shaping.

Cup seamers are used to assemble sealed edges, whereas cut edges require overlocking and neck bands may be attached using a linker. Cup seamers are designed to uncurl the sealed edges of knit fabric, which is typically the characteristic of single-jersey fabrics when they come off the knitting machine. The fabric passes between two metal cups into the stitch zone acting as a feed device and hence the name of the machine. This is considerably a more skilled process than overlocking and therefore is a more expensive production method than the fully cut method. The neck and sleeve head shaping still tends to be relatively simple and basic for cost efficiency. The design emphasis is therefore more likely to be on the textile design, employing more intricate stitch structures or use of color. Textile design and color are particularly subject to trends. More interesting shapes such as raglan and saddle shoulder styles are more likely to be used with plain jersey fabrics as the interesting style features are easily lost in textured and multicolored textile designs, which would render the styles as insufficiently cost efficient.

16.10 Fully Fashioned, a Better Quality Knitwear Production Method, and More Sophisticated Garment Styles

Electronic flatbed knitting machines and CAD packages are fully utilized for this production process, and with advances in technology, fully fashioned garment production has become a highly profitable production method. Not only are more sophisticated styles possible but also an array of knit fabrics makes this method a knitwear designer's paradise. Necklines may still be cut and sewn using an overlocker if desired, though cup seamers and linkers are more fully utilized. As linking machines join loop for loop, the process makes a very neat join possible on raglan sleeves. Fully fashioned garment blanks are shaped entirely on the knitting machine by decreasing or increasing the number of stitches across the width of the blank and thus eliminating the need to cut. This is particularly a economical method of knitting in terms of fabric waste reduction, and all edges are sealed; however, the production rate is slower than producing unshaped blanks.

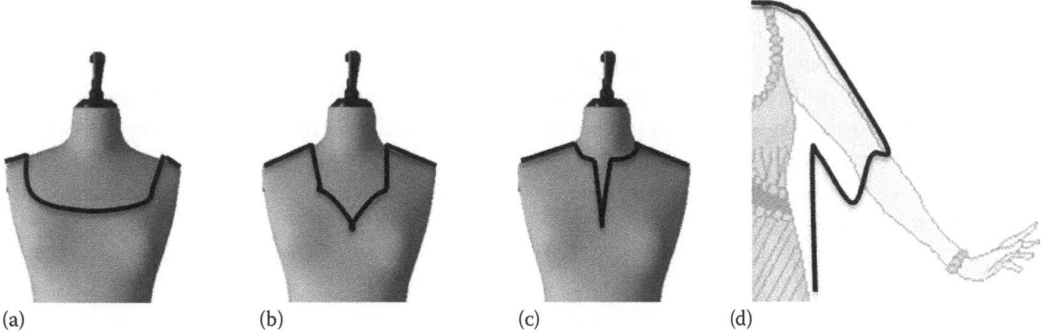

(a)   (b)   (c)   (d)

FIGURE 16.9
(a) Scoop, (b) sweetheart, (c) slot necklines, and (d) kimono sleeve.

To decrease the fabric width, one or more stitches are moved from the outer needle(s) and are placed inward onto corresponding needles; this process is known as fashioning. This forms a characteristic mark on the fabric face in which one loop can be seen to overlap another. When increasing the fabric, an additional needle is worked at the selvedge. For a smooth increase, one or more of the selvedge stitches can be removed from their needles to adjacent corresponding needles to introduce a new stitch inside the selvedge. This technique forms a hole underneath the new stitch, which may be used as a decorative feature. Alternatively, a loop from an adjacent stitch can be placed on the empty needle prior to the next course (row of knitting). Fully fashioned garments do not require overlocking; they are assembled with cup seamers and linkers. The fully fashioned method is most commonly used for more expensive yarns such as cashmere due to the reduced yarn wastage. Classics such as ladies' suits, jackets, and coats in luxurious yarns with a strong marketplace are particularly affordable to produce using this method, and designers can better exercise their creativity through more interesting design styles and fabrics.

More intricate variations of the round neck, such as scoop and standaway, and V neck variations such as sweetheart and slot are possible (Figure 16.9). The width of V necks can vary from very wide to very narrow; the slot neckline is particularly a narrow V shape. The sweetheart neckline is a variation of a hybrid V and square neckline. The slot and sweetheart styles are particularly flattering without a neckband trim and therefore highly suited to the fully fashioned technique with its characteristic neat sealed edges. Similarly, a scoop neckline can be particularly sensual with no trim. A further variation of this style is to shape the scoop in a more stylized manner, creating a horseshoe neckline or a more basic U shape. The standaway neckline is not too dissimilar from the turtle neck collar (shown previously in Figure 16.7) standing away from the neck but without a separate collar or band that looks very elegant.

Cap and kimono sleeves are subject to fashion trends. Cap sleeves are common on T-shirts in which they are often formed as a separate sleeve part. For knitwear, the cap sleeve is more likely to be knit as part of the body sections by increasing the number of needles to form the protruding cap; the style length is shown in Figure 16.3. The kimono style is particularly favored for lightweight or heavier weight outerwear. Textured yarns are often used to give added interest to the kimono style, providing a costly but stylish coat. The degree of slope or angle of the sleeve relative to the body affects the comfort and the aesthetic appeal of the style. At 90° such as a drop sleeve is full with good drape capability, depending on the softness of the yarn but will extend the shoulder line below the shoulder point, thus dropping down the arm. This can give the look of a cheaper cost garment. Whereas a slope of around 75° increases the possibility of restriction giving a poor fit unless the sleeve is large and loose such as a kimono (Figure 16.9).

A popular alternative front fastening to the single breasted is the double-breasted front fastening. Again the style is subject to trends but is considered to be a classic style. As there is a significant overlap of fabric for this style, consideration must be made to the thickness that the yarn and the fabric structure create. The style, however, provides extra warmth for the front of the body.

16.11 Integral Garments and Full Garment Knitting, New Technology for Knitwear Production, and Further Advanced Garment Styles

The original forms of integral machine knitting included the production of socks, hats, and gloves. Early use of the technology applied to garment production was restricted to more simplistic garment styles such as a waistcoat and often still incorporated some degree of cut and sew. With the aid of new state-of-the-art technology, today it is possible to knit garments entirely on the machine with little or no seams and including intricate design features such as pockets, trims, and even buttonholes all in one process and at relatively high production speeds. More complex garment styles that are possible using the fully fashioned production method can now be achieved with minimal making up using this fully automated technology. Such styles include batwing and dolman sleeve types (Figure 16.10), keyhole necklines, integral knit shirt collars, and shawl collars.

16.11.1 Fabric Designs

Different fabric structures are possible by manipulating the stitches as the fabric is constructed on the machine. However, this is not always possible when possible garments and the knitted fabrics should be unroved to accurately analyze the fabric construction and to analyze the yarns used in a garment. In this section, the stitch manipulation techniques are discussed.

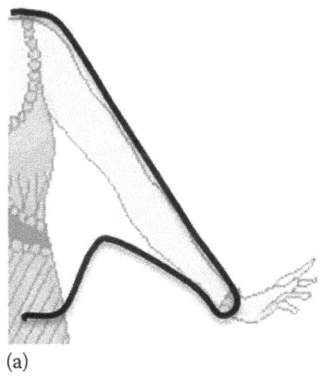
(a)

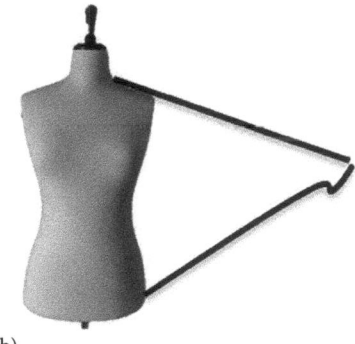
(b)

FIGURE 16.10
(a) Dolman and (b) batwing sleeves.

While the variations possible are limited in number here, in reality, when using these techniques the limitations are only in the design capability and technical realisation of the individual. The basic stitch manipulation techniques are described in this section with example fabrics in Figure 16.11. It is recommended that you fully explore stitch and fabric types through the wealth of available text and design books.

The most basic fabric is plain single jersey, which can be seen in Figure 16.1 in which the technical face has the characteristic V-shaped stitches, and the technical reverse has the U-shaped loops. To produce plain knit stitches, a new loop of yarn is pulled through the previous loop by the action of the needle; the previous loop is released from the needle and the new loop is held on the needle. An open fabric structure can be produced by simply omitting some needles in a sequence; for example, when five needles are working, one needle is out-of-work across the needle bed. Where there is a needle missing, the yarn will simply form a strand from the last needle loop to the next; the fabric has a delicate lace or laddered appearance. A variety of lace patterns can be produced using the previously described method for shaping in which the loops are transferred to adjacent needles to form holes in the fabric. Cables are also formed using the transfer method; however, rather than leaving empty needles to collect yarn for the formation of new loops as with lace patterns, cables are produced by changing the order of sets of loops periodically. For example, for a six-needle cable, stitches 1,2 and 3 are removed with a tool from their respectively numbered needles, and stitches 4, 5, and 6 are also removed from their respectively numbered needles using a separate tool. Stitches 1, 2 and 3 are replaced on the needles where stitches 4, 5 and 6 were originally. Stitches 4, 5 and 6 are crossed over stitches 1, 2 and 3 and placed on needles 1, 2 and 3. Normally, at least 6–8 rows of plain knit occur before repeating the transfer sequence. Cable sweaters were historically popular hand-knit garments using very heavy-weight yarns for warmth. Cable knits though now possible on modern knitting machines are costly to produce due to low productivity. Lace patterns are also time-consuming to produce because transfer rows are often required prior to knit rows though some machines are capable of transferring stitches and knitting in one row (or course) operation. A mock lace appearance can be achieved using tuck stitches. Tuck stitches are produced when instead of taking the yarn through the loop to produce a new loop, the needle holds the existing loop and the newly laid yarn without releasing the loop. This action may be repeated a number of times, depending on the yarn count and the needle size before knitting the loop to form a new loop. The original loop and the laid yarn strands are captured by the newly formed loop. At least one plain knit loop on either side of a tuck stitch is required. The resulting fabric is wider than plain fabric but requires more knit rows to create the same length of fabric as plain knit because the stitch formation produces flattened stitches. Tuck stitch is often used for double-bed fabrics to produce fisherman's rib or half fisherman's rib.

Miss or slip stitches are created when a needle is completely bypassed by the yarn; short strands are formed on the technical reverse of the fabric. When using only one color, the technical reverse is used as the fabric face because the strands can be strategically placed to produce patterns across the fabric surface. However, slip stitch is more likely to be used with more than one yarn to produce colored patterns on the technical face of the fabric, called single jacquard, which is more commonly known as Fair Isle after hand knit. Double jacquard is produced on double-bed machines in which the would-be strands on the technical reverse are knit on the second bed, creating a thicker fabric with no outward facing floats. Small and large motifs can also be incorporated in a knit fabric, whereas small

FIGURE 16.11

Top left to right: single-bed lace with a picot hem; double-bed tuck stitch fabric; double-bed tuck stitch with two-colored yarns; single jacquard with a plain rib welt (double bed). Bottom left to right: Two-color single jacquard with a tuck stitch rib welt (double bed); double-bed tuck stitch with transferred loops (lace stitch); intarsia knit using three-colored yarns incorporating lace pattern (transferred loops) and rose bud hem applique.

motifs are possible using the jacquard technique in which a special method and carriage are mostly employed; this process is known as intarsia knitting.

16.11.2 Garment Specs, Sample Garments (Prototyping), and Garment Finishing

Garment specifications (specs) are used in the industry to create garments exactly to the requirements of the customer/designer. The spec sheets normally contain garment sketches and/or pattern block illustrations, yarn samples and fiber content details, and knit instructions; a designer's example is given in Figure 16.12. Sample garments, or prototypes, are produced following the garment specs for inspection by the designer and/or customer (most likely to be a retailer). The prototype garments are usually made to size 12 in the United Kingdom. The garments are finished with all labels, and so on, prior to inspection that includes measuring the garments to ensure that they conform to the specs. The sample garments are often sealed and tagged with the company's seal to verify that the garment has passed the quality control inspection and now serves as the official quality for all subsequent garments that form part of that particular order. Other sizes are graded from the standard size 12.

Creases and careless staining can occur through excessive handling during the making of knitwear. At the end of the making process, garments are prepared for their journey to the retailer and subsequently to the consumer. Metal frames are used relative to the required finished garment size as inserts during the steaming process after the garment has been assembled and after the labels have been inserted. Heat, moisture—often in the form of steam, and a degree of pressure are used to deform or reform the fibers to eliminate creases. Garment is then packed for transportation to the customer's requirements, which are normally relative to how the garments will be displayed—on hangers covered with plastic wrappers or folded and boxed. Figure 16.12 also shows a mock factory layout.

16.11.3 Trend Forecast Sources for Knitwear

Once upon a time, trend forecasting information was confined to the industry sector that pays handsomely for the privilege. Trend forecasting information is more widely available today to the general public for fee, albeit however more related to what is readily available on the high street at that time rather than being months, or even a year or two in advance. Peclers Paris is one of the original trend forecasting agencies to create dedicated forecasts for knitwear and is still active today. According to expodatabase, there are currently trade shows for knitwear that are operating in 115 different countries. Many of the trade journals update the industry on-trend information from such trade fairs and shows. For students, worth global style network (WGSN) is a popular trend source accessible for free through most university design schools. Even for the general public, trend information is easy enough today to find on the Internet. Independent knitwear designers, such Debbie Bliss, give their own accounts of the current trends, and popular magazines such as *Vogue* and knit craft magazines are useful sources. Blogs and Pinterest are also becoming popular resources. Bloggers who attend catwalk shows are more likely to give a good spin on likely imminent trends.

Thus far, throughout the twenty-first century, knitwear has been predominantly plain knit, single color, and basic in style. The popularity of hand knitting as a hobby craft throughout the twentieth century influenced a number of exciting styles and yarns giving

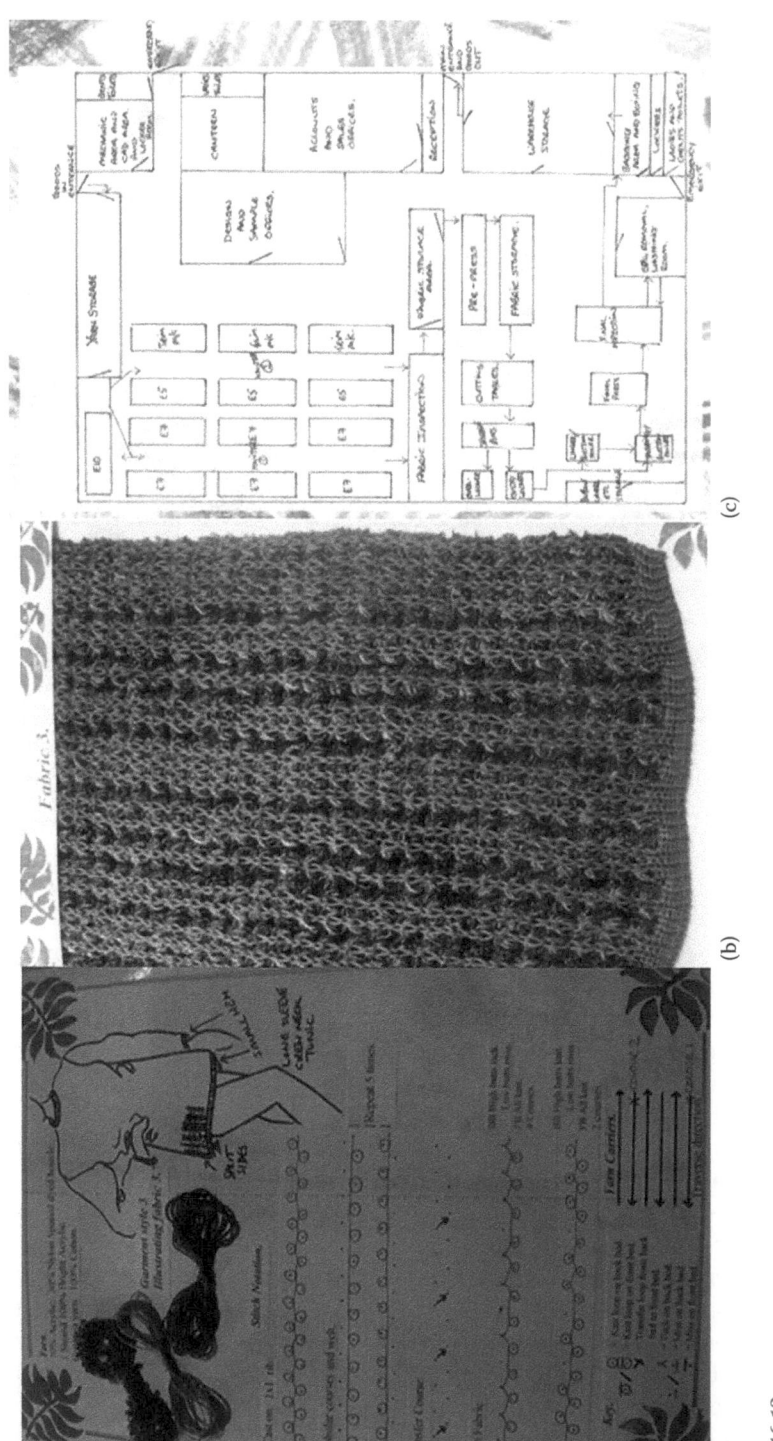

FIGURE 16.12

Example of (a) a designer's garment spec, (b) a fabric sample, and (c) a mock factory layout.

knitwear a creative edge. Even though the technology available today in the industry knitwear is still experiencing a lull; it is hoped that through the nitration of design and technology, a new generation of knitwear specialist will break free to use this highly creative medium to generate excitement again in the world of knit.

16.12 Conclusion

This chapter has introduced a rationale for encouraging knitwear specialists to develop both design and technological knowledge and skills for the betterment of the industry, education, and for the sake of readily available creative knitwear for consumers. This chapter briefly discusses the growth of the sector before focusing on the types of machines available. The four garment production processes have been discussed in relation to garment design and where relevant to trends. The reader is introduced to basic garment styles and a number of variations that have come in and out of favor for fashion over time. The reader is also introduced to some of the designers' tools such as pattern blocks and garment specs. This chapter concludes with a brief discussion of resources for knitwear trends and words of encouragement to develop as a creative knitwear specialist.

Bibliography

Brackenbury, T. (1992) *Knitted Clothing Technology*, Blackwell Publications, Oxford, UK.
Carr, H., and Latham, B. (1994) *The Technology of Clothing Manufacture*, 2nd ed., Blackwell Publications, Oxford, UK.
Donofrio-Ferrezza, L., and Hefferen M. (2008) *Designing a Knitwear Collection from Inspiration to Finished Garments*, Fairchilds Publications, New York.
Taylor, M. (1994) *Technology of Textile Properties*, 3rd ed., Forbes Publications, London, UK.

17

The Measurement of Textile Material Properties

Riikka Räisänen

CONTENTS

17.1 Introduction

This chapter describes different aspects that are connected to textile testing and measuring of textile properties. Textile material properties are tested for various reasons, for example, to obtain information that assists in material innovation and selection, to help to oversee the agreed level of production, and to secure that the final products fulfill the properties that are pledged for them. When talking about textile testing, the concept of quality is also concerned. What is quality and how can it be described? In this chapter, the aspects of quality are discussed first.

Later on, textile testing is discussed in connection to production processes and as a part of the quality control system within a company. Textile testing is used to analyze the characteristics of materials during the development of products. This chapter introduces textile standards and other methods that produce information about desired textile properties. The measurement and testing of material properties are explained under the following categories:

- Basic material properties
- Properties connected to mechanical strength

- Properties connected to aesthetic appearance and maintenance
- Properties connected to wear comfort
- Properties connected to safety

At the end, some future trends of textile testing are discussed.

17.2 Quality

Quality can be defined in various ways and can be approached from different viewpoints. There is not a single description of it. Ways to describe quality are through its objective or subjective nature. Objective quality describes the existing characteristics of an artifact or a process in measurable, quantitative way. With textile testing, the manufacturer can ensure that the garment or textile material has the properties that it is assumed to have, for example, the abrasion resistance of the furnishing material is sufficient for public premises. Subjective quality is qualitative, and it speaks about the context in which the concept of quality has been used. For example, it takes the customer's opinions into account and thus describes the quality on the personal level.

Furthermore, quality can be approached from the product base, manufacturing base, user base, and value base. Quality can also be seen as transcendent: luxury that is incomparable. Altogether, quality tries to attain positive and desirable good features. Quality indicates succeeding in processes and making. The international standard *ISO 9000 Quality management systems—Fundamentals and vocabulary* defines quality as *degree to which a set of inherent characteristics fulfils requirements.*

The product-based definition describes quality as properties that are measurable and quantitatively examined, for example, by using standardized tests. Such properties are, for example, durability and constancy. This type of quality can be considered objective. Product-based quality is connected to the pursuit of high-level characteristics, which often mean greater costs of production and sales but also a higher priced product.

The manufacturing and producing-based approach considers quality from the organizational point of view. Quality is described by manufacturing practices based on conformance to requirements or specifications. These specifications fulfill the criteria set by the producer, and the deviation from them implies a reduction in quality. The company may try to attain the optimal ratio for quality and costs by setting the level of criteria within the limits that ensure profitable production and still show characteristics of an object that are considered acceptable. Pass and fail limits divide products into *first quality* and *second quality*: the latter are products discarded as faulty or defective.

The user-based approach values quality from the personal point of view, which means that the context differs for every situation, and also the preferences vary individually from person to person. The customer may also value quality according to fitness for purpose and use and the benefit he or she gains. In the user-based approach, high quality means customer satisfaction and fulfilling his or her preferences and expectations.

The value-based description of quality takes into account not only the characteristics but also the cost and price. Thus, the customer's decision is based on the quality at an acceptable price and what could be called the *best buy*. The value-based approach also considers attributes such as ethical and sustainable production. Customers' opinions of the quality and satisfaction are affected by ethical and sustainable practices in the manufacturing processes and values into which the company has engaged, for example, the knowledge that the product is produced in developing societies to promote women's employment.

In the line of textile business, similar to any other line of businesses, public authorities lay limits to minimum quality. Quality is enacted by law. For example, the Consumer Protection Act is designed to protect consumers and to give them rights when buying goods and services, for example, manufacturers are legally obliged to put certain information on products, which in the question of textiles means fiber content, care instructions, and origin—manufacturer or country of manufacture. The Consumer Protection Act covers a range of issues including fraud, unfair business practice, and product liability. In addition to being bound by national and domestic legislation, businesses must take account of European laws and international agreements with regard to consumer protection.

17.3 Textile Testing as a Part of the Company's Quality Control System

Quality management is an important part of the operation of business, and it means comprehensive development of the company's processes in a way that demonstrates the company's ability to consistently provide products and services that enhance customer satisfaction and meet applicable statutory and regulatory requirements. Quality management aims not only to develop processes and products to achieve a company's business goals and objectives, but also to meet the needs and expectations of users and the market itself. Quality management covers different activities in the company from leadership to grassroot textile testing.

During product development, quality is evaluated in many aspects from the selection of the material till the finished product. Product development starts with ideation and defining the target group and continues with outlining and prototyping. New product proposals are valued as they take on their material, technical, and commercial forms. For example, the level and target for the quality are already set in the ideation step, and this determines, for example, the material choices, what material and fabric types are used, and what kind of in-use characteristics are emphasized. Before selecting the material for the product sample, materials are ordered from different suppliers. Materials are tested to obtain more information about their characteristics. Standardized laboratory tests and sensory and haptic evaluation may be used.

The choice of the material and fabric and the technical structure of the garment are specific and dependent on end use and situation. In addition, the final characteristics of the garment are affected by each of the materials and fabrics used and the combination of them. The properties of one fabric may be totally different when used alone than when used with a combination of other fabrics. This must be taken into account especially when designing, for example, work garments that require a high resistance against heat or flame.

In the prototyping stage, choices for materials, fabrics, and technical structures are tested and evaluated among the product development group. Solutions that produce the best compromises are selected. The final decisions are evaluated according to the appointed quality limits. To get proper information about the product in its actual use, prototypes can be tested in field trials. Such tests are common for textiles in special use, for example, technical sportswear, cold wear, and workwear.

Quality management ensures that the quality of the company's products, services, and activities are fit and as agreed between clients and customers. Higher quality allows companies to obtain higher margins. Quality increases productivity, and it leads to better performance in the marketplace. Customer satisfaction increases loyalty and sales, whereas complaints decrease them.

Customer satisfaction is one of the key factors, and well-performing companies may invest especially on this and may emphasize the need to delight customers by giving them more than what is required in the contract. These companies create a total experience for their customers, an experience that is unique in relation to the offerings of competitors. This kind of strategy is called *the experience economy*. A customer-focused company may also put a lot of effort in anticipating the future expectations of the customers. A company should listen to the customers and *real* users of the product or service in order to gain a clearer perspective on customer experiences, whereafter the company builds quality into the product, service, system, or process as upstream as is practicable.

Laboratory tests are simplified imitations of phenomena occurring when textiles are in use. Thus, the question of application and verification rises, and there is always a challenge to develop tests that correlate as much as possible with the consumer usage. Laboratory tests are of low cost and are available in short time limits, but sometimes it is necessary to perform field tests in real environments using test persons. Such tests are more subjective but give valuable information about the variation of preferences and test results.

17.4 Standards and Standardization

Textile testing is important not only from the perspective of research and product development but also from the perspective of agreements and that the product meets the requirements made of it. Textile testing can be performed not only according to standards but also according to any reliable and reported practice.

Standards are publicly available written documents that have been accepted as a solution of contract in any organization of standardization. Standards explain procedures about how to solve a problem. They aim to unify practices and courses of action. Everybody benefits from standards: manufacturers, suppliers and customers, authorities, and consumers. Standards are enacted to facilitate everyday life and to ensure that products and services are safe, reliable, and of good quality. For business, they are strategic tools that reduce costs by minimizing waste and errors and by increasing productivity. Standards help companies to access markets, level thresholds to participate and facilitate free and fair global trade. Standards are voluntary documents to use. Authorities use standards as tools, for example, to maintain law and order, as in the question of the Consumer Protection Act. Governments benefit standards in decision-making and in supporting public policy.

Materials, situations, and conditions are in constant change and development, and therefore new standards are needed all the time.

There are many organizations that develop and maintain standardization. Roughly grouped are international, regional, and local standardization organizations. When the industrialization increased production, it was necessary to draw up consistent lines in businesses. The first association for standardization was established in 1901 in Britain.

The International Organization for Standardization (ISO) is one of the most commonly known organizations. ISO is an independent, nongovernmental international organization, which began operations in 1947 in London to facilitate the international coordination and unification of industrial standards. Now, ISO's Central Secretariat is located in Geneva, Switzerland. ISO is derived from the Greek word *isos*, meaning equal. Over 160 countries are involved within ISO.

Another well-known organization is ASTM International (formerly known as the American Society for Testing and Materials). It is a voluntary, not-for-profit organization that provides a forum for the development and publication of international voluntary consensus standards for materials, products, systems, and services. More than 140 countries are involved with ASTM International.

The European Committee for Standardization (CEN), is a regional association that brings together the National Standardization Bodies of 34 European countries. CEN provides a platform for the development of European Standards and other technical documents in relation to various kinds of products, materials, services, and processes. CEN has been officially recognized by the European Union and by the European Free Trade Association (EFTA) as being responsible for developing and defining standards at European level. The standards verified in CEN are named as EN standards.

Local standardization organization exists at a national level. Most countries have their national associations. Local associations cooperate with regional and international organizations and develop and accept standards as local practices. For example, the British Standards Institution (BSI), formed in 1901, was the world's first National Standards Body. BSI represents UK economic and social interests across all European and international standards organizations. The standards verified in BSI are named as BS standards. DIN (Deutsches Institut für Normung) is the German Institute for Standardization, which was founded in 1917. American National Standards Institute (ANSI) is a coordinator of the U.S. private sector, a voluntary standardization organization. ANSI itself does not develop standards, but it represents U.S. interests in regional and international standardization activities. Nowadays, almost all national standards are regional (e.g., European) or international in origin.

Standards can be classified into groups according to International Classification of Standards (ICS), and the same standard may belong to different groups. General types of standards are as follows:

- *Basic standards*: Measures, concepts, signs, and symbols
- *Product standards*: Describe properties of products related to measures, structures, composition, durability, and safety
- *Testing standards*: Descriptions for determining raw materials and product properties
- *Service standards*: Describe quality requirements for services
- *Safety standards*: Aim at guaranteeing safety both for people and the environment

- *Terminology standards*: Concepts, descriptions, and examples
- *Designing standards*: Unifying basics for designing
- *Management standards*: For example, quality and environmental management

Standards operate with metric and International System of Units (SI unit systems).

Companies very often have their own testing laboratories, but there are also independent laboratories that offer testing services. Such laboratories operate worldwide, and they may concentrate to perform only particular tests. Part of the testing laboratories may be certified to perform specific tests that are, for example, part of the materials' or businesses' certification systems. For example, only certified laboratories may perform the official test included in the safety certification systems such as European CE marking, which sets criteria for certain products in order to let them be sold anywhere in the EU, CE stands for *Conformité Européenne*, which is French for 'European Conformity', or voluntary Oeko-Tex 100 textile standard, which is a worldwide independent testing and certification system for raw, semifinished, and finished textile products.

17.5 Textile Testing Principles

Textile testing can be performed according to standards or any trustworthily reported practice. Standardized testing provides equality and comparison of the test results and reliability and validity of procedures. Standards determine the conditions in which the testing is performed. Conditions cover, for example, laboratory conditions—temperature and humidity, sampling, apparatus and its functioning, performing the test, and reporting the test results. When using testing according to the particular standard, the results obtained from different laboratories are comparable. It is always important to report the number and the name of the standard and the year of the verification because the same standard may be developed further, and thus the year of verification is necessary to report. Results from different standards are not comparable even if they test the same or nearly the same properties.

In the case of textile testing, the question of atmospheres for conditioning and testing—laboratory conditions—is important. Textiles that absorb moisture and water content within a textile will affect its properties, for example, by increasing the mass (weight) and changing the strength properties. Therefore, testing is performed in a laboratory in which the temperature and humidity remain constant. Standard atmosphere for conditioning and testing is usually a room temperature of 20°C ± 2°C and relative humidity (RH) of 65% ± 4% (ISO 139). But there are also other conditions, for example, to apply the tropic climate. The proper testing conditions are informed in each standard. The higher the temperature, the more water air can contain. The higher the RH, the greater amount of water the textile can absorb. Before testing, the textile materials must be conditioned in the laboratory. Textiles are placed in the standard laboratory conditions in open baskets so that air can flow through and long enough that the moisture balance has been reached. This takes usually from 12 to 24 h.

Preconditioning may be necessary in cases in which it seems possible that the material would release moisture instead of absorbing it when placed in the standard laboratory conditions. The objective is that in the actual conditioning, textile would absorb moisture, not release it. Thus, preconditioning by drying the textile in the oven (60°C–70°C, RH ≤ 10%) is necessary (ISO 139).

Testing can also be performed in a regular laboratory that does not have an air-conditioner. Such testing is valid, for example, when selecting materials and fabrics within a company and among a chosen batch that has been tested in the same conditions. Unfortunately, the results are not comparable with any other results.

Before testing, it is also important to ensure that sampling is performed according to representative principles. It is not possible to study all the products, not even the whole batch, and therefore sampling must be done. A sample represents the whole textile or material under investigation, and conclusions concerning its properties can be drawn from the sample's test results. Sampling should be accomplished on a basis of randomness: samples should be taken from different parts of the studied material—along the width and the length of the fabric, for example, from the beginning, middle, and end of the roll of the fabric or the packing of yarn. A sample should be taken far enough from the end of a fabric (1 m) or a yarn, furthermore from a sufficient distance from the selvedges, and faulty areas should be avoided. A sufficient amount of parallel samples should be taken—if the material to be tested is homogenous, fewer samples are required compared to the situation where the material is very uneven. Parallel samples should contain different warp and weft yarns. Detailed instructions for sampling are not only explained in special standards concerning sampling (EN 12751, ISO 1130) but also usually in every testing standard.

17.6 Testing Material Characteristics

Testing gives information about textile properties and helps to choose proper fabrics for intended end uses. In this chapter, a few common testing procedures are explained in the following categories: basic material properties and fabric measurements, testing properties connected to mechanical strength, aesthetic appearance and maintenance, wear comfort, and safety.

17.6.1 Testing Basic Material Properties and Fabric Measures

A textile can be described by its basic structure. Fabric properties are affected by fiber type and content, yarn number and twist, fabric thickness, fabric weight, weave or knit structure, and fabric density.

17.6.1.1 Fiber Tests

There are many characteristics that can be determined from fibers, for example, fiber length, the diameter, and the fineness of the fiber. In addition, a fiber's crimp frequency—the length and the amplitude of the crimp—is critical because it determines fabric properties such as volume and elasticity. Tensile properties, breaking force, and elongation at break of a single textile fiber can also be tested.

Fiber fineness is critical when producing yarn. The finer the fiber, the thinner yarn can be produced. Fineness of the fiber also affects its stiffness; thick fibers are stiff that in turn affect the fabric's drape, stiffness, and also fabric hand such as roughness. For synthetic filament fibers, fineness is constant in the same fiber lot, but for natural fibers it varies greatly. Fiber fineness can be expressed as fiber diameter in μm (micrometer) (ISO 137) or as linear density dtex (= 0.1 tex) (ISO 1144), which means the weight of the fiber per length (g/10,000 m = 0.1 mg/m). The fiber diameter cannot usually be used to describe the fineness, because there are only a few fibers that are round. Furthermore, cross section may vary in shape along the fiber length and between the different fibers. Fiber fineness may be determined by a gravimetric and a vibroscope methods (ISO 1973) or an air permeability method using a micronaire instrument (ISO 1136, ISO 2403). In the gravimetric method, the mass of a fiber is calculated by multiplying the cross-sectional area, length, and density. The air permeability method is based on a phenomenon in which fine fibers generate more resistance against airflow, and thus less air is passed through the fibers compared to the coarse fibers. The method is relatively accurate for fibers with approximately circular cross sections and constant overall densities.

17.6.1.2 Yarn Tests

When producing yarn, the aim is that the final product is as even as possible, because yarn unevenness causes problems in production and faults in products. Yarn evenness can be examined as yarn count constancy, uniformity of fiber diameter and material content, regularity of twist, and breaking force.

Yarn count gives the ratio between the weight and length. Yarn count can be expressed with the direct or indirect count systems (ISO 1144). Direct count systems express the linear density of the yarn, and tex and denier are the most commonly used units. Tex is the yarn count recommended by the SI unit system. Tex (T_t) tells the weight in grams of the yarn that is 1 km long (g/km). Denier (T_d) gives the weight in grams of the yarn that is 9 km long; the unit is den. For determining the yarn count, a sample of fixed length, for example 100 m, is cut, and its mass is weighed. The result is then multiplied, in this case by ten, to obtain the final yarn count. In the direct count systems, the statement *the bigger the number, the thicker the yarn* is valid. Indirect count systems give the length of the yarn per the mass. Such counts are, for example, metric number Nm, which gives the length in meters of the yarn that weigh 1 g, or the English cotton count Ne_c, where Ne_c gives the number of hanks all 840 yards long that weigh 1 lb. In indirect count systems, the statement *the bigger the number the finer the yarn* is tenable.

Staple yarns have twist to keep the constituent fibers together. Twist is usually expressed as number of turns per unit length such as turns per meter. Twist direction can be denoted as S or Z according to the direction of the angle whether it turns to the left or right, respectively. The simplest method for the twist measurement is the direct counting method (ISO 2061). A fixed length of yarn is untwisted until all the fibers are parallel to the axis. The instrument shows how many turns of S or Z twist were recorded by the rotating jaw, and the final number of turns per meter can be calculated.

17.6.1.3 Basic Fabric Measurements

Dimensional properties of fabrics can be determined as length, width, and thickness.

Fabric length gives the distance from one end to the other when the fabric lays freely, having no wrinkles and when the fabric is in a nontensioned state. Total fabric gives tells the distance between the outermost edges of the fabric measured perpendicularly to the produced length of the fabric. Utility width leaves the selvedges, markings, and other irregularities in the edge areas out of the measurement. Fabric length is expressed in meters (m), and width is expressed in centimeters (cm). The width of the fabrics varies because there are different loom sizes and knitting machines, and different amounts of shrinkage occur in the finishing processes (ISO 22198).

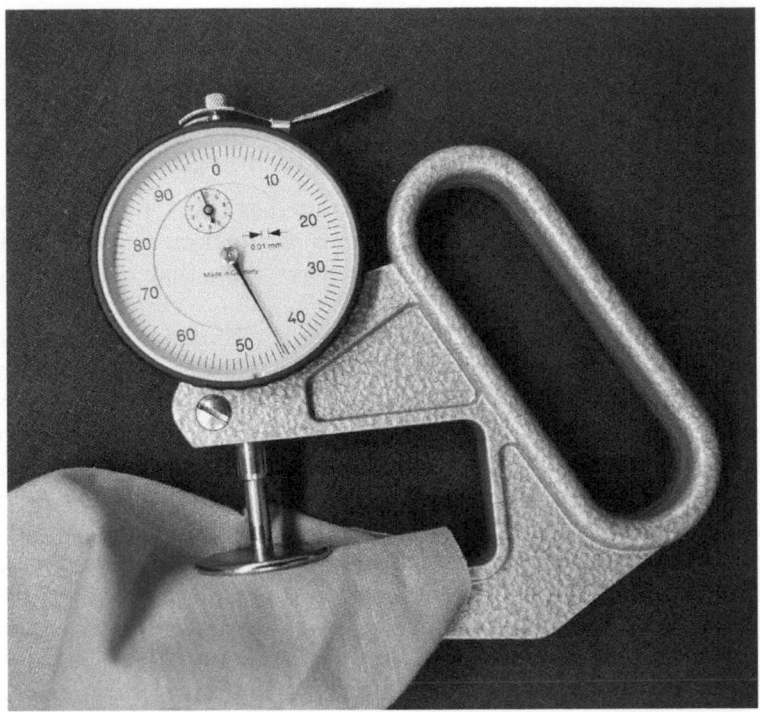

FIGURE 17.1
A thickness tester.

Thickness gives the distance between the upper and lower surfaces of the material. It can be measured using a special tool that produces a small pressure on both sides of the fabric (ISO 5084) (Figure 17.1). Thickness is expressed in millimeters (mm). Knowledge of fabric thickness is needed, for example, when measuring tear strength. In addition, thickness can be measured before and after a test and enables assessment of, for example, the fabric performance in abrasion. Fabric thickness also gives valuable information when selecting a fabric for a product. Thickness increases the warmth and bulk properties.

Fabric weight gives the mass per unit area and is calculated from a fixed area sample whose mass has been weighed (ISO 3801, EN 12127). A small sample of 100 cm^2 is usually cut, and the final measure is pronounced in grams per square meter (g/m^2). This procedure applies both for woven and knit fabrics, nonwovens, and for fabrics produced by any other technique. Fabrics are often sold with fabric weights per unit area, and it is an easy measure to compare similar materials, for example, single jerseys for T-shirts. There is a wide range of fabric weights because manufacturers try to obtain the best balance between the cost and end-use performance. Fabric weight has an effect on other fabric properties such as wear strength, fabric handle, moisture absorbance, and air permeability.

Weight can also be pronounced as mass per unit length, which means that a 1 m length of fabric as wide as the entire fabric is weighed (ISO 3801). The problem with this measure appears in comparison because fabrics have different widths.

Fabric sett tells the number of warp or weft yarns per length for woven fabrics or the number of courses or wales per length for knit fabrics (ISO 7211-2). Both the number of warps/wales and wefts/courses are counted, and according to SI units, a number per centimeter (cm) is reported. The number of yarns are usually counted from a longer distance, for example, 2–10 cm, and are divided to get the result per one unit (cm).

17.6.2 Testing Properties Connected to Mechanical Strength

The mechanical strength of a fabric can be examined as breaking, tearing and bursting strength, abrasion and strain resistance, and yarn distortion. The mechanical strength required by technical textiles is considerably more than that required by fabrics for normal use.

Breaking strength is the maximum tensile force that is recorded when extending a sample to its breaking point. Breaking elongation is the increase in length that occurs when the fabric breaks. The tensile force at the moment of rupture can be referred to as the tensile strength at break. However, the elongation of the specimen may continue after the maximum tensile force so that the tensile strength at break is actually lower than the maximum tensile force (Figure 17.2). This kind of test is valid for woven fabrics that have yarns parallel to the stretching force in a rectangular strip that is stressed along the longer side. Samples are tested parallel to warp and weft, and thus tensile force for warp and weft is obtained as newtons (N). Breaking force can also be determined from wet specimen according to these principles (ISO 13934-1). Tensile properties of fabrics and determination of the maximum force can also be tested with the grab method (ISO 13934-2).

Tearing strength by a single-tear method (ISO 13937-2) explains the force that is required to continue tearing the fabric along a cut made in it. The tear force is determined in the warp (or length) and weft (or transverse) direction, and it is expressed in newtons (N). The test specimen, which has the shape of very small trousers, is placed in the instrument and is held at each leg by a clamp. The clamps start to move in the opposite directions, tearing the sample with a constant rate of extension. The test is mainly applied for woven fabrics. If the yarns shift in a woven fabric, it is possible that the fabric does not tear directly along the direction of force. If any slippage of threads out of the fabric is observed, the result is discarded. A tearing test can be accomplished with samples of different shape, for example, there is a single-tear method for wing-shaped test specimens (ISO 13937-3) or double-tear test for tongue-shaped test specimens (ISO 13937-4). Tests for different shaped

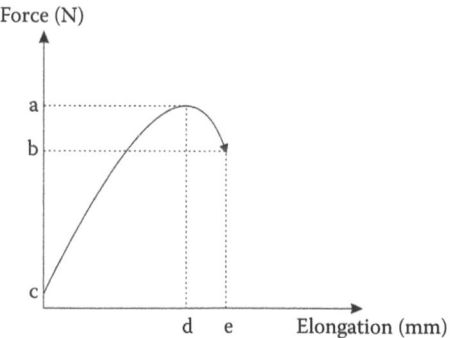

FIGURE 17.2
Force elongation curve from the determination of the tensile properties of fabrics using the strip method (ISO 13934-1), a, maximum force; b, force at rupture; c, pretension; d, elongation at maximum force; and e, elongation at rupture.

samples are needed because the trouser-shaped method does not function for all fabrics. Tear force can also be measured according to the ballistic Elmendorf method (ISO 13937-1).

Bursting strength expresses the amount of pressure required to rupture a fabric. There are two test methods that can be applied: the hydraulic method (ISO 13938-1) and the pneumatic method (ISO 13938-2) for determination of bursting strength and bursting distension. In the first method, a rubber diaphragm is expanded by liquid pressure until it pushes its way through the fabric that is clamped over it, and in the second method, a steel ball is pushed through the fabric. These test methods can be applied to all kinds of fabrics such as knitted, felted, woven, nonwoven, and laminated fabrics.

After a textile has been subjected to a force even for a short period of time, the complete removal of the force allows the specimen to recover its original dimension, rapidly at first but then more slowly, and finally a small amount of residual extension may remain. The remaining extension is known as permanent set. The *elastic recovery* can be calculated from recovered and original extension. For perfect elastic materials, the recovery is 100%. Fabric resistance to strain and elastic properties can be tested using DIN 53835-13 standard. The ISO standards for determination of elasticity of fabrics by strip tests (ISO 20932-1), multiaxial tests (ISO 20932-2), and testing of narrow fabrics (ISO 20932-3) are under development.

There are many forms of abrasion, and thus there are also many tests to measure the resistance against it. Abrasion can occur on the plain fabric, in seams and edges, and between fibers when fabric is bent. *Abrasion resistance* can be described as the resistance to the wearing away of any part of a material when it is rubbed against another material. There are several apparatus for testing abrasion resistance, for example, the Stoll Flex tester, the Taber Abraser, and the Wyzenbeek tester, and there are specific standards for each of them. The Martindale apparatus provides several methods for determining abrasion resistance, such as determination of abrasion resistance of fabrics by observing the specimen breakdown (ISO 12947-2), the mass loss (ISO 12947-3), and the appearance change (ISO 12947-4).

The main principle in the Martindale test is that a circular sample, placed in a specimen holder, is rubbed against a standard fabric (usually wool) according to a Lissajous figure, which is formed by the apparatus (Figure 17.3). The abrasion resistance is determined from the inspection interval to the breakdown of the sample (ISO 12947-2). The breakdown is reached at a point when two separate threads are broken in a woven fabric, one thread is broken in a knitted fabric, pile is fully worn off in a pile fabric, or when the first hole having a diameter of 0.5 mm appears in a nonwoven fabric. Depending on the end use of the fabric, the abrasion load can be changed to create a pressure of 12 or 9 kilopascals (kPa)—12 kPa for workwear, upholstery, bed linen, or fabrics for technical use and 9 kPa for textiles in apparel and household use. As a test result, the number of cycles on the testing machine will be reported.

17.6.3 Testing Properties Connected to Aesthetic Appearance and Maintenance

Properties connected to aesthetic appearance and maintenance of textiles can be examined from, for instance, pilling propensity, wrinkle recovery, dimensional changes in washing, and color fastness to washing, rubbing, and light. Furthermore, appearance of seams can be evaluated after predisposing them to different conditions. Tests that examine such properties give valuable information about how the product would maintain its looks in wear and under textile care operations such as washing and drying.

Surface *pilling* occurs in almost all fabrics, and it is one of the reasons why a garment is discarded even if it otherwise serves well. Pilling occurs because fibers in yarns and textile

(a)

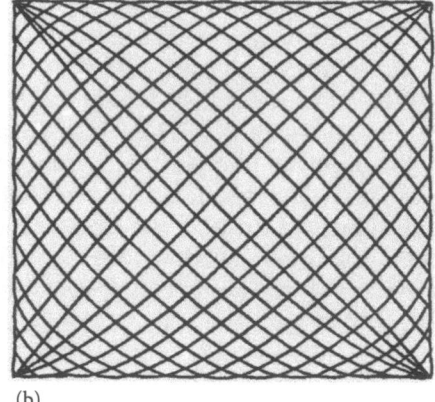
(b)

FIGURE 17.3
(a) Martindale abrasion testing apparatus (ISO 12947-1) and (b) the Lissajous figure of abrasion movement. (Courtesy of Riikka Räisänen.)

structures start to move against each other by wear and abrasion. Fiber ends stick out from the fabric surface and start to curl and tangle around each other causing small balls, pills. It is also common that when fiber ends stick out, they do not entangle and form pills, but lint comes off and fabric loses its thickness. Furthermore, it is usual when rubbing is continued long enough that fibers and pills wear off, and fabric surface appearance may become better looking.

ISO 12945 describes methods for the determination of the resistance to pilling and surface change of textile fabrics. The standard describes several methods from which the first one (ISO 12945-1) describes the pilling box method in which fabric specimens mounted on polyurethane tubes are rubbed against the walls of the cork cube and against each other. The assessment of fuzzing and/or pilling is evaluated by viewing the specimens in the viewing cabinet and grading them from 1 to 5 in accordance with the grading scheme given by the standard. Grading is subjective in nature, but reliability is enhanced by assessing specimens several times and changing specimen's position in the viewing cabinet, using several observers and additional photographic assessment. The pilling propensity can also be tested using a modified Martindale method (ISO 12945-2) and random tumble method (ISO 12945-3).

Wrinkle resistance is the property that enables a fabric to recover from being folded and from forming undesirable wrinkles. ISO 9867 describes a method for evaluating the appearance of textile fabrics after induced wrinkling. It is applicable to fabrics made from any fiber or combination of fibers. In the test, the instrument is first heated to a temperature imitating body temperature, and a damp paper is added on the lower flange. The test specimen is attached between the instrument's upper and lower flanges, and the upper one is screwed down along a groove to produce standard shaped wrinkles. A load is added to intensify wrinkling, and the specimen is allowed to remain for a time period after which the wrinkling result is evaluated by comparing the specimen to three-dimensional wrinkle recovery replicas that express the strength of shapes from 1 to 5, where 1 represents the strongest wrinkling. Wrinkle recovery is evaluated after 24 h.

Color fastness describes the properties of dyes in fabrics when enduring conditions and chemicals that they are exposed to. ISO 105 is a set of standards that explain test procedures for several color fastness tests, test conditions, and assessing procedures. Generally, one standard explains the determination of color fastness against one factor. Color fastness is assessed in most tests as a change in color and staining. A color fastness test applies not only for the specific dye examined but also for the dyeing procedure. Therefore, the same dye used in a different dyeing procedure might give a different result in a color fastness test. Furthermore, the dyed fiber type has an effect on the result.

In a color fastness test, evaluation is accomplished by assessing the change in color between the original fabric and the specimen after the test. The change in color is evaluated using a gray-assessing scale (ISO 105-A02). The five-step scale contains gray chips next to each other becoming increasingly lighter shades of gray illustrating the perceived color difference in contrasts. In the scale, the rating runs from 5 to 1, where 5 is the best value—no change in color. Staining is evaluated from the standard adjacent fabric that has been attached to the test specimen. Staining is evaluated with the gray–white assessing scale (ISO 105-A03) in which the chips start becoming white and then become increasingly darker gray. In the scale, 5 is the best value and relates to no staining, whereas 1 is the inferior value with the most staining—white–dark gray chips. A multifiber adjacent fabric or two single fiber adjacent fabrics can be used with the specimen.

Color fastness can be assessed against daylight (ISO 105-B01) or artificial light (ISO 105-B02), weathering (ISO 105-B03), domestic and commercial laundering (ISO 105-C06), rubbing (ISO 105-X12), rubbing with organic solvents (ISO 105-D02), sea water and chlorinated swimming pool water (ISO 105-E02 and E03, respectively), and also perspiration (ISO 105-E04) just to give some examples.

17.6.4 Testing Properties Connected to Wear Comfort

Wear comfort depends on a person's physical and psychological moods. Physical aspects are connected, for example, to heat equilibrium, whereas psychological aspects are connected to attitude, fashion, and personal situation. Physical aspects of wear comfort and testing procedures that provide information about them are discussed here.

There are textile properties that can be determined through their ability to increase wear comfort. Such properties are, for example, thermal insulation, breathability and air permeability, water absorbency and water removal ability, water repellency and waterproofness, and fabric hand.

For cold-weather clothing, heat comfort and heat resistance properties are important. In the stage of heat equilibrium, the heat production and heat transfer are in balance. Clothing and textiles have a great effect on how this stage of wear comfort is reached. Thickness of a fabric tells something about its thermal insulation ability, but because transmission of heat through a textile occurs in many ways, for example, by conduction through the fibers and the entrapped air and by radiation through the air spaces within the fabric, more advanced methods are needed. The ISO 5085 standard determines *thermal resistance* of textiles. The method is suitable for materials up to 20 mm thick, and it determines the material's thermal resistance and thermal conductivity.

Airflow has an effect on heat transfer. An *air permeable* fabric increases airflow, heat convection, and thus cooling within clothing. Therefore, for cold-weather clothing, windproof materials are desired as the outermost layer. The method for the determination of the permeability of fabrics to air is described in the standard ISO 9237 and is applicable to most types of fabrics, including industrial fabrics for technical purposes, nonwovens, and made-up textile articles. The method measures the velocity of an airflow that passes perpendicularly through a test specimen under specified conditions of time, test area, and pressure difference between the two sides of the fabric. The air permeability may be different when measured from different directions of the same fabric, and therefore it is important to report which side of the fabric has been tested.

The measurement of *water vapor permeability* of textiles is explained in the ISO 15496 standard. It is a simple method for testing and will provide the manufacturer with a method for quality control within the plant. It gives reliable information when comparing fabrics tested at the same time. However, this method is not applicable for classifying the water vapor resistance of textiles against values relating to physiological effects specified in product standards. Water vapor permeability (WVP) is a characteristic of a textile material that describes the amount of water vapor that diffuses through the textile per square meter, per hour (g/m^2 24 h), and per unit difference of water vapor pressure across the textile.

The increasing amount of technical textiles used in sports and workwear is designed to transfer perspiration and evaporated water from skin to the outermost textile layers. Moisture can be transferred by diffusion through pores in textile: there can be capillary transport in yarn and fiber capillaries, absorption–desorption within a fiber, and adsorption and migration on the fiber surface. The test method for the determination of surface *water absorption of microfiber fabrics* is under development (ISO 20158).

A fabric's ability to repel rain can be determined according to the *spray test* (ISO 4920), which defines the fabric's resistance to surface wetting. It is not intended for use in predicting the rain penetration resistance of fabrics, because it does not measure penetration of water through the fabric. In the test method, a certain amount of water is poured through a funnel that creates a spray on the fabric that is inclined at an angle of 45° toward the spray. The fabric's degree of wetness is assessed according to the standard rating charts based on photographs or a verbally descriptive scale explained in the standard. Evaluation runs from 1 to 5, where 5 is the best value, meaning no wetting of the sprayed surface. The method is applicable to all types of fabrics that might or might not have been given a water resistant or water repellent finish.

The Bundesmann rain-shower test (ISO 9865) determines the water repellence of fabrics by measuring the amount of water in milliliters (mL) that has penetrated the fabric after an artificial rain shower for a specific duration. In addition, the amount of water absorbed by the fabric is calculated from weighing the mass of the fabric before and after the test. Furthermore, the water runoff may be recorded. The water repellency is assessed according to the grading photographs from 1 to 5, where 5 is the best value. The test may be used to assess the effectiveness of finishing procedures for rendering textile fabrics water repellent.

Water penetration of a fabric can be tested according to the ISO 811, which explains a method to measure the hydrostatic pressure as conventional centimeters of water (cmH$_2$O) at the point in which the water penetrates the specimen in a third place. The method is primarily intended for laminated and dense fabrics, for example, waterproof but also breathable fabrics such as semipermeable fabrics used in sports, outdoor and workwear, tenting materials, and tarpaulins.

Fabric hand is probably one of the most widely used criteria for quality when evaluating and selecting textiles and materials, for example, during the product development process or when buying a textile product. Fabrics are evaluated haptically by touching. Properties that are connected to the concept of fabric hand are, for example, softness, compressibility, stretchability, resilience, density, surface structure, surface friction, and warmth.

There are guidelines compiled by the American Association for Textile Chemists and Colorists (AATCC) for the subjective evaluation of fabric hand. Certain characteristics of a fabric are evaluated by comparing the specimen to the reference materials. An assessor has his or her eyes covered, and evaluation is based on the feeling of touch and the movement of the fabric between the fingers. Evaluation is repeated several times. Evaluation by hand is subjective, but can fabric hand be objectively tested?

Obviously, there has been a need for quantitative measurement of fabric hand. In Japan, the Hand Evaluation and Standardization Committee (HESC, established in 1972) developed, under the direction of Professor Kawabata, a procedure that is based on the physical measurements of fabric qualities. The KES-FB system for the evaluation of fabric hand (Kawabata Evaluation System of Fabrics) includes several methods that use a tensile tester, a bending tester, a shear tester, a compression tester, and a surface (roughness and friction) tester. The KES-FB system preforms 6 measurements with these 4 instruments, and all together 17 different physical quantities can be obtained or calculated. As a result, a fingerprint diagram can be printed for each tested fabric in regard to these quantities. Fingerprints of different fabrics can be compared, and estimations about sewability and formability, for example, can be made.

17.6.5 Testing Properties Connected to Safety

One aim of textile testing is to increase the safety of textiles. The aspects connected to safety include, for example, chemicals that cause allergy. Dyes and finishing auxiliaries such as formaldehyde may cause irritation and allergenic reactions. Especially with workwear, the safety aspects are important because the main function of them is to protect the worker from chemicals, flame or heat, metal splashing, radiation, and so on.

Children's clothing is the kind of textiles that have special criteria for safety. Standard (EN 14682) determines the qualities of cords and drawstrings on children's clothing. In addition, children's nightwear should fulfill the burning behavior specification given by the standard (EN 14878).

The flammability of children's wear and interior textiles such as bed items, curtains, carpets, and upholstered furniture must be tested especially when they are meant for public premises. This is particularly important in fabrics that consist of cotton and other cellulosic fibers though it must be remembered that nettle fiber performs well with regard to flammability.

17.7 Conclusion

For a sustainable future, the development of materials and processes is vital. Ecological and ethical principles of textile manufacture and businesses and valuing of quality will hopefully have an increasing emphasis and become an important part of companies' objectives and courses of action. Customers and consumers are aware of environmental factors, and legislation imposes new challenges on production. Standardization is needed to give guidelines for companies on how to implement good practices. Standards and certification also give information to customers about the company's practices and engagements. Development of materials and processes is fast, and testing is needed to ensure that we all can live in a safe world in the future.

Bibliography

ANSI American National Standards Institute. https://www.ansi.org/ (accessed March 5, 2017).

ASTM International. https://www.astm.org (accessed March 5, 2017).

BSI British Standards Institution. https://www.bsigroup.com (accessed March 5, 2017).

CEN European Committee for Standardisation. http://www.cen.eu (accessed March 5, 2017).

Cohen, A. C. and I. Johnson 2010. *J.J. Pizzuto's Fabric Science*. 9th ed. New York: Fairchild Books.

Dale, B. G., T. van der Wiele, and J. van Iwaarden (Eds.) 2013. *Managing Quality*. 5th ed. Malden, MA: Blackwell.

DIN German Institute for Standardisation. http://www.din.de (accessed March 5, 2017).

Garvin, D. 1988. *Managing Quality. The Strategic and Competitive Edge*. New York: Free Press.

ISO International Organisation for Standardisation. http://www.iso.org (accessed March 5, 2017).

ISO 9000 (2015). *Quality Management Systems—Fundamentals and Vocabulary*. Geneva, Switzerland: The International Organisation for Standardisation.

ISO 9001 (2015). *Quality Management Systems—Requirements*. Geneva, Switzerland: The International Organisation for Standardisation.

Oeko-Tex Standard 100. https://www.oeko-tex.com (accessed March 5, 2017).

Räisänen, R., M. Rissanen, E. Parviainen, and H. Suonsilta 2017. *Tekstiilien Materiaalit* (Textile materials). Helsinki, Finland: Finn Lectura.

Saville, B. P. 1999. *Physical Testing of Textiles*. Manchester, UK: Woodhead Publication in association with the Textile Institute.

18

Textile and Clothing Consultancy

Tom Cassidy and Dian Li

CONTENTS

18.1 What Is a Consultant?

A consultant is a person who fits the required person specification for a client's project in terms of skills, knowledge, and experience and who has the ability to manage a project for successful completion. How does a person assess their readiness to be a consultant? This is a tough question but perhaps the person should stand in front of a mirror and ask the following questions:

- Do I have sufficient subject authority in the area of the project?
- Do I have sufficient hands-on experience?
- Am I prepared to commit the time required for the project?
- Am I prepared to travel and live on-site for the required duration of the project?
- Do I have a national/international perspective on life and respect for all cultures?
- Can I communicate well?
- Can I manage well?
- Do I understand how to plan and schedule a project?
- Do I fully understand the implications of the project?
- Am I creative and flexible when solving problem?

If the answer to all of these is yes, then you have made a start but every project will bring new questions for you to ask yourself. The questions will be less daunting as you build up experience though if you can honestly self-assess; some of them may become more daunting, but that is not a bad thing and will lead to continuous performance improvement.

18.2 The Birth of Projects

Every project starts with someone or a group of people who has an idea or who identifies a problem/challenge. Examples are as follows:

- Why are we losing stock from our cashmere and lambswool yarn stocks?
- How can we improve productivity and quality in our woolen yarn spinning mill?
- Rather than just selling all our wool overseas, should we set up a mill to process it into yarns and possibly the fabrics and garments?
- How do we change from quality control to process control?
- Should we add value to our mohair by scouring it before selling it, and therefore should we set up a scouring plant?
- What can we do with remnant warp silk filament yarns that might provide some paid work for women in poor rural areas?

At this point, we should consider the types of project that might be the domain of the design technologist by himself or herself and those in which he or she might be involved in as a part of a team. We must also consider the tools/techniques and methods that can be used by the design technology consultancy although, within the limitations of one chapter, these will be neither prescriptive nor exhaustive. In the later sections of this chapter, we will use the aforementioned examples as case studies and refer back to the following observations.

18.3 Feasibility Study

The purpose of the feasibility study is to allow the clients and/or the funders to decide the likely viability of the project as proposed by the consultant. The size and content of the feasibility study will vary, depending on the scale of the project. The days are almost over when a single consultant or even when a single agency would be asked to undertake the project management of a model new textile factory project, and so the design technology consultant is much more likely to be looking at the scale of projects given earlier or to be asked to look at specific elements of a more major project. However, it is still necessary for the consultant to provide input to the feasibility study of a major project or to prepare a suitable scale feasibility project for the more likely type of project that they can confidently take on and succeed. This does not mean that the range of projects in which design technologists will be involved in is small, in fact, the opposite. The range of projects is only limited by the consultant's subject knowledge, experience, and self-confidence.

The following is a list of the minimum contents of a feasibility study:

- *Background*: This will include the history of the clients and their own knowledge levels, experience, environmental and cultural considerations, and the qualifications and track record of the consultant.
- *The project*: This will give the reasons why the client wants this project to happen and why some of the possible routes should be considered.
- *Appropriate technology (AT) and/or management systems to be considered*: Raw materials to be used and/or acquired.
- *Market considerations*: If a new product is to be developed, then the likely market acceptance and possible costing and pricing factors must be examined. Who are the target consumers and what will be the competitive advantage provided by the new product? Similar factors have to be considered if the client wishes to make a design technology and/or systems management change.
- *Timetable scheduling*: It is important that the consultant provides a project management schedule that is realistic and that allows for possible contingencies and financial interventions that might take place.
- *Project costs*: All costs must be presented to the client and funder including consultant's fee, plant and equipment acquisition if required, new building (again if required and the design technologist will require local advice on this matter), training, and additional manpower.

The feasibility study report must be presented professionally and should not promise more than the consultant who is confident to deliver. Nondelivery is a disaster for the career of any who would be consultant. If a member of academic staff of a university carries out a consultancy, he or she will be covered by university insurance schemes as long as he or she acts through the auspices of the university consultancy arm, which is often nowadays known as enterprise and innovation. If the consultant is operating as an individual, he or she would be well advised to take personnel cover. However, insurance will only cover financial losses; reputation losses are much more difficult to recoup.

18.4 Project Management

Project management is simply the planning, organizing, and managing tasks and resources to accomplish a defined objective. It is a vital component of any consultant's toolbag and a skill that must be developed and refined. Most projects will have time and cost constraints that have to be understood and allowed for in the plan. There are typically three phases involved: (a) creating the expected schedule, (b) managing changes, and (c) communicating project information to all relevant parties.

Creating the schedule: The first and most important phase of project management consists of defining tasks and their durations, setting up relationships between tasks, tracking the use and cost of the various resources required, and assigning those resources. If this phase is carried out to an optimum level, the next two phases will benefit immensely. You will notice the use of the word *optimum* rather than *correct* as it is rarely possible to achieve the latter. The best that can be hoped for is that all possible contingencies have been considered and that the project is not subject to an *act of god*.

Managing changes: As indicated earlier, *things happen* but what counts is how the consultant reacts to them, what contingency allowances have been put in place, and how flexible the schedule tasks and resources have been designed. Remember that a design technologist is a good problem solver.

Communicating project information: A consultant has to be a good communicator and a good listener. Never keep things to yourself; your clients and project staff will not thank you for this, and the success of the project is dependent on good will, trust, and ownership by all concerned parties.

18.5 Project Management Models

Useful models for project management include the following:

- *Linear organizing method* (*calendar path*): This method involves a monthly (or weekly) calendar that shows tasks and durations. It is simple and can work for short, uncomplicated projects.
- *Critical path method* (CPM): CPM is a systematic model for calculating the total duration of a project based on the individual task durations and dependencies and identifies which tasks are critical. Many software packages are available

today to help set up a CPM and analysis. A useful calculation from CPM thinking is how to work out the time allocated to each element/task component of a project:

$$\text{Real Time} = \frac{1}{6}\left(1\,\text{PT} + 4\,\text{MT} = 1\,\text{OT}\right)$$

where:
  PT is pessimistic time
  MT is most probable time
  OT is optimistic time

- *Gantt chart*: This is a graphic way to represent activities across a timescale. It is relatively simple and adequate for most small-scale consultancy projects.

The need for such a level of sophistication in project management offered by CPM and/or PERT is determined by the size, complexity, and duration of the project. In other words, if you are asked to project manage the setting up of a mill in a particular location from equipment procurement to final plant commissioning, you will definitely need to use CPM. However, if your project is about advising on an extant process problem in an existing mill, then CPM will definitely be overkill but the task/element time requirement calculation given earlier will definitely be useful.

18.6 Intellectual Property Rights

Knowledge, skill, and experience are at the core of what a consultant offers; by applying these through the course of a project, it is more than possible that new products, designs, technologies, and systems may emerge. In the case of products and technologies, there are well trodden and established paths to protecting intellectual property rights (IPR) such as patents. However, a client may not wish to use patent laws for protection as they require public dissemination of ideas, and many clients can see this as an avenue to revealing what might become a competitive advantage for them. On the other hand, the consultant might want protection for their ideas and the perceived kudos that accompanies the obtaining of patents. In order to avoid future misunderstandings, or worse, the IPR arrangements for any new products, technologies, designs, or systems that may emerge from a project should be discussed and agreed at the beginning. Of course, the likelihood of these will depend on the nature of the consultancy. If, for example, it is simply the commissioning of a manufacturing plant, there is not likely to be such problems. If the consultant is working for an agency, then IPR will probably be written into the contract.

IPR in the United Kingdom for design and design technologists has been a topic of controversy, frustration, and complexity for many years and continues to be as such. The situation is further complicated by the different laws, affecting IPR in different countries and indeed by the total ignorance of IPR by many young designers and different countries. Indeed, there is an interesting analogy between young designers in developed countries and cultures living in developing countries and in terms of knowledge and understanding of IPR. To understand IPR, we must first alter or augment our perception of property.

We grow up to consider property as a material entity: a car, a house, a television, and so on, something tangible that an individual has paid to take possession of or has manufactured themselves and therefore has the right to charge another individual for or even to give away; the last three words are important and may become more important as we begin to discuss IPR in relation to culturally significant patterns, products, and practices. For example, many of us will give away our personal property to charity shops or any other nonprofit organization that we chose because we believe that, in doing so, we are contributing toward the good for our national and global society.

Though property is probably not semantically the right word to cover design knowledge, it is used because *property must be grounded in a public belief that it is morally right; if it is not so justified, it does not remain an enforceable claim. If it is not justified, it does not remain property.* Design knowledge can be placed in one of the two groups (Rodgers and Clarkson 1998): tacit or explicit. To put this simply, tacit knowledge is what that is in the designer's brain; it has been put there by training, education, practice, and experience. It is how the designer collects and interprets previous knowledge. The knowledge will remain tacit throughout most of the idea generation stage and even up to the stages of early sketches. Once the designer/design technologist moves onto technical drawings, models, computer simulations, and so on (i.e., they become explicit), the ideas can be tested, viewed, used, and enjoyed in the public domain and therefore should be protected, and a value should be placed on it.

18.6.1 The Protection of Design

In 2012, a substantive report was produced by the UK Intellectual Property Office. It was reported that designs are protected in the United Kingdom by five legal rights:

1. EU registered design rights
2. EU unregistered design rights
3. UK registered design rights
4. UK unregistered design rights
5. Artistic copyright

This report goes on to show that the protection of design rights presents a maze in which designers and companies have to try to navigate and that most designers and companies feel unlikely to succeed and/or that even success would be more costly than it is worth. The cost of failure when attempting to protect design rights is more or less prohibitive to SMEs, which make up the majority of design agencies. Across Europe, there has been an effort through the Community Design Regulation introduced in 2001 to bring about a unitary system and approach, but this has had limited success.

One section of the report may be considered as of significant interest to individuals, agencies, companies, and cultures—the effect of design rights on motivation and innovation. Many companies will quote their number of patents in promotional materials but very few companies will quote their design rights/copyrights, and so on, in the same manner. Are they missing a trick or is it simply that product/design life cycles are so short that it is not worthwhile? Taiwanese product/industrial design agencies have used the award of a patent to an individual designer's work for many years as a mark of esteem and as a motivational tool. Perhaps in the West, we need to consider this more diligently. In addition, may it be that cultures would be inspired, motivated, and feel important if their patterns, products, and practices were considered valuable enough to be legally protected?

It was reported that most designers, agencies, and companies were most comfortable with the concept and reality of copyright and would normally revert to that as a first choice.

18.6.2 Recommendations from the Development of Design Law Report

The recommendations of the report are briefly outlined as follows:

- The Intellectual Protection Office (IPO) could offer design rights as an insurance policy for which companies/individuals would have to pay a registration fee. The IPO should then make the information related to the protection of registered designs more accessible and clear.
- The IPO could give clear guidelines as to the likelihood of success in litigation cases brought about due to perceived design right infringement. Information regarding costs, time, and benefits of such action should be made available.
- Among the main problems perceived in the current system are time and cost. In order to alleviate this, it is recommended that a superfast track should be considered, which would be available to small-value design claims. The associated cost would then also be limited, but no precise figure is suggested. Extra expense could be attributed to any party that is wishing to step outside the superfast process.
- Problems can arise due to the judiciary that has a different value perception to design as that of the design community. In order to obviate or at least lessen the likelihood of this occurring, *lay assessors* or *informed users* could be appointed to be on the bench and could advise the judge.
- The time taken for and the cost of bringing about a judgement on IPR infringements may be reduced by introducing an IPO tribunal system. This would hear cases only where registered design rights exist and which is up to 3 h in duration.
- One final recommendation is for the formation of a design list to help accommodate an expedited process. This list would comprise cases in which it was considered by the judiciary that decisions could be reasonably achieved in 3 h or less.

These recommendations are much more expansively described in the report. The above outline gives the reader some flavor of what could be done. At this time, these recommendations remain just that, *recommendations,* and the law with regard to Design Intellectual Property Rights in the United Kingdom is still difficult to comprehend and/or to ensure firm expectations.

18.7 Soft Systems Methodology

Systems thinking was first developed by Checkland (1981) and was primarily aimed at countering the lack of user considerations being shown by the computer hardware industry. It has been applied by business analysts to many forms of activity. Since around 2005, design researchers have made sporadic use of it to address interior and product design issues. The first version that Checkland introduced had seven stages and a number of tools to express issues, ideas, and solutions. However, Checkland also made it clear that researchers could dip in and could make use of any of the stages, tools, and methods and that some might stand alone for specific purposes. Soft systems methodology (SSM) can be very useful when a project requires the study of human activities and the organization of those activities. The stages of SSM can be defined as follows:

1. *Entering the problem situation*: The process starts with the identification and evaluation of a real-world problem or design challenge. The data collected may be qualitative, quantitative, or mixed, and the researcher chooses the most appropriate tools or methods to collect this data.

2. *Expressing the problem situation*: In this stage, the problem situation is usually expressed using *rich pictures* that can reflect and examine the circumstances within the particular system that is being studied. *Rich pictures* are particularly useful and attractive to those who are visually stimulated more or equally as much as textually; they are therefore effective tools for designers to visualize their thinking and to record their insight.

   Root definition of relevant activity systems: This is the stating of a hypothesis concerning the final improvement of the problem situation by means of an implemented transformation (design solution) and whether it will be feasible or desirable. It is a structured description of a system written by the researcher from the initial research activities and the objective of the system (design solution). Therefore, it has to consider the aim of the transformation, the persons who could affect it or be affected by it, and those who would be part of or use the new system. It also identifies other key elements in the transformation that might relate to the system itself, human activity, and/or the broader environment. To be confident that a given root definition is comprehensive, several criteria need to be specified. The construction of the root definition is an iterative process, and each iteration can be tested using the CATWOE test. This test helps ensure that the necessary components of the system are addressed in the root definition. The components are as follows:

   - *C = Customers or clients*: Who (or what) benefits from this transformation.
   - *A = Actors or agents*: Who is engaged in the system activities and facilitates its operation.
   - *T = Transformation*: How is the system, practice, or product transformed? This is an essential part of the process of change.
   - *W = Weltanschauung* or Worldview.
   - *O = Owners*: Who controls the system, practice, or product/could cause it not to exist.
   - *E = Environment*: What does this system take as given from the world that surrounds and influences the system, practice, or product.

3. *Building a conceptual model*: It may occur concurrently with the formulation of the root definition. The conceptual model commonly illustrates the relationship between the system activities underpinned by the root definition. Patching (1990) suggests that the development of a conceptual model is illustrated by assembling and structuring the minimum number of verbs that is necessary to describe each component or activity in the system (commonly expressed in diagrammatic form).

The four final stages of the SSM process are more applicable to business systems but can, in some cases, help with design technology problems/solutions. They are fairly self-explanatory:

- Comparing the new system/practice or product with the existing version
- Deciding on the transformation(s) to be made
- Taking action to improve the existing system

18.8 Motion/Method Study

The purpose of motion/method study is to enable an observer to examine and analyze the motions involved in the performance of manual tasks. It must not be confused with time study that is used in many sectors of the textile and clothing industries to determine how long a task takes to perform and therefore how much time can be allocated for that task and therefore how much a person is paid for multiple performances of the task. For example, if a sewing machine operator in a clothing company has the task of inserting a sleeve into the body panel of a jacket, they are often paid for the number of times they successful complete that task, and therefore the management of the company have to determine how long it would take for the operator who is working at a reasonable speed. The author has to say that he would never be involved in this type of measurement, and motion study can be used in order to maximize the efficiency with which the task is accomplished. However, much more humane and valuable use can be made of motion study to try to allow the operator or other workers to carry out tasks comfortably and efficiently and without detriment to their well-being. Probably, the two most well-known exponents of motion/method study were Frank and Lillian Gilbreth who worked around the beginning of the last century. They developed and used a system for categorizing types of motions so that attention could be focused on improving, reducing, or eliminating motions that caused unnecessary levels of fatigue, discomfort, and wasted time often through poor design of the workplace tool or layout. The Gilbreths used a series of 18 motion icons (Figure 18.1), which could then be plotted on a simultaneous motion chart along with the time each motion took and any notes on the motion. These icons are called therbligs. Therblig is basically Gilbreth spelled backward with the exception of the *th*.

18.8.1 Using Therbligs

When examining a particular textile or clothing process, the use of a timed activity chart can help reveal what are the root causes of problems that can be linked to time wasted, energy over expenditure, operator discomfort and poor design, and/or technology usage. The most effective way of obtaining the data to be studied is by filming the operation and preferably as many iterations as possible and then studying the film using the timed activity chart technique. Of course, to do this, the explicit permission of the unit (mill, factory, studio, laboratory, workshop, etc.) and the concerned operator are required. Do not try to use any form of subterfuge to obtain permission. Work study and method study engineers of the past would use this type of technique to look at the operations on a microlevel to try to speed up or improve repetitive operations, but the technique can be employed on a more macrolevel and to try to reveal and help solve design technology issues.

Let us look at the example of a spinning operative in a traditional woolen spinning mill in Table 18.1.

You will see how this was used in Section 18.11.2.

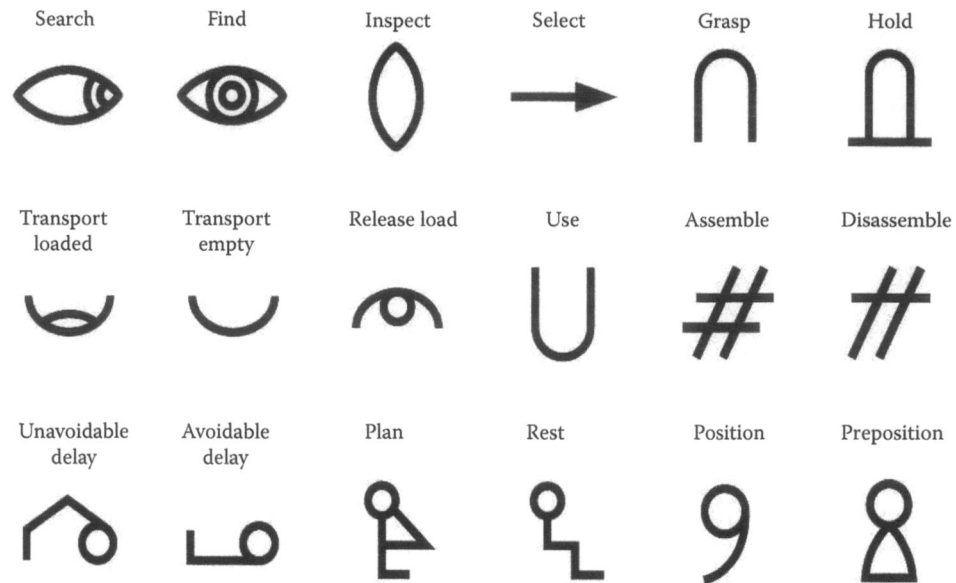

FIGURE 18.1
The Gilbreths 18 motion icons.

TABLE 18.1

Example Taken from Timed Activity Chart Using Therbligs

⌐O	Taking time to talk with friends before starting machine	10 min
⌂O	Check that all yarn ends are pieced up before starting up machine	30 s
U	Use spinning frame by switching on	10 s
◁O	Search for any ends of yarn breaking on start up	10 s
◀◉▶	Find two ends down	5 s
#	Piecen up one end	20 s
#	Piecen up second end: more troublesome	50 s
⌐L	Rest	10 min
#	Piecen up one end	20 s
⌐L	Rest	15 min
#	Peicen up two ends: troublesome because operator had not noticed ends going down because he was reading his paper???	2 min

18.9 Appropriate Technology

Schumacher (1973) developed the idea of intermediate technology, which is more productive than indigenous technology and low in capital and running costs compared to sophisticated technology used in modern industry. Intermediate technology is considered useful when workplaces are created where people live, reducing migration to urban areas. These workplaces can be created in large numbers without huge capital, and production methods can be simple—mainly using local materials. The term *intermediate technology* has been criticized for implying a technological fix for development problems that are separate from the political and social factors involved (Hollick, 1982). The term AT has been suggested as a substitute, which includes the social and the cultural dimensions of innovation.

AT can be characterized into two areas: resource localization and soft approach. Resource localization means that the appropriateness of using AT is decided by the designer/technologist according to the use of available resources in the targeted community. The soft approach is related to sensitivity toward local conditions in the development of the technology (Sianipar et al., 2013).

In this context, field study observations were used as a method to generate knowledge about specific issues in the village. There are two types of observation: structured observation and participant observation. Structured observation methods require systematic observation of behavior, without directly questioning the people who are observed. Participant observation involves asking questions that arise naturally in the course of observation. At the initial stage of the study, the researchers followed participant observation where they approached observation with a relatively open mind, in order to minimize the influence of the observers' preconceptions.

In order to make the technology used successful within a particular locality, it has to be firmly related to the technical, economical, and social conditions in existence. Culture is viewed as forming a context or background for the development of technologies. Hazeltine et al. (1998) define culture as groups, customs, and standards of taste. However, culture is not considered solely as a series of responses or adjustments to technology; rather, it is seen as an essential mediator and adversary to the noncultural, the mechanical, and artificial realm of technology. A cultural critique of technology is one in which the noncultural elements are evaluated, judged, and pushed in new directions as it fits the individual society or culture. The selection of technology that accounts for regional social values plays an important role. For example, the appropriateness of technologies should not be decided purely on economic and factor endowment grounds. AT is a technology that is appropriate to the particular situation faced by a given group of people, where consideration is given to value priorities along with the economic circumstances and the available resources. The U.S. Congress Office of Technology (1981) characterized AT as being energy efficient, small scale, environmentally sound, and labor intensive, and it can be controlled by the local community. Dunn (1978) discusses AT as rural based and as a method, which attempts to recognize the potential of a particular community and help it develop gradually. This development progressively builds the skills of the community based on local resources. The reason why AT relates well to cultures is because it can be adapted to the local needs and is controlled by those using it. The characteristics of AT such as low cash requirements, being repairable and controlled by users, match the situation of women in the developing countries. If cultural factors are not taken into consideration while introducing a new technology, then it is likely that the aim of the new technology may not be met—perhaps because of unexpected contingencies or resistance by those involved. AT tends to put participants in control so that it can be adapted to local conditions, it does not require major changes in people's lives, and it is a promising way to improve living conditions without cultural damage suggesting that a technology should be considered appropriate "when its introduction into a community creates a self-reinforcing process internal to the same community, which supports the growth of the local activities and the development of the indigenous capabilities as decided by the community itself." According to Sianipar et al. (2013), researchers believe that AT as a phenomenon emerges together from specific conditions from a local area that needs a technology that is appropriate with local people's needs and wants.

18.9.1 Why Is Appropriate Technology Suitable for Small-Scale Application?

The benefits of AT outlined by Schumacher (1973) for small-scale application are summarized as follows:

- Employment can be created in the place where the unemployed live and that can restrict migration of unemployed people to urban areas.
- Through reducing the needs for imports and creating an export market, savings from wages and profits generated can be used for investment in further capital development.
- There is a greater opportunity to use renewable resources (solar, wind, hydro, wood, and biogas).
- Small-scale industries using AT can reduce pollution and ecological imbalances that are prevalent in most concentrated large-scale industry. Furthermore, ecological problems can be remedied at much less cost.
- Growth of the industry can occur in small steps, as required by demand and made possible through new capital, which includes changes in the products through innovation.

Accordingly, AT can be defined as a technology that fits local condition and is easily and economically utilized from readily available resources in local communities to meet their needs. AT can be viewed in a broader sense in which the major concern is whether AT can produce sufficient goods and services. For developing countries, AT can lead to national development in the sense of a trained workforce. However, the problem exists whether people may accept an AT approach rather than high technology. Some leaders are understandably suspicious of AT as being a way to discourage the developing countries from industrializing and becoming competitors. The answer to this issue is that there is no other way to industrialize other than using the resources that are readily available. Many developing countries have vast resources of untrained labor; in order to channel it in the right direction, a simple technology that is easy to adapt with simple training facilities will lead to economic development of the community.

Another problem faced by the AT approach is that it is specific to locality; thus transferring expertise from one locality/country to another is a difficult process. Technology transfer is difficult between communities because of limited and weak communication between the communities and researchers, resulting in poor technological diffusion (Amiolemen et al., 2012). In this case, changes must be made in technology choice, depending on the requirements and skills of the community. AT projects are normally small-scale and done by many people independently, thus making it difficult for government officials to understand what is happening and to take action when needed. With reference to India, central and state governments have specialized departments such as The National Mission for Empowerment of Women (NMEW), The Ministry for Micro, Small & Medium Enterprise (MSME), and M. S. Swaminathan Foundation (MSSRF), and these departments aim to handle the socioeconomic development of rural areas.

18.10 Cultural Awareness and Assimilation

If a consultant is undertaking a job in a country that is foreign to them, it is wise if they can study as much as possible about the culture(s) of the people in that country including the language. Being able to say *please* and *thank you* in the native tongue are minimum requirements that will be improved if you can introduce yourself and if you can briefly ask whether the person you are talking with can be kind enough to help you by continuing in English. East Asian languages can be particularly difficult for westerners as they have vowel intonations that do not occur in the West. Any effort to be able to give them the respect and good manners of trying to use their language will almost always be received well. There are some exceptions to this general rule, but these will be exercised by those suffering from the ailment that, as a consultant, you are trying to avoid in yourself. The ailment in question is called ethnocentrism. This is perceiving of other cultures only from the perception of your own culture. Ethnocentrism is usually more attributable to inexperience or lack of knowledge about foreign cultures rather than to prejudice. The keywords are caring, understanding, and patience. The following are some of the areas to consider:

- *High-context and low-context cultures*: High-context communication is succinct and to the point. Low-context cultures value rhetoric (lots of talk).
- *Time*: Different cultures have different concepts of time. Monochromic cultures are organized, methodical and place importance on punctuality. Polychronic cultures tend to do many things at the same time and believe that punctuality is nice but not always required.
- *Long-term versus short-term thinking*: Westerners tend to make up their mind and want things to happen quickly. In other cultures, people take their time to make up their mind, and there is likely to be a time lag before anything happens.
- *Closure*: Western cultures feel that a task has to be completed, or they have wasted time. Eastern cultures are happy to let things develop slowly.
- *Individualism/collectivism*: Western cultures tend to be individualistic (look after themselves and their immediate family). Eastern and Latin American cultures are collectivist. They belong to groups who look after them in exchange for loyalty. The idea of *face* is very important in collectivist cultures. They do not insult or shame a Chinese person in front of her/his peers, preferably never do let this happen.

18.11 Case Studies

Let us now return to the projects listed at the beginning of this chapter and briefly outline what the design technologist was able to do as a consultant.

18.11.1 Why Are We Losing Stock from Our Cashmere and Lambswool Yarn Stocks?

This was a hosiery company that produces very high-quality knitted jumpers. At the time of the project, they were buying cashmere and lambswool yarns by weight (e.g., 500 kg of 2/20 nm mid-blue cashmere yarn to produce around 200 jumpers for a high-quality retailer who was one of their regular customers). At the time of the project (1985), woolen yarns were allowed to have 8% spinning oil and 17.5% regain (water content expressed as a percentage of the oven dry weight). When the knitwear production staff started to use the yarns, they were finding that they were insufficient to produce the number of jumpers ordered which was, of course, a problem for the company in terms of finance, reputation, reliability, and so on.

The management of the company started to have the yarn lots weighed before use and found that many of them were of significantly lower weight than on their *goods received documents*.

There were two possible answers for this: (1) staff were stealing valuable cones of yarn from the warehouse or (2) the yarns were losing moisture in storage. Management decided to bring in a consultant to investigate reason 2 as reason 1 would have been the remit of a security consultant rather than a textile design technologist. In order to investigate the problem, the consultant took samples of yarn cones from every lot that was already available in the warehouse and received over a period of 1 month and carried out tests. The test used was a simple weigh, dry, reweigh method. The results showed that the moisture regain of yarns that had been held in storage for a month or more had moisture regains that varied from 13% to 17%. It was therefore simple to conclude that yarns that had been held in storage for a month or more could have lost weight of 5%–7% and therefore a 500 kg lot could easily be 475 kg by the time it is used.

Three solutions were suggested:

1. Air-conditioning could be installed in the yarn warehouse.
2. An allowance could be added to the ordered weight of yarn to allow for loss in moisture content (regain).
3. Yarn could be ordered in measured length cones rather than by weight.

The third solution was a long-term solution and required much negotiation with yarn producers and suppliers. This project did not bring the solution 3 to be brought about but it did add to the weight of argument by many knitwear manufacturers, which has brought about the general adoption of this solution by most yarn suppliers and users today. So, the skills used by the design technology consultant in this case were a knowledge of woolen yarn sales methods, a knowledge of yarn moisture and regain and how these could be measured, and an ability to prepare a sound feasibility report and final management report.

18.11.2 How Can We Improve Productivity and Quality in Our Woolen Yarn Spinning Mill?

In this case study, the management of a very traditional woolen spinning mill found that the production rates of their spinning machines and the quality of their resultant fabric were poor. They called the author to have a look around and see if he could identify what the problem(s) might be. In order to investigate the spinning activity in the mill, the author asked if he could take a film of one of the spinning machines in order to examine the technology and how it was being used. Back in the office, the author applied a timed activity chart using therbligs as shown in Table 18.1. The first thing observed was that the operator was having long periods of rest between piecening up activities. This was unusual in a busy spinning mill that was producing quality yarns for nice handling fabrics. Work study engineers' approach would probably have suggested that the operator has more machines to run, but that was not the approach of this design technologist.

If a machine is running for long periods with few end breaks, it often suggests that too much twist is being inserted that will make the ends of yarn strong but also with a harsh handle in the resultant fabrics. The author tested the twist content in samples of the yarn that was around 100 tex count and found that 500 turns/m were being inserted. This meant that the twist factor was 5000, which is extremely high for a quality weaving yarn. Of course, this also meant that the machine was running slow as the delivery rollers had to be slow to allow the spindles to insert twist. Therefore by the simple expedient of lowering the twist content of the yarns, we could achieve the aims of the management at least as explained in their brief to the consultant.

18.11.3 Rather Than Just Selling All Our Wool Overseas, Should We Set Up a Mill to Process It into Yarns and Possibly Fabrics and Garments?

This project began just before the Falkland Islands conflict in 1983. A farming family on the West Falklands decided that, rather than just farming sheep and having their wool exported to brokers in the United Kingdom, they should process the wool into yarns and sell to retailers and fabric producers. They contacted one of the authors at the Textile Higher Education Institution where he taught and then visited the Institution with a suitable quantity of already scoured wool to carry out processing trials. Let us return to the earlier section in this chapter on AT. The intention of the family was to build a spinning mill on their farm and to operate the equipment themselves while providing some additional employment for a few of the farm workers.

When they arrived at the spinning workshop of the Higher Education Institution (HEI), the first thing to be decided was what would be the AT to enable them to succeed in this endeavor. The full worsted spinning system would require too many machines and the acquisition of complex skills for setting up and operating the combing process, so this was ruled out. The woolen spinning system would require less space and equipment but the complexity of the woolen card and, in particular, the tape condenser would again be a step too far. It was therefore decided to carry out trials on a semiworsted system. This was accomplished by using the first part of the woolen card and by taking the fibers off the doffer roller as sliver coiled into a can. Six of these slivers were then fed into a gill box, and a draft of six was used to give the same sliver linear density as that which had come off the carding machine. This process was repeated but with a higher draft of 8 to reduce the linear density of the sliver, which was then fed into the drafting zone of a roving machine using a draft of 10 to produce a roving of 1 ktex (1000 g/1000 m). At the spinning stage, a draft of 10 was again used to give a yarn of 100 tex, and a low twist factor of 3000 meaning 300 tpm (turns per metre) was inserted in the Z direction as it was intended for this yarn to go into the hand knitting or hand frame knitting markets.

All the aforementioned values were chosen to make life as simple as possible for the family. As time progressed and as their skills and experience increased, they could play to get a wider count range. The progress of the project was interrupted by the Falkland Islands conflict but after that it gathered momentum and a suitable range of second hand equipment was identified, refurbished, and prepared for shipment to the Falklands. This consultant and two technicians also flew over and prepared to set up the equipment. This project was fraught with many problems, though not quite as bad as the last one described, but in the end the mill was set up and the yarns began to be produced. The last comment on the mill, which was indeed a brave and bold effort by the farm family, found on a website was *although the mill produced an attractive product, it was closed due to the high cost of energy and labor.*

18.11.4 How Do We Change from Quality Control to Process Control?

A spinning mill in a Scottish Border town was producing high-quality cashmere and lambswool woolen spun yarns. They had a Quality Control Department, which did a good job of inspecting and testing materials and product at all stages of production and advising management on the rejection of substandard product. However, they had no responsibility for advising the technical staff of the mill about the possible causes of faults and possible solutions. These aspects were left in the hands of the technical staff that made senior staff feel vulnerable to the vagaries of levels of knowledge and application of the technical personnel. The company had made some investment in automated process control systems but realized that these had limitations when it came to the various technological problems that can occur at the extremely important carding stage of the woolen spinning system. They therefore wished to change the approach of the quality control staff toward that of human process control. Many other similar companies had been experimenting with this change a decade before with varying degrees of success.

It is fact that the most important issue in yarn production is the measurement and control of evenness. The company decided to invest in a sophisticated evenness measurement system costing a great deal of money and offering sophisticated data outputs concerning the overall yarn evenness but also intelligence about the wavelength of thick and thin places in the yarn. However, the mill personnel had not been given sufficient training in the use of this sophisticated technology, and there was no degree educated technologist to explain how simple calculations could be employed to reveal the importance of process control messages that are being delivered by the charts and graphs produced by the equipment after measuring the yarn evenness over a long length.

Consequently, the author was called to give lessons and demonstrations for the selection of the staff. Those were a few pleasant days spent with very nice people who did their utmost to understand how and what to do, but it was not difficult to tell that there was an air of confusion and fear of the new technology. Unfortunately, this mill did not stay in operation for a much longer time, which would have, of course, been brought about by other economic and political issues that were affecting the textile industry at that time, but these were not helped by the lack of investment in educated design technologists at the right levels of management.

18.11.5 Should We Add Value to Our Mohair by Scouring It before Selling It and Therefore Should We Set Up a Scouring Plant

This was one of the first consultancy challenges for one of the authors of this chapter and was also an example how a lack of communication, in the days before effective electronic communications became ubiquitous, could lead to a great deal of misunderstanding. The author was young and had just begun teaching at a higher institute for textile education. He had spent 9 years working in spinning plants in the United Kingdom and was now teaching yarn production. It was known to the senior management of the institution that the author also had considerable experience of hand spinning and so when a call came from a major international agency to find someone who could give advice on the hand spinning of a speciality fiber in a remote area of southern Africa, the management asked if the author would take this as a consultancy. Being young, confident, and enthusiastic to travel, the author said yes and a few weeks later found himself on a plane bound for southern Africa. You may have noticed that an important adjective that was not used earlier to describe the author, at that stage of his life, was inexperienced. He had plenty of good education and technological training but had taken the job at a face value without digging deeper into what were the main influences amongst the various players in the scenario. In other words, he did not carry out the detailed feasibility report, as discussed at the beginning of this text, before stepping onto the plane, which is a big mistake. This story could turn into a novel, so for the sake of brevity, many of the preliminary adventures will only be outlined.

The plane tickets had arrived to get, what we will now call, the consultant to a major city in the region from which he had to transfer to another plane to get to a minor city on the border of the remote country, which was his destination and locale of the job to be undertaken. I had been under the impression that I would be met at the airport of the minor city and transported by car across the border into the remote region. However, no one turned up and having no mobile phones in those days and no e-mail, I had to find my way to a hotel and find out how to get to my destination. The consultant was told that the only way I could travel was by bus or train, and that the latter was definitely safer though neither was normally used by foreigners. A few more adventures/crises/confrontations took place during this journey but the consultant finally reached the local headquarters of the international agency where he was to work via a taxi ride from the train station in a vehicle driven by a person who looked like a 13-year-old boy accompanied by a local family of 7 who decided that I could pay for them to share my ride.

When I arrived at the agency, they were expecting me but had not known how I was traveling as the main agent in charge of the project the consultant was to undertake, had left the country to get hospital treatment in the nearby relatively developed country without giving details of my arrival to anyone. At least, I was told that a hotel room was booked for me, and I was allocated an official car and driver. I also had a preliminary schedule of meetings with interested parties, so things were looking up. It turned out that the agent in charge of the project would not return for 2 weeks (I was only commissioned for three) and that another consultant was going to join me whose role would be in cooperative formation and organization. This latter news encouraged me as it made sense and would allow me to concentrate on the AT and yarn design sides. What did not encourage me was that the second consultant would not arrive for a week and a half. The author is recounting this tale as a warning to all would-be consultants to do as much background work as you can before doing a job and do not consider that large international consultancy agencies are likely to know what the right and left hand are doing and that resilience and flexibility are vital in the foreign consultant's portfolio of skills and personality.

It turned out that the international agency who had commissioned the author as a consultant did not want involvement in a hand spinning project but was actually interested in building a mohair scouring plant in the country on the basis that they had been advised by previous consultants that removing the grease and other impurities in the mohair would give added value to the exported product. I was at length to explain that mohair is normally spun on the worsted system and in that system the raw fibers were normally blended before scouring, and therefore the brokers would not welcome or pay extra for scouring to take place in the country of origin. In addition, although the farming of angora goats and subsequent selling of the mohair fiber produced was important to a small remote country's economy, it was still on a too small-scale to support a scouring plant. This all happened in the late 1970s and when the author last checked, there was still talk on the news about the possibility of a scouring plant in that country in 2016 though voices were heard to support the opinion I had given.

18.11.6 What Can We Do with Remnant Warp Silk Filament Yarns that Might Provide Some Paid Work for Women in Poor Rural Areas?

This project never became fully live though the men and women of the rural handloom weaving village in India did participate and were very interested in where it could go. In addition, two other consultancies were interested in the framework produced. The project began with a young lady (Sugathan, 2016) approaching the author with an application to carry out PhD research in mixing silk filament remnants from the handloom weaving of silk into sari. She wanted to mix the filaments with lambswool to produce a yarn that could be used in the knitting of garments. A feature of this aim was that the color of the yarns would be irreproducible and would change depending on the design of the sari being woven (different warps) and because of the craft nature of the dyeing of the silk filaments. This meant that the yarns produced would not be suitable for mainstream fashion but would be ideal for the hand knitting market where customers are always looking for something different and for a good story to go with the product.

We decided that the formation of a spinning cooperative would be a good idea to give the women somewhere to belong to outside their familial situations and a possible means of gainful employment. In considering the application of AT, it was felt that hand spinning would offer a suitable solution. The trials then commenced and we ended up cutting the remnant silk filaments into 3 in. staple lengths and mixing these with lambswool. Hand spinning, along with the one-off nature of the coloration caused by the remnant silk fibers, offered the production of a speciality yarn each lot of which would be unique. This technology also provided a therapeutic effect that would be helpful to women who lived in a stressful environment. The fiber mixing and opening could have been carried out by hand but this would have been tedious and laborious and so we decided to use a mechanized small-scale carding machine. The yarn produced was tested against a range of yarns already in the hand knitting market, and it was confirmed that the achievable price would have enabled a healthy profit to be generated.

18.12 Summary

Consultancies and consultants come in widely varying forms, and the purpose of this chapter has not been to inform the reader on the right way to go about things; indeed, if the reader wants a more detailed treatment of project management, they should refer to Ormerod (1992). This text has been aimed to allow the reader to ask themselves some pertinent questions, to discuss some of the tools that the authors found useful, and to give some anecdotal case studies. The concept of AT has been briefly discussed, and it is hoped that the reader will be sufficiently interested to read further on this subject and maybe make use of AT in some future projects. The authors hope that the reader would have learnt a lot, perhaps laughed here and there and perhaps realized that consultancy should not be taken lightly and requires high levels of knowledge, experience, resilience, and flexibility. In addition, the textile project consultancy is as far from business consultancy as it is possible to be.

References

Amiolemen, S.O., Ologeth, I.O., and Ogidan, J.A. Climate change and sustainable development: The appropriate technology concept. *International Journal of Sustainable Design* 5(5) 2012.

Checkland, P. *Systems Thinking, Systems Practice*. J. Wiley, Chichester, UK 1981.

Congress of the United States: Office of Technology Assessment 1981. An Assessment of Technology for Local Development.

Dunn, P.D. *Appropriate Technology-Technology with a Human Face*. Schoken books, New York 1978.

Hazeltine, B. and Bull, C. *Field Guide to Appropriate Technology*. Academic Press, London 2003.

Hofstede, G.H., Hofstede, G.J., and Minkov, M. *Cultures and Organisations: Intercultural Cooperation and its Importance for Survival*. McGraw-Hill, New York 2010.

Hollick, M. The appropriate technology movement and its literature: A retrospective. *Technology in Society* 4(3): 213–229, 1982.

http://www.malvinahousehotel.com/about/our-artist/

Ormerod, A. *Textile Project Management*. The Textile Institute, Manchester, UK 1992.

Patching, D. *Practical Soft Systems Analysis*. London: Pitman 1990.

Rodgers, P.A. and Clarkson, P.J. *Knowledge Usage in New Product Development (NPD)*. Loughborough University 1998.

Schumacher, E.F. Small is beautiful: A study of economics as if people mattered. Blond & Briggs, London 1973.

Sianipar, C.P.M., Yudoko, G., Adhintama, A., and Dowaki, K. Community empowerment through appropriate technology: Sustaining the sustainable environment. The third international conference on sustainable future for human security. SUSTAIN 2013.

Sugathan, M., Cassidy, T., and Carnie, B. The development and evaluation of a speciality hand knitting yarn using appropriate technology for the empowerment of women in rural India. *Research Journal of Textile and Apparel*. 20(3): 136–154. Emerald 2016.

The development of design law. (Report) Intellectual Property Office 2012.

Index

Note: Page numbers followed by f and t refer to figures and tables respectively.

Milton Keynes UK
Ingram Content Group UK Ltd.
UKHW050457071024
449327UK00015B/412